HANDBOOK FOR LINEAR REGRESSION

HANDBOOK FOR
LINEAR REGRESSION

MARY SUE YOUNGER   *The University of Tennessee*

DUXBURY PRESS
North Scituate,
Massachusetts

Handbook for Linear Regression
was edited and prepared for composition by Elizabeth DuBois.
Interior design was provided by Dorothy Booth.
The cover was designed by Oliver Kline.

Duxbury Press
A Division of Wadsworth, Inc.

Library of Congress Cataloging in Publication Data

Younger, Mary Sue.
  Handbook for linear regression.

  Bibliography
  Includes index.
  1. Regression analysis. I. Title.
QA278.2.Y68   519.5'36   78-24267
ISBN 0-87872-187-8

Printed in the United States of America
1 2 3 4 5 6 7 8 9 — 83 82 81 80 79

CONTENTS

PREFACE

With the increasing amounts of data available in our computerized society has come an increased emphasis on the use of statistical analysis by industry, government and academia. Statistical techniques were until quite recently considered to be the private domain of professional statisticians, but students in the sciences are now expected to be able to apply statistical methods to their own data and to interpret the results for themselves. Graduate students in psychology, for instance, must often design their own studies and perform a statistical analysis of the results. Undergraduate students in marketing are often expected to analyze results of studies and to present interpretations and recommendations. Even the professional of many years' experience in a given research-oriented field is often faced with the problem of using statistics to analyze an ever-increasing amount of information.

Colleges and universities require more and more courses in statistics for those fields in which the collection of data is an integral part of research, but people who are not majoring in mathematics and statistics cannot be expected to become really expert in statistical methods. Nor can the professional who received training before the advent of this emphasis on statistics be expected to become expert in its methods.

This book is intended to be both a textbook and a reference manual for simple and multiple linear regression techniques. Concepts of regression are developed for the student, and essential mathematical and statistical techniques are reviewed for those who need refreshing in them. For those using the book as a reference, example problems are given to illustrate concepts and techniques.

This text is not intended to be all things to all people. It has minimal reference to statistical theory. It does *not* cover all variations and special cases of regression. It is intended to take a student with slight training in statistics from the first basic concepts of relationships among variables through reasonable competence in using a computer to analyze complex multiple regression problems. It reviews topics such as summation notation, confidence intervals, and hypothesis tests, and thus should enable students to handle most problems in regression that they should encounter in their research, and enable them to refer to more specialized texts should they need to. For the more advanced student, this text is a handy reference and can perhaps fill in some gaps in understanding or technique.

This book came about at the request of my students, who kept threatening to publish my lecture notes. Since it grew out of lecture notes, it is very informal in style. However, my objective has been to write a readable, comfortable manuscript.

Many students have supported and encouraged this project. Primary thanks go to my students, friends, and colleagues who have been enthusiastic about my efforts. The College of Business Administration at the University of Tennessee, Knoxville has been generous in its support. Judy Henegar and Janice Savage were most helpful in preparing the manuscript. Virginia Patterson and Don Broach of the UTK Computing Center willingly helped when I came in with a problem.

I am grateful to the Literary Executor of the late Sir Ronald A. Fisher, F.R.S., to Dr. Frank Yates, F.R.S., and to Longman Group, Ltd., London, for permission to reprint Table III from their book *Statistical Tables for Biological, Agricultural, and Medical Research* (6th edition, 1974). I am also grateful to the Biometrika Trustees for portions of Tables 12 and 18 of *Biometrika Tables for Statisticians*, Vol. 1 (3rd edition, 1966), and the Biometrika Trustees and Professor J. Durbin and G. S. Watson for permission to reprint tables from *Biometrika*, Vol. 38 (1951). I would like to thank W. J. Dixon and the Biometric Society for permission to reprint portions of tables from *Biometrics*, Vol. 9 (1853), p. 74. I would also like to acknowledge the SAS Institute of the North Carolina State University, the Health Sciences Computing Facility of the University of California at Los Angeles, and SPSS Inc. of Chicago, whose packaged output I used to illustrate how to run regression programs. Finally, I would like to thank Col. William F. Skidmore, of the Center for Business and Economic Research of the University of Tennessee, Knoxville, for permission to use data published in the *Tennessee Statistical Abstract*, 1974, and Professor David W. Cravens of the Department of Marketing and Transportation at UTK for providing data previously analyzed in David W. Cravens, et al. (1972), "An Analytical Approach for Evaluating Sales Territory Performance," *Journal of Marketing Research, 36,* 31–37.

Although they remain anonymous to me, I would also like to thank the numerous reviewers of this manuscript who provided many helpful suggestions for its improvement. Finally, I must express my appreciation to the two Duxbury Press editors, Alex Kugushev and Robert West, who kept faith with me and encouraged me through the seemingly endless months of revision and more revision. I hope that the final product will justify their support.

INTRODUCTION

The purpose of this text is two-fold. (1) To show students that they really can understand what is going on and *can* use statistical techniques to interpret findings with confidence. (2) To provide a practical handbook for those who must use regression. To this end, the book discusses what we are trying to measure and how to measure it before it presents the formulas. Then, using only techniques of algebra, the book derives the measures and interprets the results specific to the problem, with an emphasis on the ways sloppy interpretation can cause misuse of statistics. Concepts that may be new to the student are carefully developed to promote motivation. The notation used is both simple and enlightening.

The book is divided into two major parts: simple linear and polynomial regression in Chapters 1–9, and multiple linear regression in Chapters 10–15. While it may seem odd to devote more space to the less complicated subject, the mathematics of the regression problem and the proper use of results provide a foundation; all these results carry over into the discussion of multiple regression.

The first nine chapters thus deal with simple regression in great detail and gear up the reader to attack the problems of multiple regression in Chapters 10–15. Chapter 1 introduces the notion of regression and gives examples of the kinds of problems that might be analyzed using regression techniques. Chapters 2–4 are devoted to methods for obtaining the regression equation: first algebraically, then by using matrices, and finally by using the computer. Use of the computer is introduced early (Chapter 4), as is use of matrices (Chapter 3) as preparation for later, more complicated work. Since the reader must analyze any sizable regression problem with the aid of a computer, the book shows how to run regression programs provided by the three most widely available packages: BMD, SPSS, and SAS. It presents instructions in such a way that a reader with no previous exposure to computing will be able to run relatively sophisticated analyses. It explains and compares printouts from the various programs. All analysis is done first by hand so that the reader may become familiar with and confident of the results of the computer programs. Chapter 5 continues the use and interpretation of the regression equation as it applies to a specific problem and points out some common misuses. Techniques of statistical inference are first used in Chapter 6, in which confidence intervals are found for mean and individual responses. Hypothesis tests for the significance of the relationship are given in Chapters 7 and 8. Throughout Chapters 1–8, a single "well behaved" problem is used to illustrate concepts and tie them together. In Chapter 9, some of the problems encountered in less well behaved data are considered.

Chapter 10, Multiple Regression, and Chapter 11, Correlation in Multiple Regression, correspond roughly to Chapters 1–6 and 8, extending the earlier results to more complicated situations. Chapters 12, 13, and 14 discuss significance tests in multiple regression, first explaining the theory, then using the computer to perform the calculations, and finally exploring some of the available variable-selection procedures. For the sake of unity and clarity, an extended version of the illustrative problem used in simple regression is continued until the middle of Chapter 13. Then a larger data problem is introduced, with further illustrations. Finally, Chapter 15 is somewhat analogous to Chapter 9, discussing how actual data may cause problems in analysis, but also introducing two relatively new model selection techniques.

So that the reader may try out new skills on problems different from the one discussed in the text, a variety of problems are included in the exercises. While each problem has a slightly different twist, the reader is not expected to work all of them. Selected answers to exercises are provided at the end of the book; detailed solutions are provided in a separate manual. For the reader's convenience, data sets used in the exercises, along with additional analyses of them, are repeated in a separate section preceding the selected answers.

An appendix contains digressions on topics that may interest the reader.

Chapter 1 SIMPLE LINEAR REGRESSION

1.1 INTRODUCTION

In this chapter, we define regression and some of its basic terms and give some examples of situations in which regression might be the appropriate method of statistical analysis to use. The hypothetical data problem introduced here will be referred to throughout most of the text. The initial analysis includes gathering data that we shall use many times as we explore various techniques of linear regression.

1.2 WHAT IS REGRESSION?

USES OF REGRESSION ANALYSIS

Regression is the study of relationships among variables. One purpose of regression may be to predict, or estimate, the value of one variable from known or assumed values of other variables related to it. For example, the Dean of Admissions of a university might use the College Board scores or high school class standing, personal recommendations, and scores on various tests to predict the college performance of applicants. Economists use such measures as the Index of Industrial Production and the Consumer Price Index to predict the unemployment rate or the Gross National Product. A psychologist working with laboratory rats might use number of hours of food deprivation to predict how long it will take a rat to learn its way through a maze to find food.

In order to make predictions, or estimates, we must identify the effective predictors of the variable of interest. Thus, one of the most crucial tasks in a regression study is to determine which variables are important indicators, which carry only a little information, and which predictors are redundant with other variables. Researchers can frequently identify many variables that may be important, but they must narrow down their predictors to small sets. They seek effective predictors that can be measured with the least cost.

Often a variable is difficult or very expensive to measure. For example, suppose a political scientist wants to measure "degree of political conservatism" among a group of people. Such an intangible concept might be measured on some arbitrary scale after extensive interviews that involve the subjective judgment of the interviewer. But the task would be simpler if the political scientist could find some easily measured variables, such as income, years of education, and age, that could be used effectively to estimate degree of political conservatism.

Note that in regression analysis we use the terms *predict* and *estimate* almost interchangeably. *Predict* normally implies the future but in regression analysis we are not trying to predict change over time. For example, our political scientist may find that one can use age, income, and education to estimate people's degree of political conservatism. But a different analysis would be needed to predict the degree of conservatism the same people will have as they grow older, increase or decrease in income, or go back to school. Predicting a change over time, such as next year's sales for a given company or next year's GNP is not achieved with regression analysis. Rather, it requires *time series* analysis. While there are some similarities between time series analysis and regression analysis, the thrust in regression analysis is to identify variables that carry information about another variable and not to extrapolate from present conditions to future conditions. Thus in regression studies, a predictor variable is one which is used to estimate some characteristic or response variable. We expect two items that have different measurements on the predictor to have different measurements on the response also.

Predictor Variable

Response Variable

Beyond merely identifying which variables can be used to estimate the value of another variable, the methods of regression analysis can also be used to describe the manner in which variables are related. One of the concerns of scientific inquiry is to be able to formulate theories, or models, of the relationships among variables. For example, an economist might hypothesize that as the price of a certain commodity rises, the number of units sold will decline. Techniques of regression analysis can be used to see if that theory is supported or refuted by empirical evidence.

Generally, we begin with an idea—an educated guess or hypothesis—about how several variables might be related to another variable and possibly about the form of the relationship also. With the techniques of regression analysis researchers can test their hypotheses by using empirical evidence and identifying relevant predictors. Once the form of the relationship is determined, methods of regression analysis can be used to estimate the value of the variable of interest. You can probably think of several examples from your own areas of interest in which it is important to formulate models of relationships and make estimates of one variable from values of other variables.

SOME DEFINITIONS AND NOTATION

Independent Variable

When setting up a problem to analyze data, it is customary to denote the predictor, or independent variable, by the symbol X. Then X_1 denotes the first predictor, X_2 denotes the second, and X_k denotes the kth, or the last. The numbering of the predictors

Dependent
Variable

is arbitrary. However, the response, or <u>dependent variable</u>—the one to be predicted from the others—is usually denoted by Y. On the other hand, an economist interested in predicting the quantity of an item sold from the price per unit might call the predictor (price) P and the response (quantity) Q. Some prefer to denote the dependent variable by X_1 and the predictors by $X_2, X_3, \ldots, X_{k+1}$, and this notation corresponds to that used in some computer programs. Still others use X_0 for the response. For most of the text, however, we shall stick to the Y and $X_1, X_2, \ldots, X_k$ notation.

We will begin with the simplest possible case: the prediction of one variable, denoted Y, from one other variable, X. A regression using only one predictor is called a <u>simple regression</u>. Though a single predictor may oversimplify reality, our results here will extend easily to more realistic situations later. When there are two or more predictors, the analysis is called a <u>multiple regression</u>. Furthermore, we are often less interested in accurate predictions than in simply finding a relationship between two variables and describing the relationship.

Simple
Regression

Multiple
Regression

A SINGLE-PREDICTOR EXAMPLE PROBLEM

Suppose a new family in a certain city wants to buy a house. Aware that property values differ considerably from city to city, the newcomers seek an estimate of the prices of various houses in their new home town. Many factors influence the price of a house: The size of the house, the size of the lot, the age and condition of the house, the part of town it is in, whether or not it is air conditioned, how badly the owner wants to sell it, local economic conditions, and so on. Let us choose just one variable, size, and see if the price can be estimated from the size. Then, in our example,

$X=$ size of the house, in thousands of square feet,
$Y=$ price of the house, in thousands of dollars.

We certainly suspect that the price of a house depends on its size to some extent. But in order to be sure that the two variables are related we must collect some data. Some preliminary research should let us determine *if* there is a relationship, and if so, *what* it is. What kind of data should we collect? Obviously, we need to look at houses of varying sizes and record their prices. Our observations then consist of a *pair* of values for each house: one value representing size and the other price. Let X_i denote a value (numerical) of the variable X and let Y_i denote a value of the variable Y. Thus, for any given house, an observation consists of the ordered pair

$$(X_i, Y_i),$$

where the first value represents size and the second denotes price. We look at several houses, say some number n, and record the size and price of each. Then for the first house, the size and price are denoted (X_1, Y_1), for the second our observations are (X_2, Y_2), and so on. Thus we obtain a set of n ordered pairs of observations

$$(X_i, Y_i) \quad \text{for } i = 1, 2, \ldots, n.$$

There are several practical questions to be answered before the actual data collection can take place.

1. How many houses should we look at? Ideally, one would collect information on as many houses as time and money allow. It is these practical considerations that make statistics so useful. Hardly anyone could spend the time, money, and effort needed to look at every house for sale. It is unrealistic to obtain information on every house of interest, or in statistical terms, on every item in the population. Thus, we can look only at a sample of houses—a subset of the population—and hope that this sample will give us reasonably accurate information about the population. Let's say we can afford to look at 20 houses.

Population

Sample

2. From what range of houses do we select the twenty? The newcomers probably have some lower limit on size, determined by how large the family is, how much furniture they have, and how much space they want. There is also probably an upper limit on size determined by much the same considerations. We will sample within this range of sizes that the newcomer would be willing to consider.

3. We would probably choose to select a simple random sample—that is, a sample in which, roughly speaking, every house in the population has equal chance of being included. Then we would expect to get a reasonably representative sample of houses throughout this size range, reflecting prices for the whole city. This sample should give us some information about all houses of all sizes within this range, since a simple random sample tends to select as many larger houses as smaller houses and as many expensive as less expensive ones.

Simple Random Sample

If 20 houses are randomly selected, both X and Y are random variables. We have no control over either and cannot know what specific (X_i, Y_i) values will be selected. It is chance only that determines them. In other situations, however, it might not be wise to sample randomly throughout the entire range of appropriate X-values. For example, a chemical engineer studying the effects of temperature on some compound may choose to look at the effects (Y) of subjecting the compound to temperatures (X), while varying X by five-degree steps throughout some range. The researcher can *choose*, or *preselect*, the X-values because his equipment enables him to apply specific temperatures. In this case, X is not a random variable, but is rather a controlled variable. However, Y, the response, is still random because the researcher cannot know its value in advance and because it is, to some degree at least, determined by chance. The chemical engineer has information about the effects on the compound of *only those temperatures that were applied*, and not about every temperature in the entire range, in contrast to the situation in predicting the price of a house from its size. Thus, whether X is random or controlled is important to the kinds of generalizations that can be made. We shall see that this distinction takes on more importance when more than one predictor is involved.

Random Variable

Controlled Variable

Suppose that the 20 houses in our random sample have the sizes and prices shown in Table 1.1. What can we tell about the relationship between size and price from our sample? (Note that these are hypothetical data chosen for the purpose of illustration and may or may not be realistic representations of prices of houses of given sizes for any particular location.) In the next section we will begin the analysis.

TABLE 1.1   *Sizes and Prices of Twenty Houses*

House Number	Size (Thousands of Square Feet)	Price (Thousands of Dollars)	House Number	Size (Thousands of Square Feet)	Price (Thousands of Dollars)
1	1.8	32	11	2.3	44
2	1.0	24	12	0.9	19
3	1.7	27	13	1.2	25
4	2.8	47	14	3.4	50
5	2.2	35	15	1.7	30
6	0.8	17	16	2.5	43
7	3.6	52	17	1.4	27
8	1.1	20	18	3.3	50
9	2.0	38	19	2.2	37
10	2.6	45	20	1.5	28

EXERCISES   *Section 1.2*

1.1   Give examples from your own field of study of cases in which the value of one variable might be predicted from values of several other variables.

1.2   In the following examples, state which is the dependent variable (Y), which is the independent variable (X).

a. A life insurance company executive wants to predict the amount of life insurance carried by heads of households from families' annual incomes.

b. An advertising agency has completed a campaign to introduce a new product in several cities. It now wants to determine whether or not sales of the product differed in the various cities according to the varying number of television commercials run in them.

c. The manager of a large supermarket wishes to estimate the average time a customer spends in a check-out lane when various numbers of lanes are open.

d. A restaurant manager wishes to estimate the number of customers to be expected on a given evening from the number of dinner reservations received by 5 PM.

e. A sociologist has assigned a "degree of femininity" index to each of several possible occupations. He wishes to know if the sizes of the families certain women came from influence the degree of femininity of their chosen occupations.

f. A psychologist has trained pigeons to peck at a stimulus formed by an illuminated red area surrounding a vertical white rectangle in order to receive food pellets. The experiment is designed to determine if the intensity of the red light has any effect on the time it takes untrained pigeons to learn the response.

g. A geneticist is studying the effect of the number of hours of exposure to radiation on the incidence of mutations in a population of fruit flies.

h. A pharmaceutical company determines the recommended dosage of a new pain-killing drug by varying the dosage on several patients and noting the number of hours relief obtained.

i. A meteorologist wants to determine if there is any relationship between the average monthly rainfall and the average monthly temperature in various sections of the United States.

j   A forester wants to determine if there is any relationship between the number of pine trees per acre and the height of ten-year-old trees.

1.3   In each situation above, state whether X is probably a random or a controlled variable.

1.3 SCATTER DIAGRAMS

PLOTTING THE POINTS

Scatter Diagram

In Table 1.1, it appears that generally speaking the larger the house, the higher the price, but there are exceptions. For example, house 16 is larger than 11, but house 11 is more expensive than 16. Also, houses 5 and 19 are the same size, but 19 is more expensive than 5. It is hard to pinpoint the relationship between size and price from just looking at the raw data. A graph of the data will let us view all 20 pairs of values at once. Such a graph, called a <u>scatter diagram</u>, consists of a set of coordinate axes on which the 20 (X, Y)-values are plotted as points. The scatter diagram for the data in Table 1.1 is shown in Figure 1.1.

    The construction of a scatter diagram is somewhat a matter of personal taste, but there are some general rules to observe. Since the purpose of a scatter diagram is to give an overall view of the problem, one should take care that it does convey information and that it does not distort things. The following four points need to be considered.

FIGURE 1.1   *Scatter Diagram for Sizes and Prices of 20 Houses*

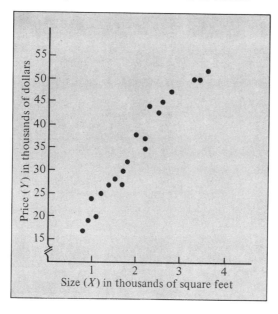

1. Labeling. It is nice to have an informative title. The axes should be labeled with the names of the variables and their units, as well. It is customary, as in algebra, to plot the predictor variable (X) on the horizontal axis and the response (Y) on the vertical axis. (A caution to students in economics: this is opposite the custom adopted by your field.)

2. Scaling. The units on both the horizontal and vertical axes must be scaled as well as labeled, and the plotting should be sufficiently precise for someone to reproduce Table 1.1 from Figure 1.1. Units on the horizontal and vertical axes need not be the same; for example, one inch need not equal one thousand units on both axes. Similarly, it is not necessary for the axes to intersect at zero; often to show a true zero on the scatter diagram means crowding the points in one corner of the quadrant, instead of centering them in the middle of the diagram. Note that on Figure 1.1 the axes do not intersect at zero. A portion of the vertical axis has been omitted in order to center the scatter diagram nicely on the axes. Although the choice of scales on the axes is arbitrary, some care is needed in choosing the vertical scale, to avoid distortion. For example, Figures 1.2(a) and (b) both show the same data plotted on the same horizontal scale; but (b) uses a larger vertical scale than (a). Note how different the two diagrams look.

3. The Points. The points stand alone. Labeling them would only obscure the lay of the data to the eye. Neither are the points connected with lines; such lines carry the eye along in a *time series* and point up the change in a single variable over time. Recall that the typical regression problem concerns a fixed point in time and not change in a variable over time.

4. The Quadrant. Since only positive X- and Y-values are usually observed, most scatter diagrams lie completely in the first quadrant. However, they can extend into other quadrants if either variable assumes negative values. For example, one can observe negative temperatures in the wintertime.

FIGURE 1.2   *Scatter Diagrams with the Same Horizontal Scale but Different Vertical Scales for the Same Set of Data*

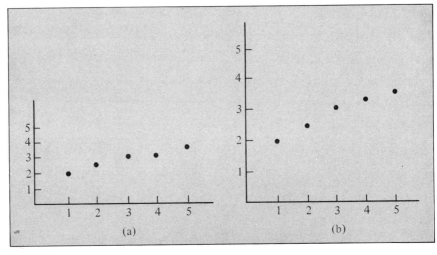

THE DIRECTION OF THE RELATIONSHIP

What can we tell from our scatter diagram? Well, first: Is there a relationship? By a "relationship" one of two things is meant: either the larger the X-value, the generally larger the Y-value; or the larger the X-value, the generally smaller the Y-value. If large X-values and large Y-values tend to occur together (large sizes and large prices) so that the mass of points tends to rise from left to right, then we say that X and Y have a <u>direct</u> relationship. Indeed, there appears to be a direct relationship between sizes and prices of houses in Figure 1.1, as we expected. On the other hand, if large X-values and small Y-values tend to occur together, so that the mass of points falls from left to right, then the relationship between X and Y is <u>inverse</u>.

Inverse relationships are as common as direct ones. For example, one expects an inverse relationship between average winter temperature and consumption of heating oil: the warmer the temperature, generally the less oil used for heating. Similarly, for many commodities studied by economists, the higher the price of an item, the lower the demand. An inverse relationship is shown in the scatter diagram of Figure 1.3(a).

Part (b) of the figure shows a typical scatter diagram indicating no relationship. During the oil shortage in the winter of 1974, the price of gasoline increased sharply but it was found that this price increase has almost no effect on the demand for gasoline. To an economist, this was a classic case of *inelastic demand*. A mathematician would say that quantity demanded was a *constant* function of the price. To a statistician, it meant that the two variables were not related: regardless of the value of X, the value of Y was generally constant. Note on the scatter diagram that the points lie more or less parallel to the horizontal axis.

<u>Direct</u>
<u>Relationship</u>

<u>Inverse</u>
<u>Relationship</u>

<u>No Relationship</u>

FIGURE 1.3   *Scatter Diagrams Indicating (a) an Inverse Relationship and (b) No Relationship Between Two Variables*

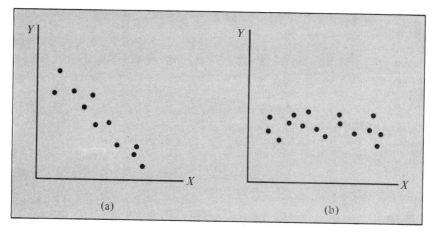

LINEAR AND CURVILINEAR RELATIONSHIPS

Linear
Relationship

Curvilinear
Relationship

If Y increases or decreases at a generally constant rate with X; so that the points tend to lie within a band formed by two parallel straight lines, then we say that there is a linear relationship between the two. If, however, the rate of change of Y with X is not generally constant, then the relationship is curvilinear, and the points on the scatter diagram lie within segments of concentric curves. Figure 1.4 shows examples of some possible linear and curvilinear relationships. The curvilinear relationships in parts (b) and (c) of Figure 1.4 show Y increasing and decreasing with X at *decreasing* rates. Of course, Y could also increase or decrease with X at an increasing rate. (See Problem 1.7.) Part (d) shows Y increasing and then decreasing with X.

To keep the mathematics simple our analyses will emphasize relationships that are considered strictly linear. In the absence of theory indicating that the relationship should be curvilinear, we can ignore any very slight curvature, attributing it to the chance variability inherent in the sampling process. For example, in Figure 1.1 you might see curvature in the range $X = 3,000$ to $X = 4,000$ square feet. But it is slight and may only indicate that we sampled some relatively low-priced houses in that size range. Thus, we begin by assuming that the relationship is linear, although it will be important for us to test this assumption later on. Even if a curvilinear relationship is seen in a given

FIGURE 1.4   *Examples of Linear and Curvilinear Relationships*

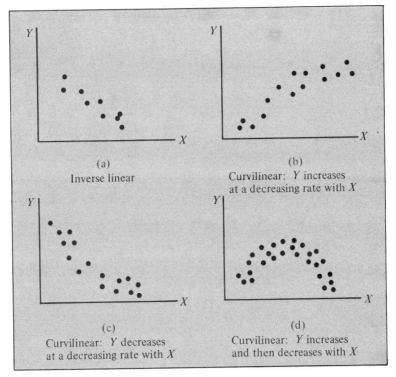

(a)
Inverse linear

(b)
Curvilinear: Y increases
at a decreasing rate with X

(c)
Curvilinear: Y decreases
at a decreasing rate with X

(d)
Curvilinear: Y increases
and then decreases with X

problem, the analysis is still considered part of *linear* regression analysis, as will be explained in Chapter 2.

THE STRENGTH OF THE RELATIONSHIP

Figure 1.1 seems to indicate a relatively strong relationship between size and price; for any small range of sizes the range of prices is also rather small. That is, the vertical *scatter* of points over a small horizontal range, is not large. This means, of course, that if two houses are close to the same size, then their prices are also relatively close. Thus, size should be a strong predictor of price. Suppose that we have two predictors of Y, X_1 and X_2, and that their scatter diagrams are as shown in Figure 1.5(a) and (b). Since for any small range of X-values, the vertical scatter of Y-values is smaller in (a) than in (b), X_1 is a stronger predictor of Y than is X_2. One can predict Y from X_1 more accurately than from X_2.

FIGURE 1.5   *Scatter Diagrams for Two Different Predictors of Y, Showing Relative Strengths of the Predictors*

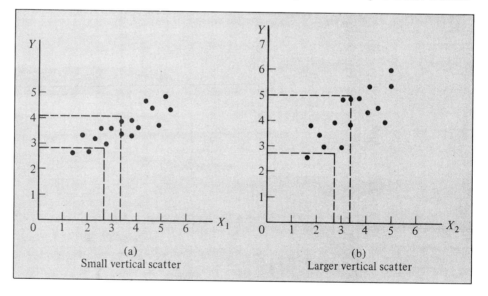

(a)
Small vertical scatter

(b)
Larger vertical scatter

IS THE SCATTER UNIFORM?

The strength of a relationship is measured not only by the amount of scatter over a small range, but also by the constancy of the vertical scatter throughout the range of X-values. The scatter diagram, again, helps us determine whether or not the scatter remains constant. As long as the points lie within parallel straight lines, we have

FIGURE 1.6   *Scatter Diagrams Showing Homogeneous and Heterogeneous Vertical Scatter*

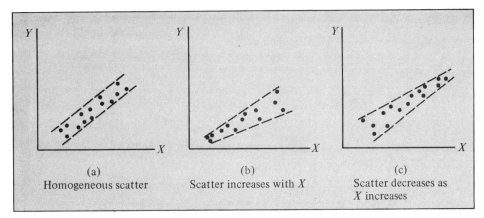

(a)
Homogeneous scatter

(b)
Scatter increases with X

(c)
Scatter decreases as
X increases

<u>Homogeneous
Scatter</u>

<u>Heterogeneous
Scatter</u>

evidence of constant, or <u>homogeneous, scatter</u>,* depicted in (a) of Figure 1.6. Two types of <u>heterogeneous</u> (nonconstant) vertical <u>scatter</u> are shown in (b) and (c).

EXERCISES   *Section 1.3*

We present here six small data sets which will form the nucleus of the exercises through Chapter 8. Some data sets are artificially constructed, while others are based on actual data; in either case n is kept relatively small. The student should pick one or two sets to analyze in parallel with the text's analysis of the house size-price problem and should compare and contrast results of these exercises to results obtained in the text.

In the section titled Analyses of Data Sets, p. 521, you will find a complete analysis of each data set. You may refer to these results in order to check your answers and to find quantities you need in order to proceed with the analysis.

1.4   For the set(s) you select to analyze:
   a.   Draw a scatter diagram.
   b.   Tell whether there is an inverse or a direct relationship, or if there appears to be none at all.
   c.   Classify the relationship(s) you find as linear or curvilinear.
   d.   Describe the strength of the relationship.
   e.   For each relationship, state whether or not the vertical scatter is uniform throughout the range of X-values.

   * Even for curvilinear relationships, as long as the points lie within arcs of concentric curves, we can consider that we have homogeneous scatter.

TEACHERS Data

Number of classroom teachers and average teacher salaries for southeastern states:

State	Number of Teachers (000) Y	Average Annual Salary ($000) X
Alabama	27	4.0
Arkansas	14	3.3
Florida	35	5.1
Georgia	32	3.9
Kentucky	23	3.3
Louisiana	24	5.0
Mississippi	17	3.3
North Carolina	36	4.2
South Carolina	20	3.4
Tennessee	26	3.9
Virginia	30	4.3
West Virginia	15	4.0

Source: Center for Business and Economic Research, The University of Tennessee, Knoxville, *Tennessee Statistical Abstract,* 1974, Table 6.11.

PRECIPITATION
Data

Average annual temperature and average annual rainfall for selected Eastern Tennessee cities, 1972:

City	Precipitation (In.) X	Temperature (°F) Y
A	69.2	57.3
B	51.2	55.8
Ch	64.5	57.7
Cl	63.8	57.7
Ga	67.4	55.1
Gr	53.6	57.6
J	62.7	56.6
Ki	54.0	57.6
Kn	58.0	58.1
L	65.9	57.0
Ne	55.0	56.3
No	63.5	58.9
O	64.9	57.3
R	65.5	54.6
S	56.0	55.1

Source: Center for Business and Economics Research, The University of Tennessee, Knoxville, *Tennessee Statistical Abstract,* 1974, Tables 9.6 and 9.7.

POLICE Data

Suppose a survey of cities of comparable size yields the following data for a given year:

City	Number of Police Officers (X)	Number of Robberies (Y)
1	64	625
2	53	750
3	67	560
4	52	690
5	82	515
6	59	680
7	67	630
8	90	510
9	50	800
10	77	550
11	88	550
12	71	525
13	58	625

SALES Data

Suppose the number of employees (X) and the average weekly retail sales in thousands of dollars (Y) of fifteen stores are:

X	Y
17	7
39	17
32	10
17	5
25	7
43	15
25	11
32	13
48	19
10	3
48	17
42	15
36	14
30	12
19	8

VALUE ADDED
Data

For a certain state, the number of persons employed in certain industries and the value added by manufacture of these industries were:

Industry	Number of Employees (000) X	Value Added (mil $) Y
Food	2.8	29
Textiles	5.5	44
Apparel	1.5	10
Lumber	3.1	42
Furniture	5.5	26
Paper	1.8	18
Printing	5.0	35
Chemicals	5.8	70
Rubber	1.1	17
Plastics	4.3	48
Stone	5.4	54
Ceramics	4.1	24
Machinery	4.9	59
Electrical	2.4	18
Transportation	3.3	31

HEATING Data

Suppose a homeowner in a small northwestern city wishes to estimate the daily winter heating oil bill from the average daily temperature. On 15 randomly selected days during December, January, and February, the homeowner records the average daily temperature and the cost of oil burned that day. The data follow.

Day	Average Daily Temperature (°F)	Daily Oil Bill ($)
1	50	0.30
2	4	0.70
3	5	0.65
4	19	0.65
5	33	0.50
6	48	0.30
7	6	0.65
8	11	0.60
9	27	0.50
10	−2	0.80
11	6	0.70
12	26	0.55
13	49	0.40
14	0	0.70
15	2	0.75

1.5 Plot the data of Table 1.1 on a scatter diagram in which the vertical and horizontal axes intersect at zero.

1.6 Use nontechnical terms to describe the relationships below. For example, a direct relationship between the size of a house and its price means that larger houses tend to cost more than smaller houses.
  a. A direct relationship between the amount of fertilizer applied per plot and the number of bushels of tomatoes harvested.
  b. An inverse relationship among college students between the average number of hours spent per week in intramural sports and grade average.
  c. No relationship between the price per jar of baby food and the number of jars sold.
  d. An inverse relationship between the time it takes to cook a vegetable and the percentage of vitamin content retained.
  e. A direct relationship between consumption of electricity and consumption of beer among families in a given location.

1.7 Sketch a scatter diagram showing:
  a. Y increasing with X at an increasing rate.
  b. Y decreasing with X at an increasing rate.

1.4 THE MODEL

Now that we have collected and plotted our data on houses and determined that there is a linear relationship between size and price, how do we describe that relationship? We know that price increases with size, but how much, and starting where? We would like to have a mathematical equation, or function, or model, which specifies what the relationship is. Since the relationship is linear, we want a <u>linear equation</u> stating Y as a function of X.

Linear
Equation

LINEAR EQUATIONS

At this point, it might be well to digress and review linear equations. The equation $Y = \alpha + \beta X$ is the equation of a straight line. The Greek letters α (alpha) and β (beta) are <u>parameters</u> of the line: once they are specified, the line is fixed. The first parameter, α, is the Y-intercept, the value of Y when $X = 0$. The parameter β is the slope of the line, the number of units and direction that Y changes for each one-unit increase in X.*

Parameters

For example, suppose that $\alpha = 4$ and $\beta = 2$, so that our equation is

$$Y = 4 + 2X.$$

It is easy to see that if $X = 0$,

$$Y = 4 + 2(0) = 4,$$

* Some writers prefer to denote the intercept by β_0 and the slope by β_1.

and $\alpha = 4$ is the value of Y when $X = 0$. Notice that as the value of X is increased by 1, the value of Y increases by $\beta = +2$:

$$\text{if } X = 1, \quad Y = 4 + 2(1) = 4 + 2 = 6,$$
$$\text{if } X = 2, \quad Y = 4 + 2(2) = 4 + 4 = 8,$$
$$\text{if } X = 3, \quad Y = 4 + 2(3) = 4 + 6 = 10,$$

and so on.

To plot the line $Y = 4 + 2X$, then, one could first plot the Y-intercept $(X = 0, Y = 4)$. Knowing that a straight line is determined by two points, we could then choose another X-value, say $X = 5$ for which $Y = 4 + 2(5) = 4 + 10 = 14$. After plotting the point $(5, 14)$, we then connect the two points with a straightedge, as shown in Figure 1.7. Note that we would not ordinarily label the coordinates of the points chosen to plot, since they are completely arbitrary but labeling draws one's attention to the labeled points.

FIGURE 1.7   *Plotting the Equation Y = 4 + 2X*

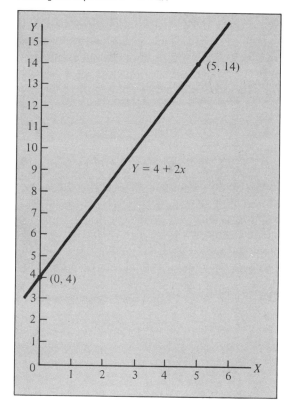

FIGURE 1.8   *Showing Large Divergence in Lines for Small Difference in Plotting Second Point When Second X-Value Chosen is Close to First One*

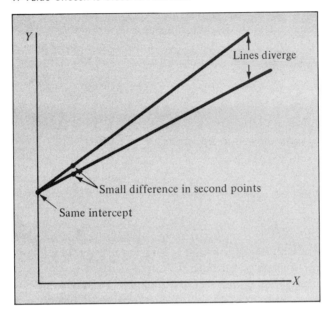

Although the points chosen to plot a straight line are arbitrarily chosen, one usually plots the Y-intercept for convenience. However if, as in Figure 1.1, the axes do not intersect at zero, the Y-intercept cannot be plotted. Another X-value, $X \neq 0$, must be selected. In addition, graphs of straight lines are more accurate if the two X-values chosen are widely spaced, perhaps at opposite ends of the range of data. If the two points are quite close together, a small, often unavoidable error in plotting can radically alter the slope of the line, as Figure 1.8 demonstrates. Conversely, any error in the slope of the line will be small if the two X-values chosen are well spaced.

Here are a few further guidelines it will pay you to remember when graphing a straight line. The Y-intercept can be any real number. If it is positive, then the line crosses the Y-axis above the origin; if it is negative, the line crosses the Y-axis below the origin; and if it is zero the line passes through the origin. As for the slope, it can also be any real number. Note carefully that the maximum possible value for the slope is *not* 1, and that a vertical line has slope equal to infinity. A line with positive slope rises from left to right; one with negative slope falls from left to right; and a horizontal line is one with slope zero. (Students of economics please note this last fact.)

DETERMINISTIC AND PROBABILISTIC MODELS

In an equation of the form

$$Y = \alpha + \beta X$$

the points all lie on a straight line. But, in our scatter diagram, all points do not lie on a straight line. How, then, can the relationship we found between house size and price be described using the equation for a straight line?

Deterministic Model

The equation $Y = \alpha + \beta X$ represents a deterministic, or *mathematical* model— once a value of X is chosen, the value of Y is automatically determined by the specific values of α and β and the rules of arithmetic. Deterministic models are often encountered in such conversion formulas as

$$F = 32 + \frac{9}{5}C,$$

which converts Celsius to Fahrenheit temperature, or

$$I = 12F,$$

which converts feet to inches. Obviously, however, no such exact relationship exists between the size of a house and its price. Many factors other than size help determine the price of a house. When some factors are disregarded, two houses of the same size tend to sell for somewhat different prices.

Furthermore, even if we considered every conceivable factor affecting price and should find two houses with all the same characteristics, we still would expect their prices to differ. Why? We can only attribute the difference to some unpredictable whim of the buyer or the seller, or to chance. This uncontrollable, unmeasurable variation is what distinguishes one identical twin from another, or one laboratory rat from its litter-mate; it's what gives a chemist slightly different results on any two runs of the same experiment. By the same token, our statistical predictions will always be

Random Error

subject to random error: they will always be, to some degree, imperfect when applied to any specific situation.

Probabilistic Model

Recognizing that we can never predict anything exactly, we describe our relationship by means of the probabilistic, or statistical, model

$$Y = \alpha + \beta X + \varepsilon.$$

Here, ε, the Greek letter epsilon, represents the error in the predictions. In this *model*, $Y = \alpha + \beta X$ says that there is a linear relationship, that generally Y increases or decreases at a constant rate with X; the ε says that this relationship is not exact for every individual pair of observations. The error term ε accounts for variables that affect Y but are not included as predictors. It accounts for chance, or random, variability as well as imprecision in the underlying relationship, which might be almost, but not exactly, linear.

We can thus say that the error term is composed of two general kinds of error:

1. *model error*, or *lack of fit*, meaning that all relevant predictors are not specified or the form of the relationship is not precisely specified; and
2. purely *random error*, which is unpredictable and uncontrollable.

ESTIMATING THE UNDERLYING RELATIONSHIP

The model $Y = \alpha + \beta X + \varepsilon$ merely states the conceptual framework of the problem. It is a shorthand way of saying that we are trying to investigate a problem in which there is an imperfect linear relationship between two variables. The exact values of α, β, and ε can never be known. Our primary goal, then, is to get a handle on this underlying relationship so that we can estimate Y from X. Thus, we shall use our data to get numerical estimates, call them a and b, and α and β, so that the estimated price $\hat{Y}$ of a given house can be obtained by simply plugging its size (X) into the equation

Estimated Price

$$\hat{Y} = a + bX.$$

Then, since the estimated price can be expected to differ somewhat from the actual price in any individual instance, our secondary concern is to estimate how far off one can expect the prediction to be. That is, we want to estimate how much error is involved in our predictions. If this error is quite large, this might be an indication that the relationship is not strong enough to bother with.

Finally, we recognize that because our information comes from a randomly chosen sample, there is a chance that it is giving us a distorted picture of how X and Y are related. It might indicate a relationship when in fact none exists. The question of whether or not a relationship exists is central to the whole study of regression: we certainly would not spend time trying to describe a nonexistent relationship and trying to make predictions and measure error from it. However, we will investigate this question last because of the many steps leading up to it. In Chapter 2 we will begin work on our primary goal: obtaining numerical estimates of the slope and the intercept of the underlying linear relationship.

EXERCISES   *Section 1.4*

1.8   Explain the difference between a deterministic and a probabilistic model.

p. 524   1.9   a.   Refer to the VALUE ADDED data. Since both the Textiles and the Furniture industries employ 5.5 thousand workers, how do you explain the fact that the values added by the two industries are not also equal?

p. 523   b.   In the POLICE data cities 3 and 7 both have the same number of police officers, but more robberies are committed in city 7 than in city 3. How do you explain this?

                c.   Suppose that two check-out lanes in a supermarket each have three customers in line. Do you expect both lanes to service their three customers equally quickly? Why or why not?

                d.   A physicist rolls a metal sphere down an incline twice and on each roll measures the time elapsed with an electronic timer. Will the physicist see identical times elapsed both times the experiment is performed?

                e.   How might the physicist in part d formulate a physical "law" relating the degree of incline to the time it takes the ball to roll down it?

1.10 Graph the following linear equations.
    a. $Y = 4 - 2X$;   b. $Y = -4 - 2X$;   c. $R = -5 + 3S$;
    d. $T = 3 + 1.1U$;   e. $Q = 160 - 2.7P$.

1.5 SUMMARY

We have made a good start at the analysis of a simple regression problem. Several concepts and terms have been defined, both generally and with reference to a specific data problem. We have plotted some data and noticed some important features of the example problem. Having decided that a simple linear model will most likely provide an adequate representation for the house size-price problem, in the next three chapters we will study methods by which to describe the underlying relationship present in these data.

Chapter 2 THE LEAST SQUARES REGRESSION EQUATION

2.1 INTRODUCTION

In this chapter we will look at the *least squares* method for fitting a straight line or a curve to bivariate data. After a brief review of summation notation, we shall explore the motivation behind use of the least squares criterion. Alternative, labor-saving methods of calculation and data transformation will also be considered.

2.2 CRITERION FOR THE BEST LINE

Recall that the model for a simple linear regression prediction problem,

$$Y = \alpha + \beta X + \varepsilon,$$

indicates that X and Y are linearly related, but imperfectly; for a given problem we need to estimate α and β in order to obtain the regression, or *prediction*, equation

Regression
Equation

$$\hat{Y} = a + bX,$$

where a and b are numerical values and $\hat{Y}$ is the predicted value of $\hat{Y}$ for a given value of X. Since the linear relationship is imperfect, we must also estimate ε, the size of the error of estimation.

THE LINEARITY ASSUMPTION

Suppose for a moment that the whole population of houses within the desired size range is at our disposal. There would be many different sizes, and for each size, there

FIGURE 2.1   *Scatter Diagram for the Population of Houses*

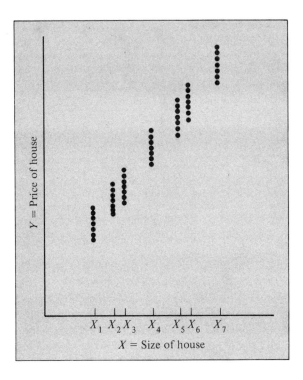

would be not one but a variety of prices. Thus there would be a subpopulation of prices for each different size house, as shown in Figure 2.1. (Note the uniform scatter of the subpopulations.) Under these circumstances, the (population) mean price for a house of a given size could be calculated. We will call this mean $\mu_{y \cdot x}$, the Greek letter *mu* representing a population mean, the subscript y indicating that the mean Y-value is being found, and the "dot-x" reading, "for a given X-value." Imagine calculating $\mu_{y \cdot x}$ for each X-value in the population. *If the relationship between X (size) and Y (price) is really linear, then all the $\mu_{y \cdot x}$-values will lie on a straight line.* That is, we assume that we have a perfect linear relationship in the population between X and the mean Y-value for each X-value. The equation of the line which passes through each mean Y-value is thus

$$\mu_{y \cdot x} = \alpha + \beta X,$$

as in Figure 2.2. If we plug in a size (X), the equation will give the *average* price $(\mu_{y \cdot x})$ for that size. By connecting the average Y-values by a straight line one can read values for α and β from the graph, or β can be determined by finding two $(X_i, \mu_{y \cdot x_i})$ points and using

FIGURE 2.2   *Determining the Underlying Linear Relationship* $\mu_{y\cdot x} = \alpha + \beta X$

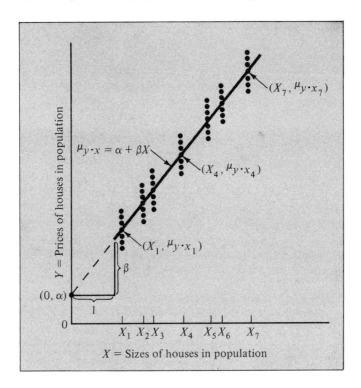

the formula

$$\beta = \frac{\mu_{y\cdot x_2} - \mu_{y\cdot x_1}}{x_2 - x_1}.$$

Thus the linear relationship between X and the average Y for each X can be determined; this is what we have been calling the underlying linear relationship. Naturally, *all* points do not lie on the line because, for a given size house, prices will deviate from the average price because of what we have called *error*: many predictors have been omitted and also random variation is present. (We do, however, have a truly linear relationship.) Since we are using only one predictor and since we cannot eliminate random variation, we shall be doing well to learn the average price for houses of a given size, even if we cannot get the exact price of any specific house. If the linear relationship is strong, the error in any specific instance will be small, and the average price will be close to the actual price.

Observed
Y-Values

Mean Values

In the scatter diagram of Figure 2.2, we notice two different kinds of Y-values corresponding to the X-values. There are observed Y-values and there are mean Y-values. The observed Y-values corresponding to any X-value are denoted Y_i and are represented by points on the scatter diagram. Then for any given X-value, if we find the

FIGURE 2.3   *Straight Lines Providing Good and Poor Descriptions of the Underlying Linear Relationships*

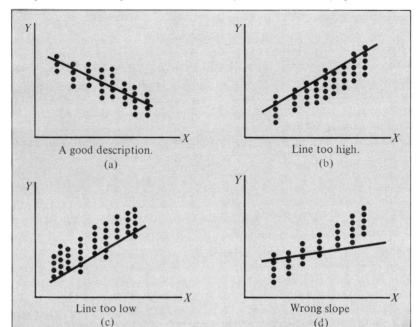

average of the associated Y_i-values we obtain $\mu_{y \cdot x_i}$, and this value lies on the line $\mu_{y \cdot x} = \alpha + \beta X$. (Note that the mean value $\mu_{y \cdot x_i}$ may or may not equal any observed Y_i.) Thus, for a specific house of size X_i, the difference between the actual price and the mean price,

$$Y_i - \mu_{y \cdot x_i},$$

is the error involved in predicting the price from the size alone. Note also that the line need not actually touch any of the points in order to give a good general description.

   Two observations that might seem obvious are nevertheless worth mentioning. If a straight line is to describe the underlying relationship apparent in a scatter diagram, it must pass through the mass of points and it must follow the lay of the points. Figure 2.3 shows some examples of freehand lines that do and do not indicate well the underlying linear relationships.

DETERMINING THE UNDERLYING RELATIONSHIP FROM A SAMPLE

We now drop the false assumption that we can consider the entire population of houses. In reality we will have only a sample of the houses within the specified range of sizes.

For any size house in the population, our sample might record only one of the various prices. Furthermore, the sample may omit some sizes altogether. In other words, the sample may omit some X-values, and it certainly will omit many Y-values corresponding to each X-value sampled. The sample will consist of a randomly selected subset of the points in Figure 2.1 and its scatter diagram will be similar to that of Figure 1.1, p. 6. Having only this sample of information, we will never know exactly what α and β are. But with our sample, we can find an estimate a of α and an estimate b of β. With these, we can fit a straight line

$$Y = a + bX$$

through the points on the scatter diagram. As before, this line should pass through the middle of the mass of points, so that we can predict the mean Y-value for a given value of X. If we denote this *predicted*, or *estimated*, mean Y-value by $\hat{Y}$, the sample regression equation will be

$$\hat{Y} = a + bX.$$

Also as before, for any X-value in the sample there are two corresponding Y-values: Y_i, the observed value of Y corresponding to the observed X_i; and $\hat{Y}_i$, the predicted mean value of Y for X_i. See Figure 2.4. Then (X_i, Y_i) is a point on the scatter diagram, and $(X_i, \hat{Y}_i)$ is a point on the line $\hat{Y} = a + bX$. The difference $Y_i - \hat{Y}_i$ is a measure of the error involved in predicting the price (Y) from the size (X) of a given house.

What has been said so far is actually after the fact. It assumes that values of a and b have been determined. We now turn to the question of how to determine these values for any particular problem; that is, how do we find the equation of the regression line?

FIGURE 2.4   *Relationship between Y_i and $\hat{Y}_i$ for a Given X_i*

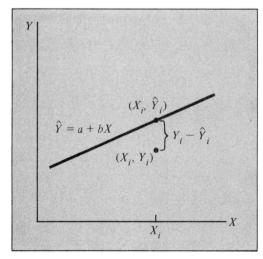

SOME PROPOSED CRITERIA FOR THE "BEST" LINE

One might simply draw a freehand straight line through the mass of points on the scatter diagram and determine the slope and intercept from points on the graph. However, we can do better. What we need is a criterion for estimating the underlying linear relationship in the sample of points.

Deviation   For every size (X_i) in our sample, we want the <u>deviation</u> $Y_i - \hat{Y}_i$ to be as near zero as possible. If $Y_i - \hat{Y}_i = 0$, this means that the line passes through the point (X_i, Y_i). Recall, however, that the line need not pass through any point on the scatter diagram, and certainly a straight line cannot pass through all the points. Thus all $Y_i - \hat{Y}_i$ values cannot be zero. What if we use the criterion that the lines pass through as many points as possible? Figure 2.5 shows the error in this reasoning. Lines A, B, and C all pass through four points and line C obviously does not describe the underlying relationship.

Another criterion that might seem reasonable is to be sure that as many points fall above the line as below. One can easily imagine that this criterion will produce several "best" lines, some such as D in Figure 2.5, having the wrong slope.

Since the objective is to predict the mean Y-value for a given X-value, the line should certainly pass through ($\overline{X}, \overline{Y}$), the mean of both X and Y. While this is desirable, it is no criterion, because an infinity of lines can pass through the point ($\overline{X}, \overline{Y}$). We need something more.

FIGURE 2.5   *Poor Descriptions Arising from Lines Passing through as Many Points as Possible (A, B, C) and from Having as Many Points above as below the Line (D)*

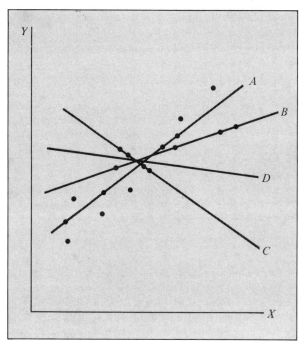

FIGURE 2.6   *Comparison of Two Lines*

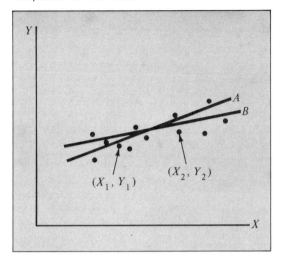

Suppose we must decide which of the two lines in Figure 2.6 better describes that set of points. Both pass through $(\bar{X}, \bar{Y})$. Note that for the point (X_1, Y_1), Line A is closer than Line B. Thus, Line A gives a more accurate prediction of Y_1 than Line B. However, Line B is closer to (X_2, Y_2) and thus Line B will better predict Y_2. We might draw a third line, Line C, which would be better in predicting both Y_1 and Y_2 than either Lines A or B, but then Line C would be further off for some other Y_i than either Line A or Line B.

Thus, we cannot concentrate on getting the line close to any individual point, but must *consider all points simultaneously*. But how can this be done?

Suppose, as in Figure 2.7, we should try to decide between Lines A and B. We might calculate the deviations $Y_i - \hat{Y}_{i(A)}$ from each point to Line A, and also the deviations $Y_i - \hat{Y}_{i(B)}$ from each point to Line B. Now suppose we average the $Y_i - \hat{Y}_{i(A)}$ values, which can be positive, negative, or zero, depending upon whether the point is above, below, or on the line. The average of these deviations from Line A would tell *how much the points deviate from Line A, on the average*. If we do the same thing for Line B, the average of $Y_i - \hat{Y}_{i(B)}$ values will tell the average amount that the points deviate from Line B. If we could repeat this procedure over and over with all possible lines, we could by elimination determine the *best line: the line which is, on the average, closest to all the points*. This seems to be a reasonable criterion for determining the line that gives the best description of the underlying linear relationship.

Of course, it is impossible to make these comparisons for all possible lines. However, as we shall see in the next section, the methods of calculus enable us to do essentially that.

FIGURE 2.7   *Deviations of Points from Two Competing Lines*

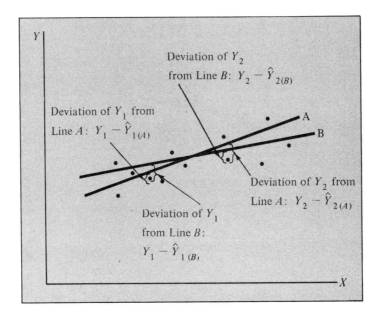

2.3 THE METHOD OF LEAST SQUARES AND NORMAL EQUATIONS

REVIEW OF SUMMATION NOTATION

Before proceeding to find the "best" line to fit a set of data, it might be well to review the use of a summation notation. The three basic rules for manipulating the $\sum$ sign that follow are important to the arguments throughout the text.

Rule 1. The summation sign distributes across the terms of a sum or difference.

$$\sum_{i=1}^{n} (X_i + Y_i - Z_i) = \sum_{i=1}^{n} X_i + \sum_{i=1}^{n} Y_i - \sum_{i=1}^{n} Z_i.$$

This equality is easily verified by letting $n = 2$. Then

$$\sum_{i=1}^{2} (X_i + Y_i - Z_i) = (X_1 + Y_1 - Z_1) + (X_2 + Y_2 - Z_2)$$
$$= (X_1 + X_2) + (Y_1 + Y_2) - (Z_1 + Z_2)$$
$$= \sum_{i=1}^{2} X_i + \sum_{i=1}^{2} Y_i - \sum_{i=1}^{2} Z_i.$$

Rule 2. A constant factors out of a sum.

$$\sum_{i=1}^{n} kX_i = k \sum_{i=1}^{n} X_i, \quad \text{for } k \text{ a constant.}$$

To verify Rule 2, let $n = 3$ and $k = 4$. Then

$$\sum_{i=1}^{3} 4X_i = 4X_1 + 4X_2 = 4X_3$$

$$= 4(X_1 + X_2 + X_3)$$

$$= 4 \sum_{i=1}^{3} X_i.$$

Rule 3. The sum of a constant over n terms is n times the constant.

$$\sum_{i=1}^{n} k = kn, \quad \text{for } k \text{ a constant.}$$

Verification: Let $n = 4$ and $k = 5$. Then

$$\sum_{i=1}^{4} 5 = \underbrace{5 + 5 + 5 + 5}_{4 \text{ terms}} = 4(5).$$

THE LEAST SQUARES CRITERION

The objective is to find the line

$$\hat{Y} = a + bX$$

in which the values a and b make the *average deviation* from the points to the line a minimum. That is, we want to find values a and b so that

$$\frac{1}{n} \sum_{i=1}^{n} (Y_i - \hat{Y}_i) = \frac{1}{n} \sum_{i=1}^{n} (Y_i - a - bX_i)$$

(since $\hat{Y}_i = a + bX_i$) is a minimum. First, we note that for a fixed sample size n, we need only consider minimizing the numerator $\sum_{i=1}^{n} (Y_i - \hat{Y}_i)$ in minimizing the left-hand side of the equation.

The problem is that, as long as the line passes through $(\bar{X}, \bar{Y})$, as we said it should, this sum will always be zero, regardless of the values of a and b. This is because if $X_i = \bar{X}$, then $\hat{Y}_i = a + bX_i = a + b\bar{X}_i = \bar{Y}$, so that

$$\sum_{i=1}^{n} (Y_i - \hat{Y}_i) = \sum_{i=1}^{n} (Y_i - \bar{Y}) = \sum_{i=1}^{n} Y_i - \sum_{i=1}^{n} \bar{Y}$$

$$= \sum_{i=1}^{n} Y_i - n\bar{Y}$$

$$= \sum_{i=1}^{n} Y_i - \sum_{i=1}^{n} Y_i = 0.$$

Thus, if the numerator is zero, the average deviation is zero. Since the average deviation is *always* zero when the line passes through $(\bar{X}, \bar{Y})$, there seems to be no advantage in trying to find a and b so that

$$\sum_{i=1}^{n} (Y_i - \hat{Y}_i) \text{ is minimized.}$$

What has happened is that if $\hat{Y}$ is the mean Y-value for a given X-value, then $\hat{Y}$ is so situated that deviations below the line (negative values) cancel deviations above the line (positive values). This must be so since a mean is a center of gravity, or balancing point. (Recall the notion of variance from introductory statistics. In finding the average deviation of the values from their mean, it was noted that $\frac{1}{n}\sum_{i=1}^{n} (Y_i - \bar{Y})$ is always zero, because the mean is a center of gravity. The same thing applies here, although we are using slightly different notation.) However, since we do not care whether any particular point is above or below the line, as long as it is reasonably close to the line, we might just look at the *distances*, or absolute values of the deviations. By averaging positive numbers and zeroes, we will find the average distance from the points to the line. Thus, we want to find a and b to minimize

$$\sum_{i=1}^{n} |Y_i - \hat{Y}_i| = \sum_{i=1}^{n} |Y_i - a - bX_i|.$$

(Recall that the quantity $\sum_{i=1}^{n} |Y_i - \hat{Y}_i|$ is analogous to the numerator of the *mean deviation* $\frac{1}{n}\sum_{i=1}^{n} |Y_i - \bar{Y}|$ from your introductory course.)

The methods of calculus used to achieve this minimization involve absolute values and thus present a problem that is not trivial. In order to simplify the minimization procedure, we will minimize, instead, the quantity

$$\sum_{i=1}^{n} (Y_i - \hat{Y}_i)^2 = \sum_{i=1}^{n} (Y_i - a - bX_i)^2.$$

By squaring the deviations we eliminate the direction of the differences, since squared real numbers are always positive or zero. (Note that the quantity $\sum_{i=1}^{n} (Y_i - \hat{Y}_i)^2$ is analogous to the numerator of the sample variance $\frac{1}{n-1} \sum_{i=1}^{n} (Y_i - \bar{Y})^2$ from your introductory course, since $\hat{Y}_i$ is also a mean.)

Squared Deviations

Our criterion has changed slightly: we now define as the "best" line the one for which the squared deviations $(Y_i - Y_i)^2$ are the smallest, on the average. This criterion is similar to our original one that (1) it considers all points simultaneously and (2) closeness is defined as a function of the vertical distance between the points and the line. We choose to use the average squared deviation because the minimization procedure for this criterion is much simpler than minimization of the mean absolute deviation. In addition, you will recall from your introductory course that a calculating formula is available for a quantity such as $\sum_{i=1}^{n} (Y_i - \bar{Y})^2$, so that this quantity is much easier to calculate for large samples than is $\sum_{i=1}^{n} |Y_i - \bar{Y}|$. With this criterion, we use the method of least squares to find the best line, the one which is, on the average, closest to all

Method of Least Squares

the points. The name *least squares* should be obvious: least = minimize; squares = squared deviations.*

NORMAL EQUATIONS

Methods of calculus can be used to find values a and b that minimize $\sum(Y_i - \hat{Y}_i)^2 = \sum(Y_i - a - bX_i)^2$ for a given set of data. Those familiar with differential calculus can refer to Appendix A for the details of the minimization procedure. Suffice it to say here that the procedure reduces to the solution of simultaneous linear equations called the normal equations:

Normal
Equations

$$\sum_{i=1}^{n} Y_i = na + b \sum_{i=1}^{n} X_i$$

$$\sum_{i=1}^{n} X_i Y_i = a \sum_{i=1}^{n} X_i + b \sum_{i=1}^{n} X_i^2.$$

Solving these equations simultaneously for a and b will give the line $\hat{Y} = a + bX$ that is, on the average, closest to all the points in terms of squared deviations.

Before proceeding to the solution of the normal equations, we might make several observations. First, the equation

$$\hat{Y} = a + bX$$

is *linear* because a and b, the *unknowns*, are raised only to the first power—not because X is raised to no power greater than 1. Thus, even should the scatter diagram indicate a curvilinear regression, an equation of the form

$$\hat{Y} = a + bX + cX^2$$

is still a *linear* equation, because a, b, and c, the constants, are raised only to the first power. Although it is the equation of a parabola, it is still referred to as *curvilinear*. In this case, we would need to solve *three* simultaneous equations in the three unknowns, a, b, and c. These normal equations also would be derived by the methods of calculus and would be simultaneous linear equations—linear in the unknowns. Any curvilinear equation in any number of unknowns can be fitted to the data points by the same procedure.

---

* We have now advanced our mathematical skill to the early Nineteenth Century. The method of least squares was first published in 1806 by the Frenchman Adrien-Marie Legendre (1752–1833). However, the method had been independently invented, although not published, some years earlier by the great German mathematician Carl Friedrich Gauss (1777–1855). After Legendre's publication of the method of least squares, Gauss commented that he had invented the method himself several years previously. This statement infuriated Legendre, who felt that a man with as many credits to his name as Gauss did not need to steal his (Legendre's) thunder.

NOTICE THE PATTERN

Writing normal equations without using calculus becomes quite easy once a memory trick is mastered. Consider the two normal equations for fitting the line $\hat{Y} = a + bX$:

$$\sum Y_i = na + b \sum X_i$$
$$\sum X_i Y_i = aX_i + b \sum X_i^2.$$

(The indices on the summation signs have been omitted for the sake of simplicity. The subscripts on X and Y are retained to help distinguish them as variables from the constants a and b.) The trick to writing down (memorizing) these two equations involves the following three steps.

Step 1.   Write down the equation you wish to fit.
We want to fit

$$\hat{Y} = a + bX.$$

Step 2.   Refer to Step 1. Find the coefficient on the first unknown. Multiply each term in the equation by this coefficient. Place a summation sign in front of every term of this new equation, and use summation rules.
In this case, the first unknown is a, and its coefficient is 1. Multiplying $\hat{Y} = a + bX$ by 1 gives

$$Y = a + bX$$

(ignore the hat on Y). Placing a summation sign in front of every term gives

$$\sum Y_i = \sum a + \sum bX_i,$$

or

$$\sum Y_i = na + b \sum X_i,$$

which is the first normal equation.

Step 3.   Repeat Step 2, applying it to the second unknown, b.
In this case, the coefficient is X. Multiplying each term by X gives

$$XY = aX + bX^2.$$

Sticking on the summation sign and applying rules of summation notation gives us the second normal equation,

$$\sum X_i Y_i = \sum aX_i + \sum bX_i^2,$$

or

$$\sum X_i Y_i = a \sum X_i + b \sum X_i^2.$$

This trick for obtaining the two normal equations is magic in that it works for any linear equation. For example, if we want to fit

$$\hat{Y} = a + bX + cX^2,$$

then the first normal equation is

$$1(Y) = 1(a) + 1(bX) + 1(cX^2),$$
$$\sum Y_i = \sum a + \sum bX_i + \sum cX_i^2,$$

or

$$\sum Y_i = na + b \sum X_i + c \sum X_i^2.$$

The second is

$$XY = aX + bX^2 + cX^3,$$

or

$$\sum X_i Y_i = a \sum X_i + b \sum X_i^2 + c \sum X_i^3,$$

and the third is

$$X^2 Y = aX^2 + bX^3 + cX^4,$$

or

$$\sum X_i^2 Y = a \sum X_i^2 + b \sum X_i^3 + c \sum X_i^4,$$

so that we would solve the following three equations simultaneously for a, b, and c:

$$\sum Y_i = na + b \sum X_i + c \sum X_i^2,$$
$$\sum X_i Y_i = a \sum X_i + b \sum X_i^2 + c \sum X_i^3,$$
$$\sum X_i Y_i^2 = a \sum X_i^2 + b \sum X_i^3 + c \sum X_i^4.$$

This trick also works for multiple regression, in which we predict Y from several X-variables, as we shall see.

NORMAL EQUATIONS FOR HOUSE SIZE-PRICE EXAMPLE

We now return to the problem of predicting house prices from sizes according to the data in Table 1.1, p. 5. From these data we must calculate the following quantities to plug into the normal equations:

$$\sum Y_i, \quad n, \quad \sum X_i, \quad \sum X_i Y_i, \quad \sum X_i^2.$$

The quantity $\sum X_i Y_i$ indicates that for each house in the sample, we multiply the size X_i by the price Y_i, and add the products. Note that

$$\sum X_i Y_i \neq \left(\sum X_i\right)\left(\sum Y_i\right),$$

since $\left(\sum X_i\right)\left(\sum Y_i\right)$ indicates that we total all the sizes $\left(\sum X_i\right)$, total all the prices $\left(\sum Y_i\right)$ and then multiply these totals. Similarly, $\sum X_i^2$ indicates that we square each size and total the squares. Note also that

$$\sum X_i^2 \neq \left(\sum X_i\right)^2,$$

since $\left(\sum X_i\right)^2$ indicates that the sizes are totaled and the total is then squared.

Necessary calculations from the data are given in Table 2.1. (Note that it includes Y^2 figures, which are not needed for solving the normal equations but will be needed later.) We find there $n = 20$ and

$$\sum Y_i = 690, \quad \sum X_i = 40, \quad \sum X_i^2 = 93.56, \quad \sum X_i Y_i = 1554.9.$$

TABLE 2.1   *Calculations for House Size and Price Problem*

House Number	X = Size (000 sq ft)	Y = Price ($000)	X^2	XY	Y^2
1	1.8	32	3.24	57.6	1024
2	1.0	24	1.00	24.0	576
3	1.7	27	2.89	45.9	729
4	2.8	47	7.84	131.6	2209
5	2.2	35	4.84	77.0	1225
6	0.8	17	0.64	13.6	289
7	3.6	52	12.96	187.2	2704
8	1.1	20	1.21	22.0	400
9	2.0	38	4.00	76.0	1444
10	2.6	45	6.76	117.0	2025
11	2.3	44	5.29	101.2	1936
12	0.9	19	0.81	17.1	361
13	1.2	25	1.44	30.0	625
14	3.4	50	11.56	170.0	2500
15	1.7	30	2.89	51.0	900
16	2.5	43	6.25	107.5	1849
17	1.4	27	1.96	37.8	729
18	3.3	50	10.89	165.0	2500
19	2.2	37	4.84	81.4	1369
20	1.5	28	2.25	42.0	784
Sum	40.0	690	93.56	1554.9	26,178

Substituting these values into the normal equations gives

$$690 = 20a + 40b$$

$$1554.9 = 40a + 93.56b.$$

These two equations must now be solved simultaneously for a and b. Any one of several equivalent methods can be used. Perhaps the easiest for now is to perform operations on the equations so that one of the unknowns has the same coefficient in both equations. This is simply done if we multiply the top equation by 2, to obtain a coefficient of 40 on a in both equations:

$$1380 = 40a + 80b$$

$$1554.9 = 40a + 93.56b.$$

(Of course, multiplying the top equation by 93.56 and the bottom by 40 would give the coefficient 40(93.56) = 3742.40 on b in both equations, should one choose to eliminate b first.) Now we simply subtract one equation from the other. Since the top equation has the smaller values, subtracting the top from the bottom will result in positive numbers, although it does not matter which is subtracted. We obtain upon subtraction

$$174.9 = 0a + 13.56b,$$

or

$$174.9 = 13.56b,$$

which is one equation in one unknown. The solution is

$$\frac{174.9}{13.56} = b = 12.8982.$$

To find a, substitute $b = 12.8982$ into one of the original equations. Since the top equation has smaller numbers,

$$690 = 20a + 40(12.8982)$$

$$690 - 515.9280 = 20a$$

$$\frac{174.072}{20} = a = 8.7036.$$

CHECKING YOUR RESULTS

Since an arithmetic error is always possible, one should substitute the values for a and b back into the original equations. Both equations must check. Checking,

$$20(8.7036) + 40(12.8982) = 690.0000$$

$$40(8.7036) + 93.56(12.8982) = 1554.8996$$

$$\approx 1554.9.$$

Noting that the second equation does not check exactly to four decimal places brings up the question of rounding error. Accuracy depends upon the number of decimal places carried in calculations; the more places carried, the more accurate the results. Obviously, then, one carries as many decimals as feasible, and this is no great burden with the wide availability of electronic calculators. (Only a masochist does regression problems by hand.) When reporting results, however, one needs to be practical. If a variable is measured in, say, thousands of dollars, then a figure such as 10.98765 means \$10,987.65. However, if the units were dollars instead of thousands of dollars, the figure 10.98765 would denote \$10.98765, which would be better expressed as \$10.99.

The number of decimal places recorded in the data often indicates the form of the result. If the house prices, for example, are recorded in round thousands of dollars, such as 32 = \$32,000, it would make little sense to predict a particular house as selling for, say, \$28,935.64. Perhaps \$28.9 thousand would be as close as you would care to estimate; as a rule of thumb, carry at least two more decimal places than shown in your data, and round results correct to one more place.

Rounding error comes about when divisions are made and when square roots are taken. Thus, error can be minimized by postponing divisions and extraction of roots as long as possible. For example, to evaluate

$$\frac{a}{b/c}$$

by first dividing b by c and then dividing a by their quotient involves two divisions and thus two sources of rounding error. It would be better done as ac/b, which involves only one division. Similarly, $\sqrt{a}\sqrt{b}$ is better calculated as $\sqrt{ab}$, since this latter form involves extracting only one square root.

GRAPHING THE REGRESSION EQUATION

For the sake of simplicity, let's round our answers in the house example to

$$a = 8.7, \qquad b = 12.9,$$

which yield the regression equation

$$\hat{Y} = 8.7 + 12.9X.$$

This is the line that, on the average, is closest to all the points in our data in terms of squared deviations. Plotting the regression equation on the original scatter diagram, as in Figure 2.8, shows that the line $\hat{Y} = 8.7 + 12.9X$ both follows the lay of the points and passes through the middle of the mass of points. Plotting the regression equation on the scatter diagram is also a good way to catch any gross errors in addition or squaring or multiplying in the original data. Note in Figure 2.8 that the regression equation is labeled, but that points chosen to plot it are not labeled.

FIGURE 2.8   *Regression Line Superimposed on Scatter Diagram*

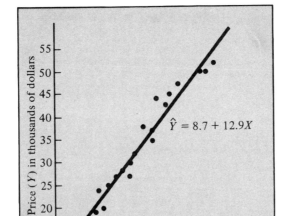

We can also verify that the least squares line passes through the point $(\bar{X}, \bar{Y})$. In our example, $\bar{X} = 40/20 = 2$ and $\bar{Y} = 690/20 = 34.5$. Substituting $\bar{X}$ for X in the regression equation produces

$$\hat{Y} = 8.7 + 12.9(2)$$
$$= 8.7 + 25.8$$
$$= 34.5 = \bar{Y},$$

so that the line does pass through $(\bar{X}, \bar{Y})$. This can also be demonstrated by locating the point (2, 34.5) in Figure 2.8.

The next section will show an alternative method for finding the regression equation.

EXERCISES   *Sections 2.2 and 2.3*

2.1   What is the difference between the points (X_i, Y_i) and $(X_i, \hat{Y}_i)$?

2.2   Use the three rules of summation notation to find equivalent forms for the following expressions. Assume all sums are from $i = 1$ to n. Example: $\sum(X_i - 4) = \sum X_i - 4n$.
   a. $\sum(X_i^2 - X_i + 4)$;
   b. $\sum(3X_i + Y_i)$;
   c. $\sum(X_i - 4)^2$;
   d. $\sum(3X_i - 4Y_i)^2$;
   e. $\sum(3X_i + Y_i)(4X_i - 5Y_i)$;
   f. $\sum(Y_i - a - bX_i)$, for a and b constants;
   g. $\sum(X_iY_i - aX_i - bX_i^2)$, for a and b constants;
   h. $\sum(Y_i - a - bX_i)(X_i)$, for a and b constants;
   i. $\sum(X_i - \bar{X})^2$, for $\bar{X}$ a constant;
   j. $\sum(X_i - \bar{X})(Y_i - \bar{Y})$, for $\bar{X}$ and $\bar{Y}$ constants.

2.3   Using rules of summation, show that
   a. $\sum(X_i - \bar{X}) = 0$;

   b. $\sum(X_i - \bar{X})^2 = \dfrac{n\sum X_i^2 - (\sum X_i)^2}{n}$;

   c. $\dfrac{\sum(X_i - \bar{X})^2}{n - 1} = \dfrac{n\sum X_i^2 - (\sum X_i)^2}{n(n - 1)}$;

   d. $\sum(X_i - \bar{X})(Y_i - \bar{Y}) = \dfrac{n\sum X_iY_i - (\sum X_i)(\sum Y_i)}{n}$;

   e. $\sum(Y_i - Y)^2 = \dfrac{n\sum Y_i^2 - (\sum Y_i)^2}{n}$.

2.4   Using the data $X : 1, 2, 3, 4, 5,$

   a.   calculate $1/n\sum_{i=1}^{n}(X_i - \bar{X})$ and show that this quantity is zero;

   b.   calculate the mean deviation

$$\text{MD} = \frac{1}{n}\sum_{i=1}^{n}|X_i - \bar{X}|;$$

   c.   calculate the variance

$$s_x^2 = \frac{1}{n - 1}\sum_{i=1}^{n}(X_i - \bar{X})^2.$$

2.5   Using $X : 1, 2, 3, 4, 5,$
   a.   show that $\sum X_i^2 \neq (\sum X_i)^2$;
   b.   calculate s_x^2 by using the formula

$$s_x^2 = \frac{n(\sum X_i)^2 - (\sum X_i)^2}{n(n - 1)}.$$

Compare your answer to that of Exercise 2.4c. Which method of calculating s_x^2 was easier and quicker?

c. True or false? $\sum X_i Y_i = (\sum X_i)(\sum Y_i)$.

2.6 Why is the regression line obtained by the method of least squares considered to be the best line that can be fitted to the data?

2.7 Write down the normal equations that one would have to solve in order to fit the following equations to a set of data.

a. $\hat{Q} = a + bP$;
b. $\hat{E} = a + bT + cT^2$;
c. $\hat{Y} = a + bX + cX^2 + dX^3$;
d. $\hat{Y} = a + bX + cZ$;
e. $\hat{Y} = b_0 + b_1 X_1 + b_2 X_2$.

2.8 Set up and solve the normal equations needed to fit a straight line of the form

$$\hat{Y} = a + bX$$

p. 521, 522, 524, 525

to the data of either the TEACHERS data, the PRECIPITATION data, the SALES data, or the HEATING data, whichever you chose to analyze. Refer to Problem 1.4, or the Analyses of Data Sets for the data to use. Check your calculations. State the regression equations.

2.9 For the data set(s) you chose to analyze in the previous problem, plot the regression equations on the appropriate scatter diagrams. Check that the lines follow the lay of the points.

2.10 For the data set(s) you are analyzing, verify that the lines pass through the points $(\overline{X}, \overline{Y})$ for the appropriate sets of data.

2.11 a. Set up and solve the normal equations necessary to fit an equation of the form

$$\hat{Y} = a + bX$$

p.523

to the POLICE data.

b. Set up and solve the normal equations necessary to fit an equation of the form

$$\hat{Y} = a + bX + cX^2$$

to the POLICE data.

c. Using a and b from part a of this exercise, calculate $\hat{Y}_i = a + bX_i$ for each of the given X-values.

d. Using a, b, and c from part b, calculate $\hat{Y}_i = a + bX_i + cX_i^2$ for each of the given X-values.

e. Using the $\hat{Y}_i$-values from part c, calculate $\frac{1}{n}\sum |Y_i - \hat{Y}_i|$.

f. Using the $\hat{Y}_i$-values from part d, calculate $\frac{1}{n}\sum |Y_i - \hat{Y}_i|$.

g. Compare your results in parts e and f. Referring to your scatter diagram, which do you think better describes the data: the straight-line regression equation $\hat{Y} = a + bX$ or the curvilinear equation $\hat{Y} = a + bX + cX^2$?

h. Plot the equations obtained in parts a and b on the scatter diagram.

2.4 SHORTCUT FORMULAS

The regression equation can always be found by using the normal equations, regardless of how many predictor variables are used and regardless of whether the relationship is linear or curvilinear. However, in simple regression, the regression line can also be found by using two shortcut formulas. These formulas eliminate the need for solving simultaneous equations. However, their greatest value lies in the fact that they involve calculating quantities which will be used extensively throughout the remainder of our analysis. Since these quantities must be calculated eventually, why not find them first and also use them to obtain values for a and b?

USEFUL NOTATION

At this point, we digress briefly to introduce some notation that will help us simplify many formulas. First recall that the rules of summation notation let us obtain a calculating formula for the quantity $\sum_{i=1}^{n}(X_i - \overline{X})^2$:

$$\sum(X_i - \overline{X})^2 = \sum(X_i^2 - 2X_i\overline{X} + \overline{X}^2) \quad \text{(expanding the square)}$$

$$= \sum_{i=1}^{n} X_i^2 - 2\overline{X}\sum_{i=1}^{n} X_i + n\overline{X}^2 \quad \text{(Rules 1, 2, 3)}$$

$$= \sum_{i=1}^{n} X_i^2 - 2\left(\frac{\sum_{i=1}^{n} X_i}{n}\right)\sum_{i=1}^{n} X_i + n\left(\frac{\sum_{i=1}^{n} X_i}{n}\right)^2 \quad \text{since} \quad \left(\overline{X} = \frac{\sum_{i=1}^{n} X_i}{n}\right)$$

$$= \sum_{i=1}^{n} X_i^2 - 2\frac{\left(\sum_{i=1}^{n} X_i\right)^2}{n} + \frac{\left(\sum_{i=1}^{n} X_i\right)^2}{n}$$

$$= \sum_{i=1}^{n} X_i^2 - \frac{\left(\sum_{i=1}^{n} X_i\right)^2}{n}$$

$$= \frac{n\sum_{i=1}^{n} X_i^2 - \left(\sum_{i=1}^{n} X_i\right)^2}{n}.$$

Let

$$S_{XX} = \sum_{i=1}^{n} (X_i - \overline{X})^2 = \frac{n\sum_{i=1}^{n} X_i^2 - \left(\sum_{i=1}^{n} X_i\right)^2}{n}.$$

The S_{XX} notation is used *only* to shorten the amount of writing which must be done. Note that $\sum_{i=1}^{n} (X_i - \bar{X})^2$ is a sum of squares (or sum of squared quantities) involving X-values; it is called a <u>corrected sum of squares</u> (corrected for the mean), as opposed to the <u>raw sum of squares</u>, $\sum X_i^2$. The S_{XX} notation reminds us of this, if we think of the letter S as denoting sum, and the XX subscript as denoting squares of quantities involving X-values $(X \cdot X = X^2)$.* Note also that

Corrected Sum of Squares

Raw Sum of Squares

$$\sum_{i=1}^{n} (X_i - \bar{X})^2 = \frac{1}{n}\left[n \sum_{i=1}^{n} X_i^2 - \left(\sum_{i=1}^{n} X_i\right)^2 \right]$$

is the numerator of the calculating formula for the sample variance of the variable X. Thus we could write the <u>sample variance</u> as

Sample Variance

$$\hat{\sigma}_x^2 = \frac{S_{XX}}{n}$$

or as

$$s_x^2 = \frac{S_{XX}}{n-1},$$

if an unbiased estimator of the population variance is required.

This sum of squares notation is quite versatile. For example, replacing X_i by Y_i gives

$$S_{YY} = \sum_{i=1}^{n} (Y_i - \bar{Y})^2 = \frac{1}{n}\left[n \sum_{i=1}^{n} Y_i^2 - \left(\sum_{i=1}^{n} Y_i\right)^2 \right].$$

In addition, if we note that

$$S_{XX} = \sum_{i=1}^{n} (X_i - \bar{X})^2 = \sum_{i=1}^{n} (X_i - \bar{X})(X_i - \bar{X}),$$

or

$$S_{XX} = \frac{1}{n}\left[n \sum_{i=1}^{n} X_i^2 - \left(\sum_{i=1}^{n} X_i\right)^2 \right]$$

$$= \frac{1}{n}\left[n \sum_{i=1}^{n} X_i X_i - \left(\sum_{i=1}^{n} X_i\right)\left(\sum_{i=1}^{n} X_i\right) \right],$$

* Some texts denote the quantity $\sum_{i=1}^{n} (X_i - \bar{X})^2$ as $\sum x_i^2$, where $x_i = X_i - \bar{X}$. Still other texts give the calculating formula as $\sum X_i^2 - (\sum X_i)^2/n$ or as $\sum X_i^2 - n\bar{X}^2$, but the form given above will be more convenient for us in most situations. Care should be taken when reading another text to get the notation straight.

then by simply replacing one of the X's in S_{XX} by a Y, we get

$$S_{XY} = \sum_{i=1}^{n} (X_i - \bar{X})(Y_i - \bar{Y})$$

$$= \frac{1}{n}\left[n \sum_{i=1}^{n} X_i Y_i - \left(\sum_{i=1}^{n} X_i\right)\left(\sum_{i=1}^{n} Y_i\right) \right].$$

Corrected Sum of Cross-Products
S_{XY} is a corrected sum of cross-products (corrected for the means of X and Y) as opposed to the *raw* sum of cross products, $\sum X_i Y_i$. Some readers might recognize that

$$S_{XY} = \sum_{i=1}^{n} (X_i - \bar{X})(Y_i - \bar{Y})$$

Covariance
is the numerator of the sample covariance between two variables X and Y and is a measure of the extent to which the two variables act alike. We will have much more to say about this later.

SHORTCUT FORMULAS

With this shorthand notation, we are ready to derive the shortcut formulas for finding a and b in the equation $Y = a + bX$. Keep in mind that these formulas work only in the case of simple linear regression, as will be apparent from their derivation.
We begin by solving the two normal equations algebraically.

$$\sum Y_i = na + b \sum X_i$$
$$\sum X_i Y_i = a \sum X_i + b \sum X_i^2.$$

Multiplying the first equation by $\sum X_i$ and the second by n will eliminate a in both equations:

$$\left(\sum X_i\right)\left(\sum Y_i\right) = na \sum X_i + b\left(\sum X_i\right)^2$$
$$n \sum X_i Y_i = na \sum X_i + nb \sum X_i^2.$$

Subtracting the top equation from the bottom,

$$n \sum X_i Y_i - \left(\sum X_i\right)\left(\sum Y_i\right) = nb \sum X_i^2 - b\left(\sum X_i\right)^2$$
$$= b\left[n \sum X_i^2 - \left(\sum X_i\right)^2\right].$$

Dividing both sides by n, we obtain

$$S_{XY} = bS_{XX},$$

or

$$b = S_{XY}/S_{XX},$$

which is the shortcut formula for b. Since we will use S_{XY} and S_{XX} many times in our analysis, we can save time by calculating S_{XY} and S_{XX} first, and then using them to find b. The result will be the same as that obtained by solving the normal equations.

Once b is found, substitution into the first normal equation yields a value for a:

$$\sum Y_i = na + b \sum X_i,$$

$$\sum Y_i - b \sum X_i = na,$$

$$\bar{Y} - b\bar{X} = a,$$

where $\bar{Y} = \sum Y_i/n$ and $\bar{X} = \sum X_i/n$ are the mean values of the dependent and independent variables, respectively.

USING SHORTCUT FORMULAS IN HOUSE SIZE-PRICE EXAMPLE

In our example about sizes and prices of houses,

$$S_{XY} = \frac{20(1554.9) - (40)(690)}{20} = \frac{3498}{20} = 174.9,$$

$$S_{XX} = \frac{20(93.56) - (40)^2}{20} = \frac{271.2}{20} = 13.56.$$

Then

$$b = \frac{174.9}{13.56} = 12.8982,$$

which is just what we found by using the normal equations. (Referring back to the calculation of b by using normal equations, we see that we found $b = 174.9/13.56 = S_{XY}/S_{XX}$. In some instances, the actual numbers in the ratio when solving normal equations would be nS_{XY}/nS_{XX}, or the *numerators* of S_{XY} and S_{XX}, because the n's in the denominators cancel. Thus, in our example, we could just as well have found b by $3498/271.2 = 12.8982$. Because nS_{XY}/nS_{XX} involves two fewer divisions than S_{XY}/S_{XX}, the use of formulas may introduce some additional rounding error.)

Continuing with our example,

$$a = \bar{Y} - b\bar{X}$$

$$= \frac{690}{20} - 12.8982\left(\frac{40}{20}\right)$$

$$= 34.5 - 25.7964$$

$$= 8.7036,$$

which is also the same value for a that we found earlier.

To summarize,

1. the same values for a and b (except for possible rounding error) will be found by solving normal equations and by using shortcut formulas;
2. shortcut formulas can be used only in simple linear regression;
3. normal equations can be used in *any* regression problem;
4. in simple regression, use of formulas often saves time, since S_{XY} and S_{XX} will have to be calculated eventually anyway.

A CLOSER LOOK

Let us pause for a moment to examine the formula for b,

$$b = \frac{S_{XY}}{S_{XX}}.$$

You will recall that b, as the slope of the line $\hat{Y} = a + bX$, measures the number of units and the direction of change in Y for a one-unit change in X. Let us see how the shortcut formula for b reflects this fact.

First, note that S_{XX}, being a sum of squares, can never be negative. It can be zero, but only if all X-values are the same; that is, if we have looked at only one X-value. In any realistic regression problem, then, because S_{XX} is always positive, the sign of b must be determined by S_{XY}.

What determines the sign of S_{XY}? First suppose that X and Y are directly related. Then large X-values tend to occur with large Y-values and small X-values with small Y-values. If the X-value is large, then $X_i - \bar{X}$ will be positive. If the Y-value is large, the difference $Y_i - \bar{Y}$ will also be positive, and so will the product $(X_i - \bar{X})(Y_i - \bar{Y})$. On the other hand, if X and Y are both small, $X_i - \bar{X}$ will be negative and so will $Y_i - \bar{Y}$; their product, however, will be positive. Thus, if X and Y are directly related, $\sum(X_i - \bar{X})(Y_i - \bar{Y})$ will be the sum of mostly positive values and will itself be positive.

Suppose now that X and Y are inversely related. Then when one is large, say X, the other tends to be small. Then if $X_i - \bar{X}$ is positive, the associated value of $Y_i - \bar{Y}$ will tend to be negative and the product will be negative. The sum $\sum(X_i - \bar{X})(Y_i - \bar{Y})$ will thus be negative if X and Y are inversely related.

Finally, if the two variables are not related, then regardless of the value of X, Y tends to remain constant. Most values of Y are thus equal to $\bar{Y}$ and the differences $Y_i - \bar{Y}$ are close to zero. Regardless of whether X is large or small, then, the products $(X_i - \bar{X})(Y_i - \bar{Y})$ will be near zero and so will the sum $\sum(X_i - \bar{X})(Y_i - \bar{Y})$. So if X and Y are not related, b will be zero (or close to it, when random variability is taken into account).

We have seen how the direction of the relationship determines the sign of S_{XY} and thus the sign of b.

The shortcut formula $b = S_{XY}/S_{XX}$ also reminds us that b measures change in Y for change in X. It does this because its units of measurement are Y-units over X-units. That this is so can be easily seen: $S_{XY} = \sum (X_i - \bar{X})(Y_i - \bar{Y})$ is in X-units times Y-units and $S_{XX} = \sum (X_i - \bar{X})^2$ is in squares of X-units. Their ratio is then in Y-units over X-units.

EXERCISES   *Section 2.4*

2.12 Write out the formulas for
       a. S_{TT}; b. S_{RR}; c. S_{RT}; d. S_{TR}.

p. 521, 522.   2.13 Use shortcut formulas to find the regression equation for either the TEACHERS data, the
524, 525   PRECIPITATION data, the SALES data, or the HEATING data. Check to see that you
                      found the same values for a and b as you did in Exercise 2.8, in which the regression
                      equation was found by solving normal equations. (Refer to the Analyses of Data Sets
                      section for preliminary calculations.)

2.5 CODING THE X-VALUES

Sometimes we can save ourselves a lot of calculating if the predictor variable is appropriately *centered* or *scaled*. For example, an economist constructing a time series across several years might have to work with the figures

1969   1970   1971   1972   1973.

It can get quite cumbersome if one must add or square these values. Life will be easier if the economist codes the years thus:

1969 $= -2$,   1970 $= -1$,   1971 $= 0$,   1972 $= 1$,   1973 $= 2$.

Nothing is really changed in the problem as long as the economist remembers that the value 0 denotes the year 1971, and so on.

An engineer testing the breaking strength of a certain metal alloy rod might subject it to pressures of 100, 150, and 200 pounds. Since the difference between the low and medium pressures is equal to the difference between the medium and high pressures—that is, the experimental levels are evenly spaced—the engineer can code the three respective pressures $-1, 0$, and $+1$, values that are much easier to work with.

CENTERING X-VALUES

In the time series example, if we denote the coded values by $U_i = X_i - \bar{X}$, then

$$\sum(U_i) = 0.$$

In effect, the coding was achieved by finding the mean of the original values and subtracting this mean from each of the original values. Thus if $X_1 = 1969$,

$$U_1 = X_1 - \bar{X} = 1969 - 1971 = -2$$

$$U_2 = X_2 - \bar{X} = 1970 - 1971 = -1$$

$$U_3 = X_3 - \bar{X} = 1971 - 1971 = 0$$

$$U_4 = X_4 - \bar{X} = 1972 - 1971 = 1$$

$$U_5 = X_5 - \bar{X} = 1973 - 1971 = 2.$$

Centering
X-Values

Since the mean is subtracted from each value, we say that such a coding scheme centers the X-values, or *corrects the X-values for their mean*.

What effect will this centering have on our values for a and b in the regression equation? There should be none. Let us check it out: If we code the X-values, then using U-values,

$$S_{XX} = S_{UU} = \frac{n \sum(U_i)^2 - (\sum U_i)^2}{n} = \sum U_i^2$$

and

$$S_{XY} = S_{UY} = \frac{n \sum U_i Y_i - (\sum U_i)(\sum Y_i)}{n} = \sum U_i Y_i,$$

since

$$\sum U_i = 0. \quad \text{Then}$$

$$b = \frac{S_{XY}}{S_{XX}} = \frac{\sum U_i Y_i}{\sum U_i^2}$$

and

$$a = \bar{Y} - b\bar{U} = \bar{Y}$$

since

$$\sum U_i/n = \bar{U} = 0.$$

Note that b really has not changed, since

$$\frac{\sum U_i Y}{\sum U_i^2} = \frac{\sum (X_i - \bar{X})Y_i}{\sum (X_i - \bar{X})^2}$$

$$= \frac{S_{XY}}{S_{XX}},$$

because $\sum (X_i - \bar{X})Y_i = \sum (X_i - \bar{X})(Y_i - \bar{Y})$ as shown below:

$$\sum (X_i - \bar{X})Y_i = \sum (X_i - \bar{X})Y_i - \sum (X_i - \bar{X})\bar{Y} + \sum (X_i - \bar{X})\bar{Y}$$

$$= \left[\sum (X_i - \bar{X})Y_i - \sum (X_i - \bar{X})\bar{Y} \right] + \sum (X_i - \bar{X})\bar{Y}$$

$$= \sum \left[(X_i - \bar{X})Y_i - \sum (X_i - \bar{X})\bar{Y} \right] + \bar{Y} \sum (X_i - \bar{X})$$

$$= \sum \left[(X_i - \bar{X})(Y_i - \bar{Y}) \right] + \bar{Y}(0)$$

$$= \sum (X_i - \bar{X})(Y_i - \bar{Y}) = S_{XY}.$$

Thus, the general underlying relationship between X and Y, as indicated by b, is not affected by centering the predictor variable. The value for a, however, is changed. To understand why: (1) recall that a is the Y-intercept, the value of Y when $X = 0$; (2) note that with the coding, the Y-axis has been shifted to the mean X-value, since $U = 0$ where $X = \bar{X}$. If the vertical axis is now located at $\bar{X}$, then the line will cross the Y-axis at $\bar{Y}$, since the line goes through the point $(\bar{X}, \bar{Y})$.

When the predictor variables are coded according to the above scheme, the model is written as

$$Y = \alpha' + \beta U + \varepsilon,$$

or

$$Y = \alpha' + \beta (X - \bar{X}) + \varepsilon.$$

CENTERING *X*-VALUES IN THE HOUSE SIZE-PRICE EXAMPLE

Now, let us apply the concept of centering X-values to our house example. Recall that $\bar{X} = 2$. The coded X-values and calculations for the problem are then as given in Table 2.2. Note there that $\sum U_i^2 = 13.56$, which is what we got for S_{XX} when the X-values were not coded; and that $\sum U_i Y_i = 174.9$, the same value we found for S_{XY}. Thus,

$$b = \sum U_i Y_i / \sum U_i^2 = \frac{174.9}{13.56} = \frac{S_{XY}}{S_{XX}} = 12.8982$$

TABLE 2.2   *Coded X-values and Calculations for the Data in Table 1.1.*

House	X	$U = X - \bar{X}$	U^2	UY	Y
1	1.8	−0.2	0.04	− 6.4	32
2	1.0	−1.0	1.00	−24.0	24
3	1.7	−0.3	0.09	− 8.1	27
4	2.8	0.8	0.64	37.6	47
5	2.2	0.2	0.04	7.0	35
6	0.8	−1.2	1.44	−20.4	17
7	3.6	1.6	2.56	83.2	52
8	1.1	−0.9	0.81	−18.0	20
9	2.0	0.0	0.00	0.0	38
10	2.6	0.6	0.36	27.0	45
11	2.3	0.3	0.09	13.2	44
12	0.9	−1.1	1.21	−20.9	19
13	1.2	−0.8	0.64	−20.0	25
14	3.4	1.4	1.96	70.0	50
15	1.7	−0.3	0.09	− 9.0	30
16	2.5	0.5	0.25	21.5	43
17	1.4	−0.6	0.36	−16.2	27
18	3.3	1.3	1.69	65.0	50
19	2.2	0.2	0.04	7.4	37
20	1.5	−0.5	0.25	−14.0	28
Sum	40.0	0.0	13.56	174.9	690

$$\text{Mean: } \bar{X} = \frac{40}{20} = 2$$

verifies that the slope, which gives us the relationship, is unaffected by centering the predictor variable.

Then

$$a' = \bar{Y} = \frac{690}{20} = 34.5,$$

which is the value of Y when $U = 0$, that is, when $X_i - \bar{X} = X_i - 2 = 0$, or $X_i = 2$. The regression equation is

$$\hat{Y} = 34.5 + 12.9U,$$

where $U = X - 2$. The intercept has changed because the vertical axis has been moved from $X = 0$ to $X = 2$. The underlying linear relationship, however, is not affected, and one can easily obtain $\hat{Y} = 8.7 + 12.9X$ from the regression equation:

$$\hat{Y} = 34.5 + 12.9(X - 2) = 34.5 + 12.9X - 25.8 = 8.7 + 12.9X.$$

FIGURE 2.9   *Effect on Slope and Intercept of Centering X*

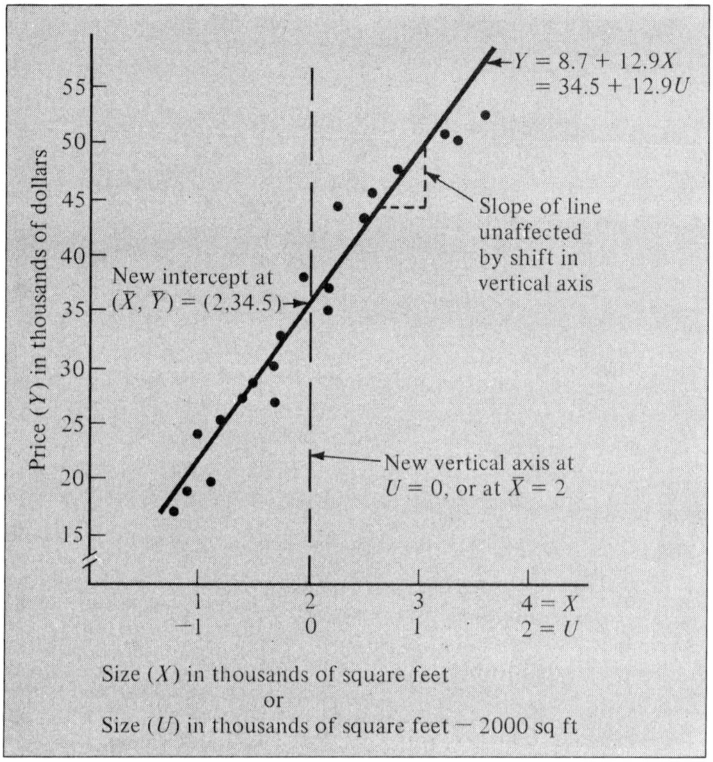

SCALING X-VALUES

In the engineering example, in which high, medium, and low pressures at equally spaced experimental levels were coded to $1, 0$, and -1, respectively, the coding was achieved by

$$Z_i = \frac{X_i - \bar{X}}{d},$$

Scaling   where d is the number of units between levels. This coding, which we shall refer to as scaling, or *standardizing*, affects the value obtained for the intercept because X has been corrected for its mean. It also affects the slope.
  In this case,

$$\sum Z_i = \frac{1}{d}\sum (X_i - \bar{X}_i) = \frac{1}{d}0 = 0,$$

but

$$\sum Z_i^2 = \sum \left(\frac{X_i - \bar{X}}{d}\right)^2 = \frac{1}{d^2} \sum (X_i - \bar{X})^2 = \frac{1}{d^2} S_{XX}$$

and

$$\sum Z_i Y_i = \sum \left(\frac{X_i - \bar{X}}{d}\right) Y_i = \frac{1}{d} \sum (X_i - \bar{X}) Y_i$$

$$= \frac{1}{d} S_{XY}.$$

Thus,

$$a = \bar{Y} - b\bar{Z} = \bar{Y},$$

but

$$b = \frac{\sum Z_i Y_i}{\sum Z_i^2} = \frac{S_{XY}/d}{S_{XX}/d^2} = d\frac{S_{XY}}{S_{XX}} = db = b',$$

and the regression model is

$$Y = \alpha' + \beta' Z + \sigma.$$

This coding has really not changed anything, either, if we compare Z's change from -1 to 0 to 1 with X's change from 100 to 150 to 200 pounds. Since a one-unit step in terms of X-values is 50 times as great as a one-unit step in terms of Z-values, b' must be 50 times as great as b. This keeps the change in Y for a one-unit change in Z proportional to the change in Y for a one-unit change in X; Exercises 2.15 and 2.16 give you some experience with this type of coding so that you can convince yourself that the only real change is in the way the scatter diagram is drawn.

This coding scheme is particularly useful when X is a controlled variable; at the outset the experimenter can select the values of the independent variable so that the X-values are equally spaced and sum to zero when coded.

Scaling can also be used when the X-values are not equally spaced and may even be random variables. For example, in psychological testing the researcher is more Z-Score comfortable working with what are referred to as Z-scores, or standard scores. A Z-score is simply a score (or some other measurement X) which has been first corrected Standard for the mean and then scaled by the standard deviation (the positive square root of the Deviation variance). That is, if a set of data has X-values $X_1, X_2, \ldots, X_n$ with mean $\bar{X}$ and standard deviation s_x, then the Z-scores are obtained by

$$Z_i = \frac{X_i - \bar{X}}{s_x}.$$

The Z-scores obtained, $Z_1, Z_2, \ldots, Z_n$, then have mean zero and standard deviation 1 and thus are easier to work with than are the original values. Of course, any regression equation using Z-scores can easily be changed into a regression equation for the original values by

$$\hat{Y} = a' + b'Z$$

$$= a' + b'\left(\frac{X - \overline{X}}{s_X}\right).$$

$$= a + bX.$$

SCALING X-VALUES IN THE HOUSE SIZE-PRICE EXAMPLE

Although you may not have noticed, we have been working with scaled values all along in the house size-price example. Note that X represents thousands of square feet, and Y, thousands of dollars. Why deal with the number 3600 when 3.6 thousand will do as well? This becomes very important when one has only a hand or desk calculator (and sometimes even if you have access to a computer).

What if we had left our X-values as $1800, 1000, \ldots, 1500$? Then $\sum X_i$ would be 40,000 and $\sum X_i^2$ would be 93,560,000. If Y were recorded in dollars instead of thousands of dollars, $\sum Y_i$ would be 690,000 and $\sum X_i Y_i$ would be 1,554,900,000. Many small calculators cannot handle such large numbers. Some might convert to scientific notation, but that can be confusing, too.

The effect of not scaling X and Y on the values in the regression equation are easily verified. The intercept would turn out to be $a = 8700$ and the slope would be $b = 12,900$. These are no different from $a = 8.7$ thousand and $b = 12.9$ thousand, which we obtained previously with much less effort. One should carefully examine data at the outset of an analysis to discover ways of saving trouble without actually affecting the results of the analysis.

EXERCISES   *Section 2.5*

p. 521, 522, 524, 525   2.14   Use either the TEACHERS data, the PRECIPITATION data, the SALES data, or the HEATING data. Refer to the Analyses of Data Sets section for preliminary calculations and previous results.
   a.   Correct X for its mean by subtracting $\overline{X}$ from each X-value.
   b.   Using corrected X-values, find a and b in order to fit the regression equation $\hat{Y} = a' + bU$. Solve normal equations.
   c.   Find a' and b by using shortcut formulas.
   d.   Compare the value of b to the one previously obtained (Exercise 2.8 or 2.13). Are they the same? Why or why not?
   e.   Compare the value of a' to the value obtained for a in Exercises 2.8 and 2.13. Are they the same? Why or why not?

FERTILIZER Data

An experiment was run in order to determine the relationship between the number of pounds of fertilizer applied per acre and the number of bushels yield for a certain crop. On three randomly selected one-acre plots, 15 pounds of fertilizer was applied; each of another three plots received 20 pounds; and 25 pounds was applied to each of three other acres. The results were

X Fertilizer (lbs)	Y Yield (bu)
15	45
15	43
15	46
20	58
20	62
20	64
25	70
25	78
25	76

2.15 For the **FERTILIZER** data:
   a. Code the X-values using the following scheme.

   Let μ = average number of pounds of fertilizer applied to all nine acres,
   d = spacing between levels of fertilizer application.

   Then $Z = \dfrac{X - \mu}{d}$.

   b. Plot the data, in terms of Z and Y, on a scatter diagram.
   c. Find the regression equation

   $$\hat{Y} = a' + b'\left(\frac{X - \mu}{d}\right) = a' + b'Z$$

   and plot it on the scatter diagram.
   d. What is the estimated average yield if 15 pounds of fertilizer is applied?
   e. Express the regression equation in terms of X and Y by substituting the values of μ and d into

   $$\hat{Y} = a' + b'\left(\frac{X - \mu}{d}\right).$$

   State the equation $\hat{Y} = a + bX$.
   f. Using the equation $\hat{Y} = a + bX$, estimate the average yield if 15 pounds of fertilizer is applied. Does your estimate agree with the one obtained in part d?
   g. Can the regression equations obtained in this problem be used to predict the average yield for acres to which 17 pounds of fertilizer is applied? Explain.

DEPRIVATION

An experiment was performed to study the relationship between the number of hours food deprivation for laboratory rats and the time required to learn a simple task for a food reward. Five different degrees of food deprivation were selected for study: slight, moderately slight, moderate, moderately great, and great. For convenience, slight deprivation is denoted by -2, moderately slight by -1, ..., great by $+2$. Results of the study were:

Degree of Deprivation	Time to Learn Task (min)
-2	73
-2	68
-1	49
-1	54
0	32
0	25
1	42
1	40
2	60
2	65

2.16 Use the **DEPRIVATION** data to work the following exercises.
   a. If the average duration of food deprivation was 72 hours, and each successive group of rats was deprived 24 hours longer than the previous group, to how many hours deprivation do the values $-2, -1, 0, 1, 2$ correspond? Use the formula

$$Z = \frac{X - \mu}{d}.$$

   b. Plot the data, in terms of Z (coded X-values) and Y, on a scatter diagram. Does it appear that time to learn task is a linear function of hours deprivation for these levels of deprivation?
   c. Fit a curvilinear equation of the form $\hat{Y} = a' + b'Z + c'Z^2$ to the data.
   d. Plot the equation obtained in c on the scatter diagram.
   e. Is it always true that if $\sum Z_i = 0$, then $\sum Z_i^3 = 0$,
      (1) if the Z-values are evenly spaced?
      (2) if the Z-values are not evenly spaced?

2.6 SUMMARY

We can now find a simple linear regression equation either by solving normal equations or by using shortcut formulas. If we are willing to solve simultaneous equations, even curvilinear regression or multiple regression models can be fitted to data, because normal equations can be written to correspond to any regression equation to be used. When more than two equations in two unknowns are involved, however, it is not easy to solve simultaneous equations. In the next chapter we shall look at the regression problem in terms of matrices with the view of obtaining a general solution for all regression problems.

Chapter 3   THE MATRIX APPROACH

3.1   A BRIEF INTRODUCTION TO MATRICES

Matrices do not save any great amount of time or labor in simple regression problems. At first, they may even appear to be more trouble than they are worth.

Try to keep in mind that when we get to more complicated problems, such as polynomial or multiple regression, our work will be greatly reduced if we can use matrices to formulate the regression problem. For one thing, matrix notation provides a shorthand that lets us reduce many cumbersome formulas down to just one. Furthermore, once one understands how the various elements of matrices combine and why, one can get a much better feeling for some of the fascinating things that happen in multiple or polynomial regressions. Finally, anyone who anticipates reading statistical regression literature in almost any field will not get past the first sentence without an understanding of the matrix form of the regression model.

We begin the matrix approach in conjunction with simple regression in order to make it as painless as possible. We will then show how the approach extends to *any* linear regression problem, regardless of the form of the model. First of all, we need some basic definitions.

3.2   SOME DEFINITIONS

MATRICES, THEIR ELEMENTS AND DIMENSIONS

Matrix   A <u>matrix</u> is merely a rectangular array of numbers, typically enclosed by square or round brackets. If $a, b, \ldots, f$ are numbers, then a matrix might look like

$$\begin{bmatrix} a & b & c \\ d & e & f \end{bmatrix}.$$

For example,

$$
\begin{bmatrix} 1 & 2 & 3 \\ 7 & 6 & 5 \\ 9 & 1 & 4 \end{bmatrix}, \quad \begin{bmatrix} 1 & 7 & 3 & 9 \\ 1 & 4 & -2 & 7 \\ 1 & 0 & 4 & -3 \end{bmatrix}, \quad \text{and} \quad \begin{pmatrix} 1 & 0 \\ 2 & 7 \\ 3 & 9 \\ 4 & 3 \end{pmatrix}
$$

are matrices. A matrix is usually denoted by a boldface capital letter, such as

$$
\mathbf{A} = \begin{bmatrix} a & b & c \\ d & e & f \end{bmatrix}.
$$

Elements

Dimension

The numbers in a matrix, called _elements_, are denoted by lowercase letters such as a, b, and so on. The _dimension_ of a matrix tells the number of rows, r, and the number of columns, c, in a given matrix. The number of rows always comes first. Thus, the matrix $\mathbf{A}$ above is a 2×3 (two-by-three) matrix since it has two rows and three columns of numbers. If $r = c$, as in the 3×3 matrix

$$
\mathbf{B} = \begin{bmatrix} b_{11} & b_{12} & b_{13} \\ b_{21} & b_{22} & b_{23} \\ b_{31} & b_{32} & b_{33} \end{bmatrix},
$$

Square Matrix

then the matrix is a _square matrix_. A square matrix with n rows and n columns is often referred to as an _n-square matrix_. Note that the elements of $\mathbf{B}$ are denoted by the symbols b_{ij}, where i indicates the row and j indicates the column. Thus, the element b_{23} is found in the second row and third column of the matrix, while b_{32} is found in the third row and second column.

In the matrix

$$
\mathbf{C} = \begin{bmatrix} 7 & 4 & -3 & 1 \\ 2.5 & -7 & 9.3 & 26 \\ -4 & 8 & 0 & 5 \end{bmatrix},
$$

$c_{13} = -3, c_{23} = 9.3, c_{32} = 8, c_{34} = 5$, and so on. Similarly, we could refer to the $(1, 3)$ element of $\mathbf{C}$ or the $(2, 3)$ element, or any other in this manner.

Vector

A matrix that has only one row or only one column is called a _vector_ and is denoted by a boldface small letter. For example,

$$
\mathbf{v}_1 = \begin{bmatrix} a & b & c & d \end{bmatrix}
$$

Row Vector   has four columns but only one row, so $\mathbf{v}_1$ is called a 1×4 <u>row vector</u>. Similarly

$$\mathbf{v}_2 = \begin{bmatrix} a \\ b \\ c \end{bmatrix}$$

Column Vector   is a 3×1 <u>column vector</u>, since it has three rows but only one column. By convention, a vector is usually written as a column vector unless otherwise specified.

Scalar   A matrix having only one row *and* only one column is called a <u>scalar</u>. An example of a scalar is

$$\underset{1 \times 1}{[3]} = 3.$$

Note that sometimes the dimension of a matrix is listed below the lower right-hand corner, for emphasis.

EQUALITY OF MATRICES

Equality   Two matrices are said to be <u>equal</u> if (1) they are of the same dimension, and (2) if they contain the same elements in the same positions. For example, if

$$\mathbf{A} = \begin{bmatrix} a & b & c \\ d & e & f \end{bmatrix} \quad \text{and} \quad \mathbf{B} = \begin{bmatrix} u & v & w \\ x & y & z \end{bmatrix},$$

$\mathbf{A} = \mathbf{B}$ if and only if $a = u$, $b = v$, $c = w$, $d = x$, $e = y$, and $f = z$.

The two matrices

$$\begin{bmatrix} 2 & 2 & 3 \\ 6 & 4 & 6 \end{bmatrix} \quad \text{and} \quad \begin{bmatrix} 2 & 2 & 4 \\ 6 & 4 & 6 \end{bmatrix}$$

are not equal because the $(1, 3)$ element of the first is 3 but the $(1, 3)$ element of the second is 4. Also,

$$\begin{bmatrix} 2 & 2 & 3 \\ 6 & 4 & 6 \end{bmatrix} \neq \begin{bmatrix} 2 & 2 \\ 6 & 4 \end{bmatrix},$$

because these two matrices are not of the same dimension: one has dimension 2×3 and the other 2×2.

TRANSPOSES

Transpose The transpose of an $r \times c$ matrix is a $c \times r$ matrix obtained by writing the rows as columns and the columns as rows, while retaining the orders of the elements in each row and column. For example, the transpose of the matrix $\mathbf{A}$ above, denoted $\mathbf{A}'$ or $\mathbf{A}^T$, is

$$\mathbf{A}' = \begin{bmatrix} a & d \\ b & e \\ c & f \end{bmatrix},$$

a 3×2 matrix. Naturally, the transpose of a column vector is a row vector and vice versa, and the transpose of a scalar is just the scalar. Since vectors are conventionally written as column vectors, a symbol $\mathbf{v}$ will denote a column vector, and $\mathbf{v}'$ will denote a row vector.

SOME SPECIAL SQUARE MATRICES

Diagonal   There are some special definitions that apply only to square matrices. The (principal) diagonal of an n-square matrix is the set of elements, in order, from the $(1, 1)$ element to the $(2, 2)$ element to the $(3, 3)$ element,..., to the (n, n) element. For example, in $\mathbf{S}$, the 4×4 matrix below, the diagonal is indicated by the dotted line. Elements on the
Diagonal   diagonal are called diagonal elements. Those that do not lie on the diagonal are called
Elements   off-diagonal elements.

Off-Diagonal
Elements

$$\mathbf{S} = \begin{bmatrix} a & b & c & d \\ e & f & g & h \\ i & j & k & l \\ m & n & o & p \end{bmatrix}.$$

Diagonal   A matrix whose off-diagonal elements are zero, but whose diagonal elements are
Matrix   not all zero, is called a diagonal matrix. For example,

$$\mathbf{D} = \begin{bmatrix} d_1 & 0 & 0 & 0 \\ 0 & d_2 & 0 & 0 \\ 0 & 0 & d_3 & 0 \\ 0 & 0 & 0 & d_4 \end{bmatrix} = \operatorname{diag}(d_1, d_2, d_3, d_4)$$

is a 4×4 diagonal matrix. (If the diagonal elements were all also zero, so that every

Zero Matrix element in the matrix is a zero, then we have a <u>zero matrix</u>, denoted by **0**.* A zero matrix need not be square, however.) A special diagonal matrix is one in which all the diagonal elements are ones:

$$\mathbf{I} = \begin{bmatrix} 1 & 0 & \cdots & 0 \\ 0 & 1 & \cdots & 0 \\ \vdots & \vdots & & \vdots \\ 0 & 0 & \cdots & 1 \end{bmatrix}$$

Identity Matrix is called the <u>identity matrix</u>, and we will see presently that it is the multiplicative identity. An $n \times n$ identity matrix is sometimes denoted $\mathbf{I}_n$.

Symmetric A <u>symmetric matrix</u> is a square matrix whose (i, j) and (j, i) elements are equal.
Matrix That is, if the matrix is folded in half along the principal diagonal, then the coincident elements will be equal. For example, the matrix **C** is symmetric:

$$\mathbf{C} = \begin{bmatrix} a & d & e \\ d & b & f \\ e & f & c \end{bmatrix}.$$

In **C**, the $(1, 2)$ element is d and the $(2, 1)$ element is also d; the $(1, 3)$ element and the $(3, 1)$ element are both e, and the $(2, 3)$ and $(3, 2)$ elements are both f.

EXERCISES   *Section 3.2*

3.1   Can any of the elements of a matrix be
    a.   negative numbers?
    b.   imaginary numbers?
    c.   zeroes?
    d.   fractions?

3.2   Give the dimensions of the following matrices.
    a.
$$\begin{bmatrix} 2 & 2 & 3 & 6 \\ 4 & 6 & 5 & 1 \\ 1 & 6 & 9 & 0 \end{bmatrix};$$

* Sometimes the zero matrix is denoted by **0** or by **0**.

b. $\begin{bmatrix} 6 & 3 & 6 \\ 5 & 5 & 2 \\ 3 & 0 & 5 \\ 5 & 1 & 2 \end{bmatrix}$;

c. $\begin{bmatrix} 9 & 7 & 4 \\ 2 & 5 & 5 \\ 6 & 10 & 4 \end{bmatrix}$;

d. $\begin{bmatrix} a_{11} & a_{12} & \cdots & a_{19} \\ a_{21} & a_{22} & \cdots & a_{29} \\ \vdots & \vdots & & \vdots \\ a_{61} & a_{62} & \cdots & a_{69} \end{bmatrix}$;

e. $\mathrm{diag}\,(a_1, a_2, \ldots, a_{12})$;

f. $[a_1, a_2, \ldots, a_{12}]$;

g. $\begin{bmatrix} a \\ b \\ c \\ d \end{bmatrix}$;

h. $[e]$.

3.3 In what row and column are the following elements found?
a. e_{47}; b. e_{74}; c. e_{77}; d. e_{ij}.

3.4 Which of the following pairs of matrices are equal?

a. $\begin{bmatrix} 9 & 7 & 4 & 2 \\ 5 & 5 & 6 & 1 \end{bmatrix}$ and $\begin{bmatrix} \frac{18}{2} & \frac{14}{2} & \frac{16}{4} & \frac{12}{6} \\ \frac{25}{5} & \frac{20}{4} & \frac{18}{3} & \frac{12}{12} \end{bmatrix}$;

b. $\begin{bmatrix} 9 & 7 & 4 & 2 \\ 5 & 5 & 6 & 1 \end{bmatrix}$ and $\begin{bmatrix} 9 & 5 \\ 7 & 5 \\ 4 & 6 \\ 2 & 1 \end{bmatrix}$;

c. $\begin{bmatrix} 9 & 7 & 4 & 2 \\ 5 & 5 & 6 & 1 \end{bmatrix}$ and $\begin{bmatrix} \frac{18}{2} & \frac{25}{5} \\ \frac{14}{2} & \frac{20}{4} \\ \frac{16}{4} & \frac{18}{3} \\ \frac{12}{6} & \frac{12}{12} \end{bmatrix}$;

d. $\begin{bmatrix} 1 & 0 & 0 \\ 0 & 1 & 0 \\ 0 & 0 & 1 \end{bmatrix}$ and $\begin{bmatrix} 5 & 0 & 0 \\ 0 & 5 & 0 \\ 0 & 0 & 5 \end{bmatrix}$;

e. $\begin{bmatrix} 1 & 0 & 0 \\ 0 & 1 & 0 \\ 0 & 0 & 1 \end{bmatrix}$ and $\begin{bmatrix} 1 & 0 \\ 0 & 1 \end{bmatrix}$.

3.5 Give the transposes of the following matrices.

a. $[1 \quad 2 \quad 3 \quad 4 \quad 5]$;

b. $\begin{bmatrix} 5 \\ 4 \\ 3 \\ 2 \\ 1 \end{bmatrix}$;

c. $\begin{bmatrix} 6 & 9 & 0 & 1 \\ 2 & 4 & 6 & 0 \end{bmatrix}$;

d. $\begin{bmatrix} a & b \\ c & d \\ e & f \end{bmatrix}$;

e. $\begin{bmatrix} a_{11} & a_{12} & a_{13} \\ a_{21} & a_{22} & a_{23} \\ a_{31} & a_{32} & a_{33} \end{bmatrix}$;

f. $\begin{bmatrix} 1 & 2 \\ 2 & 3 \end{bmatrix}$;

g. $\begin{bmatrix} 9 & 8 & 7 \\ 8 & 6 & 5 \\ 7 & 5 & 4 \end{bmatrix}$.

3.6 Indicate the (principal) diagonal of the following:

a. $\begin{bmatrix} a_{11} & a_{12} & a_{13} & a_{14} \\ a_{21} & a_{22} & a_{23} & a_{24} \\ a_{31} & a_{32} & a_{33} & a_{34} \\ a_{41} & a_{42} & a_{43} & a_{44} \end{bmatrix}$;

b. $\begin{bmatrix} 9 & 8 & 7 \\ 8 & 6 & 5 \\ 7 & 5 & 4 \end{bmatrix}$;

c. $\begin{bmatrix} 1 & 0 \\ 0 & 1 \end{bmatrix}$;

d. $\begin{bmatrix} 1 & 0 & 0 & 0 \\ 0 & 1 & 0 & 0 \\ 0 & 0 & 2 & 0 \\ 0 & 0 & 0 & 4 \end{bmatrix}$;

e. $\begin{bmatrix} 1 & 0 & 0 & 0 \\ 0 & 1 & 0 & 0 \\ 0 & 0 & 2 & 0 \end{bmatrix}$.

3.7   Write out all the elements of the following matrices:
   a.   diag$(1, 1, 2, 3, 5)$;
   b.   diag$(7, 0, 3, 4, 1, 0)$;
   c.   diag$(1, 1, 1, 1)$;
   d.   diag$(5, 5, 5)$;
   e.   diag$(a_1, a_2, \ldots, a_n)$.

3.8   Fill in the missing elements so as to make the matrices symmetric.

a.
$$\begin{bmatrix} 5 & 2 & 7 \\ & 1 & 1 \\ & & 2 \end{bmatrix};$$

b.
$$\begin{bmatrix} 9 & 9 & 2 & 9 \\ & 2 & 1 & 1 \\ & & 2 & 4 \\ & & & 0 \end{bmatrix};$$

c.
$$\begin{bmatrix} a & & \\ b & c & \\ d & e & f \end{bmatrix};$$

d.
$$\begin{bmatrix} a_{11} & a_{12} & a_{13} & a_{14} \\ & a_{22} & a_{23} & a_{24} \\ & & a_{33} & a_{34} \\ & & & a_{44} \end{bmatrix};$$

e.
$$\begin{bmatrix} 0 & & & \\ 0 & 6 & & \\ 9 & 2 & 4 & \\ 3 & 7 & 9 & 1 \end{bmatrix};$$

f.
$$\begin{bmatrix} 1 & 2 & 3 \\ & 4 & 5 \end{bmatrix}.$$

3.3   MATRIX OPERATIONS

ADDITION

Matrix Addition   Matrix addition can be performed if two matrices have the same dimension. They are added by summing corresponding elements:

$$\begin{bmatrix} a & b & c \\ d & e & f \end{bmatrix} + \begin{bmatrix} g & h & i \\ j & k & l \end{bmatrix} = \begin{bmatrix} a+g & b+h & c+i \\ d+j & e+k & f+l \end{bmatrix}.$$

Similarly, one matrix can be subtracted from another by subtracting elements of one from the corresponding elements of the other:

$$\begin{bmatrix} 4 & -3 & 7 \\ -8 & 0 & 9 \end{bmatrix} - \begin{bmatrix} 7 & 4 & -3 \\ -1 & 7 & 0 \end{bmatrix} = \begin{bmatrix} 4-7 & -3-4 & 7-(-3) \\ -8-(-1) & 0-7 & 9-0 \end{bmatrix}$$

$$= \begin{bmatrix} -3 & -7 & 10 \\ -7 & -7 & 9 \end{bmatrix}.$$

SCALAR MULTIPLICATION

Scalar Multiplication   The product of a scalar and a matrix is a matrix containing the products of the scalar times each element. If s is a scalar, then

$$s\begin{bmatrix} a & d \\ b & e \\ c & f \end{bmatrix} = \begin{bmatrix} a & d \\ b & e \\ c & f \end{bmatrix}s = \begin{bmatrix} as & ds \\ bs & es \\ cs & fs \end{bmatrix} = \begin{bmatrix} sa & sd \\ sb & se \\ sc & sf \end{bmatrix}.$$

For example,

$$-2\begin{bmatrix} 4 & -3 & 7 \\ -8 & 0 & 9 \end{bmatrix} = \begin{bmatrix} -8 & 6 & -14 \\ 16 & 0 & -18 \end{bmatrix}.$$

The same rules apply, of course, to addition of vectors and multiplication of vectors by a scalar.

MATRIX MULTIPLICATION

Conformable Matrices   *Multiplication* of matrices is a bit more complicated. Two matrices can be multiplied together only if their dimensions are conformable. This means that the number of columns of the first matrix must equal the number of rows of the second. For example, if a matrix E is of dimension 2×4 and a matrix F is of dimension 4×3, then the product EF can be formed because E has four columns and F has four rows. However, if G is 3×6 and H is 5×2, the product GH cannot be formed. If J is 3×6, the product GJ cannot be formed, either. We will see the reason for this shortly. A good way to check for conformable dimensions is to subscript the matrices with their dimensions and see that the two "inside" numbers are equal. For example,

$$E_{2 \times 4}F_{4 \times 3}$$
↑ ↑
"inside" numbers equal.

Notice that if $\mathbf{E}$ is 2×4 and $\mathbf{F}$ is 4×3, the product $\mathbf{EF}$ can be formed, but the product

$$\mathbf{F}_{4 \times 3}\mathbf{E}_{2 \times 4}$$

cannot. That is to say, matrix multiplication is *not commutative*. Indeed, even when two square matrices (having dimensions that are conformable in either order) are multiplied, such as $\mathbf{S}_{4 \times 4}\mathbf{T}_{4 \times 4}$ or $\mathbf{T}_{4 \times 4}\mathbf{S}_{4 \times 4}$, the two products will usually be different, so that

$$\mathbf{S}_{4 \times 4}\mathbf{T}_{4 \times 4} \neq \mathbf{T}_{4 \times 4}\mathbf{S}_{4 \times 4}.$$

When two conformable matrices are multiplied, the resulting product is a matrix with as many rows as the first matrix and as many columns as the second matrix. Thus, if a 2×4 matrix is multiplied on the right by a 4×3 matrix, the resulting matrix is a 2×3 matrix. For example,

$$\mathbf{E}_{2 \times 4}\mathbf{F}_{4 \times 3} = \mathbf{K}_{2 \times 3}.$$

Thus, the "outside" numbers of the subscripts of the matrices determine the dimension of the product matrix.

Matrix Multiplication

How is this matrix multiplication actually performed? First study the product below:

$$\begin{bmatrix} a & b & c \\ d & e & f \\ g & h & i \end{bmatrix}_{3 \times 3} \begin{bmatrix} j & m \\ k & n \\ l & o \end{bmatrix}_{3 \times 2} = \begin{bmatrix} aj + bk + cl & am + bn + co \\ dj + ek + fl & dm + en + fo \\ gj + hk + il & gm + hn + io \end{bmatrix}_{3 \times 2}$$

To obtain the $(1, 1)$ element of the product matrix, we work with the *first* row of the left matrix and the *first* column of the right matrix. Both have the same number of elements—in this case, three. We multiply the first elements aj, the second elements bk, and the third elements cl; then add up these products. Thus, the $(1, 1)$ element of the product matrix is $aj + bk + cl$.

To obtain the $(1, 2)$ element of the product matrix, we again use the first row of the left matrix, but we use the *second* column of the right matrix; again multiply corresponding elements, and sum. Thus the $(1, 2)$ element of the product is $am + bn + co$.

For the $(2, 1)$ element of the product, use the *second* row of the left matrix and the *first* column of the right, to get $dj + ek + fl$. For the $(3, 2)$ element of the product, use the third row of the left matrix and the second column of the right, obtaining $gm + hn + io$, and so on.

A more sophisticated way to indicate the product of the previous two matrices is

$$
\begin{bmatrix} a_{11} & a_{12} & a_{13} \\ a_{21} & a_{22} & a_{23} \\ a_{31} & a_{32} & a_{33} \end{bmatrix} \begin{bmatrix} b_{11} & b_{12} \\ b_{21} & b_{22} \\ b_{31} & b_{32} \end{bmatrix}
$$

$$
= \begin{bmatrix} a_{11}b_{11} + a_{12}b_{21} + a_{13}b_{31} & a_{11}b_{12} + a_{12}b_{22} + a_{13}b_{32} \\ a_{21}b_{11} + a_{22}b_{21} + a_{23}b_{31} & a_{21}b_{12} + a_{22}b_{22} + a_{23}b_{32} \\ a_{31}b_{11} + a_{32}b_{21} + a_{33}b_{31} & a_{31}b_{12} + a_{32}b_{22} + a_{33}b_{32} \end{bmatrix}
$$

$$
= \begin{bmatrix} \sum_{k=1}^{3} a_{1k}b_{k1} & \sum_{k=1}^{3} a_{1k}b_{k2} \\ \sum_{k=1}^{3} a_{2k}b_{k1} & \sum_{k=1}^{3} a_{2k}b_{k2} \\ \sum_{k=1}^{3} a_{3k}b_{k1} & \sum_{k=1}^{3} a_{3k}b_{k2} \end{bmatrix}
$$

and this can easily be extended to the products of matrices with different dimensions. As a numerical example, consider

$$
\begin{bmatrix} 2 & 2 & 3 \\ 6 & 4 & 6 \\ 5 & 1 & 1 \end{bmatrix}_{3 \times 3} \begin{bmatrix} 6 & 9 & 0 \\ 3 & 7 & 4 \\ 6 & 1 & 1 \end{bmatrix}_{3 \times 3}
$$

$$
= \begin{bmatrix} 2(6) + 2(3) + 3(6) & 2(9) + 2(7) + 3(1) & 2(0) + 2(4) + 3(1) \\ 6(6) + 4(3) + 6(6) & 6(9) + 4(7) + 6(1) & 6(0) + 4(4) + 6(1) \\ 5(6) + 1(3) + 1(6) & 5(9) + 1(7) + 1(1) & 5(0) + 1(4) + 1(1) \end{bmatrix}
$$

$$
= \begin{bmatrix} 36 & 35 & 11 \\ 84 & 88 & 22 \\ 39 & 53 & 5 \end{bmatrix}_{3 \times 3} .
$$

To show that matrix multiplication is generally not commutative, let us form the product in reverse order.

$$\begin{bmatrix} 6 & 9 & 0 \\ 3 & 7 & 4 \\ 6 & 1 & 1 \end{bmatrix}_{3 \times 3} \begin{bmatrix} 2 & 2 & 3 \\ 6 & 4 & 6 \\ 5 & 1 & 1 \end{bmatrix}_{3 \times 3}$$

$$= \begin{bmatrix} 6(2) + 9(6) + 0(5) & 6(2) + 9(4) + 0(1) & 6(3) + 9(6) + 0(1) \\ 3(2) + 7(6) + 4(5) & 3(2) + 7(4) + 4(1) & 3(3) + 7(6) + 4(1) \\ 6(2) + 1(6) + 1(5) & 6(2) + 1(4) + 1(1) & 6(3) + 1(6) + 1(1) \end{bmatrix}_{3 \times 3}$$

$$= \begin{bmatrix} 66 & 48 & 72 \\ 68 & 38 & 55 \\ 23 & 17 & 25 \end{bmatrix}_{3 \times 3} \neq \begin{bmatrix} 36 & 35 & 11 \\ 84 & 88 & 22 \\ 39 & 53 & 5 \end{bmatrix}_{3 \times 3}$$

The same rules apply when multiplying vectors. An $r \times 1$ column vector times a $1 \times c$ row vector will yield an $r \times c$ matrix. But a $1 \times c$ row vector times an $r \times 1$ column vector will yield a (1×1) scalar (if $r = c$). For example,

$$[2 \quad 2 \quad 3]_{1 \times 3} \begin{bmatrix} 6 \\ 4 \\ 5 \end{bmatrix}_{3 \times 1} = [12 + 8 + 15] = [35]_{1 \times 1},$$

but

$$\begin{bmatrix} 6 \\ 4 \\ 5 \end{bmatrix}_{3 \times 1} [2 \quad 2 \quad 3]_{1 \times 3} = \begin{bmatrix} 12 & 12 & 18 \\ 8 & 8 & 12 \\ 10 & 10 & 15 \end{bmatrix}_{3 \times 3},$$

and again we see that matrix multiplication is not commutative.

THE MULTIPLICATIVE IDENTITY

Multiplicative Identity

We can now verify that $\mathbf{I} = \mathrm{diag}(1, 1, \ldots, 1)$ is the multiplicative identity; that is, that for any matrix $\mathbf{A}_{r \times c}$,

$$\mathbf{A}_{r \times c} \mathbf{I}_{c \times c} = \mathbf{I}_{r \times r} \mathbf{A}_{r \times c} = \mathbf{A}.$$

For example, let

$$\mathbf{A} = \begin{bmatrix} a & b & c \\ d & e & f \end{bmatrix}_{2 \times 3}.$$

Then

$$\mathbf{A}_{2 \times 3}\mathbf{I}_{3 \times 3} = \begin{bmatrix} a & b & c \\ d & e & f \end{bmatrix} \begin{bmatrix} 1 & 0 & 0 \\ 0 & 1 & 0 \\ 0 & 0 & 1 \end{bmatrix}$$

$$= \begin{bmatrix} a(1) + b(0) + c(0) & a(0) + b(1) + c(0) & a(0) + b(0) + c(1) \\ d(1) + e(0) + f(0) & d(0) + e(1) + f(0) & d(0) + e(0) + f(1) \end{bmatrix}$$

$$= \begin{bmatrix} a & b & c \\ d & e & f \end{bmatrix} = \mathbf{A}_{2 \times 3}.$$

Also

$$\mathbf{I}_{2 \times 2}\mathbf{A}_{2 \times 3} = \begin{bmatrix} 1 & 0 \\ 0 & 1 \end{bmatrix} \begin{bmatrix} a & b & c \\ d & e & f \end{bmatrix}$$

$$= \begin{bmatrix} 1(a) + 0(d) & 1(b) + 0(e) & 1(c) + 0(f) \\ 0(a) + 1(d) & 0(b) + 1(e) & 0(c) + 1(f) \end{bmatrix}$$

$$= \begin{bmatrix} a & b & c \\ d & e & f \end{bmatrix} = \mathbf{A}_{2 \times 3}.$$

THE PRODUCTS OF A MATRIX AND ITS TRANSPOSE

Note that for any matrix $\mathbf{M}_{r \times c}$, the products $\mathbf{M}'_{c \times r}\mathbf{M}_{r \times c}$ and $\mathbf{M}_{r \times c}\mathbf{M}'_{c \times r}$ can always be formed (although they will not be equal). Also, the products $\mathbf{M}'\mathbf{M}$ and $\mathbf{M}\mathbf{M}'$ will always be square and symmetric. To illustrate, let

$$\mathbf{M} = \begin{bmatrix} 9 & 7 & 4 \\ 2 & 5 & 5 \end{bmatrix}_{2 \times 3}.$$

Then

$$\mathbf{M'M} = \begin{bmatrix} 9 & 2 \\ 7 & 5 \\ 4 & 5 \end{bmatrix} \begin{bmatrix} 9 & 7 & 4 \\ 2 & 5 & 5 \end{bmatrix}$$

$$= \begin{bmatrix} 81+4 & 63+10 & 36+10 \\ 63+10 & 49+25 & 28+25 \\ 36+10 & 28+25 & 16+25 \end{bmatrix} = \begin{bmatrix} 85 & 73 & 46 \\ 73 & 74 & 53 \\ 46 & 53 & 41 \end{bmatrix},$$

which is square and symmetric. Then

$$\mathbf{MM'} = \begin{bmatrix} 9 & 7 & 4 \\ 2 & 5 & 5 \end{bmatrix} \begin{bmatrix} 9 & 2 \\ 7 & 5 \\ 4 & 5 \end{bmatrix}$$

$$= \begin{bmatrix} 81+49+16 & 18+35+20 \\ 18+35+20 & 4+25+25 \end{bmatrix}$$

$$= \begin{bmatrix} 146 & 73 \\ 73 & 54 \end{bmatrix},$$

which is also square and symmetric.

EXERCISES   *Section 3.3*

3.9   Form the following sums:

a. $\begin{bmatrix} 2 & 4 \\ 9 & 7 \end{bmatrix} + \begin{bmatrix} 5 & 7 \\ -6 & 0 \end{bmatrix}$;

b. $\mathrm{diag}(3, 3, 1, 7) + \mathrm{diag}(6, 7, 1, 4)$,

c. $\begin{bmatrix} 1 & -3 & 2 & 7 \\ 1 & 0 & -4 & 1 \\ 1 & 5 & 2 & -2 \\ -5 & 6 & 6 & 4 \end{bmatrix} + \begin{bmatrix} 1 & -5 & 2 & 2 \\ 7 & 5 & 6 & -3 \\ 1 & -8 & 7 & 8 \\ 5 & 8 & 3 & 5 \end{bmatrix}$;

d. $\begin{bmatrix} 1 & -3 & 2 & 7 \\ 1 & 0 & -4 & 1 \\ 1 & 5 & 2 & -2 \\ -5 & 6 & 6 & 4 \end{bmatrix} - \begin{bmatrix} 1 & -5 & 2 & 2 \\ 7 & 5 & 6 & -3 \\ 1 & -8 & 7 & 8 \\ 5 & 8 & 3 & 5 \end{bmatrix}$;

e. $\begin{bmatrix} 1 & 0 & 5 \\ 4 & 0 & 9 \\ 2 & 1 & 1 \end{bmatrix} + \begin{bmatrix} 1 & 0 \\ 5 & 9 \\ 9 & 1 \end{bmatrix};$

f. $\begin{bmatrix} 1 & 0 \\ 5 & 9 \\ 9 & 1 \end{bmatrix} + \begin{bmatrix} 1 & 5 & 9 \\ 0 & 9 & 1 \end{bmatrix};$

g. $\begin{bmatrix} 2 \\ 6 \\ 1 \\ 7 \\ 2 \end{bmatrix} + \begin{bmatrix} 5 \\ 1 \\ 1 \\ 5 \\ 8 \end{bmatrix};$

h. $\begin{bmatrix} 2 \\ 6 \\ 1 \\ 7 \\ 2 \end{bmatrix} + \begin{bmatrix} 5 & 1 & 1 & 5 & 8 \end{bmatrix};$

i. $\begin{bmatrix} 2 & 6 & 1 & 7 & 2 \end{bmatrix} - \begin{bmatrix} 2 & 0 & 7 & 7 & 4 \end{bmatrix};$

j. $4 + \begin{bmatrix} 2 & 6 & 1 & 7 & 2 \end{bmatrix}.$

3.10 Form the following products:

a. $4\begin{bmatrix} 4 & -7 & 4 \\ 5 & 8 & 1 \\ -5 & 8 & 2 \end{bmatrix};$

b. $\frac{1}{4}\begin{bmatrix} 4 & -7 & 4 \\ 5 & 8 & 1 \\ -5 & 8 & 2 \end{bmatrix};$

c. $-3\begin{bmatrix} 6 & 5 \\ 7 & 4 \\ 5 & 2 \end{bmatrix};$

d. $7\begin{bmatrix} 8 & 4 & 9 & 2 & 3 \end{bmatrix};$

e. $-5\begin{bmatrix} 7 \\ 2 \\ 1 \\ -2 \\ 0 \\ 7 \end{bmatrix}.$

3.11 Which of the following pairs of matrices are conformable for multiplication in the order given? For conformable pairs, state dimension of the product.

a. $\begin{bmatrix} 8 & 4 & 9 \\ 2 & 3 & 9 \end{bmatrix}\begin{bmatrix} 6 & 7 \\ 5 & 3 \\ 7 & 8 \end{bmatrix};$

b. $\begin{bmatrix} 6 & 7 \\ 5 & 3 \\ 7 & 8 \end{bmatrix}\begin{bmatrix} 8 & 4 & 9 \\ 2 & 3 & 9 \end{bmatrix};$

c. $\begin{bmatrix} 3 & 6 & 8 \\ 4 & 0 & 3 \\ 9 & 1 & 1 \end{bmatrix}\begin{bmatrix} 2 & 2 & 2 \\ 2 & 1 & 4 \\ 3 & 9 & 5 \end{bmatrix};$

d. $\begin{bmatrix} 2 & 2 & 2 \\ 2 & 1 & 4 \\ 3 & 9 & 5 \end{bmatrix}\begin{bmatrix} 3 & 6 & 8 \\ 4 & 0 & 3 \\ 9 & 1 & 1 \end{bmatrix};$

e. $[1 \quad 2 \quad 1 \quad 1][0 \quad 6 \quad 2 \quad 3]$;

f. $[1 \quad 2 \quad 1 \quad 1]\begin{bmatrix} 0 \\ 6 \\ 2 \\ 3 \end{bmatrix}$;

g. $\begin{bmatrix} 1 \\ 2 \\ 1 \\ 1 \end{bmatrix}[0 \quad 6 \quad 2 \quad 3]$;

h. $\begin{bmatrix} 1 \\ 2 \\ 1 \\ 1 \end{bmatrix}\begin{bmatrix} 0 \\ 6 \\ 2 \\ 3 \end{bmatrix}$;

i. $[4]\begin{bmatrix} 1 \\ 2 \\ 1 \end{bmatrix}$.

3.12 Form the indicated products for the conformable pairs you found in Exercise 3.11.

3.13 Referring to Exercise 3.12, find examples to show that, in general, $\mathbf{AB} \neq \mathbf{BA}$ if
a.   either $\mathbf{A}$ or $\mathbf{B}$ or both are not square matrices;
b.   both $\mathbf{A}$ and $\mathbf{B}$ are square matrices.

3.14 Using the matrix

$$\mathbf{S} = \begin{bmatrix} 1 & 1 & 1 & 1 \\ 1 & 6 & 8 & 3 \\ 1 & 2 & 7 & 4 \end{bmatrix},$$

show that
a.   if $\mathbf{S}'$ is the transpose of $\mathbf{S}$ then the products $\mathbf{SS}'$ and $\mathbf{S}'\mathbf{S}$ can both be formed;
b.   both $\mathbf{SS}'$ and $\mathbf{S}'\mathbf{S}$ are symmetric,
c. $\mathbf{SS}' \neq \mathbf{S}'\mathbf{S}$.

3.15 Let $\mathbf{I}_n$ denote the n-square identity matrix. For each of the following matrices $\mathbf{M}$, show that $\mathbf{I}$ is the multiplicative identity by forming the products $\mathbf{I}_r\mathbf{M} = \mathbf{MI}_c = \mathbf{M}$, where r = number of rows in $\mathbf{M}$ and c = number of columns in $\mathbf{M}$.
For example, if

$$\mathbf{M} = \begin{bmatrix} 3 & 1 \\ 6 & 2 \\ 9 & 4 \\ 0 & 6 \end{bmatrix}_{4 \times 2},$$

show that

$$\mathbf{I}_4\mathbf{M}_{4\times 2} = \begin{bmatrix} 1 & 0 & 0 & 0 \\ 0 & 1 & 0 & 0 \\ 0 & 0 & 1 & 0 \\ 0 & 0 & 0 & 1 \end{bmatrix}\begin{bmatrix} 3 & 1 \\ 6 & 2 \\ 9 & 4 \\ 0 & 6 \end{bmatrix} = \begin{bmatrix} 3 & 1 \\ 6 & 2 \\ 9 & 4 \\ 0 & 6 \end{bmatrix}$$

and that

$$\mathbf{M}_{4\times 2}\mathbf{I}_2 = \begin{bmatrix} 3 & 1 \\ 6 & 2 \\ 9 & 4 \\ 0 & 6 \end{bmatrix}\begin{bmatrix} 1 & 0 \\ 0 & 1 \end{bmatrix} = \begin{bmatrix} 3 & 1 \\ 6 & 2 \\ 9 & 4 \\ 0 & 6 \end{bmatrix}.$$

a. $\begin{bmatrix} 2 & 1 \\ 1 & 9 \end{bmatrix}$;

b. $\begin{bmatrix} 1 & 2 & 7 \\ 1 & 4 & 5 \end{bmatrix}$;

c. $\begin{bmatrix} 6 & 3 & 5 \\ 7 & 2 & 9 \end{bmatrix}$;

d. $\begin{bmatrix} 5 & 1 \\ 7 & 5 \\ 2 & 6 \end{bmatrix}$;

e. $\begin{bmatrix} a_{11} & a_{12} & a_{13} & a_{14} \\ a_{21} & a_{22} & a_{23} & a_{24} \\ a_{31} & a_{32} & a_{33} & a_{34} \end{bmatrix}$;

f. $\begin{bmatrix} a & b & c \end{bmatrix}$;

g. $\begin{bmatrix} a & b & c & d \end{bmatrix}$.

3.4 DETERMINANTS

Determinant

A useful feature of a *square* matrix is its determinant. A determinant is a number (scalar) that can represent a matrix. It is determined by all elements of the matrix. For a matrix **A**, the determinant is denoted |**A**|. Straight vertical bars around an array of numbers also indicate the determinant of the array. The determinant of a 1×1 matrix, or scalar, is just the scalar itself. For example,

$$|a| = a.$$

DETERMINANT OF A 2 × 2 MATRIX

For a 2 × 2 matrix, it is also quite easy to calculate the determinant. If

$$\mathbf{A} = \begin{bmatrix} a & b \\ c & d \end{bmatrix},$$

then

$$|\mathbf{A}| = \begin{vmatrix} a & b \\ c & d \end{vmatrix} = ad - bc.$$

For example, if

$$\mathbf{A} = \begin{bmatrix} 2 & 2 \\ 3 & 6 \end{bmatrix},$$

then

$$\mathbf{A} = \begin{vmatrix} 2 & 2 \\ 3 & 6 \end{vmatrix} = 12 - 6 = 6.$$

SINGULARITY

Nonsingular Matrix

Singular Matrix

If for any square matrix $\mathbf{A}$ the determinant is nonzero, then $\mathbf{A}$ is a nonsingular matrix. If $|\mathbf{A}| = 0$, then $\mathbf{A}$ is called a singular matrix. If, for example,

$$\mathbf{A} = \begin{bmatrix} 1 & 2 \\ 2 & 4 \end{bmatrix},$$

then

$$\mathbf{A} = \begin{vmatrix} 1 & 2 \\ 2 & 4 \end{vmatrix} = 4 - 4 = 0,$$

and $\mathbf{A}$ is singular. Note that in the above matrix any row or column is a multiple of some other row or column.

Linear Combination

In general, if any row(s) or column(s) in a matrix is a linear combination of any other row(s) or column(s), the matrix is singular. If one row is a linear combination of the other rows, then that row will be obtained by adding multiples of other rows together. Arguing on columns, suppose that $\mathbf{v}_1, \mathbf{v}_2, \ldots, \mathbf{v}_c$ are the c columns of a matrix (note that they are column *vectors*) and that $k_1, k_2, \ldots, k_c$ is a set of c constants (scalars);

pick one of the columns, say v_j. If

$$v_j = k_1 v_1 + k_2 v_2 + \cdots + k_{j-1} v_{j-1} + k_{j+1} v_{j+1} + \cdots + k_c v_c,$$

then v_j is a linear combination of the remaining $c - 1$ columns; v_j can be obtained by adding together multiples of the other columns.

In the matrix

$$\mathbf{A} = \begin{bmatrix} 1 & 2 \\ 2 & 4 \end{bmatrix}$$

we can see that

$$\begin{bmatrix} 1 \\ 2 \end{bmatrix} = \frac{1}{2} \begin{bmatrix} 2 \\ 4 \end{bmatrix}$$

or that $[2 \quad 4] = 2[1 \quad 2]$. Suppose we start with two columns,

$$\mathbf{c}_1 = \begin{bmatrix} 1 \\ 3 \\ 7 \end{bmatrix} \quad \text{and} \quad \mathbf{c}_2 = \begin{bmatrix} 2 \\ 4 \\ 5 \end{bmatrix},$$

of a 3×3 matrix and combine them by

$$k_1 \mathbf{c}_1 + k_2 \mathbf{c}_2 = 2\mathbf{c}_1 + (-1)\mathbf{c}_2 = \mathbf{c}_3:$$

$$2\begin{bmatrix} 1 \\ 3 \\ 7 \end{bmatrix} + (-1)\begin{bmatrix} 2 \\ 4 \\ 5 \end{bmatrix} = \begin{bmatrix} 2 \\ 6 \\ 14 \end{bmatrix} - \begin{bmatrix} 2 \\ 4 \\ 5 \end{bmatrix} = \begin{bmatrix} 0 \\ 2 \\ 9 \end{bmatrix}.$$

Then the matrix

$$\begin{bmatrix} 1 & 2 & 0 \\ 3 & 4 & 2 \\ 7 & 5 & 9 \end{bmatrix}$$

will be singular; its determinant will be zero, as we shall see on p. 74, since the third column is a linear combination of the first two. This will have significance when we turn to the question of solving simultaneous equations, especially in the case of multiple regression.

DETERMINANTS OF LARGER MATRICES

Finding a determinant becomes progressively more complicated as the dimension of
the matrix becomes larger. We will illustrate the method of evaluating the determinant
Expansion by of a 3×3 matrix by <u>expansion by cofactors</u>. Let
Cofactors

$$\mathbf{A} = \begin{bmatrix} a_{11} & a_{12} & a_{13} \\ a_{21} & a_{22} & a_{23} \\ a_{31} & a_{32} & a_{33} \end{bmatrix}.$$

Choose an element of $\mathbf{A}$, say a_{11}. Strike out of $\mathbf{A}$ the row and column in which a_{11}
appears; this results in a 2×2 matrix,

$$\begin{bmatrix} a_{22} & a_{23} \\ a_{32} & a_{33} \end{bmatrix}.$$

The determinant

$$\begin{vmatrix} a_{22} & a_{23} \\ a_{32} & a_{33} \end{vmatrix} = a_{22}a_{33} - a_{23}a_{32}$$

Minor of this 2×2 "submatrix" is called the <u>minor</u> of the element a_{11}. Other minors
corresponding to other elements of $\mathbf{A}$ are

Element	Minor
a_{12}	$\begin{vmatrix} a_{21} & a_{23} \\ a_{31} & a_{33} \end{vmatrix} = a_{21}a_{33} - a_{23}a_{31}$
a_{23}	$\begin{vmatrix} a_{11} & a_{12} \\ a_{31} & a_{32} \end{vmatrix} = a_{11}a_{32} - a_{12}a_{31}$
a_{33}	$\begin{vmatrix} a_{11} & a_{12} \\ a_{21} & a_{22} \end{vmatrix} = a_{11}a_{22} - a_{12}a_{21}$

and so on.

Cofactor   A <u>cofactor</u> is a minor that has been multiplied by either $+1$ or -1. In order to
determine whether a given minor gets multiplied by plus or minus one, count how many
horizontal or vertical steps would be taken in order to move the corresponding element
to the $(1, 1)$ position of the matrix. If zero or an even number of steps are required, the
minor is multiplied by $+1$; if an odd number of steps must be taken, multiply the minor
by -1.

For example, since a_{11} is already in the $(1, 1)$ position, its minor,

$$\begin{vmatrix} a_{22} & a_{23} \\ a_{31} & a_{33} \end{vmatrix} = a_{22}a_{33} - a_{23}a_{31}$$

gets multiplied by $+1$ so that the cofactor corresponding to the element a_{11} is $+(a_{22}a_{33} - a_{23}a_{31})$. In order to move the element a_{12} to the $(1, 1)$ position, one horizontal step to the left must be taken, so the cofactor of a_{12} is $-(a_{21}a_{33} - a_{23}a_{31})$. Moving a_{23} to the $(1, 1)$ position takes two horizontal steps and one vertical step, so the cofactor of a_{23} is $-(a_{11}a_{32} - a_{12}a_{31})$. To move the element a_{33} to the $(1, 1)$ position, two horizontal and two vertical steps must be taken so the minor is multiplied by $+1$, and so on.*

Label cofactors with subscripted capital letters corresponding to the elements with which they are associated. For example,

Element	Cofactor
a_{11}	$A_{11} = (a_{22}a_{33} - a_{23}a_{32})$
a_{12}	$A_{12} = -(a_{21}a_{33} - a_{23}a_{31})$
a_{23}	$A_{23} = -(a_{11}a_{32} - a_{12}a_{31})$
a_{33}	$A_{33} = (a_{11}a_{22} - a_{12}a_{21})$

and so on. Of course, there is a cofactor corresponding to each element of the matrix $\mathbf{A}$.

Once the cofactors are determined, the determinant of $\mathbf{A}$ can be found by picking any row or column of $\mathbf{A}$, multiplying each element of that row or column by its corresponding cofactor, and then summing these products. For example, if we choose the first row, then

$$|\mathbf{A}| = a_{11}A_{11} + a_{12}A_{12} + a_{13}A_{13}.$$

If the third row is chosen, then

$$|\mathbf{A}| = a_{31}A_{31} + a_{32}A_{32} + a_{33}A_{33};$$

or if the second column is chosen,

$$|\mathbf{A}| = a_{12}A_{12} + a_{22}A_{22} + a_{32}A_{32}.$$

Regardless of which row or column is chosen, the numerical value of $|\mathbf{A}|$ will be the same.

Let us try a numerical example. Let

$$\mathbf{A} = \begin{bmatrix} 2 & 2 & 1 \\ 1 & 0 & 1 \\ 2 & 1 & 2 \end{bmatrix}.$$

---

* Equivalently, add the subscripts on the element whose minor has been found. If the sum is even, multiply the minor by $+1$; if the sum is odd, multiply the minor by -1. For example, the minor of a_{11} will be multiplied by $+1$ since $1 + 1 = 2$ is even, and the minor of a_{23} will be multiplied by -1 since $2 + 3 = 5$ is odd.

Suppose we decide to determine $|\mathbf{A}|$ by expansion of cofactors across the first row:

Element	Minor	Cofactor
$a_{11} = 2$	$\begin{vmatrix} 0 & 1 \\ 1 & 2 \end{vmatrix} = 0 - 1 = -1$	$+(-1) = -1 = A_{11}$
$a_{12} = 2$	$\begin{vmatrix} 1 & 1 \\ 2 & 2 \end{vmatrix} = 2 - 2 = 0$	$-(0) = 0 = A_{12}$
$a_{13} = 1$	$\begin{vmatrix} 1 & 0 \\ 2 & 1 \end{vmatrix} = 1 - 0 = 1$	$+(1) = 1 = A_{13}$

Then

$$\begin{aligned} |\mathbf{A}| &= a_{11}A_{11} + a_{12}A_{12} + a_{13}A_{13} \\ &= 2(-1) + 2(0) + 1(1) \\ &= -2 + 0 + 1 \\ &= -1. \end{aligned}$$

Expanding down the first column gives us

Element	Minor	Cofactor
$a_{11} = 2$	$\begin{vmatrix} 0 & 2 \\ 1 & 2 \end{vmatrix} = 0 - 1 = -1$	$+(-1) = -1 = A_{11}$
$a_{21} = 1$	$\begin{vmatrix} 2 & 1 \\ 1 & 2 \end{vmatrix} = 4 - 1 = 3$	$-(3) = -3 = A_{21}$
$a_{31} = 2$	$\begin{vmatrix} 2 & 1 \\ 0 & 2 \end{vmatrix} = 2 - 0 = 2$	$+(2) = 2 = A_{31}$

Then

$$\begin{aligned} |\mathbf{A}| &= a_{11}A_{11} + a_{21}A_{21} + a_{31}A_{31} \\ &= 2(-1) + 1(-3) + 2(2) \\ &= -2 - 3 + 4 = -1, \end{aligned}$$

which is the same value for $|\mathbf{A}|$ that we obtained by expanding across the first row.

Let us now verify that the matrix

$$\begin{bmatrix} 1 & 2 & 0 \\ 3 & 4 & 2 \\ 7 & 5 & 9 \end{bmatrix},$$

in which the third column was obtained by multiplying the first column by 2 and then subtracting off the second column, has zero determinant. Expanding across the first column,

Element	Minor	Cofactor
$a_{11} = 1$	$\begin{vmatrix} 4 & 2 \\ 5 & 9 \end{vmatrix} = 36 - 10 = 26$	$+(26) = 26 = A_{11}$
$a_{12} = 2$	$\begin{vmatrix} 3 & 2 \\ 7 & 9 \end{vmatrix} = 27 - 14 = 13$	$-(13) = -13 = A_{12}$

(Note that it is not necessary to find A_{13}, since the product $a_{13}A_{13}$ will be zero.) Then

$$|A| = a_{11}A_{11} + a_{12}A_{12} + a_{13}A_{13}$$
$$= 1(26) + 2(-13) + 0 = 26 - 26 = 0.$$

What if we change some element in the third column of this matrix so that the third column becomes *almost*, but not exactly, a linear combination of the first two columns? What will this do to the value of the determinant? Suppose we change the $(3, 3)$ element to 9.1, thus:

$$\begin{bmatrix} 1 & 2 & 0 \\ 3 & 4 & 2 \\ 7 & 5 & 9.1 \end{bmatrix}.$$

Then to find the determinant,

Element	Minor	Cofactor
$a_{11} = 1$	$\begin{vmatrix} 4 & 2 \\ 5 & 9.1 \end{vmatrix} = 36.4 - 10 = 26.4$	$A_{11} = +(26.4) = 26.4$
$a_{12} = 2$	$\begin{vmatrix} 3 & 2 \\ 7 & 9.1 \end{vmatrix} = 27.3 - 14 = 13.3$	$A_{12} = -(13.3) = -13.3$

and

$$|A| = a_{11}A_{11} + a_{12}A_{12} + a_{13}A_{13}$$
$$= 1(26.4) + 2(-13.3) + 0 = 26.4 - 26.6 = -0.2.$$

We see that the determinant is no longer zero, but that it is not very far away from zero, either. A consequence of this is discussed in the next section.

The method presented above extends directly to the problem of evaluating the determinant of a 4×4 or higher dimension matrix. The calculations required, however, become more and more involved. For example, if $\mathbf{A}$ is a 4×4 matrix, then the minor associated with any element is the determinant of a 3×3 matrix, and this minor itself must be evaluated by the cofactor expansion. Thus, to find $|\mathbf{A}_{4 \times 4}|$, four 3×3 determinants must be evaluated, and each involves finding three 2×2 determinants. Thus, finding such determinants by hand is a tedious process that requires quite a bit of bookkeeping. Fortunately, there are computer programs that can evaluate determinants quite efficiently.

DETERMINANTS OF DIAGONAL MATRICES

Diagonal matrices are nice because regardless of their dimension, their determinants are simply the products of their diagonal elements. Thus, if $\mathbf{D} = \mathrm{diag}(d_1, d_2, \ldots, d_n)$, then $|\mathbf{D}| = d_1 d_2 \ldots d_n$. Let us use a 3×3 diagonal matrix to see how this works. Let

$$\mathbf{D} = \begin{bmatrix} 2 & 0 & 0 \\ 0 & 6 & 0 \\ 0 & 0 & 1 \end{bmatrix}.$$

Suppose we choose to expand across the first row. Then the only cofactor we need to find is A_{11}, since the products $a_{12}A_{12}$ and $a_{13}A_{13}$ will be zero.

$$A_{11} = \begin{vmatrix} 6 & 0 \\ 0 & 1 \end{vmatrix} = 6(1) - (0)(0) = 6(1).$$

Thus $|\mathbf{D}| = a_{11}A_{11} = 2(6)(1)$, the product of the diagonal elements. The exercises that follow will help you verify for yourself that this rule applies for diagonal matrices of other dimensions.

Our primary use for determinants will come in finding inverses of matrices, which we will look at in the next section.

EXERCISES   *Section 3.4*

3.16   Find the determinants of the following matrices:

a. $\begin{bmatrix} 2 & 3 \\ 9 & 1 \end{bmatrix}$;

b. $\begin{bmatrix} 1 & 0 \\ -9 & 7 \end{bmatrix}$;

c. $\begin{bmatrix} a_{11} & a_{12} \\ a_{21} & a_{22} \end{bmatrix}$;

d. $\begin{bmatrix} \sum n & \sum x_i \\ \sum x_i & \sum x_i^2 \end{bmatrix}$;

e. $\begin{bmatrix} 1 & 2 \\ 3 & 4 \end{bmatrix}$;

f. $\begin{bmatrix} 1 & 2 \\ 2 & 4 \end{bmatrix}$;

g. $\begin{bmatrix} 6 & 0 \\ 0 & 7 \end{bmatrix}$;

h. $\begin{bmatrix} 1 & 5 & 3 \\ 0 & 0 & 8 \end{bmatrix}$;

i. $[8]$;

j. $\text{diag}(a_1, a_2)$.

3.17 Which of the matrices in Exercise 3.16 are
   a. singular?
   b. nonsingular?

3.18 Find the determinants of the following matrices:

a. $\begin{bmatrix} 1 & 0 & 7 \\ 8 & 2 & 5 \\ 0 & 6 & 7 \end{bmatrix}$;

b. $\begin{bmatrix} 1 & 0 & 0 \\ 1 & 5 & 4 \\ 9 & 2 & 4 \end{bmatrix}$;

c. $\begin{bmatrix} 5 & 7 & 0 \\ 7 & 3 & 7 \\ 0 & 7 & 3 \end{bmatrix}$;

d. $\begin{bmatrix} 1 & 2 & 3 \\ 4 & 5 & 6 \\ 7 & 8 & 9 \end{bmatrix}$;

e. $\begin{bmatrix} 1 & 2 & 3 \\ 2 & 4 & 6 \\ 4 & 8 & 12 \end{bmatrix}$;

f. $\begin{bmatrix} 4 & 5 \\ 7 & 2 \\ 3 & 5 \end{bmatrix}$;

g. $\begin{bmatrix} 9 & 1 & 5 \\ 6 & 2 & 1 \end{bmatrix}$;

h. $\begin{bmatrix} 7 & 0 & 0 \\ 0 & 5 & 0 \\ 0 & 0 & 8 \end{bmatrix}$;

i. $\text{diag}(7, 5, 3)$;

j. $\text{diag}(d_1, d_2, d_3)$.

3.19 Which of the matrices in Exercise 3.18 are
   a. singular?
   b. nonsingular?

3.20 Find the determinants of the following matrices:

a. $\begin{bmatrix} 1 & 0 & 0 & 0 \\ 6 & 0 & 6 & 2 \\ 0 & 5 & 4 & 0 \\ 6 & 6 & 8 & 2 \end{bmatrix}$;

b. $\begin{bmatrix} 5 & 6 & 2 & 4 \\ 0 & 8 & 1 & 7 \\ 0 & 4 & 1 & 5 \\ 0 & 4 & 0 & 3 \end{bmatrix}$;

c. $\begin{bmatrix} 5 & 6 & 2 & 4 \\ 0 & 8 & 1 & 7 \\ 0 & 4 & 1 & 5 \\ 0 & 12 & 3 & 15 \end{bmatrix}$;

d. $\begin{bmatrix} 1 & 2 & 3 & 4 \\ 2 & 4 & 6 & 8 \\ 4 & 8 & 12 & 16 \\ 8 & 16 & 24 & 32 \end{bmatrix}$;

e.
$$\begin{bmatrix} 1 & 5 & 9 & 6 \\ 0 & 7 & 5 & 2 \\ 4 & 6 & 1 & 7 \\ 8 & 9 & 1 & 7 \end{bmatrix};$$

f.
$$\begin{bmatrix} 3 & 0 & 0 & 0 \\ 0 & 6 & 0 & 0 \\ 0 & 0 & 7 & 0 \\ 0 & 0 & 0 & 4 \end{bmatrix};$$

g. $\text{diag}(7, 7, 8, 5)$;

h. $\text{diag}(a_1, a_2, \ldots a_n)$;

i.
$$\begin{bmatrix} 2 & 5 & 3 & 8 \\ 7 & 9 & 6 & 3 \\ 1 & 5 & 2 & 7 \\ 2 & 9 & 0 & 1 \\ 0 & 9 & 8 & 1 \end{bmatrix};$$

j.
$$\begin{bmatrix} 1 & 2 & 3 & 4 & 5 \\ 2 & 4 & 6 & 8 & 10 \\ 4 & 8 & 12 & 16 & 20 \\ 8 & 16 & 24 & 32 & 40 \\ 16 & 32 & 48 & 64 & 60 \end{bmatrix};$$

k.
$$\begin{bmatrix} 4 & 7 & 8 & 8 & 2 \\ 5 & 0 & 6 & 0 & 0 \\ 6 & 2 & 9 & 7 & 4 \\ 9 & 9 & 4 & 9 & 2 \\ 8 & 8 & 9 & 3 & 6 \end{bmatrix}.$$

3.21 Which of the matrices in Exercise 3.20 are a. singular? b. nonsingular?

3.5 INVERSES

Inverse Matrix

Determinants are useful in finding the (multiplicative) inverse of a matrix. Given a nonsingular matrix **A**, the inverse of **A** is a matrix $\mathbf{A}^{-1}$ such that

$$\mathbf{A}\mathbf{A}^{-1} = \mathbf{A}^{-1}\mathbf{A} = \mathbf{I},$$

where **I** is the identity matrix.

GENERAL METHOD

We will illustrate the method of finding an inverse using a 3×3 matrix. Let

$$\mathbf{A} = \begin{vmatrix} a_{11} & a_{12} & a_{13} \\ a_{21} & a_{22} & a_{23} \\ a_{31} & a_{32} & a_{33} \end{vmatrix},$$

and let A_{ij} denote the cofactor of the element a_{ij}. Then

$$\mathbf{A}^{-1} = \frac{1}{|\mathbf{A}|} \begin{bmatrix} A_{11} & A_{21} & A_{31} \\ A_{12} & A_{22} & A_{32} \\ A_{13} & A_{23} & A_{33} \end{bmatrix}.$$

Note that the factor $1/|\mathbf{A}|$ requires that $|\mathbf{A}|$ be nonzero. Thus, only nonsingular matrices have inverses. Note also that the matrix given in the formula for the inverse is the *transpose* of the cofactor matrix; that is, the first subscript on A_{ij} now indicates the column and the second subscript indicates the row. The transpose of the cofactor

Adjoint Matrix   matrix, often called the adjoint matrix, is denoted by adj(A).

To find $\mathbf{A}^{-1}$ where

$$\mathbf{A} = \begin{bmatrix} 2 & 2 & 1 \\ 1 & 0 & 1 \\ 2 & 1 & 2 \end{bmatrix},$$

we must first find the cofactor of every element.

Element	Minor	Sign	Cofactor
$a_{11} = 2$	$\begin{vmatrix} 0 & 1 \\ 1 & 2 \end{vmatrix} = 0 - 1 = -1$	$+$	$-1 = A_{11}$
$a_{12} = 2$	$\begin{vmatrix} 1 & 1 \\ 2 & 2 \end{vmatrix} = 2 - 2 = 0$	$-$	$0 = A_{12}$
$a_{13} = 1$	$\begin{vmatrix} 1 & 0 \\ 2 & 1 \end{vmatrix} = 1 - 0 = 1$	$+$	$1 = A_{13}$
$a_{21} = 1$	$\begin{vmatrix} 2 & 1 \\ 1 & 2 \end{vmatrix} = 4 - 1 = 3$	$-$	$-3 = A_{21}$
$a_{22} = 0$	$\begin{vmatrix} 2 & 1 \\ 2 & 2 \end{vmatrix} = 4 - 2 = 2$	$+$	$2 = A_{22}$
$a_{23} = 1$	$\begin{vmatrix} 2 & 2 \\ 2 & 1 \end{vmatrix} = 2 - 4 = -2$	$-$	$2 = A_{23}$
$a_{31} = 2$	$\begin{vmatrix} 2 & 1 \\ 0 & 1 \end{vmatrix} = 2 - 0 = 2$	$+$	$2 = A_{31}$
$a_{32} = 1$	$\begin{vmatrix} 2 & 1 \\ 1 & 1 \end{vmatrix} = 2 - 1 = 1$	$-$	$-1 = A_{32}$
$a_{33} = 2$	$\begin{vmatrix} 2 & 2 \\ 1 & 0 \end{vmatrix} = 0 - 2 = -2$	$+$	$-2 = A_{33}$

Then

$$|\mathbf{A}| = \sum_{i=1}^{3} a_{i1} A_{i1} = 2(-1) + 1(-3) + 2(2)$$
$$= -2 - 3 + 4 = -1,$$

by expanding down the first column. The matrix of cofactors is

$$
\begin{bmatrix} A_{11} & A_{12} & A_{13} \\ A_{21} & A_{22} & A_{23} \\ A_{31} & A_{32} & A_{33} \end{bmatrix} = \begin{bmatrix} -1 & 0 & 1 \\ -3 & 2 & 2 \\ 2 & -1 & -2 \end{bmatrix};
$$

and its transpose is

$$
\begin{bmatrix} A_{11} & A_{21} & A_{31} \\ A_{12} & A_{22} & A_{32} \\ A_{13} & A_{23} & A_{33} \end{bmatrix} = \begin{bmatrix} -1 & -3 & 2 \\ 0 & 2 & -1 \\ 1 & 2 & -2 \end{bmatrix},
$$

so that

$$
\mathbf{A}^{-1} = \frac{1}{-1} \begin{bmatrix} -1 & -3 & 2 \\ 0 & 2 & -1 \\ 1 & 2 & -2 \end{bmatrix} = \begin{bmatrix} 1 & 3 & -2 \\ 0 & -2 & 1 \\ -1 & -2 & 2 \end{bmatrix}.
$$

Since it is no easy job to invert a matrix, it is always a good idea to make sure that $\mathbf{A}\mathbf{A}^{-1} = \mathbf{A}^{-1}\mathbf{A} = \mathbf{I}$. Since matrix multiplication is not generally commutative, both $\mathbf{A}\mathbf{A}^{-1}$ and $\mathbf{A}^{-1}\mathbf{A}$ must be formed, because if a mistake has been made it is possible that $\mathbf{A}\mathbf{A}^{-1} = \mathbf{I}$ but $\mathbf{A}^{-1}\mathbf{A} \neq \mathbf{I}$, or vice versa. For our example,

$$
\mathbf{A}\mathbf{A}^{-1} = \begin{bmatrix} 2 & 2 & 1 \\ 1 & 0 & 1 \\ 2 & 1 & 2 \end{bmatrix} \begin{bmatrix} 1 & 3 & -2 \\ 0 & -2 & 1 \\ -1 & -2 & 2 \end{bmatrix}
$$

$$
= \begin{bmatrix} 2+0-1 & 6-4-2 & -4+2+2 \\ 1+0-1 & 3+0-2 & -2+0+2 \\ 2+0-2 & 6-2-4 & -4+1+4 \end{bmatrix} = \begin{bmatrix} 1 & 0 & 0 \\ 0 & 1 & 0 \\ 0 & 0 & 1 \end{bmatrix} = \mathbf{I};
$$

and

$$
\mathbf{A}^{-1}\mathbf{A} = \begin{bmatrix} 1 & 3 & -2 \\ 0 & -2 & 1 \\ -1 & -2 & 2 \end{bmatrix} \begin{bmatrix} 2 & 2 & 1 \\ 1 & 0 & 1 \\ 2 & 1 & 2 \end{bmatrix}
$$

$$
= \begin{bmatrix} 2+3-4 & 2+0-2 & 1+3-4 \\ 0-2+2 & 0+0+1 & 0-2+2 \\ -2-2+4 & -2+0+2 & -1-2+4 \end{bmatrix} = \begin{bmatrix} 1 & 0 & 0 \\ 0 & 1 & 0 \\ 0 & 0 & 1 \end{bmatrix} = \mathbf{I}.
$$

THE INVERSE OF A 2 × 2 MATRIX

The same method applies to inversion of a 4×4 or higher matrix; however, the process becomes even more tedious, since the cofactors involve determinants of 3×3 or higher matrices. Applying this method to a 2×2 matrix, however, results in a neat little formula. Let

$$\mathbf{A} = \begin{bmatrix} a & b \\ c & d \end{bmatrix}.$$

Then

Element	Minor	Sign	Cofactor
a	d	$+$	d
b	c	$-$	$-c$
c	b	$-$	$-b$
d	a	$+$	a

and

$$|\mathbf{A}| = ad + b(-c) = ad - bc,$$

as we have already noted. Then

$$\mathbf{A}^{-1} = \frac{1}{|\mathbf{A}|} \begin{bmatrix} d & -c \\ -b & a \end{bmatrix}' = \frac{1}{|\mathbf{A}|} \begin{bmatrix} d & -b \\ -c & a \end{bmatrix} = \frac{1}{ad - bc} \begin{bmatrix} d & -b \\ -c & a \end{bmatrix}$$

$$= \begin{bmatrix} \dfrac{d}{ad - bc} & \dfrac{-b}{ad - bc} \\ \dfrac{-c}{ad - bc} & \dfrac{a}{ad - bc} \end{bmatrix}.$$

Note that the denominator of each term is $|\mathbf{A}| = ad - bc$; that a and d switch places but retain the same sign; and that b and c retain their positions but change sign. To check that an inverse has been found,

$$\mathbf{A}\mathbf{A}^{-1} = \begin{bmatrix} a & b \\ c & d \end{bmatrix} \begin{bmatrix} \dfrac{d}{ab - bc} & \dfrac{-b}{ad - bc} \\ \dfrac{-c}{ad - bc} & \dfrac{a}{ad - bc} \end{bmatrix}$$

$$= \begin{bmatrix} \dfrac{ad}{ad - bc} - \dfrac{bc}{ad - bc} & \dfrac{-ab}{ad - bc} + \dfrac{ab}{ad - bc} \\ \dfrac{cd}{ad - bc} - \dfrac{cd}{ad - bc} & \dfrac{-bc}{ad - bc} + \dfrac{ad}{ad - bc} \end{bmatrix} = \begin{bmatrix} 1 & 0 \\ 0 & 1 \end{bmatrix} = \mathbf{I};$$

and

$$\mathbf{A}^{-1}\mathbf{A} = \begin{bmatrix} \dfrac{d}{ad-bc} & \dfrac{-b}{ad-bc} \\[2ex] \dfrac{-c}{ad-bc} & \dfrac{a}{ad-bc} \end{bmatrix} \begin{bmatrix} a & b \\ c & d \end{bmatrix}$$

$$= \begin{bmatrix} \dfrac{ad-bc}{ad-bc} & \dfrac{bd-bd}{ad-bc} \\[2ex] \dfrac{-ac+ac}{ad-bc} & \dfrac{-bc+ad}{ad-bc} \end{bmatrix} = \begin{bmatrix} 1 & 0 \\ 0 & 1 \end{bmatrix} = \mathbf{I}.$$

THE INVERSE OF A DIAGONAL MATRIX

If the matrix to be inverted is a diagonal matrix, the procedure is very simple. If $\mathbf{D} = \text{diag}(d_1, d_2, \ldots, d_n)$, or

$$\mathbf{D} = \begin{bmatrix} d_1 & 0 & 0 & \cdots & 0 \\ 0 & d_2 & 0 & \cdots & 0 \\ 0 & 0 & d_3 & \cdots & 0 \\ \vdots & \vdots & \vdots & & \vdots \\ 0 & 0 & 0 & \cdots & d_n \end{bmatrix},$$

then

$$\mathbf{D}^{-1} = \text{diag}\left(\frac{1}{d_1}, \frac{1}{d_2}, \ldots, \frac{1}{d_n}\right)$$

$$= \begin{bmatrix} 1/d_1 & 0 & 0 & \cdots & 0 \\ 0 & 1/d_2 & 0 & \cdots & 0 \\ 0 & 0 & 1/d_3 & \cdots & 0 \\ \vdots & \vdots & \vdots & & \vdots \\ 0 & 0 & 0 & \cdots & 1/d_n \end{bmatrix}.$$

For example, if

$$\mathbf{D} = \begin{bmatrix} a & 0 \\ 0 & d \end{bmatrix},$$

then $|\mathbf{D}| = ad$ and

$$\mathbf{D}^{-1} = \begin{bmatrix} \dfrac{d}{ad} & \dfrac{0}{ad} \\ \dfrac{0}{ad} & \dfrac{a}{ad} \end{bmatrix} = \begin{bmatrix} \dfrac{1}{a} & 0 \\ 0 & \dfrac{1}{d} \end{bmatrix}.$$

The reader can verify that this is always the case either by applying the inversion rule to a 3×3 or larger diagonal matrix, or by completing the following check that

$\mathbf{DD}^{-1} = \mathbf{D}^{-1}\mathbf{D} = \mathbf{I}$:

$$\mathbf{DD}^{-1} = \begin{bmatrix} d_1 & 0 & \cdots & 0 \\ 0 & d_2 & \cdots & 0 \\ \vdots & \vdots & & \vdots \\ 0 & 0 & \cdots & d_n \end{bmatrix} \begin{bmatrix} 1/d_1 & 0 & \cdots & 0 \\ 0 & 1/d_2 & \cdots & 0 \\ \vdots & \vdots & & \vdots \\ 0 & 0 & \cdots & 1/d_n \end{bmatrix}$$

$$= \begin{bmatrix} 1+0+\cdots+0 & 0+0+\cdots+0 & \cdots & 0+0+\cdots+0 \\ 0+0+\cdots+0 & 0+1+\cdots+0 & \cdots & 0+0+\cdots+0 \\ \vdots & \vdots & & \vdots \\ 0+0+\cdots+0 & 0+0+\cdots+0 & \cdots & 0+0+\cdots+1 \end{bmatrix}$$

$$= \begin{bmatrix} 1 & 0 & \cdots & 0 \\ 0 & 1 & \cdots & 0 \\ \vdots & \vdots & & \vdots \\ 0 & 0 & \cdots & 1 \end{bmatrix} = \mathbf{I}.$$

In the next section, we will use what we have learned about matrices to solve systems of simultaneous equations.

EXERCISES   *Section 3.5*

3.22 Referring to the exercises for Section 3.4, p. 78, find the inverse of each nonsingular matrix in the following exercises. Check your answers.
a. Exercise 3.16;
b. Exercise 3.18;
c. Exercise 3.20, parts a–j.

3.23 a. Find the inverse of the matrix

$$\begin{bmatrix} 1 & 0 \\ 0 & 6 \end{bmatrix}.$$

b. Find the inverse of the matrix

$$\begin{bmatrix} 1 & 0 & 0 \\ 0 & 2 & 3 \\ 0 & 9 & 1 \end{bmatrix}.$$

Compare to the answer to Exercise 3.16a. Note that the matrix above can be written

$$\left[\begin{array}{c|cc} 1 & 0 & 0 \\ \hline 0 & & \\ & \mathbf{M} & \\ 0 & & \end{array}\right], \quad \text{where} \quad \mathbf{M} = \begin{bmatrix} 2 & 3 \\ 9 & 1 \end{bmatrix}.$$

Can you make a generalization?

c. Find the inverse of the matrix

$$\begin{bmatrix} 1 & 0 & 0 & 0 \\ 0 & 1 & 0 & 7 \\ 0 & 8 & 2 & 5 \\ 0 & 0 & 6 & 7 \end{bmatrix}.$$

Compare to the answer to Exercise 3.18a. Does this answer support the generalization in part b?

d. Complete the rule:

$$\left[\begin{array}{c|c} 1 & 0 \cdots 0 \\ \hline 0 & \\ \vdots & \mathbf{M} \\ 0 & \end{array}\right]^{-1} = ?$$

e. Complete the rule:

$$\left[\begin{array}{c|c} d & 0 \cdots 0 \\ \hline 0 & \\ \vdots & \mathbf{M} \\ 0 & \end{array}\right]^{-1} = ?$$

3.6 APPLICATION TO SOLVING SIMULTANEOUS EQUATIONS

WRITING SIMULTANEOUS EQUATIONS IN MATRIX FORM

Matrix theory evolved from the study of simultaneous equations. Let us see how this relationship works. Suppose we have the following system of equations to solve for X

and Y, where a_i, b_i, and c_i are constants:

$$a_1 X + b_1 Y = c_1$$
$$a_2 X + b_2 Y = c_2.$$

This system can be more compactly written as

$$\mathbf{A}\mathbf{x} = \mathbf{c},$$

where

$$\mathbf{A}_{2 \times 2} = \begin{bmatrix} a_1 & b_1 \\ a_2 & b_2 \end{bmatrix},$$

$$\mathbf{x}_{2 \times 1} = \begin{bmatrix} X \\ Y \end{bmatrix},$$

and

$$\mathbf{c}_{2 \times 1} = \begin{bmatrix} c_1 \\ c_2 \end{bmatrix},$$

since

$$\underset{2 \times 2}{\begin{bmatrix} a_1 & b_1 \\ a_2 & b_2 \end{bmatrix}} \underset{2 \times 1}{\begin{bmatrix} X \\ Y \end{bmatrix}} = \underset{2 \times 1}{\begin{bmatrix} c_1 \\ c_2 \end{bmatrix}},$$

or

$$\underset{2 \times 1}{\begin{bmatrix} a_1 X + b_1 Y \\ a_2 X + b_2 Y \end{bmatrix}} = \underset{2 \times 1}{\begin{bmatrix} c_1 \\ c_2 \end{bmatrix}},$$

so that

$$a_1 X + b_1 Y = c_1$$

and

$$a_2 X + b_2 Y = c_2.$$

SOLVING SIMULTANEOUS EQUATIONS

Solving the matrix equation

$$\mathbf{Ax} = \mathbf{c}$$

for $\mathbf{x}$ (that is, for X and Y), gives

$$\mathbf{A}^{-1}\mathbf{Ax} = \mathbf{A}^{-1}\mathbf{c}$$
$$\mathbf{Ix} = \mathbf{A}^{-1}\mathbf{c}$$
$$\mathbf{x} = \mathbf{A}^{-1}\mathbf{c}.$$

Pre-multiplication

Post-multiplication

Note that we have multiplied both sides of the equation by $\mathbf{A}^{-1}$ *on the left*. In this case, we say that $\mathbf{A}^{-1}\mathbf{Ax}$ premultiplies the vector $\mathbf{Ax}$ by $\mathbf{A}^{-1}$. If we had had an expression such as $\mathbf{BA}$, then the matrix $\mathbf{B}$ would be postmultiplied by the matrix $\mathbf{A}$.

Note that in solving $\mathbf{Ax} = \mathbf{c}$, both sides of the equation are premultiplied by $\mathbf{A}^{-1}$; it would not work to form

$$\mathbf{AxA}^{-1} = \mathbf{cA}^{-1},$$

because it is not the case that $\mathbf{AxA}^{-1} = \mathbf{x}$, even if the dimensions had been conformable for multiplication in that order. Remember when solving matrix equations that if one side is premultiplied by a matrix (or vector), then the other side must also be premultiplied by that same matrix in order to preserve the equality. The same reasoning applies to postmultiplication.

LARGER SYSTEMS

Note also that while we started out with two equations in two unknowns,

$$a_1 X + b_1 Y = c_1$$
$$a_2 X + b_2 Y = c_2,$$

matrix notation reduced the problem to that of solving only *one* equation in one unknown:

$$\mathbf{Ax} = \mathbf{c}.$$

Furthermore, any set of n equations in n unknowns can be expressed in the same form. For example, for four equations in four unknowns,

$$a_{11}X_1 + a_{12}X_2 + a_{13}X_3 + a_{14}X_4 = c_1$$
$$a_{21}X_1 + a_{22}X_2 + a_{23}X_3 + a_{24}X_4 = c_2$$
$$a_{31}X_1 + a_{32}X_2 + a_{33}X_3 + a_{34}X_4 = c_3$$
$$a_{41}X_1 + a_{42}X_2 + a_{43}X_3 + a_{44}X_4 = c_4,$$

we get

$$\mathbf{A} = \begin{bmatrix} a_{11} & a_{12} & a_{13} & a_{14} \\ a_{21} & a_{22} & a_{23} & a_{24} \\ a_{31} & a_{32} & a_{33} & a_{34} \\ a_{41} & a_{42} & a_{43} & a_{44} \end{bmatrix},$$

$$\mathbf{x} = \begin{bmatrix} X_1 \\ X_2 \\ X_3 \\ X_4 \end{bmatrix} \quad \text{and} \quad \mathbf{c} = \begin{bmatrix} c_1 \\ c_2 \\ c_3 \\ c_4 \end{bmatrix};$$

and the system can be written simply as

$$\mathbf{Ax} = \mathbf{c}.$$

The solution to the matrix equation,

$$\mathbf{x} = \mathbf{A}^{-1}\mathbf{c}$$

holds regardless of the dimension of the system. Thus this single equation is the solution to *all* systems of simultaneous linear equations. We need not study as separate topics the solutions to two equations in two unknowns, three equations in three unknowns, or 140 equations in 140 unknowns: we have already solved them all!

NUMERICAL EXAMPLES

For example, if the system of equations is

$$2X + 5Y = -5$$
$$X - 2Y = 5,$$

then

$$\mathbf{A} = \begin{bmatrix} 2 & 5 \\ 1 & -2 \end{bmatrix}, \qquad \mathbf{x} = \begin{bmatrix} X \\ Y \end{bmatrix}, \qquad \mathbf{c} = \begin{bmatrix} -5 \\ 5 \end{bmatrix},$$

and

$$|\mathbf{A}| = -4 - 5 = -9$$

and

$$\mathbf{A}^{-1} = \frac{1}{9} \begin{bmatrix} -2 & -5 \\ -1 & 2 \end{bmatrix} = \begin{bmatrix} \frac{2}{9} & \frac{5}{9} \\ \frac{1}{9} & -\frac{2}{9} \end{bmatrix},$$

so that

$$\mathbf{x} = \mathbf{A}^{-1}\mathbf{c}$$

$$= \begin{bmatrix} \frac{2}{9} & \frac{5}{9} \\ \frac{1}{9} & -\frac{2}{9} \end{bmatrix} \begin{bmatrix} -5 \\ 5 \end{bmatrix}$$

$$= \begin{bmatrix} -\frac{10}{9} + \frac{25}{9} \\ -\frac{5}{9} - \frac{10}{9} \end{bmatrix} = \begin{bmatrix} \frac{15}{9} \\ -\frac{15}{9} \end{bmatrix} = \begin{bmatrix} \frac{5}{3} \\ -\frac{5}{3} \end{bmatrix};$$

thus,

$$X = \frac{5}{3}, \qquad Y = -\frac{5}{3}.$$

Checking,

$$2\left(\frac{5}{3}\right) + 5\left(-\frac{5}{3}\right) = \frac{10}{3} - \frac{25}{3} = -\frac{15}{3} = -5$$

$$1\left(\frac{5}{3}\right) - 2\left(-\frac{5}{3}\right) = \frac{5}{3} + \frac{10}{3} = \frac{15}{3} = 5.$$

Note that the same solutions could be reached by multiplying the second equation by 2, subtracting, and thus eliminating X:

$$2X + 5Y = -5$$
$$\underline{2X - 4Y = 10}$$
$$9Y = -15$$

$$Y = -\frac{15}{9} = -\frac{5}{3}$$

and

$$2X + 5\left(-\frac{5}{3}\right) = -5$$

$$2X = -5 + \frac{25}{3} = \frac{-15 + 25}{3} = \frac{10}{3}$$

$$X = \frac{10/3}{2} = \frac{5}{3}.$$

Matrices may not save much time if there are only two equations in two unknowns, but for three or more simultaneous equations to be solved, the matrix approach is as efficient as any other method, and often more efficient. For example, suppose we are to solve

$$2X_1 + 2X_2 + X_3 = 1$$
$$X_1 \qquad + X_3 = -2$$
$$2X_1 + X_2 + 2X_3 = 0,$$

where X_1, X_2, and X_3 are unknown.

Then

$$\mathbf{A} = \begin{bmatrix} 2 & 2 & 1 \\ 1 & 0 & 1 \\ 2 & 1 & 2 \end{bmatrix}, \qquad \mathbf{x} = \begin{bmatrix} X_1 \\ X_2 \\ X_3 \end{bmatrix}, \qquad \text{and} \quad \mathbf{c} = \begin{bmatrix} 1 \\ -2 \\ 0 \end{bmatrix}.$$

Recognizing that $\mathbf{A}$ is the matrix we inverted earlier,

$$\mathbf{A}^{-1} = \begin{bmatrix} 1 & 3 & -2 \\ 0 & -2 & 1 \\ -1 & -2 & 2 \end{bmatrix},$$

we have

$$\mathbf{x} = \mathbf{A}^{-1}\mathbf{c} = \begin{bmatrix} 1 & 3 & -2 \\ 0 & -2 & 1 \\ -1 & -2 & 2 \end{bmatrix}\begin{bmatrix} 1 \\ -2 \\ 0 \end{bmatrix}$$

$$= \begin{bmatrix} 1 - 6 + 0 \\ 0 + 4 + 0 \\ -1 + 4 + 0 \end{bmatrix} = \begin{bmatrix} -5 \\ 4 \\ 3 \end{bmatrix} = \begin{bmatrix} X_1 \\ X_2 \\ X_3 \end{bmatrix}.$$

Checking,

$$2(-5) + 2(4) + 1(3) = -10 + 8 + 3 = \quad 1$$
$$1(-5) + 0(4) + 1(3) = \quad -5 + 0 + 3 = -2$$
$$2(-5) + 1(4) + 2(3) = -10 + 4 + 6 = \quad 0.$$

WHEN SYSTEMS ARE AND ARE NOT SOLVABLE

If there are n unknowns, then n different equations are required to solve the system. By n different equations, we mean that each equation must carry different information about the n unknowns. In the foregoing example, all three equations carry different information:

$$2X_1 + 2X_2 + X_3 = 1$$

carries different information about X_1, X_2, and X_3 than does

$$X_1 + 0X_2 + X_3 = -2;$$

and both carry information different from

$$2X_1 + X_2 + 2X_3 = 0.$$

An example of two equations in X_1, X_2, and X_3 that do not carry different information is

$$2X_1 + 2X_2 + X_3 = 1$$

and

$$4X_1 + 4X_2 + 2X_3 = 2.$$

We note that the second equation is obtained by multiplying the first equation by 2, so it gives the same information as the first. (For example, we would not gain any information on the height of an object if we had first measured its height in feet and then proceed to measure its height in inches, since the second measurement will just be 12 times the first.) Without different information about the three unknowns, we cannot solve for them. For example, if our set of simultaneous equations were

$$2X_1 + 2X_2 + \quad X_3 = 1$$
$$4X_1 + 4X_2 + 2X_3 = 2$$
$$2X_1 + \quad X_2 + 2X_3 = 0,$$

then

$$
\mathbf{A} = \begin{bmatrix} 2 & 2 & 1 \\ 4 & 4 & 2 \\ 2 & 1 & 2 \end{bmatrix},
$$

and $|\mathbf{A}|$ will be zero, because one row is a linear combination of the others: row 2 = 2(row 1) + 0(row 3). Thus $\mathbf{A}^{-1}$ will not exist and the three equations cannot be solved. We essentially have only two equations in the three unknowns.

Suppose our system of equations is

$$
2X_1 + 2X_2 + X_3 = 1
$$
$$
4X_1 + 4X_2 + 2X_3 = 3
$$
$$
2X_1 + X_2 + 2X_3 = 0.
$$

The only difference between these equations and the previous ones is that the constant on the right-hand side in the second equation has been changed from 2 to 3. In this case, two of the equations carry *contradictory*, rather than redundant, information. The first equation implies that

$$
4X_1 + 4X_2 + 2X_3 = 2,
$$

and this contradicts the second equation, which says that

$$
4X_1 + 4X_2 + 2X_3 = 3.
$$

Thus, when any row of a matrix is a linear combination of the other rows, this means that the associated system of simultaneous equations contains either redundant or contradictory information, and thus cannot be solved.

What if some column of the coefficient matrix is a linear combination of some other columns? For example, consider the system

$$
2X_1 + 2X_2 + X_3 = 1
$$
$$
4X_1 + 5X_2 + 2X_3 = 4
$$
$$
6X_1 + X_2 + 3X_3 = 6.
$$

We can see that in the coefficient matrix

$$
\mathbf{A} = \begin{bmatrix} 2 & 2 & 1 \\ 4 & 5 & 2 \\ 6 & 1 & 3 \end{bmatrix}
$$

the first column is twice the third column. Essentially, this means that we have more equations than unknowns: any information given about X_1 is the same as the information given about X_3, and we will reach a contradiction when trying to solve the equations. For example, the second equation implies that

$$2X_1 + \frac{5}{2}X_2 + X_3 = 2$$

while the first two together imply

$$(2 - 2)X_1 + \left(2 - \frac{5}{2}\right)X_2 - (1 - 1)X_3 = 2 - 1,$$

or that

$$-\frac{5}{2}X_2 = 1 \quad \text{and} \quad X_2 = -\frac{2}{5}.$$

But the third equation implies

$$2X_1 + \frac{1}{3}X_2 + X_3 = 2,$$

so that the first and third equations taken together imply

$$(2 - 2)X_1 + \left(2 - \frac{1}{3}\right)X_2 + (1 - 1)X_3 = 1 - 2 = -1.$$

or $\quad \frac{5}{3}X_2 = -1, \quad$ or $\quad X_2 = -\frac{3}{5}.$

In the next section, we shall apply what we have learned about systems to solving normal equations.

EXERCISES   *Section 3.6*

3.24 Using matrices, solve the following systems of equations. Check your answers. (*Hint:* refer to Exercise 3.22a, p. 85.)

a. $2a + 3b = 1$
    $9a + b = 5.$

b. $x + = 1$
    $-9x + 7y = 2.$

c. $m + 2n - 2 = 0$
    $3m + 4n + 2 = 0.$

d. $x + 2y = 1$
    $2x + 4y = 3.$

e. $6p = 7$
    $7q = 3.$

f. $x_1 + 5x_2 + 3x_3 + 3 = 0$
    $x_3 = 0.$

g. $8a = 5.$

3.25 Solve the following systems of equations using matrices. Check your answers. (*Hint:* refer to Exercise 3.22b.)

a. $1x_1 + 0x_2 + 7x_3 = 3$
    $8x_1 + 2x_2 + 5x_3 = -3$
    $0x_1 + 6x_2 + 7x_3 = 9.$

b. $x = 3$
    $x + 5y + 4z = 7$
    $9x + 2y + 4z = 6.$

c. $5a + 7b \qquad - 1 = 0$
$7a + 3b + 7c + 8 = 0$
$\qquad 7b + 3c - 4 = 0.$

d. $m + 2n + \ \ 3p = 3$
$2m + 4n + \ \ 6p = 3$
$4m + 8n + 12p = 9.$

e. $4r + 5s = 2$
$7r + 2s = 5$
$3r + 5s = 5.$

f. $7a_1 = 5$
$5a_2 = 5$
$8a_3 = 5.$

3.26 Using matrices, solve the following systems of equations. Check your answers. (*Hint:* refer to answers to Exercise 3.22c.)

a. $a \qquad\qquad\qquad = 7$
$6a \qquad + 6c + 2d = 0$
$\qquad 5b + 4c \qquad = 3$
$6a + 6b + 8c + 2d = 7.$

b. $5x_1 + 6x_2 + 2x_3 + 4x_4 - 4 = 0$
$8x_2 + \ \ x_3 + 7x_4 - 9 = 0$
$4x_2 + \ \ x_3 + 5x_4 - 5 = 0$
$4x_2 \qquad\quad + 3x_4 - 2 = 0.$

c. $m + \ \ 2n + \ \ 3p + \ \ 4q = 1$
$2m + \ \ 4n + \ \ 6p + \ \ 8q = 5$
$4m + \ \ 8n + 12p + 16q = 2$
$8m + 16n + 24p + 32q = 4.$

d. $x_1 + 5x_2 + 9x_3 + 6x_4 = -6$
$7x_2 + 5x_3 + 2x_4 = \ \ 7$
$4x_1 + 6x_2 + \ \ x_3 + 7x_4 = \ \ 7$
$8x_1 + 9x_2 + \ \ x_3 + 7x_4 = -4.$

e. $2x_1 + 5x_2 + 3x_3 + 8x_4 = 2$
$7x_1 + 9x_2 + 6x_3 + 3x_4 = 7$
$x_1 + 5x_2 + 2x_3 + 7x_4 = 3$
$2x_1 + 9x_2 \qquad\quad + \ \ x_4 = 6$
$9x_2 + 8x_3 + \ \ x_4 = 7.$

3.7 APPLICATION TO SIMPLE LINEAR REGRESSION

Since finding the values a and b to substitute into the regression equation $Y = a + bX$ requires solving the normal equations, which are simultaneous equations, we can reformulate the regression problem in terms of matrices.

WRITING THE MODEL IN MATRIX FORM

Recall that the model for a simple linear regression,

$$Y = \alpha + \beta X + \varepsilon,$$

says that each Y-value is linearly related to an X-value, but imperfectly. Thus, for every one of the N items in the population,

$$Y_1 = \alpha + \beta X_1 + \varepsilon_1$$

$$Y_2 = \alpha + \beta X_2 + \varepsilon_2$$

$$\vdots$$

$$Y_N = \alpha + \beta X_N + \varepsilon_N,$$

and we have a system of equations which we can write in matrix form as

$$\mathbf{y} = \mathbf{X}\boldsymbol{\beta} + \boldsymbol{\varepsilon},$$

where

$$\mathbf{y}_{N \times 1} = \begin{bmatrix} Y_1 \\ Y_2 \\ \vdots \\ Y_N \end{bmatrix}, \quad \boldsymbol{\beta}_{2 \times 1} = \begin{bmatrix} \alpha \\ \beta \end{bmatrix}, \quad \boldsymbol{\varepsilon}_{N \times 1} = \begin{bmatrix} \varepsilon_1 \\ \varepsilon_2 \\ \vdots \\ \varepsilon_N \end{bmatrix}$$

and

$$\mathbf{X}_{N \times 2} = \begin{bmatrix} 1 & X_1 \\ 1 & X_2 \\ \vdots & \vdots \\ 1 & X_N \end{bmatrix}.$$

Note that in order to be conformable for multiplication with the vector $\boldsymbol{\beta}$, the matrix $\mathbf{X}$ must have two columns, and the column of ones in X is necessary to pick up the α in the vector $\boldsymbol{\beta}$:

$$\mathbf{X}\boldsymbol{\beta} = \underset{N \times 2}{\begin{bmatrix} 1 & X_1 \\ 1 & X_2 \\ \vdots & \vdots \\ 1 & X_N \end{bmatrix}} \underset{2 \times 1}{\begin{bmatrix} \alpha \\ \beta \end{bmatrix}} = \underset{N \times 1}{\begin{bmatrix} \alpha + \beta X_1 \\ \alpha + \beta X_2 \\ \vdots \\ \alpha + \beta X_N \end{bmatrix}}$$

Then we can see that

$$\underset{N \times 1}{\begin{bmatrix} Y_1 \\ Y_2 \\ \vdots \\ Y_N \end{bmatrix}} = \underset{N \times 1}{\begin{bmatrix} \alpha + \beta X_1 \\ \alpha + \beta X_2 \\ \vdots \\ \alpha + \beta X_N \end{bmatrix}} + \underset{N \times 1}{\begin{bmatrix} \varepsilon_1 \\ \varepsilon_2 \\ \vdots \\ \varepsilon_N \end{bmatrix}}$$

gives the set of N equations we started with.

NORMAL EQUATIONS IN MATRIX FORM

Since a sample of n observations (X_i, Y_i) drawn from the population of N items produces the estimating equation

$$\hat{Y} = a + bX,$$

the sample information is written in the following form:

$$\mathbf{y} = \begin{bmatrix} Y_1 \\ Y_2 \\ \vdots \\ Y_n \end{bmatrix}_{n \times 1}, \quad \mathbf{X} = \begin{bmatrix} 1 & X_1 \\ 1 & X_2 \\ \vdots & \vdots \\ 1 & X_n \end{bmatrix}_{n \times 2}, \quad \hat{\boldsymbol{\beta}} = \begin{bmatrix} a \\ b \end{bmatrix}_{2 \times 1}.$$

Recall that the "hat" on $\boldsymbol{\beta}$ indicates that *estimates* are being made and in order to find the value of $\hat{\boldsymbol{\beta}}$ (the values of a and b), we must come up with a system of normal equations. Recall that the normal equations are

$$\sum Y_i = na + b \sum X_i$$
$$\sum X_i Y_i = a \sum X_i + b \sum X_i^2.$$

How do we write these equations in matrix form? (Remember, the *unknowns* here are a and b.) In matrices we get

$$\begin{bmatrix} n & \sum X_i \\ \sum X_i & \sum X_i^2 \end{bmatrix}_{2 \times 2} \begin{bmatrix} a \\ b \end{bmatrix}_{2 \times 1} = \begin{bmatrix} \sum Y_i \\ \sum X_i Y_i \end{bmatrix}_{2 \times 1}.$$

In terms of our data, $\mathbf{y}$, $\mathbf{X}$, and $\hat{\boldsymbol{\beta}}$, obviously, $\begin{bmatrix} a \\ b \end{bmatrix} = \hat{\boldsymbol{\beta}}$. Also, the vector on the right-hand side involves both X's and Y's. Since $\mathbf{X}$ is $n \times 2$ and $\mathbf{y}$ is $n \times 1$, the two matrices are not conformable for multiplication. However, $\mathbf{X}'$ is $2 \times n$ and the product $\mathbf{X}'\mathbf{y}$ can be formed, and this product is 2×1, the same dimension as $\hat{\boldsymbol{\beta}}$. Forming the product $\mathbf{X}'\mathbf{y}$,

$$\mathbf{X}'\mathbf{y} = \begin{bmatrix} 1 & 1 & \cdots & 1 \\ X_1 & X_2 & \cdots & X_n \end{bmatrix} \begin{bmatrix} Y_1 \\ Y_2 \\ \vdots \\ Y_n \end{bmatrix} = \begin{bmatrix} Y_1 + Y_2 + \cdots + Y_n \\ X_1 Y_1 + X_2 Y_2 + \cdots + X_n Y_n \end{bmatrix}$$

$$= \begin{bmatrix} \sum Y_i \\ \sum X_i Y_i \end{bmatrix},$$

and this is the vector on the right-hand side of the equation. The matrix

$$\begin{bmatrix} n & \sum X_i \\ \sum X_i & \sum X_i^2 \end{bmatrix}$$

involves the square of the X-variable. It is also symmetric. Note that since $\mathbf{X}$ is $n \times 2$, $\mathbf{X}'$ is $2 \times n$ and $\mathbf{X}'\mathbf{X}$ is 2×2, which is conformable for multiplication on the right by $\hat{\boldsymbol{\beta}}$. Then

$$\mathbf{X}'\mathbf{X} = \begin{bmatrix} 1 & 1 & \cdots & 1 \\ X_1 & X_2 & \cdots & X_n \end{bmatrix} \begin{bmatrix} 1 & X_1 \\ 1 & X_2 \\ \vdots & \vdots \\ 1 & X_n \end{bmatrix}$$

$$= \begin{bmatrix} 1 + 1 + \cdots + 1 & X_1 + X_2 + \cdots + X_n \\ X_1 + X_2 + \cdots + X_n & X_1^2 + X_2^2 + \cdots + X_n^2 \end{bmatrix}$$

$$= \begin{bmatrix} n & \sum X_i \\ \sum X_i & \sum X_i^2 \end{bmatrix}.$$

Thus, the normal equations can be written in matrix notation as

$$(\mathbf{X}'\mathbf{X})\hat{\boldsymbol{\beta}} = \mathbf{X}'\mathbf{y}.$$

This formulation of normal equations is the same for every regression problem, regardless of the number of predictors and regardless of whether we are fitting a straight line or a curve. The matrix $\mathbf{X}$ for multiple regression or curvilinear regression will be defined differently, as will the vector $\hat{\boldsymbol{\beta}}$; but as long as $\mathbf{X}$ and $\hat{\boldsymbol{\beta}}$ are appropriately defined the normal equations can always be written as

$$(\mathbf{X}'\mathbf{X})\hat{\boldsymbol{\beta}} = \mathbf{X}'\mathbf{y}.$$

For example, if we want to fit an equation of the form

$$\hat{Y} = a + bX + cX^2,$$

then we need to solve the normal equations

$$\sum Y_i = na + b\sum X_i + c\sum X_i^2$$
$$\sum X_i Y_i = a\sum X_i + b\sum X_i^2 + c\sum X_i^3$$
$$\sum X_i^2 Y_i = a\sum X_i^2 + b\sum X_i^3 + c\sum X_i^4.$$

We first define

$$
\mathbf{y} = \begin{bmatrix} Y_1 \\ Y_2 \\ \vdots \\ Y_n \end{bmatrix}, \quad \hat{\boldsymbol{\beta}} = \begin{bmatrix} a \\ b \\ c \end{bmatrix}, \quad \text{and} \quad \mathbf{X} = \begin{bmatrix} 1 & X_1 & X_1^2 \\ 1 & X_2 & X_2^2 \\ \vdots & \vdots & \vdots \\ 1 & X_n & X_n^2 \end{bmatrix}.
$$

Note that there is a column in $\mathbf{X}$ corresponding to the coefficient of each unknown in the equation to be fitted. Then

$$
\mathbf{X'y} = \begin{bmatrix} 1 & 1 & \cdots & 1 \\ X_1 & X_2 & \cdots & X_n \\ X_1^2 & X_2^2 & \cdots & X_n^2 \end{bmatrix} \begin{bmatrix} Y_1 \\ Y_2 \\ \vdots \\ Y_n \end{bmatrix} = \begin{bmatrix} \sum Y_i \\ \sum X_i Y_i \\ \sum X_i^2 Y_i \end{bmatrix}
$$

and

$$
\mathbf{X'X} = \begin{bmatrix} 1 & 1 & \cdots & 1 \\ X_1 & X_2 & \cdots & X_n \\ X_1^2 & X_2^2 & \cdots & X_n^2 \end{bmatrix} \begin{bmatrix} 1 & X_1 & X_1^2 \\ 1 & X_2 & X_2^2 \\ \vdots & \vdots & \vdots \\ 1 & X_n & X_n^2 \end{bmatrix}
$$

$$
= \begin{bmatrix} n & \sum X_i & \sum X_i^2 \\ \sum X_i & \sum X_i^2 & \sum X_i^3 \\ \sum X_i^2 & \sum X_i^3 & \sum X_i^4 \end{bmatrix};
$$

and we can see that

$$(\mathbf{X'X})\hat{\boldsymbol{\beta}} = \mathbf{X'y}$$

gives the normal equations.

In general, $\mathbf{y}$ will simply be a vector containing the n Y-values. $\hat{\boldsymbol{\beta}}$ will be a vector containing the unknowns. For p unknowns, in the regression equation, $\hat{\boldsymbol{\beta}}$ will be a $p \times 1$ vector and $\mathbf{X}$ will be an $n \times p$ matrix containing columns corresponding to the coefficients of the unknowns in the regression equation. Once $\mathbf{y}$, $\hat{\boldsymbol{\beta}}$, and $\mathbf{X}$ are appropriately defined for any problem, the normal equations are always given by

$$(\mathbf{X'X})\hat{\boldsymbol{\beta}} = \mathbf{X'y}.$$

And their solutions are given by

$$(\mathbf{X}'\mathbf{X})^{-1}(\mathbf{X}'\mathbf{X})\hat{\beta} = (\mathbf{X}'\mathbf{X})^{-1}\mathbf{X}'\mathbf{y}$$
$$\mathbf{I}\hat{\beta} = \hat{\beta} = (\mathbf{X}'\mathbf{X})^{-1}\mathbf{X}'\mathbf{y},$$

as long as $(\mathbf{X}'\mathbf{X})$ is nonsingular so that its inverse, $(\mathbf{X}'\mathbf{X})^{-1}$, exists.

THE SOLUTION TO NORMAL EQUATIONS FOR SIMPLE REGRESSION

In the simple regression problem, we have

$$(\mathbf{X}'\mathbf{X}) = \begin{bmatrix} n & \sum X_i \\ \sum X_i & \sum X_i^2 \end{bmatrix},$$

and

$$|\mathbf{X}'\mathbf{X}| = n\sum X_i^2 - (\sum X_i)^2 = nS_{XX},$$

so that

$$(\mathbf{X}'\mathbf{X})^{-1} = \begin{bmatrix} \dfrac{\sum X_i^2}{nS_{XX}} & \dfrac{-\sum X_i}{nS_{XX}} \\ \dfrac{-\sum X_i}{nS_{XX}} & \dfrac{n}{nS_{XX}} \end{bmatrix}.$$

(Note that both $\mathbf{X}'\mathbf{X}$ and $(\mathbf{X}'\mathbf{X})^{-1}$ are symmetric. It will always be true that the inverse of a symmetric matrix is also symmetric.) Then

$$(\mathbf{X}'\mathbf{X})^{-1}\mathbf{X}'\mathbf{y} = \begin{bmatrix} \dfrac{\sum X_i^2}{nS_{XX}} & \dfrac{-\sum X_i}{nS_{XX}} \\ \dfrac{-\sum X_i}{nS_{XX}} & \dfrac{n}{nS_{XX}} \end{bmatrix} \begin{bmatrix} \sum Y_i \\ \sum X_i Y_i \end{bmatrix}$$

$$= \begin{bmatrix} \dfrac{\sum X_i^2 \ \sum Y_i - \sum X_i \ \sum X_i Y_i}{nS_{XX}} \\ \dfrac{-\sum X_i \ \sum Y_i + n\sum X_i Y_i}{nS_{XX}} \end{bmatrix}.$$

The numerator of the bottom element is easily recognizable as nS_{XY}. Thus, the bottom element is $nS_{XY}/nS_{XX} = S_{XY}/S_{XX} = b$, which is what we obtained before. It is a little more difficult to see that the top value is $a = \bar{y} - b\bar{X}$:

$$\frac{\sum X_i^2 \sum Y_i - \sum X_i \sum X_i Y_i}{nS_{XX}}$$

$$= \frac{\sum X_i^2 \sum Y_i - \sum X_i \sum X_i Y_i - (\sum X_i)^2 \bar{Y} + (\sum X_i)^2 \bar{Y}}{nS_{XX}}$$

$$= \frac{\sum X_i^2 \sum Y_i - (\sum X_i)^2 \bar{Y} - [\sum X_i \sum X_i Y_i - (\sum X_i)^2 \bar{Y}]}{nS_{XX}}$$

$$= \frac{\dfrac{n\sum X_i^2 \sum Y_i - (\sum X_i)^2 \sum Y_i}{n} - \dfrac{n\sum Y_i \sum X_i Y_i - (\sum X_i)^2 \sum Y_i}{n}}{nS_{XX}}$$

$$= \frac{\left[\dfrac{n\sum X_i^2 - (\sum X_i)^2}{n}\right]\sum Y_i - \left[\dfrac{n\sum X_i Y_i - (\sum X_i)(\sum Y_i)}{n}\right]\sum X_i}{nS_{XX}}$$

$$= \frac{S_{XX}\sum Y_i - S_{XY}\sum X_i}{nS_{XX}}$$

$$= \frac{\sum Y_i}{n} - \frac{S_{XY}}{S_{XX}}\frac{\sum X_i}{n}$$

$$= \bar{Y} - b\bar{X} = a.$$

Thus, the matrix approach gives the same values for a and b as we obtained before.

THE HOUSE SIZE-PRICE PROBLEM USING MATRICES

Let us return to the house size-price example, using matrices for the problem we worked out in Chapter 2, pp. 34ff.
We find

$$\mathbf{y} = \begin{bmatrix} 32 \\ 24 \\ 27 \\ 47 \\ \vdots \\ 37 \\ 28 \end{bmatrix}_{20 \times 1}, \qquad \hat{\boldsymbol{\beta}} = \begin{bmatrix} a \\ b \end{bmatrix}, \qquad \mathbf{X} = \begin{bmatrix} 1 & 1.8 \\ 1 & 1.0 \\ 1 & 1.7 \\ 1 & 2.8 \\ \vdots & \vdots \\ 1 & 2.2 \\ 1 & 1.5 \end{bmatrix}_{20 \times 2},$$

$$\mathbf{X'X} = \begin{bmatrix} 20 & 40 \\ 40 & 93.56 \end{bmatrix}_{2 \times 2}, \qquad \mathbf{X'y} = \begin{bmatrix} 690 \\ 1554.9 \end{bmatrix}_{2 \times 1},$$

The normal equations are $\mathbf{X'X\beta} = \mathbf{X'y}$, or

$$\begin{bmatrix} 20 & 40 \\ 40 & 93.56 \end{bmatrix} \begin{bmatrix} a \\ b \end{bmatrix} = \begin{bmatrix} 690 \\ 1554.9 \end{bmatrix},$$

which yields

$$20a + 40b \quad = 690$$
$$40a + 93.56b = 1554.9,$$

as before (p. 34).

To solve the normal equations,

$$|\mathbf{X'X}| = 20(93.56) - (40)^2 = 271.2 = nS_{XX},$$

so that

$$(\mathbf{X'X})^{-1} = \begin{bmatrix} \dfrac{93.56}{271.2} & \dfrac{-40}{271.2} \\[2ex] \dfrac{-40}{271.2} & \dfrac{20}{271.2} \end{bmatrix}$$

and

$$\hat{\boldsymbol{\beta}} = (\mathbf{X'X})^{-1}\mathbf{X'y} = \begin{bmatrix} \dfrac{93.56}{271.2} & \dfrac{-40}{271.2} \\[2ex] \dfrac{-40}{271.2} & \dfrac{20}{271.2} \end{bmatrix} \begin{bmatrix} 690 \\ 1554.9 \end{bmatrix}$$

$$= \begin{bmatrix} \dfrac{93.56(690) - 40(1554.9)}{271.2} \\[2ex] \dfrac{-40(690) + 20(1554.9)}{271.2} \end{bmatrix} = \begin{bmatrix} \dfrac{2360.4}{271.2} \\[2ex] \dfrac{3498}{271.2} \end{bmatrix}$$

$$= \begin{bmatrix} 8.7035 \\ 12.8982 \end{bmatrix} = \begin{bmatrix} a \\ b \end{bmatrix}.$$

Note that these results are almost identical to what we found in Chapter 2. The slight difference in the value for a (8.7035 here vs. 8.7036 before) is due to rounding error. Note also that the fractions in $(\mathbf{X'X})^{-1}$ were not converted to decimal fractions prior to multiplication with $\mathbf{X'y}$; this was done to reduce rounding error.

CORRECTING X FOR ITS MEAN

Suppose X is corrected for its mean, so that our model is

$$\hat{Y} = \alpha' + \beta(X - \bar{X}) + \varepsilon.$$

Let us denote the **X**-matrix for centered X-values by $\mathscr{X}$. Then

$$\mathscr{X} = \begin{bmatrix} 1 & X_1 - \bar{X} \\ 1 & X_2 - \bar{X} \\ \vdots & \vdots \\ 1 & X_n - \bar{X} \end{bmatrix};$$

the vector $\hat{\boldsymbol{\beta}}$ is

$$\hat{\boldsymbol{\beta}} = \begin{bmatrix} a' \\ b \end{bmatrix},$$

and **y** remains unchanged. Then

$$\mathscr{X}'\mathbf{y} = \begin{bmatrix} 1 & 1 & \cdots & 1 \\ X_1 - \bar{X} & X_2 - \bar{X} & \cdots & X_n - \bar{X} \end{bmatrix} \begin{bmatrix} Y_1 \\ Y_2 \\ \vdots \\ Y_n \end{bmatrix}$$

$$= \begin{bmatrix} Y_1 + Y_2 + \cdots + Y_n \\ Y_1(X_1 - \bar{X}) + Y_2(X_2 - \bar{X}) + \cdots + Y_n(X_n - \bar{X}) \end{bmatrix}$$

$$= \begin{bmatrix} \sum Y_i \\ \sum Y_i(X_i - \bar{X}) \end{bmatrix}.$$

Recall from Section 2.4, that $\sum Y_i(X_i - \bar{X}) = \sum (Y_i - \bar{Y})(X_i - \bar{X}) = S_{XY}$, so that if X is corrected for its mean,

$$\mathscr{X}'\mathbf{y} = \begin{bmatrix} \sum Y_i \\ S_{XY} \end{bmatrix}.$$

Now

$$\mathscr{X}'\mathscr{X} = \begin{bmatrix} 1 & 1 & \cdots & 1 \\ X_1 - \bar{X} & X_2 - \bar{X} & \cdots & X_n - \bar{X} \end{bmatrix} \begin{bmatrix} 1 & X_1 - \bar{X} \\ 1 & X_2 - \bar{X} \\ \vdots & \vdots \\ 1 & X_n - \bar{X} \end{bmatrix}$$

$$= \begin{bmatrix} 1 + 1 + \cdots + 1 & (X_1 - \bar{X}) + (X_2 - \bar{X}) + \cdots + (X_n - \bar{X}) \\ (X_1 - \bar{X}) + (X_2 - \bar{X}) + \cdots + (X_n - \bar{X}) & (X_1 - \bar{X})^2 + (X_2 - \bar{X})^2 + \cdots + (X_n - \bar{X})^2 \end{bmatrix}$$

$$= \begin{bmatrix} n & \sum (X_i - \bar{X}) \\ \sum (X_i - \bar{X}) & \sum (X_i - \bar{X})^2 \end{bmatrix} = \begin{bmatrix} n & 0 \\ 0 & S_{XX} \end{bmatrix}.$$

Note that in this case, $\mathscr{X}'\mathscr{X}$ is diagonal. The normal equations are still

$$\mathscr{X}'\mathscr{X}\hat{\boldsymbol{\beta}} = \mathscr{X}'\mathbf{y},$$

so that

$$\begin{bmatrix} n & 0 \\ 0 & S_{XX} \end{bmatrix} \begin{bmatrix} a' \\ b \end{bmatrix} = \begin{bmatrix} \sum Y_i \\ S_{XY} \end{bmatrix},$$

or

$$na' = \sum Y_i$$
$$S_{XX}b = S_{XY}.$$

These are two separate equations in two separate unknowns, so

$$a' = \frac{\sum Y_i}{n} = \bar{Y}$$

$$b = \frac{S_{XY}}{S_{XX}},$$

as before. Should one desire to solve the normal equations as

$$\hat{\boldsymbol{\beta}} = (\mathscr{X}'\mathscr{X})^{-1}\mathscr{X}'\mathbf{y},$$

inversion of $\mathcal{X}'\mathcal{X}$ is simplified quite a bit because $\mathcal{X}'\mathcal{X}$ is diagonal:

$$(\mathcal{X}'\mathcal{X})^{-1} = \begin{bmatrix} \dfrac{1}{n} & 0 \\ 0 & \dfrac{1}{S_{XX}} \end{bmatrix}$$

and

$$\hat{\boldsymbol{\beta}} = \begin{bmatrix} \dfrac{1}{n} & 0 \\ 0 & \dfrac{1}{S_{XX}} \end{bmatrix} \begin{bmatrix} \sum Y_i \\ S_{XY} \end{bmatrix} = \begin{bmatrix} \dfrac{\sum Y_i}{n} \\ \dfrac{S_{XY}}{S_{XX}} \end{bmatrix}$$

$$= \begin{bmatrix} \bar{Y} \\ \dfrac{S_{XY}}{S_{XX}} \end{bmatrix} = \begin{bmatrix} a' \\ b \end{bmatrix}.$$

Note the great reduction in labor achieved if X is corrected for its mean, or centered, so that $\mathcal{X}'\mathcal{X}$ is diagonal. A similar reduction is achieved if X is scaled to $Z = (X - \bar{X})/d$. The reduction in labor is even greater in some multiple regression problems, where $\mathcal{X}'\mathcal{X}$ is 3×3 or greater.

EXERCISES   *Section 3.7*

p. 521, 522
524, 525

3.27   Refer to one of the data sets: TEACHERS, PRECIPITATION, SALES, or HEATING presented in Chapter 1 or in the Analyses of Data Sets section. For your selected data set,
   a. display the matrix $\mathbf{X}$ and the vectors $\mathbf{y}$ and $\boldsymbol{\beta}$ necessary to fit a simple linear equation $\hat{Y} = a + bX$ to the data,
   b. find $\mathbf{X'X}$ and $\mathbf{X'y}$,
   c. find $(\mathbf{X'X})^{-1}$,
   d. find $\hat{\boldsymbol{\beta}}$. State the regression equation and compare to that found in Exercise 2.8,
   e. repeat parts a–d, but use X-values corrected for their mean. Compare your regression equation to that found in Exercise 2.14.

p. 523

3.28   For the POLICE data in Chapter 1 or in the Analyses of Data Sets section,
   a. display the matrix $\mathbf{X}$ and the vectors $\mathbf{y}$ and $\hat{\boldsymbol{\beta}}$ necessary to fit a curvilinear equation $\hat{Y} = a + bX + cX^2$ to the data,
   b. find $\mathbf{X'X}$ and $\mathbf{X'y}$,
   c. find $(\mathbf{X'X})^{-1}$,
   d. find $\hat{\boldsymbol{\beta}}$. State the regression equation. Compare to answer to Exercise 2.11b.

p. 526

3.29   Referring to the FERTILIZER data in Chapter 2 or the Analyses of Data Sets section, and using given X-values,

      a.   display the matrix $\mathbf{X}$ and the vectors $\mathbf{y}$ and $\hat{\boldsymbol{\beta}}$ necessary to fit a linear equation $\hat{Y} = a + bX$ to the data,

      b.   find $\mathbf{X'X}$ and $\mathbf{X'y}$,

      c.   find $(\mathbf{X'X})^{-1}$,

      d.   find $\hat{\boldsymbol{\beta}}$. State the regression equation. Compare answer to that of Exercise 2.15e.

3.30   Repeat Exercise 3.29, but scale the X-values. Compare the regression equation to the one found above.

p. 527   *3.31*   Refer to the DEPRIVATION data of Chapter 2 or the Analyses of Data Sets section.

      a.   Display the matrix $\mathbf{Z}$ and the vectors $\mathbf{y}$ and $\hat{\boldsymbol{\beta}}$ necessary to fit a curvilinear equation $\hat{Y} = a' + b'Z + c'Z^2$ to the data.

      b.   Find $\mathbf{Z'Z}$ and $\mathbf{Z'y}$.

      c.   Find $(\mathbf{Z'Z})^{-1}$.

      d.   Find $\hat{\boldsymbol{\beta}}$. State the regression equation. Compare to answer to Exercise 2.16c.

3.8 SUMMARY

We have now explored three equivalent methods for finding a least squares regression equation. Normal equations have been solved as simultaneous equations, by using algebraic formulas, and finally by using matrices. All three methods produce the same results, although at first they may appear to be different.

Since the methods presented in this chapter have universal applicability, they lay the foundation for the study of multiple regression. Understanding some of the problems involved in solving simultaneous equations will also provide insight into the interrelationships among variables in multiple regression.

While we have in theory solved all solvable sets of simultaneous equations, there remains the tedious job of actually obtaining a solution for a real data problem. We shall next see how to make a computer do the dirty work.

Chapter 4 USE OF COMPUTER PROGRAMS IN SIMPLE REGRESSION

4.1 INTRODUCTION

So far, our simple regression problems have been small enough to do with the aid of only a calculator. The larger n becomes, however, the more time-consuming and error-prone the calculations become. You probably are not very eager to calculate $\sum X_i Y_i$ even if $n = 20$: imagine your feelings if n were 175! When we get to multiple regression, the computations become enormous. Thus, it is important to be able to use a computer for the calculations, in the interest of both time and accuracy. While it may seem like overkill to run a simple regression problem through the computer, one must begin with relatively simple problems.

In our discussions of some computer programs, instructions for obtaining access to the program are outlined, with major emphasis on interpretation of the output. The programs discussed will be various BMD P-series, SPSS, and SAS packages. Most large computing centers offer at least one of these libraries. All the programs use punched cards (or data stored on magnetic tape) and produce printed output. (The student is assumed to know keypunch.) Since all regression programs give essentially the same output, the student who has easier access to programs other than the ones presented here should be able to interpret the output by relating it to the output discussed here.

4.2 SIMPLE LINEAR REGRESSION USING BMD P-SERIES PACKAGES

BMD
P-Series

The BMD packages are Biomedical Computer Programs, put out by the Health Sciences Computing Facility at UCLA. The original BMD programs required a very rigid input format, and while they are still used extensively, they are being replaced by the P-series which uses a parameter language control rather than a fixed format program control. (Those familiar with programming will know what this means.) The P-series is now available in most large computing centers.*

The extensive BMD offerings can perform a variety of statistical analyses. Among the several regression (R) programs available, we shall look at the first and second catalogued.

BMDP1R

BMDP2R

The BMDP1R program is entitled Multiple Linear Regression and the BMDP2R is called Stepwise Regression. While both are intended for use in multiple regression problems, both can perform simple linear regressions as a special case and can be made to fit curvilinear regression equations with relative ease.

SYSTEM CARDS

System Cards

Since every computing center is set up a bit differently from every other, you are on your own to obtain appropriate information about system cards. These cards give one access to the particular computing system, providing such information as the user's name, account number, time requirements for the program, and so on. They also ask the computer to refer to the selected program, which it keeps in memory, ready for use. The system cards always go, in a specified order, at the first of the deck of cards punched. In some facilities, system cards follow as well as precede the program cards.

A SAMPLE PROGRAM

Once the system cards have been punched, the following steps will allow the user to run the BMDP1R and the BMDP2R programs.† Figure 4.1 shows a worksheet for the cards to be punched, using data in Table 1.1. Note that for a simple linear regression analysis, the program in Figure 4.1 can be run either by the BMDP1R or the BMDP2R. A system card tells the computer which program to run.

LETTERS AND NUMBERS

Before going on to discuss the program, we note that the symbol $\emptyset$ denotes the letter "oh", not to be confused with the number 0 (zero). (When proofreading your cards, note that $\emptyset$ is more squarish than the number 0. Similarly, the letter I and the number 1 should be carefully distinguished for the computer to understand what you are telling it to do.

* Those who have access only to the original series are referred to the 1975 manual, *BMD Biomedical Computer Programs*, The University of California Press, Berkeley.

† Refer to the manual *BMDP Biomedical Computer Programs*, W. J. Dixson, ed, The University of California Press, Berkeley, 1977, for a detailed explanation.

FIGURE 4.1 *Worksheet for Fitting a Simple Linear Regression Equation to House Size-Price Data Using Either BMDP1R or BMDP2R*

```
         1-10      11-20     21-30     31-40
1234567890123456789012345678901234567890
    {SYSTEM CARDS
PRØBLEM   TITLE='SIMPLE REGR  HØUSES'./
INPUT     VARIABLES=2.
          FØRMAT='(F3.1,F3.0)'.
          CASES=20./
VARIABLE  NAMES=SIZE,PRICE./
PRINT     DATA./
PLØT      VARIABLE=SIZE./
REGRES    DEPENDENT=PRICE./
END/
1.8 32
1.0 24
1.7 27
2.8 47
2.2 35
 .8 17
3.6 52
1.1 20
2.0 38
2.6 45
2.3 44
 .9 19
1.2 25
3.4 50
1.7 30
2.5 43
1.4 27
3.3 50
2.2 37
1.5 28
FINISH/
```

FORMATS

Format

A format statement tells the computer how to read the data (numbers) you have punched on cards for a particular problem. Consider, for example, the format statement (which is always enclosed in parentheses)

(F6.2, F3.0).

This says that:

1. there are values of *two* variables punched on each card, since two F's (formats) are specified;

2. the first six columns of each card contain values of the first variable, and the next three columns contain values of the second variable;
3. thus, all numbers are punched in columns 1–9;
4. values of the first variable can have as many as two digits to the right of the decimal point; values of the second variable are whole numbers.

(Those familiar with formats should note that an F-type format statement must always be used with BMD regression programs, even for data containing only integer values. The programs will not run if you use I-type format statements.)

Suppose we have a pair of values

$$X = 223.64 \quad \text{and} \quad Y = 65.1.$$

How do we write a format statement to handle these variables? First, there are eight numbers to be punched, if we do not punch the decimal point. The first value has five numbers, two of which are to the right of the decimal; and the second has three numbers with one decimal place. A format statement such as

(F5.2, F3.1)

will handle this pair of values and we would punch, beginning in column 1 on the data card and ending in column 8:

column	1 2 3 4 5 6 7 8
punch	2 2 3 6 4 6 5 1

Proofreading the punched cards is easier if the decimal point is indicated. For this, we can come up with other format statements. To punch the decimal points in both values requires 10 spaces, so we need this format statement

(F6.2, F4.1).

We would punch the data card, beginning in column 1 and ending in column 10:

column	1 2 3 4 5 6 7 8 9 10
punch	2 2 3 . 6 4 6 5 . 1

To leave a space between the first and second values and still punch the decimal points would require

(F6.2, F5.1),

and we would punch

column	1	2	3	4	5	6	7	8	9	10	11
punch	2	2	3	.	6	4		6	5	.	1

Note that column 7 is left blank. Finally, we could space between the two values and *not* punch the decimal point by using

    (F5.2, F4.1)

and punching

column	1	2	3	4	5	6	7	8	9
punch	2	2	3	6	4		6	5	1

You are free to set up the format to your own liking, *as long as your data are punched to conform to the format statement.* For example, suppose the format statement were

    (2F10.4),

indicating that ten spaces have been allotted to both values, with the last four of each ten being decimal places. How would one punch in

    $X = 123.456,$ $Y = 7.8?$

To punch the decimals, the number 1 would be placed in column 3, the decimal point in column 6, and the number 6 in column 9; then 7 would appear in column 15, the decimal point in column 16, and the 8 in column 17:

column	1	2	3	4	5	6	7	8	9	10	11	12	13	14	15	16	17	18	19	20
punch			1	2	3	.	4	5	6						7	.	8			

If the decimal points are not punched, then the data would be entered as

column	1	2	3	4	5	6	7	8	9	10	11	12	13	14	15	16	17	18	19	20
punch			1	2	3	4	5	6							7	8				

Note that if the value 123456 began in column 1 and ended in column 6, the computer, following the format statement (2F10.4), would read the number as 123,456.0000 instead of 123.456. Thus, extreme care must be taken to make the data conform to the format statement.

Right Justify | This last example leads us to the consideration of what it means to right justify. Right-justifying is simply positioning a value so that the decimal point is in its proper place. If you were to add up the three numbers 10, 243, and 7 by hand, you might write them in a column thus:

```
  10
 243
   7
 ___
```

You have lined up the three numbers so that the understood decimal points are one above the other. Suppose that in columns 9–13 of a card, you are to enter the number 20, with a corresponding format of F5.2, so that the decimal point is understood to stand after column 11. Then the 2 must be punched in column 10 and the zero in column 11. For example,

		in column						the computer reads
7	8	9	10	11	12	13		
			2	0				20
				2	0			2
if you punch		2	0					200
	2	0						0
			2					20

EXAMINATION OF THE BMDP1R AND BMDP2R PROGRAMS

We are now ready to discuss the BMDP1R and BMDP2R programs for running simple linear regressions. First note that BMD programs are divided into "sentences" and "paragraphs." Just as in English grammar, a paragraph is made up of one or more sentences, and all sentences end with periods. For example, in Figure 4.1,

FØRMAT = '(F3.1, F3.0)'.

is a sentence. The three sentences

VARIABLES = 2.

FØRMAT = '(F3.1, F3.0)'.

CASES = 20./

make up the INPUT paragraph. Note that all paragraphs end with slashes.* Since the END/ and FINISH/ statements are paragraphs that contain no sentences, there are no periods before their slashes. It does not matter in what order paragraphs are listed (except for PRØBLEM, END, and FINISH), or in what order sentences occur within paragraphs.

Data Cards   The cards that contain only numbers are <u>data cards</u>. The numerical values for our two variables, size and price, are punched according to the format statement (F3.1, F3.0). Note that the information for each house is punched on a separate card; also the sizes are first and the prices second, to correspond with the statement

    NAMES = SIZE, PRICE.

It does not matter in which order the values are entered, as long as the data you enter comply with the NAMES sentence and the FØRMAT statement. For example, if you punched your cards with prices entered first, beginning in column 1 and ending in column 6,

    32   1.8
    24   1.0
        ⋮
    28   1.5,

then the NAMES statement should be

    NAMES = PRICE, SIZE.

and the format becomes

    (F2.0, F4.1).

           The first paragraph of a BMD P-series program must always be a "problem" paragraph, indicated by

    PRØBLEM.

This gives your program a title, as indicated in Figure 4.1. Note the apostrophes that enclose your title. If you choose not to name your program, then just write

    PRØBLEM/

on the first card.

           It does not matter in what column any sentence or paragraph begins, or how many spaces you leave between words. We have begun all paragraphs in column 1 for convenience, and all sentences in column 10 because this makes an easy layout to read. Also, you can abbreviate words by shortening them or leaving out the vowels.

_____

* These slashes may be placed after the period on the same card, as indicated in Figure 4.1, or may precede the first character on the subsequent card.

PRØBLEM, for example, could also be written PRØBL or PRØB or PRBLM. However, one peculiarity of the BMD program is that VARIABLE cannot be abbreviated as VAR. In the interest of clarity, we shall not abbreviate much. Also, the equal sign (=) can be replaced by IS or ARE, if you prefer to write in English.

The INPUT paragraph tells the computer how to read your data cards. In our example,

VARIABLES = 2.

says that we are working with two variables (size and price). We have already explained the format statement—just note that apostrophes surround its parentheses. Then

CASES = 20.

says that $n = 20$ for this problem.

The VARIABLES paragraph also identifies the variables to the computer, assigning each a name and denoting their order on the data cards: SIZE and PRICE, respectively, in this case. The names can be anything you like, as long as they do not begin with a number and are no more than eight spaces long. We could have called these variables just X and Y, for example. If you do not specify names, then the computer will name them X(1), X(2), etc.

The PRINT paragraph can be used to ask that some optional output be printed. All we need in simple regression is for the computer to reproduce the data. This helps us check for keypunching errors or mistakes in constructing the format.

The PLØT paragraph will give us two plots, one a scatter diagram of the data with the regression line superimposed, and the other, a scatter diagram of the *residuals*, or values of $Y_i - \hat{Y}_i$. The sentence

VARIABLE = SIZE.

says that the observed and predicted values and residuals are to be plotted against SIZE = X; in other words, the size variable is on the horizontal axis. (This becomes less obvious in a multiple regression problem.)

The REGRES paragraph instructs the computer to perform a regression analysis. The only sentence needed in a simple regression is the one that specifies the dependent, or response variable (Y). In this case,

DEPENDENT = PRICE.

The END/ card says that this is all you want to do with this problem. (To do further analyses on the same data, you'd start over again with a new PRØBLEM paragraph and additional program instructions.) Then come the data cards, which we have already discussed. Note, however, that no period or slash follows the information on a data card. The FINISH/ card signals that you are through giving program commands and feeding in data cards. In some computing facilities, additional system cards might be needed after the FINISH card.

READING THE PRINTOUT FROM BMDP1R

While the program in Figure 4.1 can be used with either the BMDP1R or the BMDP2R to perform a simple linear regression analysis, the two programs do give results in different forms. We will look first at the printout from the P1R and then show how to find the same results using the P2R.

The BMDP1R output consists of four pages of printout. Most of the first page reproduces program information. You should look this over to check on your keypunching. First, the program statements are reproduced as you punched them, and then some tables show how the computer interpreted what you told it. The first results of the REGRES command are the simple statistics for the two variables, and we see in Figure 4.2 some familiar numbers, such as $X = 2$, $Y = 34.49994$ (≈ 34.5). In addition, the standard deviations for the two variables are given, and the ratios of the standard deviations to the means for both variables also appear. This ratio is called the

Coefficient of Variation

coefficient of variation. It is used to correct a measure of variability (the standard deviation) for the sizes of the values involved. That is, it is used to answer the question, "Is the standard deviation large because there is a lot of variability in the data, or is it large because the numbers in the data set are large?" More will be said about coefficients of variation in Chapter 6.

FIGURE 4.2   *Simple Statistics from BMDP1R Simple Regression Analysis*

VARIABLE	MEAN	STANDARD DEVIATION	ST.DEV/MEAN	MINIMUM	MAXIMUM
1 SIZE	2.00000	0.84480	0.42240	0.80000	3.60000
2 PRICE	34.49994	11.17563	0.32393	17.00000	52.00000

We are as yet unable to interpret the wealth of information on the second page of printout. The entire analysis is shown in Figure 4.3, however, for later reference. In the last two lines we can find *a* and *b*: look under "CØEFFICIENT" for "INTERCEPT" and "SIZE 1" and see $a = 8.704$ and $b = 12.898$. The 1 after SIZE indicates that size was the first variable entered.

FIGURE 4.3   *Analysis from BMDP1R Simple Regression Analysis*

MULTIPLE R	0.9750	STD. ERROR OF EST.	2.5506
MULTIPLE R-SQUARE	0.9507		

ANALYSIS OF VARIANCE

	SUM OF SQUARES	DF	MEAN SQUARE	F RATIO	P(TAIL)
REGRESSION	2255.899	1	2255.899	346.765	0.00000
RESIDUAL	117.100	18	6.506		

VARIABLE	COEFFICIENT	STD. ERROR	STD. REG COEFF	T	P(2 TAIL)
INTERCEPT	8.704				
SIZE 1	12.898	0.693	0.975	18.622	0.0

Figure 4.4 shows the third page of the BMDP1R printout. Here we see the results of the PRINT command. From left to right across the page, we find columns listing values of the residuals $Y_i - \hat{Y}_i$, predicted values $\hat{Y}_i$, the predictor SIZE, and the response PRICE, for each of the 20 houses in the sample. The predicted values are of course obtained by plugging the corresponding X-value into the regression equation

Residuals

$\hat{Y} = 8.704 + 12.898X$. The residuals are the differences between actual and predicted prices for each size of house, what we have called error. Note that the term residual implies "left over" and reminds us that factors other than size alone affect price. There is much more to say about residuals, but the most important thing for now is to use this listing to check the accuracy of our keypunching.

Finally, the PLØT command produces the graphs shown in the fourth page of the printout. Part a of Figure 4.5 looks familiar: it is a scatter diagram of the data with the regression equation drawn on. The data points are plotted with O's (for observed values) and the points on the line with P's (for predicted values). If you connect the P's with a straight line, the graph should look pretty much like Figure 2.8, p. 37. Where observed and predicted values coincide or are close enough to overlap, the symbol printed is an asterisk.

The plot in part b in Figure 4.5 graphs the residuals from Figure 4.4 against X-values. Since residuals can be positive or negative, we see the plotted points varying around the horizontal line $Y_i - \hat{Y}_i = 0$. All the points here are plotted as 1. If two points had coincided, they would have been plotted as 2, three coincident points would be indicated with a 3, and so on.

FIGURE 4.4   *BMDP1R List of Data, Predicted Values, and Residuals*

| | | PREDICTED | VARIABLES | |
NO.	RESIDUAL	VALUE	1 SIZE	2 PRICE
1	0.0797	31.9203	1.8000	32.0000
2	2.3982	21.6018	1.0000*	24.0000
3	-3.6305 *	30.6305	1.7000	27.0000
4	2.1814	44.8186	2.8000	47.0000*
5	-2.0796	37.0796	2.2000	35.0000
6	-2.0221	19.0221	0.8000*	17.0000*
7	-3.1371 *	55.1371	3.6000*	52.0000*
8	-2.8916 *	22.8916	1.1000*	20.0000*
9	3.5000 *	34.5000	2.0000	38.0000
10	2.7611 *	42.2389	2.6000	45.0000
11	5.6305 **	38.3695	2.3000	44.0000
12	-1.3120	20.3120	0.9000*	19.0000*
13	0.8186	24.1814	1.2000	25.0000
14	-2.5575 *	52.5575	3.4000*	50.0000*
15	-0.6305	30.6305	1.7000	30.0000
16	2.0509	40.9491	2.5000	43.0000
17	0.2389	26.7611	1.4000	27.0000
18	-1.2677	51.2677	3.3000*	50.0000*
19	-0.0796	37.0796	2.2000	37.0000
20	-0.0509	28.0509	1.5000	28.0000

NOTE — NEGATIVE CASE NUMBER DENOTES A CASE WITH MISSING VALUES.
    THE NUMBER OF STANDARD DEVIATIONS FROM THE MEAN IS DENOTED BY UP TO 3
      ASTERISKS TO THE RIGHT OF EACH RESIDUAL OR VARIABLE.
    MISSING VALUES ARE DENOTED BY MORE THAN THREE ASTERISKS.

SERIAL CORRELATION OF RESIDUALS = 0.0324

FIGURE 4.5   *Plots from BMDP1R Simple Regression Analysis (a) Scatter Diagram with Regression Equation Drawn On*

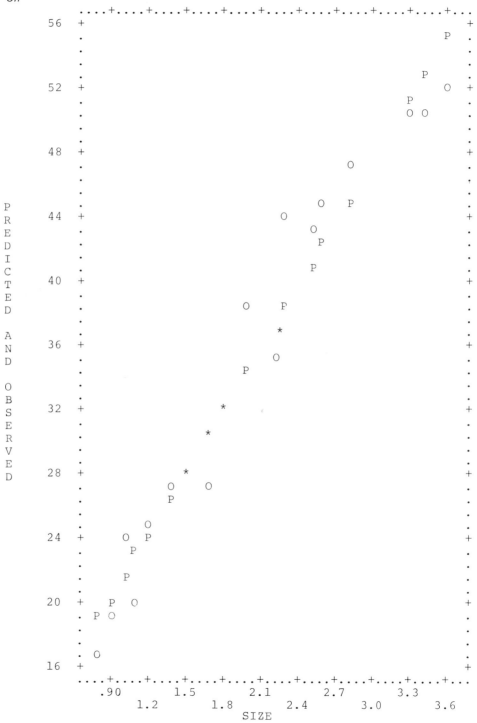

FIGURE 4.5   *(b) Residuals from Figure 4.4 Plotted Against X-values*

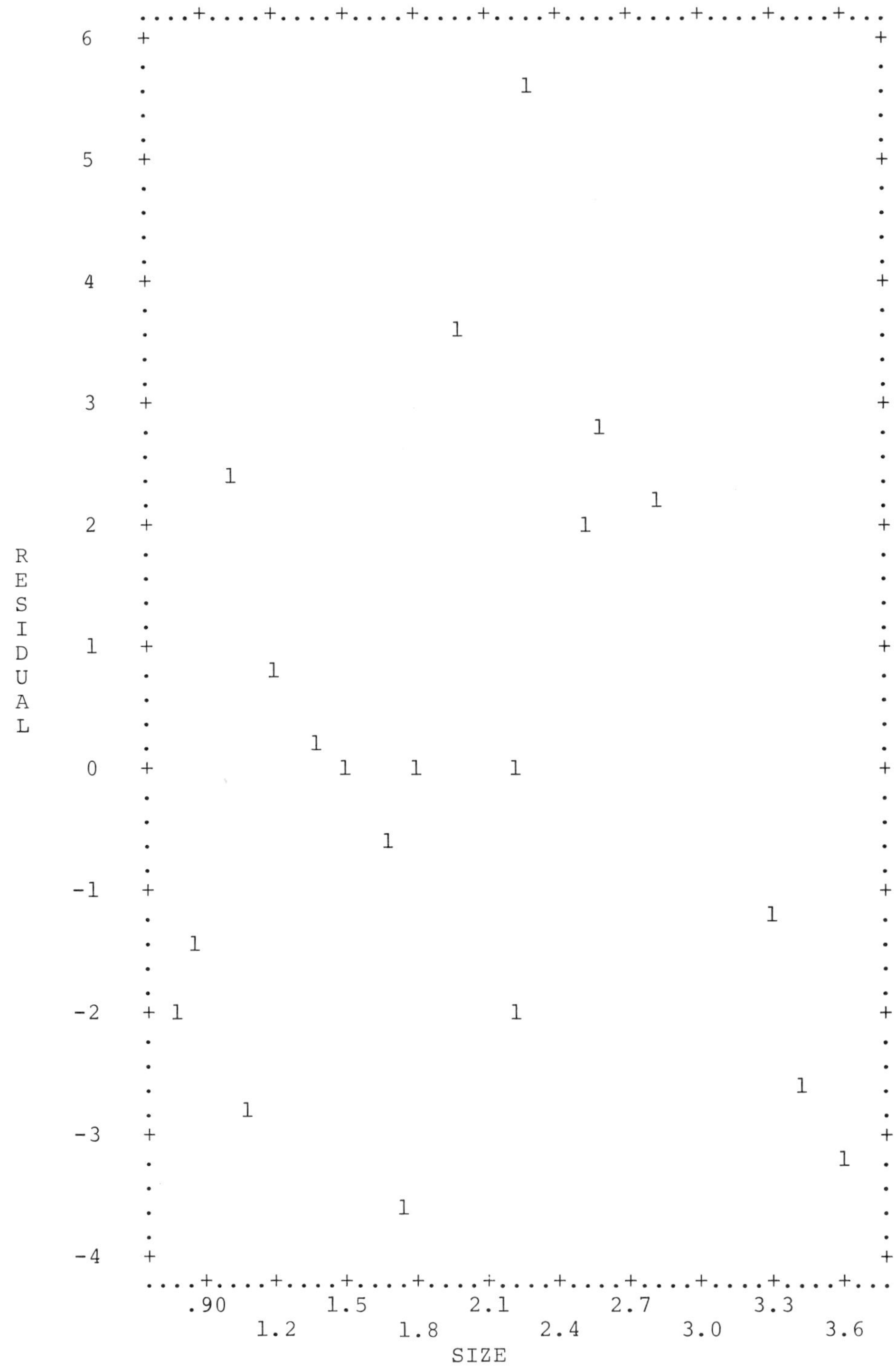

THE PRINTOUT FROM BMDP2R

The first page of the printout from the BMDP2R analysis of these data reproduces the program and shows how the computer interpreted its statements.

The second page of the P2R printout (Figure 4.6) is similar to that in Figure 4.2, but gives some additional information. Recall from elementary statistics that the *Skewness* measure of skewness tells whether or not a distribution is symmetric or lopsided (skewed), and the kurtosis measure tells whether the distribution is short and fat or tall *Kurtosis* and skinny. Also, a standardized score is of the form

$$Z_i = \frac{X_i - \overline{X}}{s_X},$$

as we saw in Chapter 2. The warning printed should be heeded when we come to tests of hypotheses: if a kurtosis value is greater than 3, the analysis given by the computer might not be trustworthy.

FIGURE 4.6   *Simple Statistics from BMDP2R*

VARIABLE NO.	NAME	MEAN	STANDARD DEVIATION	COEFFICIENT OF VARIATION	SKEWNESS	KURTOSIS
1	SIZE	2.0000	0.8448	0.4224	0.3508	1.9317
2	PRICE	34.4999	11.1756	0.3239	0.0744	1.5512

	SMALLEST VALUE	LARGEST VALUE	SMALLEST STD SCORE	LARGEST STD SCORE
	0.8000	3.6000	-1.4205	1.8939
	17.0000	52.0000	-1.5659	1.5659

NOTE — KURTOSIS VALUES GREATER THAN THREE INDICATE A DISTRIBUTION WITH HEAVIER
TAILS THAN NORMAL DISTRIBUTION

The third page of the BMDP2R printout reminds us that this program is intended for a much more complicated problem. Our problem is not complex enough for the program to "do its thing." The entire analysis is reproduced in Figure 4.7 for later reference; however. Suffice it here to note that in Step 0 is an analysis of the dependent variable, price, by itself, unrelated to X. Then in Step 1 we find the simple regression analysis and $a = 8.704$ and $b = 12.898$.

The summary tables from the regression analysis in Figure 4.8 also will be more meaningful when we analyze a more complicated problem.

The BMDP2R also lists the data, predicted values, and residuals, all nearly identical to those given in the P1R printout. The plots obtained by this program are identical to those in the earlier program.

FIGURE 4.7   *BMDP2R Simple Regression Analysis*

STEP NO. 0

MULTIPLE R   0.0
MULTIPLE R-SQUARE   0.0
STD. ERROR OF EST.   11.1756

ANALYSIS OF VARIANCE

	SUM OF SQUARES	DF	MEAN SQUARE	F RATIO
REGRESSION	0.0	0	0.0	0.0
RESIDUAL	2372.9993	19	124.8947	

VARIABLES IN EQUATION

VARIABLE	COEFF	STD. ERROR OF COEFF	STD REG COEFF	F TO REMOVE	LEVEL
(Y-INTERCEPT	34.500)				

VARIABLES NOT IN EQUATION

VARIABLE	PARTIAL CORR.	TOLERANCE	F TO ENTER	LEVEL
SIZE 1	0.97501	1.00000	346.768	1

STEP NO. 1
VARIABLE ENTERED 1 SIZE

MULTIPLE R   0.9750
MULTIPLE R-SQUARE   0.9507
STD. ERROR OF EST.   2.5506

ANALYSIS OF VARIANCE

	SUM OF SQUARES	DF	MEAN SQUARE	F RATIO
REGRESSION	2255.8997	1	2255.900	346.766
RESIDUAL	117.09950	18	6.505527	

VARIABLES IN EQUATION

VARIABLE	COEFF	STD. ERROR OF COEFF	STD REG COEFF	F TO REMOVE	LEVEL
(Y-INTERCEPT	8.704)				
SIZE 1	12.898	0.693	0.975	346.767	1

VARIABLES NOT IN EQUATION

VARIABLE	PARTIAL CORR.	TOLERANCE	F TO ENTER	LEVEL

FIGURE 4.8   *Summary Tables from BMDP2R*

STEPWISE REGRESSION COEFFICIENTS

VARIABLES STEP	0 Y-INTCPT	1 SIZE
0	34.4999*	12.8982
1	8.7036*	12.8982*

NOTE—
1) REGRESSION COEFFICIENTS FOR VARIABLES IN THE EQUATION ARE INDICATED BY AN ASTERISK
2) THE REMAINING COEFFICIENTS ARE THOSE WHICH WOULD BE OBTAINED IF THAT VARIABLE WERE TO ENTER IN THE NEXT STEP

SUMMARY TABLE

STEP NO.	VARIABLE ENTERED	REMOVED	MULTIPLE R	RSQ	INCREASE IN RSQ	F-TO-ENTER	F-TO-REMOVE	NUMBER OF INDEPENDENT VARIABLES INCLUDED
1	1 SIZE		0.9750	0.9507	0.9507	346.7666		1

EXERCISES

p. 521, 522,
524, 525

Section 4.2

4.1 Refer to the TEACHERS data, the PRECIPITATION data, the SALES data or the HEATING data found in Chapter 1 or in the Analyses of Data Sets section. For the data set you choose, use either the BMDP1R or the BMDP2R computer program to perform a simple linear regression analysis.
   a. Find the regression equation and compare it to the one obtained by your by-hand analysis.
   b. Compare the scatter diagram and plot of the regression equation made by the computer to the one you drew by hand.

4.2 Repeat Exercise 4.1, but use X corrected for its mean. Compare your results to those obtained in your by-hand analysis, and also to those obtained in Exercise 4.1.

4.3 POLYNOMIAL REGRESSION USING BMD PACKAGES

Of course, one would not work a regression problem both by hand and by computer; we do so only in order to see that the same results are obtained. In practice, one might not know at the outset whether a straight line or a curve should be fitted to a given set of data. The BMD programs allow us to investigate what kind of equation should be fitted to a set of data. For example, suppose that we did not know whether a straight-line or curved-line equation would be more appropriate. If the scatter diagram for a given set of data indicates that there might be a curvilinear relationship between two variables, we might want to fit a second-degree polynomial equation of the form

Second-Degree
Polynomial
Equation

$$\hat{Y} = a + bX + cX^2.$$

Either the BMDP1R or the BMDP2R can be used, but the instructions differ according to which program is used and the output also differs.

POLYNOMIAL REGRESSION USING BMDP1R

Figure 4.9 shows a worksheet for fitting a second-degree polynomial equation to the house size-price data, using BMDP1R. It looks very much like the one for fitting a simple linear regression. The only difference is that we ask the computer to create a new variable X^2 (or SIZE squared), to the data set, and to perform a regression analysis using both X and X^2 to predict Y. This is accomplished by the sentence

ADD = 1.

in the VARIABLE paragraph, which simply says that we are going to add one more variable to the data set. Then the TRANSFØRM paragraph instructs the computer to

FIGURE 4.9   *Worksheet for Fitting a Second-Degree Polynomial to House Size-Price Data Using BMDP1R*

```
            1-10            11-20              21-30
1 2 3 4 5 6 7 8 9 10 11 12 13 14 15 16 17 18 19 20 21 22 23 24 25 26 27 28 29 30 31 32 33 34
   {SYSTEM      CARDS
PRØBLEM        TITLE='QUADRATIC  P1R'./
INPUT          VARIABLES=2.
               CASES=20./
VARIABLE       NAMES=SIZE,PRICE.
               ADD=1./
TRANSFØRM      X(3)=SIZE**2./
PRINT          DATA./
PLØT           VARIABLE=SIZE./
REGRES         TITLE='QUADRATIC'.
               DEPENDENT=PRICE./
END/
1.8  32
1.0  24
  :
1.5  28
FINISH/
```

create the new variable. The sentence

$$X(3) = SIZE ** 2.$$

says to define a new variable, $X(3)$, by squaring the values of the SIZE variable. (The symbol $**$ is a standard one indicating that a value is to be raised to a power. We can also write X(3) = SIZE*SIZE., to indicate that the values of the SIZE variable are to be multiplied by themselves.) Note that a peculiarity of the BMDP1R program is that a variable added by a transformation must be denoted X( ). Should you try to give X^2-values a name such as XSQR = SIZE**2, XSQR would not be added to the data set.

No other changes are necessary to the program in Figure 4.1, but you might want to give your problem an appropriate title. It is also possible to give a title to the regression analysis portion of the output, but again this is not mandatory.

Figure 4.10 shows the analysis from fitting a second-degree polynomial and requesting a plot of the data, regression line, and residuals. On the last three lines of the first page of printout, we find the coefficients

INTERCEPT =   1.776

SIZE   = 20.469

X(3)   = −1.756

so that the least squares equation obtained from fitting a second-degree polynomial equation is

$$\hat{Y} = 1.776 + 20.469X - 1.756X^2.$$

FIGURE 4.10   *Worksheet for Fitting a Second-Degree Polynomial to House Size-Price Data Using BMDP1R*

MULTIPLE R	0.9811	STD.ERROR OF EST.	2.2862
MULTIPLE R-SQUARE	0.9626		

ANALYSIS OF VARIANCE

	SUM OF SQUARES	DF	MEAN SQUARE	F RATIO	P(TAIL)
REGRESSION	2284.147	2	1142.073	218.513	0.00000
RESIDUAL	88.852	17	5.227		

VARIABLE		COEFFICIENT	STD.ERROR	STD. REG COEFF	T	P(2 TAIL)
INTERCEPT		1.776				
SIZE	1	20.469	3.315	1.547	6.175	0.000
X(3)	3	-1.756	0.755	-0.583	-2.325	0.033

FIGURE 4.10   *(Part 2)*

CASE LABEL NO.	RESIDUAL	PREDICTED VALUE	VARIABLES 1 SIZE	2 PRICE	3 X(3)
1	-0.9310	32.9310	1.8000	32.0000	3.2400
2	3.5112 *	20.4888	1.0000*	24.0000	1.0000
3	-4.4986 *	31.4986	1.7000	27.0000	2.8900
4	1.6766	45.3234	2.8000	47.0000*	7.8400
5	-3.3093*	38.3093	2.2000	35.0000	4.8400
6	-0.0272	17.0272	0.8000*	17.0000*	0.6400*
7	-0.7093	52.7093	3.6000*	52.0000*	12.9600**
8	-2.1670	22.1670	1.1000*	20.0000*	1.2100
9	2.3096 *	35.6904	2.0000	38.0000	4.0000
10	1.8741	43.1259	2.6000	45.0000	6.7600
11	4.4339 *	39.5661	2.3000	44.0000	5.2900
12	0.2244	18.7756	0.9000*	19.0000*	0.8100*
13	1.1900	23.8100	1.2000	25.0000	1.4400
14	-1.0735	51.0735	3.4000*	50.0000*	11.5600*
15	-1.4986	31.4986	1.7000	30.0000	2.8900
16	1.0256	41.9744	2.5000	43.0000	6.2500
17	0.0092	26.9908	1.4000	27.0000	1.9600
18	-0.2030	50.2030	3.3000*	50.0000*	10.8900*
19	-1.3093	38.3093	2.2000	37.0000	4.8400
20	-0.5285	28.5285	1.5000	28.0000	2.2500

NOTE – NEGATIVE CASE NUMBER DENOTES A CASE WITH MISSING VALUES.
THE NUMBER OF STANDARD DEVIATIONS FROM THE MEAN IS DENOTED BY UP TO 3
ASTERISKS TO THE RIGHT OF EACH RESIDUAL OR VARIABLE.
MISSING VALUES ARE DENOTED BY MORE THAN THREE ASTERISKS.

SERIAL CORRELATION OF RESIDUALS = - 0.2475

FIGURE 4.10   *(Part 3)*

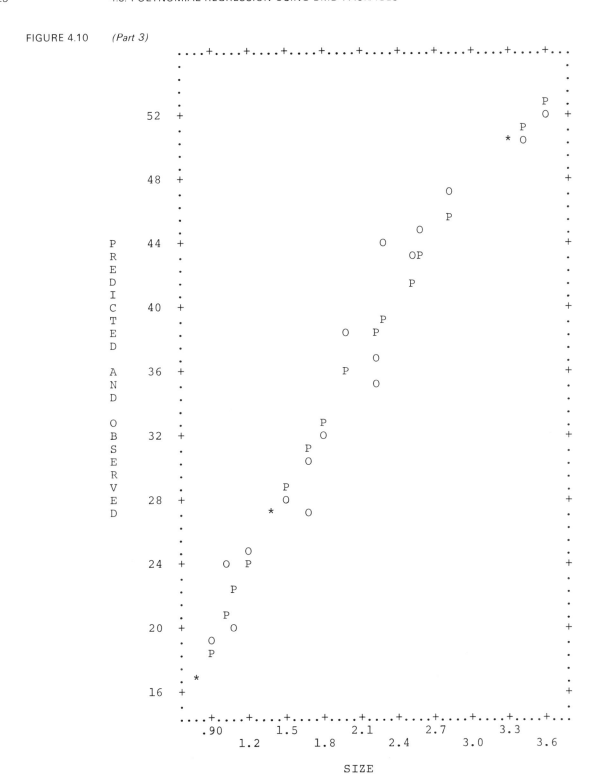

FIGURE 4.10   *(Part 4)*

```
              ....+....+....+....+....+....+....+....+....+....+...
      4.5   +                                  1                        +
            .                                                           .
            .                                                           .
            .                                                           .
            .                                                           .
      3.6   +        1                                                  +
            .                                                           .
            .                                                           .
            .                                                           .
            .                                                           .
      2.7   +                                                           +
            .                                                           .
            .                              1                            .
            .                                                           .
            .                                                           .
      1.8   +                                      1                    +
            .                                          1                .
            .                                                           .
            .           1                                               .
            .                                      1                    .
R      .90  +                                                           +
E           .                                                           .
S           .                                                           .
I           .                                                           .
D           .      1                                                    .
U     0.0   + 1              1                                          +
A           .                                          1                .
L           .                                                           .
            .                   1                                       .
            .                                                    1      .
      -.90  +                        1                                  +
            .                                                 1         .
            .                             1                             .
            .                      1                                    .
            .                                                           .
      -1.8  +                                                           +
            .        1                                                  .
            .                                                           .
            .                                                           .
      -2.7  +                                                           +
            .                                                           .
            .                             1                             .
            .                                                           .
      -3.6  +                                                           +
            .                                                           .
            .                                                           .
            .                                                           .
            .                                                           .
      -4.5  +                   1                                       +
              ....+....+....+....+....+....+....+....+....+....+...
                  .90      1.5       2.1       2.7       3.3
                      1.2       1.8       2.4       3.0       3.6
                                     SIZE
```

The second page of printout shows values of X, Y, and X^2, as well as the predicted values and residuals using the quadratic regression equation. The scatter diagram with regression equation and plot of residuals is shown on the third page. From the scatter diagram, we can see a gentle curvature in the plotted regression equation.

Let us pause now to investigate the meaning of the values a, b, and c in the second-degree equation. As in the linear regression equation, a is the Y-intercept, the value of Y when $X = 0$. It is senseless, however, to talk about a slope when one has a curve, since the slope changes at each point. The value b is a kind of slope, but it is the slope (of the tangent to the curve) at a *single point*: at $X = 0$. That b is positive says that the curve is rising, so that Y increases with X. Then c tells us how much the *slope* changes per one unit increase in X. That c is negative says that the curve is *decelerating*, or in this case, increasing at a *decreasing* rate. The value $c = -1.75566$ is *one-half* of the amount that the slope changes per one-unit increase in X. Thus, for each one-unit increase in X, the slope becomes $2(1.75566) = 3.51132$ units *less* steep.*

A comparison of the plots in Figures 4.5 and 4.10 reveals little difference in the two graphs. The curvature in Figure 4.10 is so slight that the graph at first glance appears to be a straight line. Thus we have evidence that a straight-line description of the data is probably preferable to the curvilinear description—at least, the linear description is easier to work with. In the absence of a theoretical framework which dictates that the relationship is curvilinear, we would choose the straight-line description and say that the regression equation is

$$\hat{Y} = 8.7 + 12.9X.$$

This assumption must be verified later, however. If we referred to the Tables of Residuals in Figures 4.4 and 4.10 and calculated the average absolute deviation

$$\frac{\sum |y_i - \hat{y}_i|}{n}$$

for both sets of residuals, we would find that the curved line is, on the average, closer to the data points than is the straight line. However, this slight difference does not seem to justify using the curvilinear equation.

The higher the degree of the polynomial fitted, the better the fit to any set of data, but how does one interpret the meaning of a high-order polynomial equation? What one loses in accuracy of prediction with a linear equation is made up by ease of interpretation. This is not to say that a straight line should always be used; but if the curvature is very slight, the straight-line description may be preferred over a curvilinear equation in the absence of any theory to the contrary. If a polynomial equation is to be fitted to a set of data, it is certainly handy to have access to a computer. In a designed experiment, with evenly spaced treatment levels, one can avoid using a computer to fit a curvilinear equation if orthogonal polynomials are used. The interested reader should refer to Appendix C.

Orthogonal
Polynomials

---

* This doubling comes about because acceleration is defined as the second derivative of the function. If $\hat{Y} = a + bX + cX^2$, then the first derivative is $d\hat{Y}/dX = b + 2cX$, and the second derivative is $d^2\hat{Y}/dX^2 = 2c$; acceleration $= 2c$.

If it is desired to fit a still higher-degree polynomial using BMDP1R, one proceeds as indicated in Figure 4.9, but with the appropriate alterations. To fit a third-degree polynomial

$$\hat{Y} = a + bX + cX^2 + dX^3,$$

we must add two variables, X^2 and X^3, to the data set:

ADD = 2.

and ask the computer to calculate X^3 as well as X^2:

TRANSFORM   X(3) = SIZE**2

X(4) = SIZE**3./

Then, as we can see from Figure 4.11, the computer will use X, X^2, and X^3 to predict Y. Residuals and plots have been deleted from Figure 4.11. It should now be obvious how to fit a fourth-degree or higher equation to the data. If this were done, we would find closer and closer fit, but at the expense of more and more complicated descriptions.

The analysis in Figure 4.11 shows a strange message printed next to where we expected to find the value for c (the coefficient on $X(3) = X^2$):

REDUNDANT VARIABLE—NØT USED.   TØLERANCE = 0.000358.

This means that the equation

$$Y = a + bX + dX^3$$

should be used instead of

$$\hat{Y} = a + bX + cX^2 + dX^3.$$

FIGURE 4.11   *Third-Degree Polynomial Fitted Using BMDP1R*

| MULTIPLE R | 0.9819 | STD. ERROR OF EST. | 2.2371 |
| MULTIPLE R-SQUARE | 0.9641 | | |

ANALYSIS OF VARIANCE

	SUM OF SQUARES	DF	MEAN SQUARE	F RATIO	P(TAIL)
REGRESSION	2287.923	2	1143.962	228.589	0.00000
RESIDUAL	85.076	17	5.004		

VARIABLE		COEFFICIENT	STD. ERROR	STD. REG COEFF	T	P(2 TAIL)
INTERCEPT		3.475				
SIZE	1	17.261	1.828	1.305	9.440	0.000
X(3)	3					
X(4)	4	-0.285	0.113	-0.350	-2.530	0.022

REDUNDANT VARIABLE—NOT USED.

TOLERANCE = 0.000358

Exactly what is meant by "redundant variable" and "tolerance" will have to wait until we study multiple regression. Note that the $\hat{Y}$-values, residuals, and plots, had they been shown in Figure 4.11, would all be based on the equation

$$\hat{Y} = 3.475 + 17.261X - 0.285X^3.$$

POLYNOMIAL REGRESSION USING BMDP2R

When the BMDP1R is used to analyze polynomial regressions, all variables in the model are used to predict Y. But suppose we want to decide whether to use a simple linear regression equation or a second-degree polynomial equation to describe the house size-price data. We would want to see analyses for both the simple regression case and the second-degree case. This could be accomplished using BMDP1R only by running two programs (although both could be done in one run by inserting the appropriate regression paragraph for the second-degree analysis after the regression paragraph on the simple linear regression program).

The BMDP2R is more appropriate for obtaining separate analyses for equations of the first, second, third, etc. degree. Consider the worksheet in Figure 4.12, which requests a fit of a third-degree equation. Instead of using X, X^2, and X^3 all together to predict Y, as the previous program did, however, the BMDP2R will first show an

FIGURE 4.12 *BMDP2R Worksheet for Third-Degree Polynomial Equation Fitted to House Size-Price Data*

```
         1-10                11-20               21-30               31-40
1 2 3 4 5 6 7 8 9 10 ...
  {SYSTEM   CARDS
PRØBLEM     TITLE='QUAD AND CUBIC 2R'./
INPUT       VARIABLES=2.
            FØRMAT='(F3.1,F3.0)'.
            CASES=20./
VARIABLE    NAMES=SIZE,PRICE
            ADD=2./
TRANSFØRM   X(3)=SIZE**2.
            X(4)=SIZE**3./
PRINT       DATA./
PLØT        VARIABLE=SIZE./
REGRES      DEPENDENT=PRICE.
            LEVELS=1,0,2,3./
END/
1.8 32
1.0 24
  :
  .
1.5 28
FINISH/
```

FIGURE 4.13   BMDP2R Stepwise Analysis for Polynomial Regression

STEP NO. 0

MULTIPLE R	0.0
MULTIPLE R-SQUARE	0.0
STD. ERROR OF EST.	11.1756

ANALYSIS OF VARIANCE

	SUM OF SQUARES	DF	MEAN SQUARE	F RATIO
REGRESSION	0.0	0	0.0	0.0
RESIDUAL	2372.9993	19	124.8947	

VARIABLES IN EQUATION

VARIABLE	COEFFICIENT	STD. ERROR OF COEFF	STD REG COEFF	F TO REMOVE	LEVEL
(Y-INTERCEPT	34.500)				

VARIABLES NOT IN EQUATION

VARIABLE		PARTIAL CORR.	TOLERANCE	F TO ENTER	LEVEL
. SIZE	1	0.97501	1.00000	346.768	1
. X(3)	3	0.93733	1.00000	130.252	2
. X(4)	4	0.88102	1.00000	62.429	3

STEP NO. 1
VARIABLE ENTERED   1 SIZE

MULTIPLE R	0.9750
MULTIPLE R-SQUARE	0.9507
STD. ERROR OF EST.	2.5506

ANALYSIS OF VARIANCE

	SUM OF SQUARES	DF	MEAN SQUARE	F RATIO
REGRESSION	2255.8997	1	2255.900	346.766
RESIDUAL	117.09950	18	6.505527	

VARIABLES IN EQUATION

VARIABLE		COEFFICIENT	STD. ERROR OF COEFF	STD REG COEFF	F TO REMOVE	LEVEL
(Y-INTERCEPT		8.704)				
SIZE	1	12.898	0.693	0.975	346.767	1

VARIABLES NOT IN EQUATION

VARIABLE		PARTIAL CORR.	TOLERANCE	F TO ENTER	LEVEL
. X(3)	3	-0.49117	0.03507	5.405	2
. X(4)	4	-0.52294	0.11040	6.399	3

STEP NO. 2
VARIABLE ENTERED   3 X(3)

MULTIPLE R   0.9811
MULTIPLE R-SQUARE   0.9626

STD. ERROR OF EST.   2.2861

ANALYSIS OF VARIANCE

	SUM OF SQUARES	DF	MEAN SQUARE	F RATIO
REGRESSION	2284.1492	2	1142.074	218.517
RESIDUAL	88.849945	17	5.226467	

VARIABLES IN EQUATION

VARIABLE		COEFFICIENT	STD. ERROR OF COEFF	STD REG COEFF	F TO REMOVE	LEVEL
(Y-INTERCEPT		1.776)				
SIZE	1	20.469	3.315	1.547	38.127	1
X(3)	3	-1.756	0.755	-0.583	5.405	2

VARIABLES NOT IN EQUATION

VARIABLE	LEVEL	PARTIAL CORR.	TOLERANCE	F TO ENTER	LEVEL
X(4)	4	-0.38962	0.00113	2.864	3

analysis for a simple regression, and then an analysis for a second-degree polynomial regression, and then an analysis for the third-degree regression. This is accomplished by assigning

LEVELS $= 1, 0, 2, 3$.

to the variables, in the order they are entered. Our NAMES statement said that we first entered $X = $ SIZE and then $Y = $ PRICE. Then by the TRANSFORM statements, we added X^2 and X^3, in that order. The level 1 is thus assigned to X, the level 0 to Y, the level 2 to X^2, and the level 3 to X^3. These assignments of levels determine the order in which the variables are entered as predictors. Since Y is not a predictor, it is assigned the level 0. Variables with nonzero levels are entered into the model with those with lower levels entering first.

Figure 4.13 shows the results. We see steps labeled 0, 1, and 2 in the analysis. In Step 0, no predictors are used, and we see only an analysis of the dependent variable $Y = $ price. Listed under VARIABLES IN EQUATION, under Y-INTERCEPT we see just $Y = 34.500$. Other quantities in this printout will be discussed later, especially in Chapter 14 in which the BMDP2R is used to analyze multiple regression problems. Of interest to us here are the levels shown under VARIABLES NOT IN EQUATION. Since the variable SIZE is assigned level 1, it is used as a predictor in Step 1.

In Step 1 we see the simple regression analysis using SIZE to predict price. As before, we find $a = 8.704$ and $b = 12.898$. Under VARIABLES NOT IN EQUATION in Step 1, since $X(3) = X^2$ is assigned level 2, it is added to X in Step 2 to predict Y. Comparing the results of Step 2 to those in Figure 4.6 in which the BMDP1R was used to fit a second-degree polynomial, we again find that the best equation is

$$\hat{Y} = 1.776 + 20.469X - 1.756X^2.$$

Under VARIABLES NOT IN EQUATION we see $X(4) = X^3$ with level assignment of 3. In some other problem, the program would continue on to a third step and show the coefficients for the equation

$$\hat{Y} = a + bX + cX^2 + dX^3.$$

However, as we saw before (Figure 4.12), both the quadratic and cubic terms are not needed to describe these data.

Thus, the analysis procedure terminates with only the linear and quadratic terms in the equation. The important thing is that the stepwise procedure allows us to see first a linear analysis, then a quadratic, then a cubic, and so on, all in the output from one program. Thus, it lets us choose the model that best describes our data. The graph of the regression equation (not shown here) looks almost the same as that in Figure 4.10; it is based on the terms in the model at the *last* step in the analysis, that is, the quadratic equation $\hat{Y} = a + bX + cX^2$.

EXERCISES   *Section 4.3*

4.3   a.   Find the average of the absolute values of the residuals in Figure 4.4.
b.   Find the average of the absolute values of the residuals in Figure 4.10.
c.   Show that the higher the degree of the polynomial, the better the regression equation fits the data.

4.4   a.   Using the residuals, $Y_i - \hat{Y}_i$, in Figure 4.4, calculate the sample variance

$$\frac{\sum (Y_i - \hat{Y}_i)^2}{n - 2}.$$

(The reason for the $n - 2$ in the denominator will be explained in Chapter 6.)
b.   Find where this value is given on the printout (refer to Figure 4.3).

p. 523   4.5   Run the POLICE data through either the BMDP1R or BMDP2R program, fitting a simple linear regression equation.
a.   Compare the values for a and b to those obtained in Exercise 2.11a, in which the simple linear regression equation was found by hand.
b.   Compare the residuals to those found by hand.
c.   Calculate $1/n \sum |Y_i - \hat{Y}_i|$ and compare to the value obtained in the by-hand analysis.
d.   Compare the plot of the data and the regression line to that made by hand in Exercise 2.11h.

4.6   Run the POLICE data through the BMDP1R program, fitting a second-degree polynomial regression.
a.   Compare the values of a, b_1, and b_2 to the values of a, b, and c found by hand in Exercise 2.11b.
b.   Compare the residuals to those found by hand.
c.   Calculate $1/n \sum |Y_i - \hat{Y}_i|$ and compare to the value found by hand.
d.   Compare the plot of the data and the regression equation to those made by hand.

4.7   Comparing the results of Exercises 4.5 and 4.6, does a straight line or a curve give the better description of the POLICE data? Which regression equation would you use, and why?

4.4 SIMPLE LINEAR REGRESSION USING SPSS

SPSS   The Statistical Package for the Social Sciences, or SPSS, developed by the University of Chicago, provides many programs for various statistical analyses.* We will outline here the procedure for running a simple linear regression and a polynomial regression. The reader should note the similarities and differences between the output from the SPSS and the BMD packages. One's choice of which program to run depends on (1) which packages are available and (2) which program most easily produces the kind of information wanted from the data.

———————
* For a detailed explanation of the use of the SPSS programs, refer to the manual *SPSS: Statistical Package for the Social Sciences*, by Norman H. Nie, McGraw-Hill, 1975.

OBTAINING ACCESS TO THE SPSS

As in the case of the BMD package, this program needs to be preceded by system cards, set up to allow access to the particular package in the particular computing center. Figure 4.14 shows a worksheet for an SPSS simple regression with scatter diagram for the house size-price data. Note that columns 1–15 contain general instructions to the computer, and that the specifics for a given problem begin in column 16.

Let us consider the first five cards, PAGESIZE, RUN NAME, DATA LIST, INPUT MEDIUM, and N ØF CASES. The first two are optional: PAGESIZE...NØEJECT saves paper in the printout; if it is not included, each different analysis will appear on its own page. This can waste a lot of paper. On the other hand, the NØEJECT option will cause some parts of the analysis to be split between two sheets of paper. The RUN NAME option allows one to give a useful title to the analysis.

The fourth card,

INPUT MEDIUM...CARD,

simply tells the computer that the data are entered on punched cards, rather than magnetic tape or whatever. The DATA LIST and N ØF CASES cards correspond to the format card in the BMD programs. N ØF CASES...20 is self-explanatory.

DATA LIST...FIXED/1 X 1–3 Y 5–6

FIGURE 4.14 *Worksheet for SPSS Simple Regression*

```
     }SYSTEM CARDS
PAGESIZE       NØEJECT
RUN NAME       HØUSES,SIMPLE
DATA LIST      FIXED/1 X 1-3 Y 5-6
INPUT MEDIUM   CARD
N ØF CASES     20
PRINT FØRMATS  X(1),Y(0)
LIST CASES     CASES=20/VARIABLES=ALL
READ INPUT DATA
1.8 32
1.0 24
  :

1.6 28
SCATTERGRAM    Y WITH X
STATISTICS     ALL
REGRESSIØN     VARIABLES=Y,X/
               REGRESSIØN=Y WITH X (1) RESID=0/
STATISTICS     2,4
FINISH
```

tells the computer that on each of the 20 data cards, X-values will be entered in columns 1–3 and Y-values, in columns 5–6. Referring to the data cards, we see that each pair of (X, Y)-values is punched according to the DATA LIST and that the decimal points are punched.

SPSS allows you to give your variables more interesting titles than X and Y. For example, our DATA LIST statement could say

... FIXED/1   SIZE 1–3   PRICE 5–6.

Then on our output, we would see SIZE instead of X, and PRICE instead of Y. Of course, X would have to be changed to SIZE and Y to PRICE everywhere they appear in Figure 4.14.

The first five cards simply prepare the computer to receive data. The READ INPUT DATA card says, "Here come the data," and the data cards supply the numbers to be used. But we have not yet asked the computer to do anything with the data. In order to get the printout from SPSS to look as much as possible like that from the BMD, we first ask the computer to read the data back, so that we can check for keypunching errors. LIST CASES does this. CASES = 20/ says that you want to see all 20 data cards, and VARIABLES = ALL says you want to see both X and Y. (The implication is that you need not look at all 20 cards or at both variables if you don't want to.) But, if only the LIST CASES statement is included, the computer will print *only* values to the left of the decimal point. In order to see the values to the right of the decimal, the PRINT FØRMATS statement must be used. X(1) says you want to see one value to the right of the decimal in the X-values, and Y(0) says you don't care about numbers to the right of the decimal in Y.

The SCATTERGRAM procedure card tells the computer to run the SPSS routine that constructs a scatter diagram of the data. Y WITH X says that Y is the dependent variable, to be plotted on the vertical axis.

Several statistics can be calculated in conjunction with the scatter diagram. While we are not interested in all of them at this point, we call for all of them in Figure 4.14 for purposes of comparison to the BMD programs and for later reference. The SPSS manual lists the statistics available.

EXAMINING THE PRINTOUT

Let us now examine the output corresponding to the program to this point. Most of the first page (not shown here) has to do with the internal workings of the computer. The statements in the program are reproduced. After READ INPUT DATA comes a listing of the values punched on the data cards.

In Figure 4.15, we see the results of the SCATTERGRAM routine. The observed values of (X, Y) are printed as asterisks. The graph is divided into nine approximately equal squares, primarily for ease in locating individual points. One can ask the computer not to divide the scatter diagram in this manner. Note that no regression line is plotted. Instead, the statistics requested are listed. While most of these do not interest

FIGURE 4.15   *SCATTERGRAM Routine for SPSS*

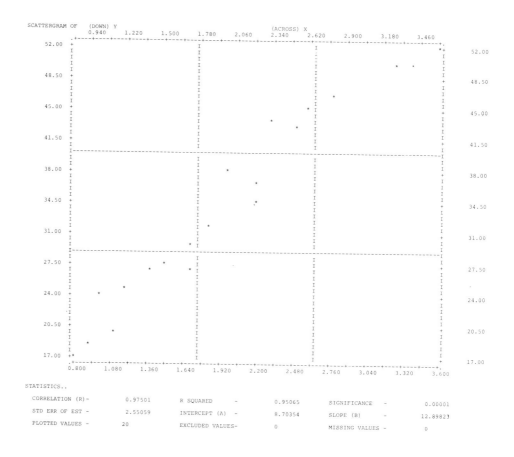

us yet, we see the familiar

INTERCEPT (A) = 8.70354   and   SLOPE (B) = 12.89823.

Note that the **SCATTERGRAM** routine does not give us means, standard deviations, residuals, or an analysis of variance table, as did the **BMD** programs. In order to get output from the **SPSS** as similar as possible to the previous program, we call for the **REGRESSIØN** routine. The statement

REGRESSIØN = Y WITH X(1) RESID=0/

in Figure 4.14 says that Y is the dependent variable, X is the first (in this case, only) variable to be analyzed with Y, and that we want a list of residuals to be printed. The statement

STATISTICS ... 2,4

asks for a table of means and variances (2) and the list of residuals (4). Then follows the FINISH card to tell the computer it is through with this problem.

The first portion of printout corresponding to these last statements is shown in Figure 4.16. As before, most of the first part of the REGRESSIØN printout reproduces the program statements and gives data on the machine's space requirements. Then follows the table of means and standard deviations and again we see that $\bar{X} = 2$, $\bar{Y} = 34.5$. Figure 4.16 shows only the output from the REGRESSIØN routine. While we are not ready to analyze all the output, we can see that this analysis is very similar to what we obtained using the BMDP2R (Figure 4.7). Note also that the output is titled Multiple Regression, so that an equation of the form

$$\hat{Y} = a + b_1 X_1 + b_2 X_2 + \cdots + b_k X_k$$

is being fitted to the data. Our simple regression is just a special case of the multiple regression, with $k = 1$. Thus, we are fitting

$$\hat{Y} = a + b_1 X_1, \quad \text{or} \quad \hat{Y} = a + bX,$$

to our data. Since this is a multiple regression program, however, several of the quantities computed are not relevant to the simple regression analysis.

Beta Coefficient

BETA is the *standardized regression coefficient*, or beta coefficient. This is the regression coefficient, b, with its units removed by multiplication by the ratio s_x/s_y, where s_x is the standard deviation of the sample X-values and s_y is the sample standard deviation of the Y-values. This is of interest, mostly to sociologists, in multiple regression. (Refer to Chapter 13.) We see that

$$s_x = 0.8448, \quad s_y = 11.1756, \quad \text{and} \quad b = 12.89823.$$

Thus

$$\text{BETA} = \frac{0.8448}{11.1756}(12.89823) = 0.97501.$$

Following the analysis of variance and a summary table, the residuals are calculated, as shown in Figure 4.17. Note that these are approximately the same as in Figure 4.5 (the differences are due to rounding error). Note also that the computer sometimes converts to scientific notation, so that

$$0.7965207E - 01 = 0.07965207.$$

Standardized Residuals

That is, $E - 01$ means: times 10^{-1}. Similarly, E05 means times 10^5, and so on. Also in Figure 4.17 is a plot of the standardized residuals —that is, the residuals divided by their standard deviation, or scaled down. In this way, the standardized residuals will usually lie in the range $(-2, 2)$ and thus the plot will not take up much space. This plot gives us a visual picture of the relative sizes and positions of the residuals; note that the residuals are plotted in the same order that the data values were given, and not in order of magnitude of X. We will discuss plots of the residuals more fully later on.

FIGURE 4.16   REGRESSIØN Routine for SPSS *Simple Regression*

DEPENDENT VARIABLE. .   Y

VARIABLE(S) ENTERED ON STEP NUMBER 1. .   X

				ANALYSIS OF VARIANCE	DF	SUM OF SQUARES	MEAN SQUARE	F
MULTIPLE R	0.97501			REGRESSION	1.	2255.90041	2255.90041	346.76645
R SQUARE	0.95065			RESIDUAL	18.	117.09959	6.50553	
ADJUSTED R SQUARE	0.94791							
STANDARD ERROR	2.55059							

VARIABLES IN THE EQUATION

VARIABLE	B	BETA	STD ERROR B	F
X	12.89823	0.97501	0.69265	346.766
(CONSTANT)	8.70354			

VARIABLES NOT IN THE EQUATION

VARIABLE	BETA IN	PARTIAL	TOLERANCE	F

MAXIMUM STEP REACHED

SUMMARY TABLE

VARIABLE	MULTIPLE R	R SQUARE	RSQ CHANGE	SIMPLE R	B	BETA
X	0.97501	0.95065	0.95065	0.97501	12.89823	0.97501
(CONSTANT)					8.70354	

FIGURE 4.17 *List of Residuals, SPSS Simple Regression*

SEQNUM	OBSERVED Y	PREDICTED Y	RESIDUAL
1	32.00000	31.92033	.7965207E-01
2	24.00000	21.60176	2.398228
3	27.00000	30.63052	-3.630531
4	47.00000	44.81857	2.181419
5	35.00000	37.07964	-2.079648
6	17.00000	19.02213	-2.022124
7	52.00000	55.13716	-3.137168
8	20.00000	22.89159	-2.891587
9	38.00000	34.50000	3.499995
10	45.00000	42.23892	2.761064
11	44.00000	38.36946	5.630535
12	19.00000	20.31195	-1.311948
13	25.00000	24.18141	.8185841
14	50.00000	52.55751	-2.557524
15	30.00000	30.63052	-.6305320
16	43.00000	40.94911	2.050879
17	27.00000	26.76105	.2389401
18	50.00000	51.26768	-1.267696
19	37.00000	37.07964	-.7964820E-01
20	28.00000	28.05089	-.5088806E-01

PLOT OF STANDARDIZED RESIDUAL

-2.0	-1.0	0.0	1.0	2.0

EXERCISES   *Section 4.4*

p. 521, 522,
524, 525   4.8   Refer to either the TEACHERS data, the PRECIPITATION data, the SALES data, or the HEATING data, shown in Chapter 1 or in the Analyses of Data Sets section. For the selected data, use the SPSS to
a.   find the regression equation,
b.   plot a scatter diagram,
c.   produce an analysis of variance table,
d.   list the residuals.
Compare the regression equation and the scatter diagram to the ones found by hand.

4.9   Repeat Exercise 4.8 using X-values corrected for their means.

4.5 FITTING A POLYNOMIAL EQUATION USING SPSS

To fit a quadratic regression using the SPSS, recall that the REGRESSIØN routine fits an equation

$$\hat{Y} = a + b_1 X_1 + b_2 X_2 + \cdots + b_k X_k$$

to the data. In order to fit

$$\hat{Y} = a + bX + cX^2,$$

we equate

$$b_1 = b, \quad X_1 = X, \quad b_2 = c, \quad X_2 = X^2.$$

If we enter a program such as shown in the worksheet in Figure 4.18, the computer will give the scatter diagram, table of means and standard deviations, and analysis of variance for the simple regression just as in Figures 4.15–4.16. In addition, it will fit a second-degree equation and give a list of residuals. But it will not plot the second-degree equation on the scatter diagram.

Note that there are only a few changes to be made from the simple regression program. First, the computer is instructed to find the X^2-values for

CØMPUTE...XSQR $= X**2.$

The PRINT FØRMATS also includes the X^2-values, for which two decimal places are required. When the LIST CASES card is encountered, if VARIABLES $=$ ALL, then values of X, Y, and X^2 will be displayed. If only X- and Y-values are required, then write

VARIABLES $=$ X,Y.

FIGURE 4.18   *Worksheet for SPSS: Fitting a Second Degree Polynomial*

1-10	11-20	21-30	31-40	41-50	51-
}SYSTEM	CARDS				
PAGESIZE	NØEJECT				
RUN NAME	HØUSES, SIMPLE, QUAD				
DATA LIST	FIXED/1 X 1-3 Y 5-6				
INPUT MEDIUM	CARD				
N ØF CASES	20				
CØMPUTE	XSQR=X**2				
PRINT FØRMATS	X(1),Y(0),XSQR(2)				
LIST CASES	CASES=20/VARIABLES=ALL				
READ INPUT DATA					
1.8 32					
1.0 24					
⋮					
1.5 28					
SCATTERGRAM	Y WITH X				
STATISTICS	ALL				
REGRESSIØN	VARIABLES=Y,X,XSQR/				
	REGRESSIØN=Y WITH X(4),XSQR(2) RESID=O/				
STATISTICS	4,2				
FINISH					

In the REGRESSIØN card, the XSQR variable is included, and the REGRESSIØN card beginning in column 16 also includes the XSQR variable. The (4) after X, being larger than the (2) after XSQR forces the computer to first fit the equation

$$\hat{Y} = a + bX$$

and then in the second step to fit

$$\hat{Y} = a + bX + cX^2.$$

The residuals will be computed using the second-degree equation. Figure 4.19 shows the results.

High-order polynomials may be fitted in a like manner. For example, to fit

$$\hat{Y} = a + bX + cX^2 + dX^3,$$

one would insert

CØMPUTE…XCUBE=X**3

after

CØMPUTE…XSQR=X**2,

FIGURE 4.19   SPSS REGRESSIØN Routine to Fit a Second-Degree Equation (Part 1)

DEPENDENT VARIABLE.. Y

VARIABLE(S) ENTERED ON STEP NUMBER 1.. X

		ANALYSIS OF VARIANCE	DF	SUM OF SQUARES	MEAN SQUARE	F
MULTIPLE R	0.97501	REGRESSION	1.	2255.90041	2255.90041	346.7664
R SQUARE	0.95065	RESIDUAL	18.	117.09959	6.50553	
ADJUSTED R SQUARE	0.94791					
STANDARD ERROR	2.55059					

VARIABLES IN THE EQUATION

VARIABLE	B	BETA	STD ERROR B	F
X	12.88823	0.97501	0.69265	346.766
(CONSTANT)	8.70354			

VARIABLES NOT IN THE EQUATION

VARIABLE	BETA IN	PARTIAL	TOLERANCE	F
XSQR	-0.58257	-0.49115	0.03507	5.405

VARIABLE(S) ENTERED ON STEP NUMBER 2.. XSQR

		ANALYSIS OF VARIANCE	DF	SUM OF SQUARES	MEAN SQUARE	F
MULTIPLE R	0.98110	REGRESSION	2.	2284.14806	1142.07403	218.51250
R SQUARE	0.96256	RESIDUAL	17.	88.85194	5.22658	
ADJUSTED R SQUARE	0.95815					
STANDARD ERROR	2.28617					

VARIABLES IN THE EQUATION

VARIABLE	B	BETA	STD ERROR B	F
X	20.46853	1.54728	3.31500	38.125
XSQR	-1.75566	-0.58257	0.75519	5.405
(CONSTANT)	1.77591			

VARIABLES NOT IN THE EQUATION

VARIABLE	BETA IN	PARTIAL	TOLERANCE	F

MAXIMUM STEP REACHED

SUMMARY TABLE

VARIABLE	MULTIPLE R	R SQUARE	RSQ CHANGE	SIMPLE R	B	BETA
X	0.97501	0.95065	0.95065	0.97501	20.46853	1.54728
XSQR	0.98110	0.96256	0.01190	0.93733	-1.75566	-0.58257
(CONSTANT)					1.77591	

SEQNUM	OBSERVED Y	PREDICTED Y	RESIDUAL
1	32.00000	32.93091	-.9309214
2	24.00000	20.48878	3.511221
3	27.00000	31.49855	-4.498557
4	47.00000	45.32343	1.676567
5	35.00000	38.30928	-3.309292
6	17.00000	17.02710	-.2710872E-01
7	52.00000	52.70927	-.7092978
8	20.00000	22.16693	-2.166935
9	38.00000	35.69034	2.309661
10	45.00000	43.12582	1.874162
11	44.00000	39.56609	4.433908
12	19.00000	18.77550	.2244996
13	25.00000	23.80998	1.190005
14	50.00000	51.07350	-1.073512
15	30.00000	31.49855	-1.498557
16	43.00000	41.97437	1.025624
17	27.00000	26.99075	.9244107E-02
18	50.00000	50.20294	-.2029457
19	37.00000	38.30928	-1.309292
20	28.00000	28.52847	-.5284729

PLOT OF STANDARDIZED RESIDUAL

```
        -2.0      -1.0       0.0       1.0       2.0

                          *I
                         * I-
                     *    I- *
                        * I- *
                          *I *-
                         * *I-
                        * *I *-
                          *I *- *
                         * I- *
                         *I- *
                          *I *-  *
                       *  I
                        *I-
                        *I- *
                        * *I-
                        *I- *
                        *I- *
                       *I *-  *
```

FIGURE 4.20   *SPSS REGRESSIØN Routine to Fit a Third-Degree Equation*

DEPENDENT VARIABLE.. Y

VARIABLE(S) ENTERED ON STEP NUMBER 1.. X

MULTIPLE R	0.97501
R SQUARE	0.95065
ADJUSTED R SQUARE	0.94791
STANDARD ERROR	2.55059

ANALYSIS OF VARIANCE	DF	SUM OF SQUARES	MEAN SQUARE	F
REGRESSION	1.	2255.90041	2255.90041	346.76645
RESIDUAL	18.	117.09959	6.50553	

VARIABLES IN THE EQUATION

VARIABLE	B	BETA	STD ERROR B	F
X	12.89823	0.97501	0.69265	346.766
(CONSTANT)	8.70354			

VARIABLES NOT IN THE EQUATION

VARIABLE	BETA IN	PARTIAL	TOLERANCE	F
XSQR	-0.58257	-0.49115	0.03507	5.405
XCUBE	-0.34964	-0.52296	0.11040	6.399

VARIABLE(S) ENTERED ON STEP NUMBER 2.. XSQR

MULTIPLE R	0.98110
R SQUARE	0.96256
ADJUSTED R SQUARE	0.95815
STANDARD ERROR	2.28617

ANALYSIS OF VARIANCE	DF	SUM OF SQUARES	MEAN SQUARE	F
REGRESSION	2.	2284.14806	1142.07403	218.51250
RESIDUAL	17.	88.85194	5.22658	

VARIABLES IN THE EQUATION

VARIABLE	B	BETA	STD ERROR B	F
X	20.46853	1.54728	3.31500	38.125
XSQR	-1.75566	-0.58257	0.75519	5.405
(CONSTANT)	1.77591			

VARIABLES NOT IN THE EQUATION

VARIABLE	BETA IN	PARTIAL	TOLERANCE	F
XCUBE	-2.25162	-0.39029	0.00113	2.875

VARIABLE(S) ENTERED ON STEP NUMBER 3.. XCUBE

MULTIPLE R	0.98400
R SQUARE	0.96826
ADJUSTED R SQUARE	0.96231
STANDARD ERROR	2.16964

ANALYSIS OF VARIANCE	DF	SUM OF SQUARES	MEAN SQUARE	F
REGRESSION	3.	2297.68286	765.89429	162.70279
RESIDUAL	16.	75.31714	4.70732	

VARIABLES IN THE EQUATION

VARIABLE	B	BETA	STD ERROR B	F
X	-3.08160	-0.23295	14.24033	0.047
XSQR	10.22132	3.39168	7.09958	2.073
XCUBE	-1.83347	-2.25162	1.08127	2.875
(CONSTANT)	15.36197			

VARIABLES NOT IN THE EQUATION

VARIABLE	BETA IN	PARTIAL	TOLERANCE	F

and change the PRINT FØRMATS card to read

PRINT FØRMATS...X(1),Y(0),XSQR(2),XCUBE(3).

The two REGRESSIØN cards would then be

REGRESSIØN...VARIABLES=Y,X,XSQR,XCUBE/

REGRESSIØN=Y WITH X(6),XSQR(4),XCUBE(2) RESID=0/

The output in Figure 4.20 shows the results of fitting a third-degree equation. Note that SPSS allows the cubic term to enter (BMDP2R did not).

EXERCISES   *Section 4.5*

p. 523   *4.10*   Run the POLICE data using SPSS for simple linear regression.
- a. Compare the values of a and b to those obtained in Exercise 2.11a.
- b. Compare the residuals to those found by hand.
- c. Calculate $\frac{1}{n}\sum|Y_i - \hat{Y}_i|$ and compare the value obtained in the previous analysis.
- d. Compare the plot of the data and the regression line to those made by hand.

     4.11   Fit a second-degree polynomial to the POLICE data, using SPSS.
- a. Compare the values of a, b_1, and b_2 to the values of a, b, and c found in Exercise 2.11b.
- b. Compare the residuals to those found by hand.
- c. Calculate $\frac{1}{n}\sum|Y_i - \hat{Y}_i|$ and compare to the value found by hand.
- d. Compare the plot of the data and the regression equation to that made by hand. Does the straight line or the curve give a better description of the data? Which regression equation would you use, and why?

4.6   SIMPLE LINEAR REGRESSION USING SAS

SAS   A third widely available set of programs is the Statistical Analysis System (SAS).* As with the BMD and SPSS programs, we will outline how to run a simple regression and obtain a scatter diagram for the data of our house size-price problem. Those who have access to SAS also automatically have access to the BMD P-series programs. Thus, any BMD program discussed in this book can be run through SAS, if one refers to the SAS manual for instructions.

---

* For a detailed description of the use of these programs, refer to *A User's Guide to SAS 76* and the *SAS Supplemental Library User's Guide*, by Anthony J. Barr et al., SAS Institute, Inc., Raleigh, N.C., 1976 and 1977.

OBTAINING ACCESS TO THE PROGRAM

Figure 4.21 shows a worksheet to produce a scatter diagram, a simple regression analysis, and a plot of residuals. Note first that, with the exception of the data cards, every statement in the program ends with a semicolon. Not all statements need to be on separate cards; the computer finds the end of a statement by looking for the semicolon.

When data are read into the computer, a *data set* is created. The statement DATA; tells the computer that you are creating a data set. Entering data is easy with SAS. The statement INPUT Y   X; says simply that we are entering two variables, Y and X (or we could have named them PRICE and SIZE). The data cards are then punched with the Y-values entered first, and separated from the X-values by a space. Decimal points are punched. (We switched the order of entering X and Y just for variety.) It is not always necessary to specify a format when using SAS, but one can do it. Section 13.4 shows how. CARDS; says that the data cards follow. The first three statements and the data cards enter the data set into the computer. We now ask the computer to perform some analyses.

PRØC PRINT; asks the computer to print the data so that we may check for keypunching errors. Following the command to print, we can ask the computer to title the printout. Titles are not necessary, but they are nice. When desired, they are placed after the command.

FIGURE 4.21 *Worksheet for SAS Simple Regression, Scatter Diagram, and Plot of Residuals*

```
{SYSTEM CARDS
DATA;
INPUT Y  X;
CARDS;
32  1.8
24  1.0
  :
28  1.5
PRØC  PRINT;
TITLE  LIST  ØF  DATA;
PRØC  GLM;
MØDEL  Y=X/P;
TITLE  SIMPLE  REGRESSIØN;
ØUTPUT  PREDICTED=YHAT  RESIDUAL=R;
PRØC  PLØT;
PLØT  Y*X  YHAT*X='*'/ØVERLAY;
TITLE  SCATTER  DIAGRAM  WITH  REGRESSIØN  EQU;
PRØC  PLØT;
PLØT  R+X;
TITLE  PLØT  ØF  RESIDUALS;
```

To find the regression equation, we ask the computer to proceed with the *general linear model* analysis by **PRØC GLM**;. Then we must specify *which* linear model to use. For a simple linear regression, use

MØDEL   Y = X;

indicating that Y is a linear function of the variable X. The slash followed by P asks the computer to print observed and predicted values of Y and residuals. This portion of the analysis is titled **SIMPLE REGRESSION**.

So far, we have not requested a scatter diagram. This would be easily done by adding two statements:

PRØC PLØT;

PLØT Y*X;

We want a bit more, however. If we obtain a plot of the regression equation on the scatter diagram, the SAS plot will be comparable to the BMD plot. We can then see how the regression line fits the data.

Recall that all points $(X_i, \hat{Y}_i)$ lie on a straight line if the $\hat{Y}_i$ values are calculated by

$$\hat{Y}_i = a + bX_i,$$

which fits a straight line to the data. Thus, in order to plot the regression line, the computer must be able to plot $\hat{Y}$-values versus X-values. To do this, it needs a data set containing $\hat{Y}$-values. The **ØUTPUT** statement allows us to create this new data set.

The **PRØC GLM**; statement, followed by the print option, calculates predicted values and residuals. We want to output these quantities from the **GLM** procedure and add them to our data set. This "outputted" data set will be a new data set containing not only X and Y, the variables used in performing the **GLM** procedure, but also the predicted values, $\hat{Y}$ (by **PREDICTED = YHAT**) and the residuals, R (by **RESIDUAL = R**).

Then the computer is asked to produce a scatter diagram using this new data set:

PRØC PLØT;.

The SAS program will always use the most recently created data set. We first ask the computer to plot the (X, Y)-points (X horizontal, Y vertical) in

PLØT Y*X;

and then to plot the $(X, \hat{Y})$-values

YHAT*X;

on the same graph, by adding the overlay option

/ ØVERLAY.

As we shall see, the (X, Y)-points are plotted with letters of the alphabet, with A denoting one point, B denoting two coincident points, and so on. In order to make the regression line distinguishable from the data points, we asked the computer to print the $(X, \hat{Y})$-values as asterisks. This is done by

YHAT*X = '*'.

Noting that the residuals $(Y_i - \hat{Y}_i)$ are also included in the new data set, we can request that the residuals be plotted. The way the statement is written in Figure 4.21, the residuals will be shown as deviations about a horizontal line. Recalling that the SPSS output plotted deviations around a vertical line, if we want to make the two plots more alike, we can change the SAS statement to read

PLØT X*R;

and then R will be plotted horizontally and X, vertically.

Note that no finish card is necessary with SAS programs. Also SAS is not picky about spacing, as long as neighboring items in a statement are separated from one another by blanks or by special symbols (such as =).

THE OUTPUT

The first page of the SAS printout reproduces the entire program, checks for errors, and tells you how much time and memory were used at each step.

The third page (Figure 4.22) gets on with the simple regression analysis. There, the analysis of variance table looks very much like those in the BMD and the SPSS. The SAS analysis contains several additional items that we will save for later. The printout tells us that $a = 8.70353982$ and $b = 12.89823009$, under the columns labeled PARAMETER and ESTIMATE. Although our SAS program did not request that $\bar{X}$ and $\bar{Y}$ be found, we could have asked for it by adding

PRØC MEANS;

VAR X Y;

which simply ask the computer to print means (and other simple statistics) for the two variables. A list of observed and predicted Y-values and residuals is given. It is shown that $\sum(Y_i - \hat{Y}_i) = 0$ and that $\sum(Y_i - \hat{Y}_i)^2 = 117.09955752$.

Figure 4.23 shows the scatter diagram of the data with the regression line superimposed. Compare Figure 4.23 to the BMDP1R plot in Figure 4.5.

FIGURE 4.22   *SAS Simple Linear Regression Analysis*

SIMPLE REGRESSION
GENERAL LINEAR MODELS PROCEDURE

DEPENDENT VARIABLE: Y

SOURCE	DF	SUM OF SQUARES	MEAN SQUARE	F VALUE	PR > F	R-SQUARE	C.V.
MODEL	1	2255.90044248	2255.90044248	346.77	0.0001	0.950653	7.3930
ERROR	18	117.09955752	6.5055097			STD DEV	Y MEAN
CORRECTED TOTAL	19	2373.00000000				2.55059424	34.50000000

SOURCE	DF	TYPE I SS	F VALUE	PR > F
X	1	2255.90044248	346.77	0.0001

SOURCE	DF	TYPE IV SS	F VALUE	PR > F
X	1	2255.90044248	346.77	0.0001

PARAMETER	ESTIMATE	T FOR H0: PARAMETER=0	PR > |T|	STD ERROR OF ESTIMATE
INTERCEPT	8.70353982	5.81	0.0001	1.49810288
X	12.89823009	18.62	0.0001	0.69264632

OBSERVATION	OBSERVED VALUE	PREDICTED VALUE	RESIDUAL
1	32.00000000	31.92035398	0.07964602
2	24.00000000	21.60176991	2.39823009
3	27.00000000	30.63053097	-3.63053097
4	47.00000000	44.81858407	2.18141593
5	35.00000000	37.07964602	-2.07964602
6	17.00000000	19.02212389	-2.02212389
7	52.00000000	55.13716814	-3.13716814
8	20.00000000	22.89159292	-2.89159292
9	38.00000000	34.50000000	3.50000000
10	45.00000000	42.23893805	2.76106195
11	44.00000000	38.36946903	5.63053097
12	19.00000000	20.31194690	-1.31194690
13	25.00000000	24.18141593	0.81858407
14	50.00000000	52.55752212	-2.55752212
15	30.00000000	30.63053097	-0.63053097
16	43.00000000	40.94911504	2.05088496
17	27.00000000	26.76106195	0.23893805
18	50.00000000	51.26769912	-1.26769912
19	37.00000000	37.07964602	-0.07964602
20	28.00000000	28.05088496	-0.05088496

SUM OF RESIDUALS	-0.00000000
SUM OF SQUARED RESIDUALS	117.09955752
SUM OF SQUARED RESIDUALS - ERROR SS	-0.00000000
FIRST ORDER AUTOCORRELATION	0.03240865
DURBIN-WATSON D	1.93510889

FIGURE 4.23   *SAS Scatter Diagram and Regression Line*

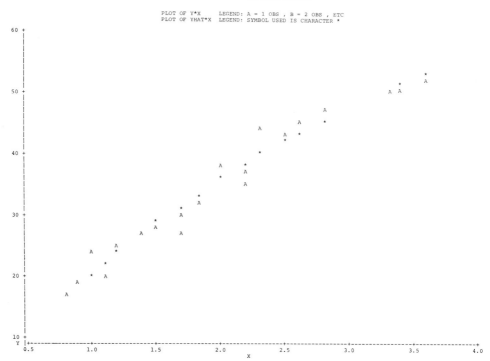

FIGURE 4.24   *SAS Plot of Residuals*

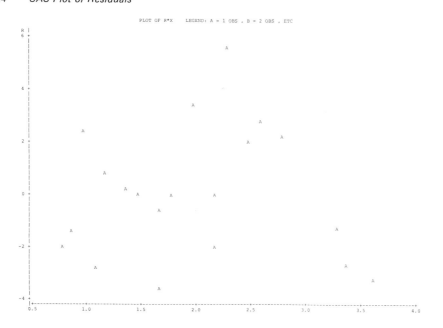

Finally, the residuals are plotted in Figure 4.24. Note that they vary around a horizontal line $R = 0$. If you draw in this horizontal line and compare the plot of residuals to the plot of (X, Y)-values, you will see that the pattern of variation of R-values around the line $R = 0$ is the same as that of the Y-values around the regression line. Except for scale, the plot of residuals looks the way Figure 4.23 would look if the picture were rotated, so that the regression line is horizontal.

EXERCISES *Section 4.6*

p. 521, 522, 524, 525

4.12 Using the TEACHERS data, the PRECIPITATION data, the SALES data, or the HEATING data in Chapter 1 or the Analyses of Data Sets section, use SAS to perform a simple linear regression analysis with scatter diagram, plot of the regression equation, and plot of residuals. Compare the regression equation and scatter diagram found here to the ones found by hand.

4.13 Repeat Exercise 4.12, but use X-values corrected for their means.

4.7 POLYNOMIAL REGRESSION USING SAS

It is very easy to have SAS fit a second-, third-, or higher-degree equation to our data. If we want to fit an equation of the form

$$\hat{Y} = a + bX + cX^2,$$

all we need to change in the worksheet of Figure 4.21 is the MØDEL statement. We change it to

MØDEL Y = X X*X;

which says that Y is a linear function of X and X^2 (X*X is the notation for X^2). One would probably change the titles also, to reflect a quadratic regression, instead of a simple regression. To fit a third-degree polynomial equation, the model statement would be

MØDEL Y = X X*X X*X*X;

and so on. The print option, denoted by /P, can be included if you want a list of $\hat{Y}$-values and residuals.

The printout from fitting the second-degree polynomial is shown in Figure 4.25. We see that the regression equation is given as

$$\hat{Y} = 1.77590491 + 20.46852858X - 1.75565671X^2$$

FIGURE 4.25   SAS Quadratic Equation (Part 1)

QUADRATIC REGRESSION
GENERAL LINEAR MODELS PROCEDURE

DEPENDENT VARIABLE: Y

SOURCE	DF	SUM OF SQUARES	MEAN SQUARE	F VALUE	PR > F	R-SQUARE	C.V.
MODEL	2	2284.14810434	1142.07405217	218.51	0.0001	0.962557	6.6266
ERROR	17	88.85189566	5.22658210			STD DEV	Y MEAN
CORRECTED TOTAL	19	2372.00000000				2.28617193	34.50000000

SOURCE	DF	TYPE I SS	F VALUE	PR > F	DF	TYPE IV SS	F VALUE	PR > F
X	1	2255.90044248	431.62	0.0001	1	199.26157592	38.12	0.0001
X*X	1	28.24766186	5.40	0.0327	1	28.24766186	5.40	0.0327

PARAMETER	ESTIMATE	T FOR H0: PARAMETER=0	PR > |T|	STD ERROR OF ESTIMATE
INTERCEPT	1.7759049	0.54	0.5939	3.2684761
X	20.46852858	6.17	0.0001	3.31500152
X*X	-1.75565671	-2.32	0.0327	0.75519175

OBSERVATION	OBSERVED VALUE	PREDICTED VALUE	RESIDUAL
1	32.00000000	32.93092863	-0.93092863
2	24.00000000	20.48877678	3.51122322
3	27.00000000	31.49855561	-4.49855561
4	47.00000000	45.32343636	1.67656364
5	35.00000000	38.30928933	-3.30928933
6	17.00000000	17.02710748	-0.02710748
7	52.00000000	52.70929690	-0.70929690
8	20.00000000	22.16694173	-2.16694173
9	38.00000000	35.69033525	2.30966475
10	45.00000000	43.12583989	1.87416011
11	44.00000000	39.56609667	4.43390333
12	19.00000000	18.77549870	0.22450130
13	25.00000000	23.80999355	1.19000645
14	50.00000000	51.07351057	-1.07351057
15	30.00000000	31.49855561	-1.49855561
16	43.00000000	41.97437195	1.02562805
17	27.00000000	26.99075778	0.00924222
18	50.00000000	50.20294770	-0.20294770
19	37.00000000	38.30928933	-1.30928933
20	28.00000000	28.52847019	-0.52847019

SUM OF RESIDUALS   -0.00000000
SUM OF SQUARED RESIDUALS   88.85189566
SUM OF SQUARED RESIDUALS - ERROR SS   -0.00000000
FIRST ORDER AUTOCORRELATION   -0.24754441
DURBIN-WATSON D   2.47899672

FIGURE 4.25   *(Part 2)*

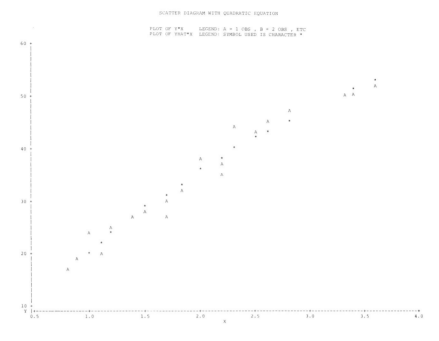

and that the predicted values are found and the residuals are calculated using this second-degree equation. There is a slight curvature in the plot of the regression equation.

The printout resulting from fitting the third-degree polynomial equation is shown in Figure 4.26.

We have seen that computer programs certainly reduce the amount of labor involved in fitting regression equations. The next section shows how easily a computer can invert a matrix.

EXERCISES   *Section 4.7*

p. 523   *4.14*   Use SAS to fit a simple linear regression equation to the POLICE data.
   a.   Compare the values of a and b to those found in the by-hand analysis.
   b.   The residuals found in the by-hand analysis were

− 12.5000	29.3476
46.1414	78.0436
− 59.4022	− 9.0762
− 19.8912	57.2824
− 13.9132	− 70.2718
12.3370	− 48.6956
10.5978	

How do the residuals from the SAS analysis compare to these?

FIGURE 4.26   SAS Third-Degree Polynomial Regression

(1)

THIRD DEGREE POLYNOMIAL REGRESSION
GENERAL LINEAR MODELS PROCEDURE

DEPENDENT VARIABLE: Y

SOURCE	DF	SUM OF SQUARES	MEAN SQUARE	F VALUE	PR > F	R-SQUARE	C.V.
MODEL	3	2297.68284888	765.8942896	162.70	0.0001	0.968261	6.2888
ERROR	16	75.31715112	4.70732194			STD DEV	Y MEAN
CORRECTED TOTAL	19	2373.00000000				2.16963636	34.50000000

SOURCE	DF	TYPE I SS	F VALUE	PR > F	DF	TYPE IV SS	F VALUE	PR > F
X	1	2255.90044248	479.23	0.0001	1	0.22044241	0.05	0.8314
X*X	1	28.24766186	6.00	0.0262	1	9.7571516	2.07	0.1692
X*X*X	1	13.53474454	2.88	0.1093	1	13.53474454	2.88	0.1093

PARAMETER	ESTIMATE	T FOR H0: PARAMETER=0	PR > |T|	STD ERROR OF ESTIMATE
INTERCEPT	15.36199038	1.79	0.0927	8.59175543
X	-3.08164241	-0.22	0.8314	14.24038123
X*X	10.22134091	1.44	0.1692	7.09960373
X*X*X	-1.83347354	-1.70	0.1093	1.08127601

OBSERVATION	OBSERVED VALUE	PREDICTED VALUE	RESIDUAL
1	32.00000000	32.23936091	-0.23936091
2	24.00000000	20.66821534	3.33178466
3	27.00000000	30.65501801	-3.65501801
4	47.00000000	46.62029321	0.37970679
5	35.00000000	38.53084083	-3.53084083
6	17.00000000	18.49959619	-1.49959619
7	52.00000000	51.19411439	0.80588561
8	20.00000000	21.89965295	-1.89965295
9	38.00000000	35.41628088	2.58371912
10	45.00000000	44.22085372	0.77914628
11	44.00000000	40.03723369	3.96276631
12	19.00000000	19.53119614	-0.53119614
13	25.00000000	23.21450812	1.78549188
14	50.00000000	50.98026307	-0.98026307
15	30.00000000	30.65501801	-0.65501801
16	43.00000000	42.89324098	0.10675902
17	27.00000000	26.05046780	0.94953220
18	50.00000000	50.61343431	-0.61343431
19	37.00000000	38.53084083	-1.53084083
20	28.00000000	27.54957062	0.45042938

```
SUM OF RESIDUALS                              -0.00000000
SUM OF SQUARED RESIDUALS                      75.31715112
SUM OF SQUARED RESIDUALS - ERROR SS            0.00000000
FIRST ORDER AUTOCORRELATION                   -0.23167475
DURBIN-WATSON D                                2.45909450
```

    c.   Calculate $\frac{1}{n}\sum|Y_i - \hat{Y}_i|$ and compare to the value found by hand.

    d. Compare the plot of the data and regression line to those made previously.

4.15   Use SAS to fit a second-degree polynomial equation to the POLICE data.

    a.   Compare the values of a, b_1, and b_2 to those we found in the by-hand analysis.

    b. How do the residuals from the SAS analysis compare to the residuals found previously?

    c.   We previously found $\frac{1}{n}\sum|Y_i - \hat{Y}_i|$ to be 27.7418. What do you get if you calculate this quantity from the SAS residuals?

    d. Compare the plot of the data and regression equation to those made by hand.

4.16   Comparing the results of Exercises 4.14 and 4.15, do you think a straight line or a curve gives the better description of the POLICE data? Which regression equation would you use, and why?

4.8 INVERTING A MATRIX

As we have seen, inverting a matrix larger than 2×2 is surely a job for a computer. Actually, the regression packages we have discussed all invert matrices to find the regression equation, but they do not all display this inversion as an essential part of the output. Seeing the inverted matrix, however, helps one understand the workings of the regression problem.

IMSL PACKAGES

IMSL

Almost every center using an IBM/360-370 or comparable computer has access to the packages of the International Mathematical and Statistical Libraries, Inc. (IMSL). These contain a great variety of standard mathematical and statistical calculations.* If these are not available at your computing center, then surely some other program for matrix inversion is available.

    Recall that in solving the normal equations to fit the straight line $\hat{Y} = a + bX$ to the data of the house size-price problem, we had to invert the matrix

$$X'X = \begin{bmatrix} 20 & 40 \\ 40 & 93.56 \end{bmatrix}.$$

Figure 4.27 shows a worksheet using the IMSL for inverting this matrix. To discuss this program in detail would constitute a course in Fortran programming. But we can point out a few of its limitations and other pertinent features.

    The statement

REAL*4 MATRIX(8,8)

*For a detailed description of what is available, refer to the *Library 1 Reference Manual*, available from IMSL, Suite 510, 6200 Hillcraft, Houston, Texas 77036.

157 INVERTING A MATRIX

FIGURE 4.27   *Worksheet to Invert **X'X** for the House Size-Price Problem, Calling IMSL INVERT Routine*

```
        } SYSTEM CARDS
          REAL*4  MATRIX(8,8)
          LOGICAL    SING
          INTEGER    ORDER,DIM
          READ(5,100)  N
          ORDER = N
          DIM = 8
          DO 20  I=1,N
          READ(5,10) (MATRIX(I,J),J=1,N)
     20   CONTINUE
          WRITE(6,40)
          DO 30  I=1,N
          WRITE(6,50)(MATRIX(I,J),J=1,N)
     30   CONTINUE
          CALL    INVERT(MATRIX,ORDER,DIM,SING)
          IF(SING)    GO TO 88
          WRITE(6,80)
          DO 60  I=1,N
          WRITE(6,50)(MATRIX(I,J) J=1,N)
     60   CONTINUE
          GO TO 99
     88   WRITE(6,70)
     10   FORMAT(8F10.4)
     40   FORMAT(1H1,T20,'MATRIX GIVEN')
     50   FORMAT(1H0,8(5X,F15.7))
     70   FORMAT(1H0,T10,'MATRIX GIVEN IS SINGULAR')
     80   FORMAT(1H0,/////,T20,'INVERSE COMPUTED')
    100   FORMAT(I3)
     99   STOP
          END
        } SYSTEM CARD(S)
  002
          20.0        40.0
          40.0        93.56
```

says that this particular program can be used to invert a matrix of dimension 8×8 or smaller. Those familiar with FØRTRAN can easily modify the program to handle larger matrices, but the 8×8 should be sufficient for most of our purposes. The statement numbered 10,

      10 . . . FØRMAT(8F10.4)

says that any data card is divided into eight groups of 10 columns each, the last four columns of each group containing decimal places. Thus, the first value in any row of the matrix is punched in columns 1–10, the second value in columns 11–20, and so on. One data card is entered for each *row* of the matrix to be inverted.

The order, or dimension, of the matrix to be inverted, is punched on the first data card, in the first three columns, right justified. With the maximum dimension of 8×8, the largest number that can be punched on the first data card is 008.

The computer will read in the data and then reproduce the matrix given, under the title MATRIX GIVEN. This lets you check on keypunching. Should the matrix be singular, the computer will also print out this message. If it is nonsingular, the computer will find the inverse and will display it under the heading INVERSE COMPUTED.

The results of inverting

$$\mathbf{X'X} = \begin{bmatrix} 20 & 40 \\ 40 & 93.56 \end{bmatrix}$$

are shown in Figure 4.28. Note that even though the value 93.56 was punched as the $(2, 2)$ element of the matrix, the value of the $(2, 2)$ position of MATRIX GIVEN is 93.5599976. This sort of thing sometimes occurs because the computer converts decimal system values to the binary number system.* In such a conversion and reconversion to the decimal system, some rounding error occasionally creeps in, but it is nothing to worry about. The IMSL matrix inversion program called for in Figure

Single-Precision

4.28 is a single-precision program. This means that any number entered occupies one storage location in the computer and has eight significant digits. Had we used a double-precision matrix inversion program, each number would occupy two adjacent storage

Double-Precision

locations and would have 16 significant digits. Obviously, we could obtain greater accuracy with a double-precision program. We use the single-precision program here because a few regression programs are only single-precision, and you should realize that some rounding error can creep into them. However, any unusual or unexpected value should always be investigated, and not attributed automatically to rounding error.

Comparing the results of the printout to what we found in Section 3 of Chapter 3, note that

$$(\mathbf{X'X})^{-1} = \begin{bmatrix} \dfrac{93.56}{271.2} & \dfrac{-40}{271.2} \\ \dfrac{-40}{271.2} & \dfrac{20}{271.2} \end{bmatrix} = \begin{bmatrix} 0.344985 & -0.147493 \\ -147493 & 0.073746 \end{bmatrix}.$$

FIGURE 4.28   *Output of a Program Inverting a 2 × 2 Matrix*

MATRIX GIVEN

20.0000000	40.0000000
40.0000000	93.5599976

INVERSE COMPUTED

0.3449844	−0.1474922
−0.1474922	0.0737461

* Those interested in the binary number system should refer to Appendix B.

This is identical to the values on the printout up to the sixth decimal place, where some rounding error occurs.

The particular program given will work for any matrix up to dimension 8×8. Since $X'X$ is always symmetric in regression problems, it might be better to use one of the IMSL programs designed expressly for inverting symmetric matrices. However, this is a more general procedure.

To invert a matrix of dimension greater than 2×2, the only necessary changes are in the data cards. Enter the dimension of the matrix on the first card; then enter the appropriate number (same as the dimension) of data cards, each containing one row of the matrix, in order. If it is desired to invert a matrix larger than 8×8, the student who is familiar with Fortran should have no difficulty amending the given program to suit his needs.

SAS MATRIX PROCEDURE

Should you have access to a computing center that offers the SAS system, matrix inversion is so easy it is almost sinful. In order to invert our matrix for the house size-price problem, all one needs are the appropriate system cards and the following three statements:

PRØC MATRIX PRINT;
A = 20 40 / 40 93.56;
B = INV(A);

and you will see the printout of Figure 4.29.

SAS allows you to perform many manipulations of matrices. Adding the word PRINT will insure that all steps are printed out. After calling the MATRIX procedure, simply name the matrix (**A** in this case) and list the elements in the matrix, separating the rows by a slash. If you take up all the columns on the card before entering all the elements in the matrix, simply continue on to the next card. The semicolon tells the computer when you are through listing elements. To ask for the inverse of the matrix **A** to be printed, simply write INV(A). We have named this matrix **B**. Note that the SAS inversion procedure is a double-precision procedure. If the matrix is singular, SAS will print a message stating so.

FIGURE 4.29   *SAS Matrix Inversion*

A	COL1	COL2
ROW1	20	40
ROW2	40	93.56

B	COL1	COL2
ROW1	0.344985	-0.147493
ROW2	-0.147493	0.0737463

INVERSION USING SAS GLM PROCEDURE

One of the nicest features of the SAS general linear model procedure is that you can request that the $\mathbf{X'X}$ and $(\mathbf{X'X})^{-1}$ matrices for the model to be fitted be printed right along with the rest of the regression analysis. The procedure is quite simple. Refer back to Figure 4.21, and recall that in order to run the general linear models procedure, one first writes

PRØC GLM;

and follows this by a model statement. Options follow the slash in the model statement. For example, P asks for the optional print of residuals to be made. The model statement

MODEL Y = X/P XPX I;

not only asks for the print option, but also for $\mathbf{X'X}$ to be printed (XPX) and for $(\mathbf{X'X})^{-1}$ to be shown (the I—for inverse—option).

INVESTIGATING MATRICES

Now that matrix inversion has been simplified through the availability of computer packages, let us investigate some properties of matrices a bit more fully.

Recall that a singular matrix is one whose determinant is zero. Also to invert a matrix, every element of the adjoint matrix must be divided by the determinant of the matrix to be inverted. Since division by zero is not defined, a singular matrix has no inverse. Let us consider for a moment the matrix

$$\begin{bmatrix} 1 & 2 & 3 \\ 2 & 4 & 6 \\ 4 & 8 & 12 \end{bmatrix},$$

from Exercise 3.18e. It is obviously singular, since the second row is twice the first row and the third row is twice the second row. In Chapter 3, we found that this determinant is zero and that the inverse does not exist. Suppose this matrix is inverted using the IMSL program in Figure 4.27. The results are shown in Figure 4.30.

The computer should have informed us that the matrix is singular. However, because of rounding error in a single-precision program, the computer found the determinant to be not exactly zero. Note how large all the entries in the matrix are. The asterisks in the (1, 1) position of the inverse indicate that the number which belongs there is so large that it cannot be fitted into the print format. This is a clue that the given matrix is singular.

FIGURE 4.30   *IMSL Inversion of a 3 × 3 Singular Matrix*

MATRIX GIVEN

1.0000000	2.0000000	3.0000000
2.0000000	4.0000000	6.0000000
4.0000000	8.0000000	12.0000000

INVERSE COMPUTED

**************	1118481.0000000	-279620.1875000
1118481.0000000	-69905.0000000	-244667.6875000
-372826.9375000	-326223.6875000	256318.5625000

Now let us consider the matrix

$$\begin{bmatrix} 1 & 2 & 3 & 4 \\ 2 & 4 & 6 & 8 \\ 4 & 8 & 12 & 16 \\ 8 & 16 & 24 & 32 \end{bmatrix},$$

from Exercise 3.20d. We know that this matrix is singular, and so does the IMSL program, as seen in Figure 4.31. Let us play with it enough to make the rows/columns not *quite* linear combinations of each other. We want to construct a matrix that is *almost*, but not exactly, singular. For example, let

$$\mathbf{A} = \begin{bmatrix} 1 & 2 & 3 & 4 \\ 2.001 & 4 & 6 & 7.999 \\ 4 & 7.999 & 12 & 16 \\ 8 & 16 & 24.001 & 32 \end{bmatrix}.$$

The result of running this matrix through the SAS inversion program, where rounding error is minimized, is shown in Figure 4.32. Note the very large and very small values in the inverse matrix. We can conclude that if a matrix is nearly singular, which means that some row(s) or column(s) gives essentially the same information as some other

FIGURE 4.31   *IMSL Recognizes a Singular Matrix*

MATRIX GIVEN

1.0000000	2.0000000	3.0000000	4.0000000
2.0000000	4.0000000	6.0000000	8.0000000
4.0000000	8.0000000	12.0000000	16.0000000
8.0000000	16.0000000	24.0000000	32.0000000

MATRIX GIVEN IS SINGULAR

FIGURE 4.32   *SAS Inversion of a Nearly Singular Matrix*

A	COL1	COL2	COL3	COL4
ROW1	1	2	3	4
ROW2	2.001	4	6	7.999
ROW3	4	7.999	12	16
ROW4	8	16	24.001	32

B	COL1	COL2	COL3	COL4
ROW1	1600.2	800	400	-600
ROW2	4000	-889039E-16	-1000	-265920E-15
ROW3	-8000	3552621E-16	-710395E-15	1000
ROW4	3600.2	-200	400	-600

row(s) or column(s), the elements of its inverse will be very large or very small, in absolute value, in relation to the elements in the original matrix.

    If the matrix of coefficients from a system of simultaneous equations turns out to be nearly singular, then we say that the system of equations, or the matrix of coefficients, **Ill-Conditioned** is ill-conditioned. Then, we should look for <u>linear dependencies</u> in the system—that is, for rows or columns that are linear combinations of other rows or columns, or are very nearly so. If for example, one equation in a system is redundant, we should try to replace **Linear** it by another equation that gives unique information; otherwise the system of **Dependencies** equations will have no solution. In multiple regression, the problem of ill-conditioned matrices is very important.

EXERCISES   *Section 4.8*

4.17 Use an available matrix inversion program to invert these square matrices. Compare results to those found in Exercise 3.22a.

a. $\begin{bmatrix} 1 & 0 & 7 \\ 8 & 2 & 5 \\ 0 & 6 & 7 \end{bmatrix}$;
b. $\begin{bmatrix} 1 & 0 & 0 \\ 1 & 5 & 4 \\ 9 & 2 & 4 \end{bmatrix}$;
c. $\begin{bmatrix} 5 & 7 & 0 \\ 7 & 3 & 7 \\ 0 & 7 & 3 \end{bmatrix}$;

d. $\begin{bmatrix} 1 & 2 & 3 \\ 4 & 5 & 6 \\ 7 & 8 & 9 \end{bmatrix}$;
e. $\begin{bmatrix} 7 & 0 & 0 \\ 0 & 5 & 0 \\ 0 & 0 & 8 \end{bmatrix}$
f. diag(7, 5, 3).

4.18 Use an available matrix inversion program to invert these square matrices. Compare results with those to the results of Exercise 3.22c.

a. $\begin{bmatrix} 1 & 0 & 0 & 0 \\ 6 & 0 & 6 & 2 \\ 0 & 5 & 4 & 0 \\ 6 & 6 & 8 & 2 \end{bmatrix}$;
b. $\begin{bmatrix} 5 & 6 & 2 & 4 \\ 0 & 8 & 1 & 7 \\ 0 & 4 & 1 & 5 \\ 0 & 4 & 0 & 3 \end{bmatrix}$;

c.
$$\begin{bmatrix} 5 & 6 & 2 & 4 \\ 0 & 8 & 1 & 7 \\ 0 & 4 & 1 & 5 \\ 0 & 12 & 3 & 15 \end{bmatrix};$$

d.
$$\begin{bmatrix} 1 & 5 & 9 & 6 \\ 0 & 7 & 5 & 2 \\ 4 & 6 & 1 & 7 \\ 8 & 9 & 1 & 7 \end{bmatrix};$$

e.
$$\begin{bmatrix} 3 & 0 & 0 & 0 \\ 0 & 6 & 0 & 0 \\ 0 & 0 & 7 & 0 \\ 0 & 0 & 0 & 4 \end{bmatrix};$$

f. diag(7, 7, 8, 5);

g.
$$\begin{bmatrix} 1 & 2 & 3 & 4 & 5 \\ 2 & 4 & 6 & 8 & 10 \\ 4 & 8 & 12 & 16 & 20 \\ 8 & 16 & 24 & 32 & 40 \\ 16 & 32 & 48 & 64 & 60 \end{bmatrix};$$

h.
$$\begin{bmatrix} 4 & 7 & 8 & 8 & 2 \\ 5 & 0 & 6 & 0 & 0 \\ 6 & 2 & 9 & 7 & 4 \\ 9 & 9 & 4 & 9 & 2 \\ 8 & 8 & 9 & 3 & 6 \end{bmatrix}$$

4.19 Invert the following matrix, using an available program.

$$\begin{bmatrix} 20 & 40 & 430 \\ 40 & 93.56 & 844.7 \\ 430 & 844.7 & 10{,}990 \end{bmatrix}$$

4.9 CHOICE OF THE PACKAGE

Often the choice of which computer package to use for a regression or to invert a matrix is restricted by what is available. Most large computing centers, however, have all the packages we have discussed. Then the choice becomes largely a matter of personal preference.

Since one can execute a BMD program through the SAS system, those who have access to SAS also have access to the BMD P-series library. The *SAS '76 User's Guide* gives instructions for using BMD programs.

As we go along, you may want to refer back to the three packages in this chapter to see which gives the assortment of measurements you require. The reader who has none of the three available should run our house size-price data through whatever package is available and compare results to those presented here. Since all regression packages give essentially the same output, you should be able to figure out what yours gives by comparing it to one or more of the programs discussed here.

Insofar as the matrix inversion programs are concerned, most large computing centers have the IMSL library and the SAS packages. While it certainly seems easier to use the SAS matrix inversion procedure, note that once the program cards for IMSL are punched, these same cards can be used to invert any matrix (of dimension 8 × 8 or smaller) by simply adding the appropriate data cards. Thus, in the long run, the IMSL program may be as easy to use as the SAS program.

4.10 SUMMARY

The choice of which computer package to use boils down to three considerations:

1. what is available through your computing center;
2. what graphs, statistics, and tests you need; and
3. personal preference as to layout of results, ease of writing program, and so on.

We have now solved simultaneous equations and fitted regression equations to data in a large variety of ways. Very little of this is new to the reader who has had reasonably good training in mathematics or some exposure to computers. It is time to get back to statistics and see what the regression equation tells us about our variables of interest.

Chapter 5 USE AND INTERPRETATION OF THE REGRESSION EQUATION

5.1 INTRODUCTION

We turn now to consider what the regression equation we have found actually means with reference to a specific set of data. We return to our house size-price problem, this time to illustrate how to use the regression equation for making estimates and, perhaps even more important, how not to use it.

5.2 PREDICTING AN AVERAGE

Until now our primary emphasis has been on the use of mathematical techniques to fit a straight line $\hat{Y} = a + bX$ through a sample of points. We now turn to the practical interpretation of our mathematical results.

ESTIMATING PRICE FROM SIZE

For describing the relationship between sizes and prices of houses, we found the equation

$$\hat{Y} = 8.7 + 12.9X.$$

It is simple enough to estimate the price of a house given its size; just plug the size in for X and solve the equation. For example, we estimate the price of a house 2000 square feet in size to be

$$\hat{Y}_2 = 8.7 + 12.9(2) = 34.5 \text{ thousand dollars},$$

or \$34,500. Note that we substitute $X = 2$, not $X = 2000$, into the equation, since the units on X are given as thousands of square feet (Table 1.1). Similarly, we estimate that a house 2200 square feet in size will sell for

$$\hat{Y}_{2.2} = 8.7 + 12.9(2.2) = 37.08 \text{ thousand dollars},$$

or \$37,080.

Recall that the regression equation predicts or estimates the *average* price for houses of a given size. Thus, $\hat{Y}_2 = \$34,500$ means that the average price of houses 2000 square feet in size is \$34,500; or, equivalently, the typical house 2000 square feet in size will sell for that amount. The prediction says nothing about any specific house. In Table 1.1, House No. 9 is 2000 square feet in area, and its price is \$38,000. Thus, House No. 9 is more expensive than the typical house that size, due to (1) other factors which we have not taken into account and (2) random variation.

Likewise, Houses No. 5 and 19, both having 2200 square feet, cost \$35,000 and \$37,000, respectively. And both cost less than the average house of their size, which we found to be \$37,080. One might ask why the estimated average value obtained by using $\hat{Y} = 8.7 + 12.9(2.2)$ is not somewhere between the actual sample values, \$35,000 and \$37,000. After all, doesn't the regression equation give an average? The scatter diagram of Figure 2.8, reproduced in Figure 5.1, helps answer the question. Locate the points (2.2, 35) and (2.2, 37), corresponding to Houses No. 5 and 19, respectively. Note that the line passes above both points. Slightly to the left of these points lies the point (2, 38), which corresponds to House No. 9. These three houses are nearly the same size. The fact that House No. 9 is relatively more expensive than the other two is what raises the line above the points for Houses No. 5 and 19; in other words, the line tends to average all three houses, not just Houses No. 5 and 19.*

THE RELEVANT RANGE

Since a random sample of houses was taken and the size and price were recorded for each sampled item, the independent variable, size, is a random variable. Thus we consider that we have a representation of all sizes within the relevant range (800 to 3600 square feet). Therefore any size within this range can be plugged into the regression equation to obtain a predicted price. The size need not be one of those sampled.

One must be extremely cautious, however, in using this sample to predict the price of a house that has less than 800 square feet or more than 3600 square feet. Why? Does not the straight line $\hat{Y} = 8.7 + 12.9X$ extend infinitely in both directions?

Although mathematically speaking, a straight line can be extended as far as one likes, we must remember that we are using mathematics only as a tool to describe a real-world problem, and the interpretations are limited by what makes sense in the problem.

---

* The term *regression* actually came about because only averages can be predicted. Studying Darwin's theory of evolution, Sir Francis Galton (1822–1911) figured that if tall parents tend to produce tall children and short parents tend to produce short children, then humans should evolve into one race of giants and one race of midgets. Actual observations showed, however, that his conclusion would not hold; both unusually tall and unusually short sets of parents tend to produce children of roughly average height. Thus the children's heights tend to converge, or regress, to the average height for that generation.

FIGURE 5.1   *Regression Line Superimposed on Scatter Diagram, from Figure 2.8*

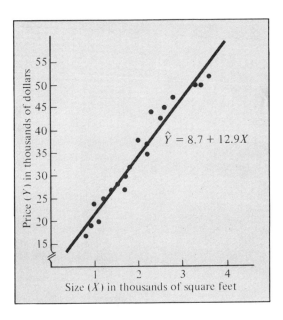

Extending the line indiscriminately can lead to some absurdities. Suppose we extend the line to $X = 0$. Then $\hat{Y}_0 = 8.7 + 12.9(0) = 8.7$, and the interpretation would be that a house zero square feet in size typically sells for \$8700. Or if $X = -5$,

$$\hat{Y}_{-.5} = 8.7 + 12.9(-.5) = 2.25 = \$2250,$$

so that a house of minus 500 square feet size will cost on the average \$2250. Obviously these interpretations are nonsense.

But what about a house 500 square feet, or one 4000 square feet? Why could one not predict their average price? The reason is that *we have information about prices of houses only for houses between 800 and 3600 square feet in size. We do not know anything about prices of larger or smaller houses.* How do you know that the same relationship that exists for houses between 800 and 3600 square feet also exists for larger or smaller houses? On the other hand, how do you know that this same relationship does not exist? The point is, we have no information, so we just don't know if this same relationship holds or not. Our equation was determined as a good general description of the relationship *in this range.* It is not intended to be some law about the relationship between all sizes and prices. The situation might be as shown in Figure 5.2, in which very small houses cost very little but great mansions are fantastically expensive. To describe the relationship between size and price for *all* houses, an S-shaped curve might be appropriate. However, in the range from 800 to 3600 square feet, a straight line is adequate. Using the regression equation to predict outside of the sampled range can result in some very gross errors, as is illustrated in Figure 5.2.

FIGURE 5.2   *The Straight Line Provides a Good Description Only over a Limited Portion of the Entire Range*

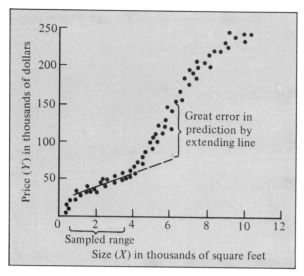

What about going just a little bit outside the range? For example, suppose the newcomer to town were particularly interested in a house of 4000 square feet. Since this is only a little bit larger than the largest house in the sample, could one not use the regression equation to estimate the price of this house? Maybe. Usually one would not expect the relationship to change very drastically very fast, at least not without some indication of a change beginning at the end of the scatter diagram, so maybe the newcomer could get a ball-park estimate of the price. But then again, maybe not. We just don't know anything about houses other than those in the sampling range. Anyone interested in houses larger than 3600 square feet should include them in the sample.

Sometimes there can be a theoretical structure to the problem that dictates that the same relationship found in the sample must continue past the sampled range. If this is the case, then it is safe (provided the theory is correct) to make predictions outside the range using the regression equation. In the absence of a theoretical framework, however, it is best to stick to situations that you know something about. Note that the regression line in Figure 5.1 is not extended past the points, this showing that the relationship has been found only for those data.

One of the most common misuses of a regression is extrapolating beyond the data. But certainly there is no reason to believe that a relationship between IQ and some other variable found for mentally retarded children with IQ's between, say, 60 and 80, should also hold for children with IQ's between 90 and 120. The two groups of children can be thought of as two different populations. Laws of science, whether they be laws of physics, chemistry, or economics, are established on the basis of much more than the evidence from a single sample from a restricted range.

5.3 PREDICTING THE AVERAGE IN A DESIGNED EXPERIMENT

CONTROLLED VERSUS RANDOM VARIABLES

Recall from Chapter 1 the distinction between the regression study in which X is a random variable and one in which X is a preselected, or controlled, variable. In our example about house size and prices, a range of house sizes of interest was defined and a random sample of houses in this size range was taken. Thus, X, as well as Y, was a random variable, since we had no way of foretelling what X-values we would end up looking at. Our sample of houses from throughout the relevant range is considered to give information about *all* sizes within the range. Therefore, we can use our regression equation to predict the price of any house in the range.

Controlled Variable
    The situation is slightly changed, however, if the predictor variable values are controlled, or preselected, as often happens in a controlled experiment. The laboratory scientist, the agricultural experimenter, the experimental psychologist, and others have control over the values of the independent variable (the "experimental level," the "treatment level," the "stimulus intensity"); they can run their experiments under whatever conditions they choose. Often the levels of the independent variable are chosen at evenly spaced intervals, as was illustrated in Chapter 2 in which we discussed scaling the X-values. This might be done for the sake of reducing computational complexity, but is more often done according to the dictates of experimental design.

Experimental Design
Briefly, an experimental design is a method for running an experiment in order to obtain the most, and the least ambiguous, information possible at the least cost.

This gives the laboratory scientist a distinct advantage over the social scientist, because an experimental design lets the laboratory scientist know the separate effects of each of several variables on the response. The social scientist rarely enjoys such a situation. The advantage held by the laboratory scientist, however, is not universal. The laboratory scientist who preselects values of the independent variable chooses specific X-values from his range of interest to study. One might say, then, that the laboratory scientist is actually interested in only those levels of the predictor variable that he chooses to use in his experiment—not in any X-value between any two experimental levels.

ESTIMATES IN DESIGNED EXPERIMENTS

Since the laboratory scientist works with only a few X-values, the resulting predictions of the average response levels (Y-values) can be made *only* for those X-values studied. For example, let's suppose an agronomist applies 20, 25, and 30 pounds of fertilizer to various plots of soil, measures the yield of a certain crop, and finds a regression equation relating intensity of fertilization to yield. He can predict the average yield for the respective plots to which 20, 25, and 30 pounds of fertilizer are applied. He cannot make a valid prediction of yield for any amount of fertilizer between 20 and 25 pounds or between 25 and 30 pounds unless he is absolutely certain that his model is correct.

Thus, by preselecting his experimental levels, the laboratory scientist loses a degree of generality in the results. The social scientist has the advantage in this regard through the ability to generalize for any value in the range.*

In practice the laboratory scientist does generalize for any value in his range of X-values; this is usually the purpose of his study. But he runs the risk of making bad estimates and missing important features of the relationship. Suppose, for example, that the actual relationship is as shown in Figure 5.3.

FIGURE 5.3   *Picking Points at Evenly Spaced Intervals Might Miss Interesting Features of the Relationship*

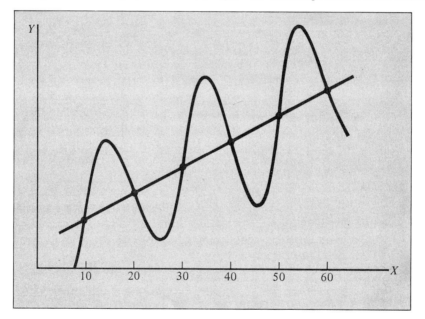

A researcher who conducts an experiment at $X = 10, 20, \ldots, 60$, would certainly feel justified in concluding that Y is linearly related to X. An estimate of Y for $X = 35$, however, would not be a very accurate prediction. While an underlying relationship such as this one might be relatively unlikely, and while it might be still more unlikely that the experimenter would be so unlucky as to choose the X-levels shown, such an instance is not impossible. Whenever one generalizes between preselected levels, there is a risk that the predictions will be off.

_____

* The reader who is familiar with fixed and random effects models in the analysis of variance can recognize the analogy between our discussion and the generalizations that can be made in these models.

5.4 MEANING OF THE INTERCEPT

THE INTERCEPT MIGHT NOT BE INTERPRETABLE

Regression
Constant

We noted earlier that the intercept, or <u>regression constant</u>, in our house size-price example has no physical meaning: it makes no sense to say that the average price of houses of zero square feet is $\hat{Y}_0 = 8.7 = \$8700$.

How, then, do we interpret the meaning of $a = 8.7$ in the equation? Obviously, the value is needed to get a reasonably accurate prediction of price from size. It is a reference, or starting point, but where does it come from? We don't have enough information to answer this question exactly. Perhaps it is some kind of base, or minimum, price, or a realtor's fee, or some such. It is a real value affecting price determined by *something outside the relationship between size and price.*

WHEN THE INTERCEPT CAN BE INTERPRETED

Can the intercept ever be meaningful to the relationship? Suppose $X =$ winter average daily temperature and $Y =$ amount of electricity used for heating. Then the value a in the regression equation will give the average amount of electricity used on days when the average temperature is zero. The intercept will have physical meaning to the problem and will be the average value of Y when X is zero provided that *both* of two conditions hold:

1. It must be physically possible for X to equal zero.
2. Data must be collected around $X = 0$.

When $X =$ winter average daily temperature and $Y =$ amount of electricity used for heating, X satisfies both conditions, since (1) it is possible for the average daily temperature in winter to be $0°$ and (2) we suppose that data have been collected for such days. If, however, the data were collected around Miami, Florida during the summer, then condition (2) would fail, and *a would not* be the average amount of electricity used for heating on zero-degree days; instead it would be a constant determined by something besides temperature. For the house size and price example, condition (1) fails because it is impossible to have a house of no size. If condition (1) fails, then, of course, condition (2) also fails.

FITTING THE REGRESSION EQUATION THROUGH THE ORIGIN

Suppose that one recognizes in a given problem that it is not physically possible for X to be zero, so that it seems logical that if $X = 0, Y$ should also be zero. For example, if the size of a house is zero, should not the price also be zero? Or, if $X =$ number of items purchased $= 0$ and $Y =$ total amount paid, then Y should also be zero. In such cases, it seems reasonable that the intercept should be zero and that the regression equation should be of the form

$$\hat{Y} = 0 + b^*X, \quad \text{or} \quad \hat{Y} = b^*X.$$

Fitting through the Origin

We denote the slope by b^* to remind ourselves that it will not be the same as if the line has a nonzero intercept. A regression line with zero intercept is said to be fitted through the origin. How would such be accomplished, in our house size-price example?

To find the line passing through $(0, 0)$, which is on the average closest to all the points (in terms of squared deviations), we need to find the value of b^* that minimizes

$$Q_1 = \sum(Y_i - \hat{Y}_i)^2 = \sum(Y_i - b^*X_i)^2.$$

With calculus it can be shown that

$$b^* = \frac{\sum X_i Y_i}{\sum X_i^2}.$$

For the interested reader, Appendix D gives the details of obtaining this result. Note that this value for b^* is of the same form as $b = S_{XY}/S_{XX}$, although the two quantities are not equal unless X has been corrected for its mean. Inspection of the scatter diagram in Figure 5.1 reveals that if the line is forced to pass through the origin, then the slope must change in order for the line to pass through the mass of dots.

FITTING THROUGH THE ORIGIN IN THE HOUSE SIZE-PRICE EXAMPLE

In our example, $\sum X_i Y_i = 1554.9$ and $\sum X_i^2 = 93.56$, so that the regression equation fitted through the origin is

$$\hat{Y} = 16.62X.$$

How well does this equation describe the underlying relationship? In Figure 5.4, we see that this line does not follow the lay of the dots as well as our original regression line did. Thus, even though it might seem reasonable for Y to be 0 if $X = 0$, such a strong assumption should be made only after careful consideration. Otherwise the equation obtained might not provide a very good description of the relationship. One problem in this example is that zero square feet is probably not the lower limit on house size; perhaps a dwelling must be at least, say, 300 square feet in area before it can be classified as a house. In addition, by fitting the line through the origin, we are assuming that the same relationship that is found for houses between 800 and 3600 square feet in size also holds for houses between 0 and 800 square feet in size. As we saw in Section 5.1, this assumption is probably not justified.

FIGURE 5.4

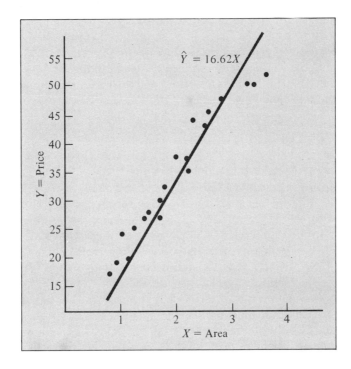

5.5 MEANING OF THE SLOPE

THE SLOPE DESCRIBES THE RELATIONSHIP

The intercept of the regression equation gives us one specific piece of information: it is either the value of Y when $X = 0$, or it is a starting point determined outside the relationship. As such, the intercept really reveals no more about the relationship than any other $(X, \hat{Y})$-value would. What really informs us about the overall relationship is the slope of the line; it tells how much and in what direction Y changes with X, i.e., what the relationship *is*.

Regression
Coefficient

The value b is usually referred to as the <u>regression coefficient</u>, rather than slope, since we are trying to leave the realm of pure mathematics and return to a real-world problem. We will have to change some of our terminology and be careful of our language when interpreting the value of the regression coefficient in a specific problem.

CROSS-SECTIONAL VERSUS LONGITUDINAL DATA

Cross-Sectional
Study

Note first that the sample was taken at one point in time, and thus our data are *not time-dependent*. That is, one day we looked at the sizes and prices of twenty houses; we did not look at one house on twenty different days. In comparison to psychological development and similar studies, we have a <u>cross-sectional study</u> rather than a longitudinal one. Thus, *we have not looked at how either the price or the size of any house has changed over time*. Rather, we have looked at twenty houses which are at one time *different* from each other in size and also different in price. This is entirely different from a time series study, in which an economic variable, GNP for example, is recorded for several successive years and one is interested in the growth or decline in GNP over a period of time. Both the assumptions and interpretations in regression studies differ from those in time series studies.

A COMMON MISINTERPRETATION

It is important to keep in mind that in a regression study, we are observing existing differences and not changes over time. Otherwise, our results might be misinterpreted. To illustrate why this is so, suppose a statistician has been hired by a group of local merchants to analyze sales. As part of his analysis, the statistician has obtained a regression equation $\hat{Y} = 79 + 8X$, relating Y = weekly sales in hundreds of dollars to X = number of employees in retail stores. When explaining to the merchants what the regression equation means, the statistician might say that as the number of employees in a store increases by 1, sales increase \$800 per week.

Now, what the statistician is really saying is that if he has observed one store with one more employee than another store, he has also observed a difference in sales of \$800, the larger store having the greater sales. Note carefully that the relationship between sales and number of employees is probably a function of the *size* of the store, the larger store having both more employees and greater sales.

This is what the statistician is saying, but it is probably not what the untrained observer is hearing. The merchant could easily reason, "Well, if sales increase by \$800 for every additional employee, I'll hire 25 or 30 more employees and my sales will really increase"! He could easily fill his store so full of employees that there would be no room for customers.

You might think that this is an extreme instance, but in fact it is not all that unusual for people to try to use a regression equation in this way. The problem is in the language. The words *increase, decrease, change,* and even *predict* carry a connotation of a time-dependent activity. Thus, to be strictly proper, and to avoid misleading an untrained listener, we should learn how to interpret the regression coefficient without using terms that have a time connotation.

INTERPRETING THE REGRESSION COEFFICIENT
IN THE HOUSE SIZE-PRICE EXAMPLE

It would be easy to interpret $b = 12.9$ in our regression equation by saying that as the size increases by one thousand square feet, the price increases by 12.9 thousand dollars. So long as we recognize that we really do not expect any house to grow and increase its size by 1000 square feet, there is nothing wrong with such an interpretation. However, the word *increase* implies some kind of a change over time, and this is what we often need to avoid. How can we interpret the regression coefficient without using the word *increase*?

Rather than talk about a change, we need to talk about two things which are, at a given point in time, different from each other. Thus, *if two houses differ in size by 1000 square feet* (1 X-unit), *then on the average the larger house will sell for $12,900* (12.9 Y-units) *more* (since $b = 12.9$ is a positive number) *than the smaller house.* Note that, rather than saying that an increase in size causes an increase in price, we can only report that houses of different sizes sell for different amounts of money, larger houses tending to cost more than smaller houses.

Note also that we can talk about an *average* difference in price for a 1000-square foot difference in size, because not every pair of houses 1000 square feet different in size will have prices differing by exactly $12,900. For example, House 13 has 1200 square feet and sells for $25,000 while House 19 has 2200 square feet and costs $37,000. Although the size difference is 1000 square feet, the price difference is $12,000 instead of $12,900, because we do not have a perfect linear relationship between the two variables.

5.6 RELATIONSHIPS AND CAUSAL RELATIONSHIPS

Cause-Effect Relationship

As illustrated in the example about the relationship between the number of employees in a store and the weekly sales, the careless interpretation of the meaning of a relationship often leads to the most common misuse of the results of a regression study: namely, the assumption that the existence of a relationship implies a cause-effect relationship.

As we have seen, the fact that stores with large numbers of employees also have large sales volumes does not necessarily mean that a change in the number of employees will cause a change in sales. Instead, the size of the store probably determines both the size of the sales force and the sales volume.

Is the size of a house the cause of its price? Does cigarette smoking cause lung cancer? Does depriving a rat of food cause it to learn to press a bar? Does fertilizer cause corn to grow? What is meant by "cause," anyway?

HOW CAUSE-EFFECT RELATIONSHIPS ARE ESTABLISHED

"Everyone knows" that cigarette smoking causes lung cancer in humans. But if you ask "everyone" how he "knows" this, you are likely to get one of two answers. The first is that, "The Surgeon-General says so." This is all well and good, except that proof by appeal to authority has not been an accepted method of proof since the Dark Ages. The other typical answer is that it has been established that the more cigarettes people smoke, the greater their incidence of lung cancer. Here, an appeal is made to the relationship between the number of cigarettes smoked and the incidence of lung cancer at a given age. But is this proof that smoking *causes* cancer? How would you go about setting up a scientific experiment, following the rules of the scientific method, in order to prove that smoking causes cancer? One method might be to form four groups of children, evenly matched as to a variety of physiological attributes so that one group is as much like another as possible. (Sorry, monkeys or rats won't do—who cares about the incidence of lung cancer among monkeys or rats?) The children of one group are prevented from smoking entirely, or from even breathing someone else's smoke during their entire lifetimes. The second group can be exposed to a little smoke, from someone else's cigarette or a very occasional smoke of their own. The third group is forced to smoke between one-half and $1\frac{1}{2}$ packs per day, and the fourth group is forced—whether the individuals want to or not—to smoke between $1\frac{1}{2}$ and 4 packs a day as long as they live.

Periodically, individuals from all four groups are "sacrificed" and studies made of changes in the lung tissue. For those individuals not sacrificed, their ages at time of death and causes of death are recorded. All deceased are autopsied and studies made of lung tissue.

From a controlled experiment such as that described, scientists should be able to determine whether or not smoking causes lung cancer. Obviously, however, such a study is not likely to be made.

Yes, almost everybody does believe that smoking causes lung cancer, but for more reasons than just that a statistical relationship has been found to exist. Physicians can observe changes in the cilia lining of the throat and lungs resulting from exposure to cigarette smoke, smokers themselves can feel the effects of the smoke, and so on.* Our objective is not to dispute the unhealthy effects of cigarette smoking but to point out that *a casual relationship cannot be inferred from only knowing that two variables are related*. The existence of a statistical relationship may—and often does—provide an indication that a cause and an effect are present. However, *experimentation, and not the existence of the statistical relationship, is necessary to establish a cause.*

SITUATIONS IN WHICH *X* AND *Y* WILL BE RELATED

Let us explore under what conditions we might observe that two variables are related. There are six.

1. *X* and *Y* will be related if *X* causes *Y*.

   Suppose X = the number of checkout counters open in a supermarket, and Y = the number of customers per lane. Then we will see that X and Y are inversely related because should the store manager open up an additional checkout lane, some customers will leave their lanes to go to the newly opened one, thus causing the number of customers per lane to decrease.

2. *X* and *Y* will be related if *Y* causes *X*.

   Using the same variables as before, how do you know that the situation is not just the opposite? If the number of customers per lane becomes large, this will cause the store manager to open up additional lanes, and we will see that X and Y are directly related.

---

* For the story of how it has been "determined that cigarette smoking is dangerous to your health" the reader is referred to the essay, "Statistics, Scientific Method, and Smoking," by B. W. Brown, Jr., in *Statistics: A Guide to the Unknown*, edited by Judith M. Tanur et al., Holden-Day, Inc., 1972.

3. *X* and *Y* will be related if *X* and *Y* interact with each other.

   If we carry the two examples a bit further, we see that an increase in the number of customers per lane will cause more lanes to be opened up. But opening additional lanes will cause the number of customers per lane to decrease, and this decrease in the sizes of the lines will result in some counters being closed down. Then the fewer lanes being open, the more customers per lane, and the cycle begins anew. Thus, because *X* and *Y* interact with each other, we observe that they are related.

4. *X* and *Y* will be related if both are caused by a third variable, *Z*.

   Perhaps both the number of lanes open and the number of customers per lane are determined by the number of customers in the store. Thus, when the store is busy, there will be many lanes open and a large number of customers in each lane—just the opposite when there are few shoppers. Still, we will see a relationship between *X* and *Y*.

5. *X* and *Y* will be related if they act alike just by chance.

   Suppose the growth rates of baby chicks and piglets happen to follow the same pattern. Obviously, this does not mean that pig growth influences chick growth or vice versa. In fact, all growth curves look very similar, regardless of whether an organism or an investment of interest is what is doing the growing. Suppose we have two variables for which the following values have been observed:

	X	*Y*
Sunday	2	200
Monday	5	500
Tuesday	6	600
Wednesday	3	300
Thursday	1	100
Friday	7	700
Saturday	2	200

   Obviously, *X* and *Y* are related; they are even perfectly related, since $Y = 100X$. What if you are now told that *X* = the number of pencils sold by a street vender in New York City and *Y* = the number of butterflies born in a laboratory in Calcutta? It is hard to see how any causal relationship exists between the two variables; it is just chance that the two patterns are alike. This is what we mean by a relationship—merely that two variables act alike, to some extent at least. The two variables need not have anything to do with each other; still, we can use one to predict the other.

6. *X* and *Y* can be related spuriously because of a nonrepresentative (biased) sample.

   Sometimes, even though two variables are not related in the population, the sample drawn makes it appear that a relationship exists. Suppose, as in Figure 5.5, there is no relationship between *X* and *Y*; the average *Y*-value $\mu_{y.x}$ for each

FIGURE 5.5   *How a Simple Random Sample That Is Biased Can Indicate a Relationship That Is Only Spurious*

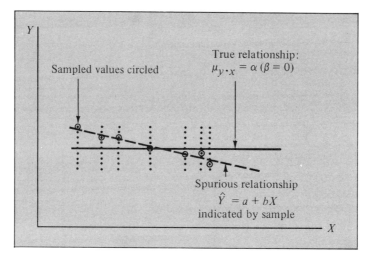

X-value lies on the horizontal line $\mu_{y.x} = \alpha$ for which $\beta = 0$. It is possible that a simple random sample could choose those items circled in Figure 5.5. Since the only available information comes from the sample, one could conclude that an inverse relationship exists, when in fact there is no relationship. This seeming relationship, caused by a nonrepresentative sample, is called a **spurious relationship**. It is the only one of the six reasons for relationship that can be handled by statistics alone.

Spurious Relationship

    We will see in Chapter 7 that one can test for the *significance* of the relationship, so that we can be reasonably confident that the relationship seen in the sample is not spurious. In each of the other five cases, further exploration is necessary, beyond statistical analysis on the given data, in order to determine whether or not a causal relationship exists and if so, the kind of relationship it is. Note carefully that we are not claiming that there exists no such thing as cause and effect. What we are saying is that causes and effects are established by investigation and experimentation, and not by statistics calculated from sample data.

EXERCISES   *Chapter 5*

p. 521   *5.1*   For the TEACHERS data, we found the regression equation $Y = -4.1414 + 7.3102X$, where X = average annual salary ($ thousands) and Y = number of thousands of teachers in southeastern states; salaries ranged from $3300 to $5000.
       a.   Estimate the number of classroom teachers for states in which the average annual teacher salary is $4000.
       b.   Interpret the meaning of the estimate in a.
       c.   Compare the estimate in a to the actual number of teachers in Alabama (27,000) and West Virginia (15,000) and explain any differences. (Both states paid average annual salary of $4000.)

d. Suppose that for the given period, the average annual teacher salary in Pennsylvania was $7000. Can the regression equation be used to estimate the number of teachers in Pennsylvania? Why or why not?

e. Suppose that ten years later, the average salary in Florida had risen to $5500. Can the regression equation be used to estimate the number of teachers in Florida at this later time? Why or why not?

f. Interpret the meaning of the regression constant for this problem.

g. Does it seem reasonable for this problem that the regression line should be fitted through the origin? Why or why not?

h. Find the regression line fitted through the origin for this problem. Plot the line on the scatter diagram, and comment on the appropriateness of the fit to the data.

i. Interpret the meaning of the regression coefficient in this problem.

j. From your analysis of the data, can you say that if a southeasthern state wants to increase its number of teachers, it should pay higher salaries? Explain.

p. 522 5.2 The PRECIPITATION data set was concerned with X = average annual number of inches precipitation and Y = average annual temperature in 15 Tennessee cities. The regression equation was $Y = 56.4928 + 0.0058X_1$ and X ranged from 51.2 to 69.2 inches.

a. Estimate the average annual temperature for a city whose average annual precipitation is 58 inches.

b. Interpret the meaning of the estimate in part a.

c. Compare the estimate in a to the actual temperature of city "Kn," whose average precipitation is 58 inches and temperature is 58.1 degrees, and explain the differences.

d. Suppose that in another Tennessee city the average annual rainfall is 40 inches. Can the regression equation be used to estimate the average temperature in this city? Explain.

e. Suppose that another city had average annual rainfall of 50 inches. Explain why the regression equation can or cannot be used to estimate the temperature in that city.

f. Interpret the meaning of the regression constant for this problem.

g. Does it seem reasonable to use a regression line fitted through the origin for this problem? Why or why not?

h. Find the regression line fitted through the origin for this problem. Plot the line on the scatter diagram for these data and comment on the fit of the line to the points.

i. Interpret the meaning of the regression coefficient in this problem.

j. Criticize the following statement: "Since the regression coefficient is a positive number, although a small one, one can see that an increase in rainfall will result in a small increase in temperature for these cities."

p. 523 5.3 In the POLICE data analysis set we found the regression equation $Y = 2010.2296 - 35.3156X + 0.2096X^2$. Here X = number of police officers (between 50 and 90) and Y = number of robberies in cities of comparable size.

a. Estimate the number of robberies in cities employing 67 police officers.

b. Interpret the meaning of the estimate in part a.

c. Compare the estimate in a to the actual number of robberies, 560 and 630, respectively, in cities numbered 3 and 7, which both employed 67 policemen. Explain any differences.

d. Suppose that another city of comparable size employed 107 policemen. Can the regression equation be used to estimate the number of robberies in that city? Why or why not?

e. Suppose that some other city of comparable size employed 45 police officers. Can the regression equation be used to estimate the number of robberies in that city? Why or why not?

f. If another city employed 60 policemen, can the regression equation be used to predict the number of robberies there? Explain.

g. Can it be said that if a city wants to reduce the number of robberies, then it should employ more policemen? Why or why not?

h. Interpret the meaning of the regression constant in this problem.

p. 524   5.4   We found the regression equation $Y = -0.3256 + 0.3842X$ to describe the relationship between X = number of employees and Y = average weekly retail sales ($ thousands) in the SALES data. The number of employees ranged from 10 to 48 in the 15 stores sampled.

a. Estimate the average weekly sales of a store which employs 17 people.

b. Interpret the meaning of the estimate in part a.

c. Compare the estimate in a to the actual sales of the two stores in the sample which had 17 employees, and explain any differences.

d. Suppose that another store employed 3 people. Can the regression equation be used to estimate sales for that store? Why or why not?

e. If another store employed 40 people, can the regression equation be used to estimate its sales? Explain.

f. If another store employed 75 people, can the regression equation be used to estimate the sales for that store? Explain.

g. Interpret the meaning of the regression constant in this problem.

h. Does it seem reasonable in this problem to use a regression line fitted through the origin? Explain.

i. Find the regression line fitted through the origin. Plot this line on the scatter diagram for these data and comment on how well the line follows the lay of the data points.

j. Interpret the meaning of the regression coefficient.

k. The manager of one of the sampled stores has been unhappy that his sales have not been larger. From looking at the results of the analysis of these data, he concludes that he can increase his sales by hiring more employees. Comment.

p. 525   5.5   We found an equation $Y = 0.7348 - 0.008X$ to describe the relationship for the HEATING data. Here, X = average daily temperature, which ranged from -2 to 50 degrees, and Y = daily heating oil bill, in dollars.

a. Estimate the daily heating oil bill for days on which the average temperature was 33 degrees.

b. Interpret the meaning of the estimate in part a.

c. Compare the estimate in a to the actual bill, 50¢, on day 5, on which the temperature averaged 33 degrees. Explain any differences.

d. Suppose that one day in January the temperature dropped to -15 degrees. Can the regression equation be used to estimate the bill on that day? Explain.

e. In April, the temperature was 76 degrees on several days. Can the regression equation be used to estimate the oil bill on those days? Why or why not?

f. Although the sample taken did not include any days for which the average temperature was 20 degrees, can the regression equation be used to estimate the bill on days in which the temperature was 20 degrees? Why or why not?

g. Interpret the meaning of the regression constant.

h. Would it be appropriate to use a regression line fitted through the origin in this problem? Explain.

   i.   Find the regression line fitted through the origin for this problem. Plot the line on the scatter diagram for these data, and comment on how well the line follows the lay of the data points.
   j.   Interpret the meaning of the regression coefficient.

5.6   Exercise 2.15 concerned an experiment in which different plots of ground received 15, 20, or 25 pounds of fertilizer in order to determine the effect on crop yield. Using X-values as given, the regression equation $\hat{Y} = 0.2222 + 3X$ was obtained; if X was scaled, the regression equation was $\hat{Y} = 60.2222 + 15Z$.
   a.   Estimate the yield per acre of plots to which 25 pounds of fertilizer was applied.
   b.   Interpret the meaning of the estimate in a.
   c.   Compare the estimate in a to the actual yield to the three plots which received 25 pounds of fertilizer. Explain any differences.
   d.   Can the regression equations $\hat{Y} = a' + b'Z$ or $\hat{Y} = a + bX$ found in Exercise 2.15 be used to estimate the yield if no fertilizer is applied? Why or why not?
   e.   Can the regression equation be used to estimate the yield from plots receiving 17 pounds of fertilizer? Explain.
   f.   Interpret the meaning of the regression constant a' in $\hat{Y} = a' + b'Z$.
   g.   Interpret the meaning of the regression constant a in $\hat{Y} = a + bX$.
   h.   Would it seem reasonable in this problem to fit a line $\hat{Y} = b*X$ through the origin? Comment.
   i.   Interpret the meaning of the regression coefficient b' in $\hat{Y} = a' + b'Z$.
   j.   Interpret the meaning of the regression constant b in $\hat{Y} = a + bX$.

5.7   For each of the relationships described below, tell whether X and Y are probably related because
          (i) X causes Y,
          (ii) Y causes X,
          (iii) X and Y interact with each other,
          (iv) X and Y are both caused by Z,
          (v) X and Y are related by chance,
          (vi)   the relationship between X and Y is spurious.
   a. X = number of television commercials run per week,
        Y = weekly sales of advertised product,
        X and Y are directly related.
   b. X = dosage level of pain-killing drug,
        Y = number of hours relief obtained by patients,
        X and Y are directly related.
   c. X = inches growth per month of pine saplings,
        Y = inches growth per month of cedar saplings,
        X and Y are directly related.
   d. X = number of employees in a store,
        Y = sales volume of store,
        X and Y are directly related.
   e. X = number of flat tires by motorists on a certain turnpike,
        Y = number of shoes thrown by race horses at a certain track,
        X and Y are directly related.

5.7 SUMMARY

In this chapter we began examination of the interpretation of the regression equation found for a given set of data. We noticed that the meaning of the regression equation depends to a great extent upon the data from which it is obtained. Only a careful examination of the data can tell the researcher what information the equation can accurately convey. Perhaps the most important concept for any student to grasp is that just because two variables are found to be related, one cannot infer that a cause-effect relationship exists without more information than just a list of X- and Y-values.

If the regression equation is to be used to estimate values of one variable from values of another, a question immediately arises: How accurately can such estimates be made? The next chapter addresses this question.

Chapter 6 MEASURING ERROR IN ESTIMATION

6.1 INTRODUCTION

Now that we know how to predict the average Y-value for any given value of X (within the range of X-values sampled, of course), the question arises as to how much variability about this mean we can expect. That is, while we can estimate the average price for houses of a specified size, how close to this average price can we expect most of the actual prices to be?

What is needed is a measure of the scatter, or *error*. In other words, we need to determine how much the actual Y-values differ from the mean Y-value because of omitted predictors, random variation, and inaccuracy of the form of the model.

6.2 ASSUMPTIONS

Let us return to the conceptualization of the population as illustrated in Figures 2.1 and 2.2, reproduced here in Figures 6.1 and 6.2. We note that for any X-value in the population, there is a subpopulation of Y-values (Figure 6.1) and the underlying linear relationship,

$$\mu_{y \cdot x} = \alpha + \beta X,$$

is a straight line passing through the mean Y-value for each X-value (Figure 6.2). Note also in both figures that the scatter of the subpopulation is homogeneous, or uniform.

THE VARIANCE OF THE Y-VALUES FOR ANY GIVEN X-VALUE

If we had the entire population at our disposal, then for any X-value we could measure the variability in Y-values by simply finding the variance of the subpopulation of Y-values associated with each X-value. For example, consider houses of size X_i. There are

FIGURE 6.1   *Scatter Diagram for the Population of Houses, from Figure 2.1*

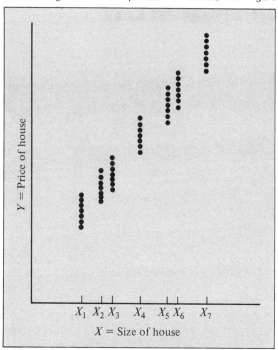

FIGURE 6.2   *Determining the Underlying Linear Relationship $\mu_{y \cdot x} = \alpha + \beta X$, from Figure 2.2*

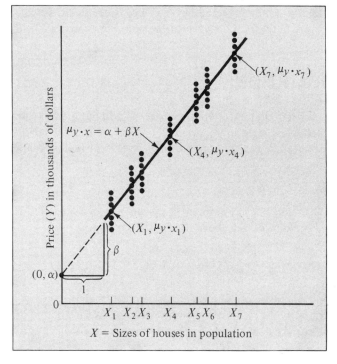

several different prices, $Y_{i1}, Y_{i2}, \ldots, Y_{iN_i}$, for which houses of size X_i sell (N_i is the number of houses of size X_i). Consider these Y-values as a small population of interest. The mean of this population is $\mu_{y \cdot x_i}$, as we have previously determined. We denote the variance of this subpopulation by

$$\sigma^2_{y \cdot x_i}.$$

To measure this variance, we proceed as usual and find the average squared deviation of the values from their mean:

$$\sigma^2_{y \cdot x_i} = \frac{\displaystyle\sum_{j=1}^{N_i} (Y_{ij} - \mu_{y \cdot x_i})^2}{N_i}.$$

That is, there are N_i Y-values associated with the X-value X_i, and we denote these as Y_{ij}, $j = 1, 2, \ldots, N_i$, the first subscript indicating that all Y-values are associated with the ith X-value.

  If the entire population were available, we could measure $\sigma^2_{y \cdot x_i}$ for each X-value and thus could know the variability of prices for every different house. This could be done regardless of whether or not the scatter of the dots is homogeneous for all X-values. If the scatter is not uniform, then each $\sigma^2_{y \cdot x_i}$-value will be different, but if scatter is homogeneous, all $\sigma^2_{y \cdot x_i}$-values will be the same. In either case, we would have a measure of how much each price differs from the average for houses of a specified size.

ESTIMATING THE VARIANCE FROM A SAMPLE

All of the above assumes that the entire population is available for our inspection, but in reality, we will have only a sample. For any X-value in our sample, we will have at best only a few observed Y-values, and in most cases only one Y-value. In addition, we will have an *estimate* of the mean Y-value for that X-value, given by

$$\hat{Y}_i = a + bX_i.$$

Recall from the scatter diagram of Figure 1.1 that for any X-value we have at best only two observed Y-values, and in all but one instance, we have only one observed price for each size house. Pick any X-value for which there is only one Y-value. How can we measure the variability in observed Y-values for that X-value if there is only *one* Y-value? This can be done only if we make a very important assumption—that the scatter of the points around the line is the same (homogeneous) for *every* X-value. This

Homoscedasticity

assumption is often referred to as the assumption of homoscedasticity. (This is a term coined by economists, whose data in many cases does not display uniform scatter, and

Heteroscedasticity

who must work in situations in which heteroscedasticity or nonhomogeneous scatter is present.)

  How will the assumption of homoscedasticity allow an estimate of the variance to be made from the sample? Recall that if scatter is uniform in the population, then all

$\sigma^2_{y \cdot x_i}$-values are equal, regardless of which X_i is being considered. Thus, we may regard our sample as a collection of subsamples from populations having equal variances, and may *pool* these subsamples to get the best possible estimate of the common variance.

The pooling of sample variances in a regression problem is a relatively straightforward procedure. For each subsample, find how far the observed value, Y_i, differs from the estimated mean value, $\hat{Y}_i$, and average these (squared) deviations across all X-values. Referring to the scatter diagram and regression line in Figure 2.8, reproduced in Figure 6.3, we find the vertical distance of each point from the line, square these distances, and average. Then $s^2_{y \cdot x}$ is the sample variance of the points around the line (their mean). Note that $s^2_{y \cdot x}$ does not measure total variability in the response variable Y, but measures variability in Y *after it has been taken into account* (by the line) *that Y varies with X.* Simply put, it measures variability around the line. Note also that this pooling could not be accomplished unless the assumption of homoscedasticity is met. For example, if $\sigma^2_{y \cdot x_1}$ were not equal to $\sigma^2_{y \cdot x_2}$, then $(Y_i - \hat{Y}_i)^2$ at $X = X_1$ would give information about $\sigma^2_{y \cdot x_1}$ and $(Y_i - \hat{Y}_i)^2$, and $X = X_2$ would give information about a *different* quantity, namely $\sigma^2_{y \cdot x_2}$. It would then make no sense to put these two pieces of information together, since they do not give information about the same thing.

THE VARIANCE OF THE POINTS AROUND THE LINE

The mathematical definition of the sample variance of Y, given X, is

$$s^2_{y \cdot x} = \frac{\sum_{i=1}^{n} (Y_i - \hat{Y}_i)^2}{n - 2}.$$

FIGURE 6.3   *Regression Line Superimposed on Scatter Diagram, from Figure 2.8*

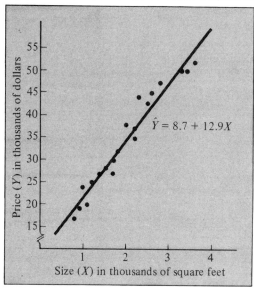

Unbiased
Estimator

Note the analogy of $s_{y \cdot x}^2$ to $\sigma_{y \cdot x}^2$. The reason for $n - 2$ in the denominator, instead of n, is that we want $s_{y \cdot x}^2$ to be an <u>unbiased estimator</u> of $\sigma_{y \cdot x}^2$. This means that $s_{y \cdot x}^2$ tends neither to overestimate nor to underestimate $\sigma_{y \cdot x}^2$. It is accurate on the average. For any particular sample, $s_{y \cdot x}^2$ may be larger or smaller than $\sigma_{y \cdot x}^2$ because of chance, but not because of anything inherent in $s_{y \cdot x}^2$ itself. Were the denominator just n, then we would tend to underestimate $\sigma_{y \cdot x}^2$, and this underestimation would not be a function of chance alone.

DEGREES OF FREEDOM

Degrees of
Freedom

The quantity $n - 2$ in the denominator $s_{y \cdot x}^2$ is called <u>degrees of freedom</u>. For an intuitive understanding of what is meant by degrees of freedom, consider first the population variance of a random variable Y:

$$\sigma_y^2 = \frac{\sum_{i=1}^{N} (Y_i - \mu_y)^2}{N}.$$

In the population, the true mean, μ_y, is known; and a variance is measured by finding the average squared deviation from the mean, which is considered a typical value. In a sample, however, the true mean can only be *estimated* using the sample itself. We would calculate a sample variance using $\sum_{i=1}^{n} (Y_i - \bar{Y})^2$ as the numerator. From this estimation procedure, we *lose a degree of freedom*. One might think of the situation as having to pick one sample value to stand in for the typical value, leaving only $n - 1$ other sample values to compare to this typical value. Thus the unbiased estimator of the sample variance is calculated as

$$s_y^2 = \frac{\sum_{i=1}^{n} (Y_i - \bar{Y})^2}{n - 1}.$$

In the regression problem, we are trying to measure the same thing—how much the various values differ from a typical value. Only in this case, the typical value is a point on the line. In order to find a line, we need two points, thus leaving only $n - 2$ other values to compare to the typical one. We have used up *two* points in order to find a typical value. Or, in order to find the line, two parameters, α and β, must be estimated by a and b from the sample. Since two estimates had to be made, two degrees of freedom were lost.

The general rule is: you lose one degree of freedom every time you have to estimate a parameter. If μ_y must be estimated by $\bar{Y}$ in order to find s_y^2, then one degree of freedom is lost; if $\mu_{y \cdot x}$ must be estimated by $Y = a + bX$, where a estimates α and b estimates β, then two degrees of freedom are lost. This rule continues to work in multiple regression, where we predict Y using some number, say k, of X-variables. The denominator of $s_{y \cdot x}^2$ will be $n - (k + 1)$, since there will be a β associated with each of the k X's and also an α. Generally, the denominator is $n - p$, where p is the number of parameters, $p = k + 1$.

THE STANDARD ERROR OF ESTIMATE

Returning to the problem at hand, any variance is a rather artificial measure, in that it averages *squared* deviations. Thus, the units of a variance are the squares of the units of the variable; as such, they often have no physical meaning or are not relevant to the problem under study. For example, if Y is price of house in thousands of dollars, then $\hat{Y}$ is also in thousands of dollars, the difference $Y - \hat{Y}$ is in thousands of dollars, and $(Y - \hat{Y})^2$ has units "square thousands of square dollars." The variance is then the average of square thousands of square dollars. To get this measure of variability back into meaningful terms, the square root is taken to form a standard deviation, whose units are the same as the units on the original variable. The standard deviation of the Y-values around the line is called the standard error of estimate and is denoted $s_{y \cdot x}$:

Standard Error of Estimate

$$s_{y \cdot x} = \sqrt{s_{y \cdot x}^2}.$$

Note that the terminology here is poor, because the standard error of estimate is *not* the standard error of any estimate. It is rather an estimate of the standard deviation of the points around the regression line and is a measure of error variation.

EXERCISES   *Section 6.2*

p. 524   *6.1*   Referring to the VALUE ADDED data, explain why the methods presented in the text for estimating error cannot be used.

p. 523   *6.2*   Referring to the POLICE data and the second-degree polynomial regression line found, how many degrees of freedom will the standard error of estimate have? Why?

6.3 CALCULATING FORMULA FOR THE STANDARD ERROR OF ESTIMATE

DERIVATION OF THE FORMULA

It is too laborious to calculate $s_{y \cdot x}$ by taking the square root of

$$s_{y \cdot x}^2 = \frac{\sum\limits_{i=1}^{n} (Y_i - \hat{Y}_i)^2}{n - 2},$$

since this formula requires that we first find $\hat{Y}_i$ for every X-value in the sample, subtract $\hat{Y}_i$ from Y_i for all n observed Y-values, square, add, and divide. Since $s_{y \cdot x}^2$ is a variance, there is a shortcut calculating formula for it, as there are for all variances. Let us derive a calculating formula. Consider the numerator:

$$\sum_{i=1}^{n} (Y_i - \hat{Y}_i)^2 = \sum_{i=1}^{n} (Y_i - a - bX_i)^2.$$

Recall that $a = \bar{Y} - b\bar{X}$, so we can write

$$\sum_{i=1}^{n} [Y_i - (\bar{Y} - b\bar{X}) - bX_i]^2 = \sum_{i=1}^{n} [(Y_i - \bar{Y}) - b(X_i - \bar{X})]^2$$

$$= \sum_{i=1}^{n} [(Y_i - \bar{Y})^2 - 2b(Y_i - \bar{Y})(X_i - \bar{X})$$

$$+ b^2(X_i - \bar{X})^2]$$

$$= \sum_{i=1}^{n} (Y_i - \bar{Y})^2 - 2b \sum_{i=1}^{n} (Y_i - \bar{Y})(X_i - \bar{X})$$

$$+ b^2 \sum_{i=1}^{n} (X_i - \bar{X})^2$$

$$= S_{YY} - 2bS_{XY} + b^2 S_{XX}.$$

Now recall that $b = S_{XY}/S_{XX}$, so that

$$b^2 S_{XX} = \left(\frac{S_{XY}}{S_{XX}}\right)^2 S_{XX} = \frac{(S_{XY})^2}{(S_{XX})^2} S_{XX} = \frac{(S_{XY})^2}{S_{XX}}$$

$$= \frac{S_{XY}}{S_{XX}} S_{XY} = bS_{XY}.$$

Then we have that

$$\sum_{i=1}^{n} (Y_i - \hat{Y}_i)^2 = S_{YY} - 2bS_{XY} + bS_{XY}$$

$$= S_{YY} - bS_{XY}.$$

Formula for $s_{y \cdot x}^2$ The calculating formula for $s_{y \cdot x}^2$ is

$$s_{y \cdot x}^2 = \frac{S_{YY} - bS_{XY}}{n - 2}.$$

STANDARD ERROR OF ESTIMATE FOR HOUSE SIZE-PRICE PROBLEM

Let us illustrate the use of this formula by calculating the standard error of estimate for our problem concerning the relationship between house size and price. One of the beauties of this formula is that most of the quantities involved have already been calculated: we found in Chapter 2 that

$$S_{XY} = 174.9, \quad b = 12.8982 \quad \text{and} \quad n = 20.$$

We must now find

$$S_{YY} = \frac{n \sum Y_i^2 - (\sum Y_i)^2}{n}.$$

In Table 2.1 we calculated $\sum Y_i$ and $\sum Y_i^2$, although we have not yet used $\sum Y_i^2$. Thus

$$S_{YY} = \frac{20(26{,}178) - (690)^2}{20} = 2373.$$

Then

$$s_{y \cdot x}^2 = \frac{2373 - 12.8982(174.9)}{20 - 2} = \frac{117.1048}{18} = 6.5058,$$

and the standard error of estimate is

$$s_{y \cdot x} = \sqrt{6.5058} = 2.5506,$$

or approximately \$2550.

EXAMINATION OF THE CALCULATING FORMULA

Obviously, the calculating formula requires much less effort than the formula

$$s_{y \cdot x}^2 = \frac{\sum\limits_{i=1}^{n} (Y_i - \hat{Y}_i)^2}{n - 2},$$

since this definitional formula requires the calculation of 20 $\hat{Y}_i$ values, 20 subtractions $Y_i - \hat{Y}_i$, 20 squares $(Y_i - \hat{Y}_i)^2$, and then finally an averaging of these values. Additionally, the calculating formula helps explain exactly what is being calculated when a standard error of estimate is found. Note that S_{YY} is a measure of the total amount of variation in the Y-values, since $s_y^2 = S_{YY}/(n-1) = \sum(Y_i - \bar{Y})^2/(n-1)$ is the variance of Y and thus tells the average squared amount that the Y-values differ from their mean. In the standard error of estimate, we do not want to measure total variability in Y, but we want to measure how much the Y-values vary *around the line*, or how much Y-values vary *after we have taken into account that Y varies partly because it is related to X, and X also varies*. Naturally, then, we expect $s_{y \cdot x}^2$ to be smaller than s_y^2. So from S_{YY}, the total variability in Y, we subtract bS_{XY}, a measure of how much Y varies because of X. This latter quantity involves the slope of the line, b, and the numerator of the covariance of X and Y, S_{XY}. The units of the product bS_{XY} are

$$\frac{Y\text{-units}}{X\text{-units}} \cdot (X \text{ units} \times Y \text{ units}),$$

or Y-units squared, and thus measure variance of Y because of relationship with X. The difference $S_{YY} - bS_{XY}$ thus measures variance of Y, having adjusted for, or taken out, variation in X. Much more will be made of this concept when we study correlation in Chapter 8.

6.4 INTERPRETATION OF THE STANDARD ERROR OF ESTIMATE

We have seen that $s_{y \cdot x}$ measures the variation in Y after we have accounted for variation because of the relationship with X. What does this mean in terms of the problem of relating prices of sizes of houses?

The relationship of Y with X is described by the regression equation $\hat{Y} = a + bX$, which gives the estimated average price of houses of a given size. Naturally one does not expect all prices to be equal to the mean price, because of random variation and such other characteristics as location and age, which have not been considered. The $2550 figure found for $s_{y \cdot x}$ is a measure of how much the prices in the sample, on the average, differ from the mean price because of the factors we call error. Given our assumption that scatter is homogeneous around the line, then regardless of the size of house being considered, the easiest way to interpret $s_{y \cdot x} = 2.55$ is that *the average amount that the actual prices differ from the estimated mean price is $2550*. Some houses will differ more than $2550 from the predicted mean price, some houses will be closer to the estimated mean price than $2550; some houses may even cost exactly as much as the estimated mean price. But if we should calculate how much each house in the sample differs from the predicted mean price and then average these differences, this average difference would be about $2550. (Actually, the average amount that the actual prices differ from the estimated mean price is measured by $\sum |Y_i - \hat{Y}_i| / n$, and not by $s_{y \cdot x} = [\sum (Y_i - \hat{Y}_i)^2 / (n - 2)]^{1/2}$.) The correct interpretation of $s_{y \cdot x}^2$ is the unbiased estimate of the average squared deviation (variance) about the line, but this technically correct interpretation requires reasonably extensive training in statistics to understand.

THE SIZE OF THE ERROR

Is this amount of variation excessive? Is it so large that the estimates are essentially useless? Is $2550 a large amount of error, or is it small enough that we can have some measure of confidence that our estimates are close enough to be useful? Obviously, too large an error in estimation indicates a very weak relationship between the sizes and the prices of the houses under consideration, and we will not gain much information concerning price from knowing the size of a house.

An error of $2550 can be relatively small or quite large. For example, an error of $2550 in estimating the cost of putting a man on the moon would be almost negligible, since the total cost would run to millions or billions of dollars. But an error of $2550 in estimating the cost of a hamburger is so large as to be ridiculous. Thus, the error is large or small depending upon the magnitude of the values involved. If the houses in our sample cost between $17,000 and $52,000, is $2550 a large or a small amount of error?

We might compare the size of the error to the average, or typical, price from our sample. We found

$$\bar{Y} = \frac{690}{20} = 34.5 \text{ thousand dollars.}$$

Thus, as compared to the typical price, the error in estimation is

$$\frac{2.55}{34.5} = 0.0739$$

Coefficient of Variation

or $7\frac{1}{3}\%$. You may notice that this measure of the percentage error, $s_{y \cdot x}/\bar{Y}$, is analogous to the coefficient of variation, $s_y/\bar{Y}$ from your introductory course, where it was computed to determine the amount of variability relative to the sizes of the values.

Now, whether $7\frac{1}{3}\%$ error in estimation is excessive or not depends upon the nature and intended use of the data at hand. If for a given problem the percentage error in estimation is considered too large, can we find another predicting equation for which the standard error of estimate is smaller? No, because the method of least squares gives the line which is, on the average, closest to all the data points in terms of squared deviations; it gives the smallest average squared errors and the most accurate predictions possible. Recall that in Chapter 2 we found the regression line by minimizing the quantity

$$Q = \sum (Y_i - \hat{Y}_i)^2 = \sum (Y_i - a - bX_i)^2.$$

We recognize this now as the line that minimizes error in predictions. It is the line for which $s_{y \cdot x}$ is smallest.

EXERCISES   *Sections 6.3 and 6.4*

p. 521   6.3   For the TEACHERS data, refer to the Analyses of Data Sets section and
                    a.   find the standard error of estimate and state its units,
                    b.   interpret the meaning of the standard error of estimate,
                    c.   calculate the coefficient of variation; does the error seem large?

6.5 INTERVAL ESTIMATES: A REVIEW

INTERVAL ESTIMATES

The equation

$$\hat{Y} = a + bX$$

Interval
Estimate

estimates the mean Y-value for a given value of X, and $s_{y \cdot x}$ gives us the average amount that the Y-values differ from $\hat{Y}$ at each X-value. We now want to put these two measures together and obtain an interval estimate of the mean price of houses of a given size or of the actual price of a house of a specified size. Recall that interval estimates are often used to obtain a measure of the confidence we have in our estimate. We have no confidence that $\hat{Y}$ is exactly accurate for the true mean price or for any individual price, since any different sample would give a different value for $\hat{Y}$. But an interval estimate gives a range of values in which we can be reasonably confident that the true mean price or some individual price lies.

THE NORMALITY ASSUMPTION

Confidence
Interval

In order to obtain a confidence interval estimate of price, given size, we must return to the population and make another important assumption. Recall our assumption that there is a perfect linear relationship between X and the average Y-value for each X, described by

$$\mu_{y \cdot x} = \alpha + \beta X.$$

We have also assumed that the scatter of Y-values around $\mu_{y \cdot x}$ is uniform, regardless of the value of X. That is, we assumed several subpopulations of Y-values, each with possibly different means $\mu_{y \cdot x_i}$, but all with the same variance, $\sigma^2_{y \cdot x}$.

Normal
Population

In addition, we will now assume that each subpopulation of Y-values is a normal population. That is, for any X-value, say X_i,

$$Y \sim N(\mu_{y \cdot x_i}, \sigma^2_{y \cdot x}),$$

which is read, "Y is distributed as a normal random variable with mean $\mu_{y \cdot x_i}$ and variance $\sigma^2_{y \cdot x}$." This assumption is illustrated in Figure 6.4. Is this a realistic

FIGURE 6.4   *Values of Y Are Normally Distributed about the Line at Each X-Value*

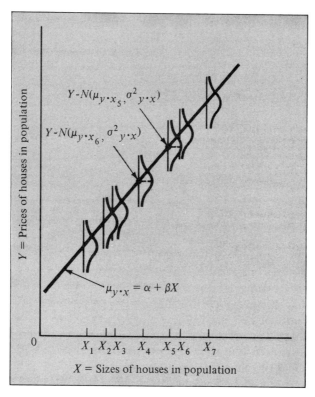

assumption? In many cases, it can be justified: First, it might seem reasonable that for a particular size of house we would expect a concentration of prices around the mean price and fewer prices observed farther away from the mean. Later we shall check the reasonableness of this assumption for a given set of data. For now, it is enough to know that if this assumption is justified, then the distribution of Y-values is bell-shaped and centered on the line (at $\mu_{y \cdot x}$).

In addition, it might be reasonable to assume that prices larger than the mean and those less than the mean behave in the same way, with their frequency decreasing at the same rate as we move farther from the mean. This assumption will make the distribution of Y-values symmetrical about the line. From the assumption of symmetric bell-shaped distributions of Y-values at each X-value, then, it is not too great a leap to call the distributions normal.

Recall the relationships that exist for all normal distributions:

approximately 68% of the values lie within one standard deviation of their mean;

approximately 95% of the values lie within two standard deviations of their mean;

approximately 99% of the values lie within three standard deviations of their mean.

Restating these relationships in symbols, approximately 68% of all normal values lie in the interval

$$\mu_{y \cdot x_i} - \sigma_{y \cdot x} \quad \text{to} \quad \mu_{y \cdot x_i} + \sigma_{y \cdot x};$$

approximately 95% of all normal values lie in the interval

$$\mu_{y \cdot x_i} \pm 2\sigma_{y \cdot x};$$

and

$$P(\mu_{y \cdot x_i} - 3\sigma_{y \cdot x} < Y < \mu_{y \cdot x_i} + 3\sigma_{y \cdot x}) > 99\%,$$

where P represents probability.

Exact probabilities can be found in tables of the normal distribution. For example, *exactly* 95% of all normal values lie within 1.96 standard deviations of their mean.

From the relationships above we can say that in the population, the probability that the price of any house of a specified size, X_i, lies in the interval

$$\mu_{y \cdot x_i} \pm 1.96\sigma_{y \cdot x}$$

is 95%. Equivalently, 95% of all prices of houses of size X_i are between $\mu_{y \cdot x_i} - 1.96\sigma_{y \cdot x}$ and $\mu_{y \cdot x_i} + 1.96\sigma_{y \cdot x}$. Or, we are 95% confident that the price of a house of size X_i is between $\mu_{y \cdot x_i} - 1.96\sigma_{y \cdot x}$ and $\mu_{y \cdot x_i} + 1.96\sigma_{y \cdot x}$. Only a very few houses of size X_i (about $2\frac{1}{2}\%$) cost less than $\mu_{y \cdot x_i} - 1.96\sigma_{y \cdot x}$; and only $2\frac{1}{2}\%$ of all houses of size X_i cost more than $\mu_{y \cdot x_i} + 1.96\sigma_{y \cdot x}$. The extreme values are attributed to random variation and features we have not taken into account.

THE CENTRAL LIMIT THEOREM

Now suppose we take a simple random sample of n_i houses, all of size X_i, and calculate the mean price of these houses, call it $\overline{Y}_i$. We hope that $\overline{Y}_i$ will be close to $\mu_{y \cdot x_i}$, but we do not expect $\overline{Y}_i$ to be exactly equal to $\mu_{y \cdot x_i}$. Someone else taking another sample of the same size from this subpopulation will get a different value of $\overline{Y}_i$ from the one we obtained, and the $\overline{Y}_i$ of this sample will probably not be exactly equal to $\mu_{y \cdot x_i}$ either. Imagine a great many people taking samples of size n_i from this subpopulation of houses of size X_i. When we put all the $\overline{Y}_i$-values together, we have a large collection of different values, some of which will actually be equal to $\mu_{y \cdot x_i}$. Many of them will be very close, and there will be about as many over-estimates of $\mu_{y \cdot x_i}$ as under-estimates. Most values of $\overline{Y}_i$ will tend to cluster around $\mu_{y \cdot x_i}$.

Central Limit
Theorem

You may recognize that we are describing the Central Limit theorem, which says that the collection of sample means tends to be normally distributed around the population mean, but that the variance of the $\overline{Y}_i$-values will not be as great as the variance of the original Y_i-values. That is, if the collection of Y_i-values is normally distributed with mean $\mu_{y \cdot x_i}$ and variance $\sigma_{y \cdot x}^2$, then the collection of $\bar{Y}_i$-values based on samples of size n_i is also normally distributed with mean $\mu_{y \cdot x_i}$ but with variance $\sigma_{y \cdot x}^2 / n_i$. In symbols, if

$$Y_i \sim N(\mu_{y \cdot x_i}, \sigma_{y \cdot x}^2),$$

then

$$\bar{Y}_i \sim N(\mu_{y \cdot x_i}, \sigma_{y \cdot x}^2 / n_i),$$

where $\bar{Y}_i$ is the mean of a sample of n_i values drawn from houses of size X_i.

Now, if we could know the values of $\mu_{y \cdot x_i}$ and $\sigma_{y \cdot x}^2$, then we could say that 95% of all $\bar{Y}_i$-values will lie within $1.96 \sigma_{y \cdot x} / \sqrt{n_i}$ of the true mean value $\mu_{y \cdot x_i}$. That is,

$$P(\mu_{y \cdot x_i} - 1.96 \sigma_{y \cdot x} / \sqrt{n_i} < \bar{Y}_i < \mu_{y \cdot x_i} + 1.96 \sigma_{y \cdot x} / \sqrt{n_i}) = 0.95.$$

Similar statements could be made about the intervals containing 68% or 99% of the $\bar{Y}_i$-values.

The foregoing theoretical argument assumed that (1) we were interested in only one subpopulation, say houses of size X_i and their mean price; (2) we had access to this entire subpopulation and thus could know the true values of $\mu_{y \cdot x_i}$ and $\sigma_{y \cdot x}^2$; and (3) we could draw many samples of size n_i and calculate the sample mean value $\bar{Y}_i$ for each.

This is very far from our actual situation, in which (1) we are interested in many mean values, not just in the mean price of houses of only one size; (2) we do not have access to the entire population or even to any one subpopulation and thus cannot know $\mu_{y \cdot x_i}$ or $\sigma_{y \cdot x}^2$; and (3) we have only *one* sample of size n, drawn from houses of many sizes, and we can calculate $\hat{Y}_i = a + bX_i$ to estimate the mean price of houses of size X_i. How can we make the theory relevant to our problem?

A CONFIDENCE INTERVAL FOR ONE SUBPOPULATION

First, we note that our objective is to estimate $\mu_{y \cdot x_i}$ using an interval estimate of the form

number $< \mu_{y \cdot x_i} <$ number,

to which we can attach a measure of confidence. From the statement

$$P(\mu_{y \cdot x_i} - 1.96 \sigma_{y \cdot x} / \sqrt{n_i} < \bar{Y}_i < \mu_{y \cdot x_i} + 1.96 \sigma_{y \cdot x} / \sqrt{n_i}) = 0.95,$$

we can obtain such an interval. Take the inequality in parentheses and play around with it until $\mu_{y \cdot x_i}$ is by itself in the middle, as shown below.

Subtract $\mu_{y \cdot x_i}$ from each term:

$$P\left(\frac{-1.96\sigma_{y \cdot x}}{\sqrt{n_i}} < \overline{Y}_i - \mu_{y \cdot x_i} < \frac{1.96\sigma_{y \cdot x}}{\sqrt{n_i}}\right) = 0.95.$$

Now subtract $\overline{Y}_i$ from each term:

$$P\left(-\overline{Y}_i - \frac{1.96\sigma_{y \cdot x}}{\sqrt{n_i}} < -\mu_{y \cdot x_i} < -\overline{Y}_i + \frac{1.96\sigma_{y \cdot x}}{\sqrt{n_i}}\right) = 0.95.$$

Now multiply all three terms by -1, to get rid of the minus on $-\mu_{y \cdot x_i}$. This will reverse the direction of the inequalities:

$$P\left(\overline{Y}_i + \frac{1.96\sigma_{y \cdot x}}{\sqrt{n_i}} > \mu_{y \cdot x_i} > \overline{Y}_i - \frac{1.96\sigma_{y \cdot x}}{\sqrt{n_i}}\right) = 0.95.$$

We now have $\mu_{y \cdot x_i}$ by itself in the middle, but the larger bound is on the left and the smaller on the right. Rewriting the inequality so as to lead off with the smaller number,

$$P\left(\overline{Y}_i - \frac{1.96\sigma_{y \cdot x}}{\sqrt{n_i}} < \mu_{y \cdot x_i} < \overline{Y}_i + \frac{1.96\sigma_{y \cdot x}}{\sqrt{n_i}}\right) = 0.95.$$

This then is a *95% confidence interval* on $\mu_{y \cdot x_i}$. If $\sigma_{y \cdot x}$ were known, then 95% of all samples taken from the subpopulation of houses of size X_i will yield $\overline{Y}_i$-values such that the interval

$$\overline{Y}_i \pm \frac{1.96\sigma_{y \cdot x}}{\sqrt{n_i}}$$

covers $\mu_{y \cdot x_i}$. Therefore the probability that any *one* sample will give an interval of this form which contains $\mu_{y \cdot x_i}$ is 0.95. Note that this is *still* for only *one* subpopulation, and we are still assuming that $\sigma_{y \cdot x}^2$ is known.

IF $\sigma_{y \cdot x}^2$ IS UNKNOWN

Working yet with houses of only *one* size, suppose that, as is more realistic; $\sigma_{y \cdot x}^2$ is not known, but that the only information we have about the subpopulation comes from our sample of size n_i. We could then estimate $\sigma_{y \cdot x}^2$ by s_y^2. Note that since we are working with only one subpopulation, we estimate the variance by just

$$s_y^2 = \frac{\sum(Y_i - \overline{Y}_i)^2}{n_i - 1}.$$

This procedure introduces more variability into our estimate of $\mu_{y \cdot x_i}$, because we must consider not only how much $\overline{Y}_i$ varies from sample to sample, but also how much s_y^2 varies from sample to sample. The interval estimate of $\mu_{y \cdot x_i}$ must then be wider than

$$\overline{Y}_i \pm \frac{1.96\sigma_{y \cdot x}}{\sqrt{n_i}},$$

which means that the coefficient on the standard deviation must be larger than 1.96. How large should it be?

6.6 THE t-DISTRIBUTION

COMPARISON OF THE t TO THE STANDARD NORMAL

Student's
t-Distribution

Since $\sigma_{y \cdot x}$ is no longer known but must be estimated by s_y, we can no longer refer to the normal distribution, but must instead refer to the Student's t-distribution.* The t-distribution looks very much like the standard normal distribution. It is also bell-shaped, symmetrical, and centered at zero. However, a t-distribution is shorter and fatter, and its variance is larger than that of the standard normal (which is 1). Further, while there is only one standard normal distribution, there are many t-distributions, one for each different positive integer. In this case, the integers are called *degrees of freedom*. The smaller the number of the degrees of freedom, the shorter and fatter (more variable) the distribution. The greater the number of degrees of freedom, the taller and skinnier the distribution, until for an infinite number of degrees of freedom, the t and the normal are the same. Figure 6.5 compares some selected t-distributions with the standard normal distribution.

FIGURE 6.5   *Comparison of Two t-Distributions with the Standard Normal*

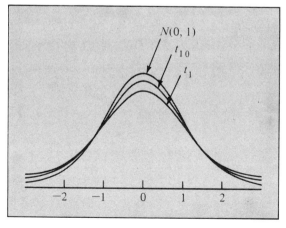

*The t-distribution was invented by W. S. Gossett, who was forced to use the pseudonym Student because his employer, the Guinness Brewery of Dublin, Ireland, would not permit him to publish his research under his own name.

We can see in the figure that the distribution of a t with one degree of freedom (denoted t_1) has more area in its "tails" than does the standard normal. Thus, to find the points that cut off $2\frac{1}{2}\%$ of the area under the upper and lower tails we shall have to go farther out than ± 1.96. But we shall not have to go as far out on t_{10} as on t_1. Let $t_{\alpha,\nu}$ denote that value of t with ν degrees of freedom which cuts off $100\alpha\%$ of the area under the upper tail of the distribution. That is, by definition,

$$P(t_\nu > t_{\alpha,\nu}) = \alpha.$$

Refer to Figure 6.6. By symmetry around zero,

$$P(t_\nu < -t_{\alpha,\nu}) = \alpha,$$

or if $t_{\alpha,\nu}$ cuts off the $100\alpha\%$ of the area under the upper tail, then $-t_{\alpha,\nu}$ cuts off $100\alpha\%$ of the area under the lower tail. (Do not confuse this α with the intercept in the model. Here, α represents a small probability, or a small percentage. We believe the context will make clear which α is being referred to.)

TABLES OF THE t-DISTRIBUTION

Table I, p. 517 gives values of the t-distribution. Across the top row are values of α, and down the outside column are listed degrees of freedom. From the example given on the tables, we see that

$$P(t_{10} > 2.228) = 0.025.$$

By symmetry around zero,

$$P(t_{10} < -2.228) = 0.025$$

FIGURE 6.6   *Definitions of $t_{\alpha,\nu}$*

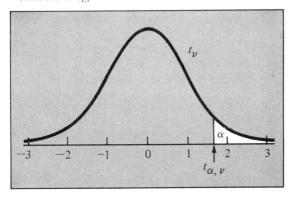

also. (The reader should be careful when using other tables—not all of them are set up like this one. It is necessary to look closely at the table headings in order to determine exactly what areas are being given.) As another example, the t-values which cut off the upper and lower 5% of the area under a distribution with 20 degrees of freedom are

$$t_{.05,20} = 1.725 \quad \text{and} \quad -t_{.05,20} = -1.725,$$

respectively. For degrees of freedom not given in the tables, one can usually use the t-value corresponding to the next lowest number. Note that for degrees of freedom equal to infinity, the values ± 1.96 cut off the upper and lower $2\frac{1}{2}\%$ of the area. Thus, the numbers corresponding to the standard normal numbers are given in the bottom row of the t-tables. Note also that for any of the percentages indicated in the table, all t-values are larger than the standard normal values. Using t-values in a confidence interval will thus give a wider interval than will using standard normal values.

CONFIDENCE INTERVALS USING THE t-DISTRIBUTION

Recall that if $\sigma_{y \cdot x}$ were known, a 95% confidence interval on $\mu_{y \cdot x_i}$ would be given by

$$\bar{Y}_i \pm \frac{1.96\sigma_{y \cdot x}}{\sqrt{n_i}}.$$

If $\sigma_{y \cdot x}$ is not known but must be estimated from the sample of size n_i by s_y, then the 95% confidence interval on $\mu_{y \cdot x_i}$ is

$$\bar{Y}_i - \frac{t_{.025,n_i - 1}s_y}{\sqrt{n_i}} < \mu_{y \cdot x_i} < \bar{Y}_i + \frac{t_{.025,n_i - 1}s_y}{\sqrt{n_i}}.$$

The degrees of freedom for t are $n_i - 1$, and this comes from the *denominator of s_y^2*. The degrees of freedom for a t always equal the denominator of the estimate of the variance.

Confidence Coefficient   Interval estimates with other <u>confidence coefficients</u> can be found by picking the t with the appropriate percentage value.

In general, a $100(1 - \alpha)\%$ confidence interval on the mean of *one* subpopulation, based on a sample of size n_i from that one subpopulation, is given by

$$\bar{Y}_i \pm \frac{t_{\alpha/2,n_i - 1}s_y}{\sqrt{n_i}}.$$

This is still not what we really want to find, however. What we have is a sample of size n which has drawn observations from many subpopulations (from houses of many different sizes). We have found a line $\hat{Y} = a + bX$ which will allow us to make a point estimate $\hat{Y}_i$ of $\mu_{y \cdot x_i}$ for *any* selected value of X_i. What we want then is an interval estimate of $\mu_{y \cdot x_i}$ based on $\hat{Y}_i = a + bX_i$.

EXERCISES   *Section 6.6*

6.8   Using Table I, find the indicated *t*-values.
   a. $t_{.05,20}.$
   b. $-t_{.01,27}.$
   c.   That value of *t* with 12 degrees of freedom which cuts off the smallest 10% of the values.
   d.   That value of *t* with 25 degrees of freedom which is exceeded by only 0.5% of the values.
   e.   Those values of *t* with 6 degrees of freedom which cut off the smallest $2\frac{1}{2}\%$ and the largest $2\frac{1}{2}\%$ of the values.

6.7 CONFIDENCE INTERVAL ON $\mu_{y \cdot x_i}$ BASED ON $\hat{Y}_i$

THE FORM OF THE INTERVAL

Notice that the general form of a confidence interval based solely on the information provided in a sample (where variance is unknown) is

point estimate $\pm$ (*t*-value)(estimate of variance of point estimate).

In order to estimate $\mu_{y \cdot x_i}$, then, we know that the point estimate is $\hat{Y}_i$. What we need now is to find an estimate of the variance of $\hat{Y}_i = a + bX_i$. Note that the values of both *a* and *b* vary from sample to sample. It can be shown that the variance of $\hat{Y}_i$ is

$$s_{y \cdot x}^2 \left[\frac{1}{n} + \frac{(X_i - \bar{X})^2}{S_{XX}} \right].$$

Note that X_i in the formula is that value of X whose mean we want to estimate; it denotes a particular chosen size.

We have that a $100(1 - \alpha)\%$ confidence interval on $\mu_{y \cdot x_i}$, using $\hat{Y}_i$, is given by

$$\hat{Y}_i - t_{\alpha/2, n-2} s_{y \cdot x} \sqrt{\frac{1}{n} + \frac{(X_i - \bar{X})^2}{S_{XX}}} < \mu_{y \cdot x_i} < \hat{Y}_i + t_{\alpha/2, n-2} s_{y \cdot x} \sqrt{\frac{1}{n} + \frac{(X_i - \bar{X})^2}{S_{XX}}},$$

where the $n - 2$ degrees of freedom for *t* comes from the denominator of $s_{y \cdot x}^2$.

CONFIDENCE INTERVAL FOR HOUSE SIZE–PRICE EXAMPLE

Let us work an example, using our house size-price data. Suppose we want a 95% confidence interval on the mean price of houses 2200 square feet in size. We have found

$$\hat{Y}_i = 8.7036 + 12.8982(2.2) = 37.0796;$$

$$s_{y \cdot x} = 2.5506, \quad n = 20, \quad S_{XX} = 13.56;$$

$$t_{.025,18} = 2.101, \quad \bar{X} = 2.$$

Then

$$\hat{Y}_i \pm t_{.025,18} s_{y \cdot x} \sqrt{\frac{1}{n} + \frac{(X_i - \bar{X})^2}{S_{XX}}}$$

$$= 37.0796 \pm 2.101(2.5506)\sqrt{\frac{1}{20} + \frac{(2.2 - 2)^2}{13.56}}$$

$$= 37.0796 \pm 1.2331.$$

We are 95% sure that the mean price of houses 2200 square feet in size is between

$$37.0796 - 1.2331 = 35.8465 \approx \$35,846$$

and

$$37.0796 + 1.2331 = 38.3127 \approx \$38,313.$$

EXAMINING THE CONFIDENCE LIMITS

Note that the farther X_i is from $\bar{X}$, the larger the interval will become because of the term $(X_i - \bar{X})^2$ in the numerator of the variance of $\hat{Y}_i$. For example, if $X_i = 2.2$, the interval is of width $2(1.2331) = 2.4662$; but if $X_i = 3.5$, then the interval is of width

$$2(2.101)(2.5506)\sqrt{\frac{1}{20} + \frac{(3.5 - 2)^2}{13.56}} = 2(2.101)(2.5506)\sqrt{0.05 + 0.16}$$

$$= 2(2.4557) = 4.9114.$$

Table 6.1 shows confidence limits for several selected values of X_i. Thus, if we plotted $100(1 - \alpha)\%$ confidence limits for each X_i in the sample, these limits would not lie on

TABLE 6.1   *Confidence Limits for $\mu_{y \cdot x_i}$ for Selected Values of X_i.*

X_i	$(X_i - \bar{X})^2$	$\dfrac{(X_i - \bar{X})^2}{S_{XX}}$	$\left[\dfrac{1}{n} + \dfrac{(X_i - \bar{X})^2}{S_{XX}}\right]^{1/2}$	$ts_{y \cdot x}\left[\dfrac{1}{n} + \dfrac{(X_i - \bar{X})^2}{S_{XX}}\right]^{1/2}$	$\hat{Y}_i$	*Lower Limit*	*Upper Limit*
0.8	1.44	0.1062	0.3952	2.1178	19.0222	16.9044	21.1400
1.0	1.00	0.0737	0.3517	1.8847	21.6018	19.7171	23.4865
1.5	0.25	0.0184	0.2615	1.4013	28.0509	26.6496	29.4522
2.0	0.00	0.0000	0.2236	1.1982	34.5000	33.3018	35.6982
2.2	0.04	0.0030	0.2302	1.2336	37.0796	35.8460	38.2030
2.5	0.25	0.0184	0.2615	1.4013	40.9491	39.5478	42.3504
3.0	1.00	0.0737	0.3517	1.8847	47.3982	45.5135	49.2829
3.5	2.25	0.1659	0.4646	2.4897	53.8473	51.3576	56.3370
3.6	2.56	0.1888	0.4887	2.6188	55.1371	52.5183	57.7559

Confidence
Bands
straight lines parallel to the regression line; rather, they would curve, as illustrated in Figure 6.7. We call these lines <u>confidence bands</u>. Note that the vertical distance between the upper band and the lower band is smallest at $X_i = \bar{X}$, and this distance becomes greater the farther X_i is from $\bar{X}$. What this says, essentially, is that we have less trustworthy information about the relationship between size and price for houses at the lower and upper ranges of sizes in our sample; we trust the regression line better for those houses in the middle range. Imagine rotating the regression line shown in Figure 6.7 clockwise, pivoting at the point $(\bar{X}, \bar{Y})$, until the line touches the confidence bands. Then rotate the line counterclockwise until it touches the bands again. We feel that the true line can lie anywhere between these two extremes. Thus we can pinpoint $\mu_{y \cdot x_i}$ fairly accurately as long as X_i is near $\bar{X}$, but not so accurately if X_i is rather far away from $\bar{X}$.

FIGURE 6.7   *Confidence Bands for $\mu_{y \cdot x_i}$*

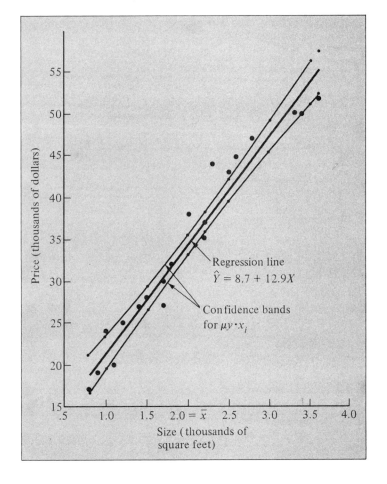

6.9 a. Refer to the TEACHERS data in the Analyses of Data Sets section. Give a 95% confidence interval for the average number of teachers in states paying average annual salary of $3900. Interpret the meaning of this interval, in terms specific to the problem.
b. Refer to the PRECIPITATION data. Give a 99% confidence interval on the average temperature for cities receiving 55 inches of precipitation. Interpret the meaning of this interval, in terms specific to the problem.
c. Using the SALES data, graph the 95% confidence bands for mean sales on the scatter diagram with regression line drawn in Exercise 2.9.
d. Graph the 99% confidence bands for the mean daily bill on the scatter diagram with regression line drawn in Exercise 2.9 from the HEATING data.
e. Confidence bands on $\mu_{y \cdot x_i}$ are the shortest distance apart at $X_i = \bar{X}$, and the farther X_i is from $\bar{X}$ the wider the bands become, for a specified confidence coefficient. Discuss the implication of this to the use of the regression equation to estimate Y-values for X-values outside the range of data sampled.

6.8 CONFIDENCE INTERVAL FOR AN INDIVIDUAL Y-VALUE

We can use the regression equation $\hat{Y} = a + bX$ to estimate the price of a single house of any size within our sampled range. While we realize that our estimated price, $\hat{Y}_i$, is an estimate of the mean price of houses of size X_i, we would like to obtain an interval in which we expect some large percentage of *all* prices to lie for houses that size, not just the mean price. Let Y_i^* denote the actual price of a house of size X_i. We want to find an interval of the form

number $< Y_i^* <$ number

Prediction
Interval
in which we can be, say, 95% sure that actual price will lie. Some people call this a prediction interval.
This interval will be wider than the one for $\mu_{y \cdot x_i}$, because sample means tend to cluster much closer around the true mean value than do individual observations. It can be shown that if we use $\hat{Y}_i$ to estimate the true value of Y for a given $X = X_i$, then the variance of $\hat{Y}_i$ is

$$s_{y \cdot x}^2 \left[1 + \frac{1}{n} + \frac{(X_i - \bar{X})^2}{S_{XX}} \right].$$

Again, the closer X_i is to $\bar{X}$, the smaller this interval will be. If we choose $X_i = 2.2$, then a

95% confidence interval on an individual Y_i-value, Y_i^*, is given by

$$\hat{Y}_i \pm t_{.025, n-2} s_{y \cdot x} \sqrt{1 + \frac{1}{n} + \frac{(X_i - \bar{X})^2}{S_{XX}}}$$

$$= 37.0796 \pm 2.101(2.5506)\sqrt{1 + \frac{1}{20} + \frac{(2.2 - 2)^2}{13.56}}$$

$$= 37.0796 \pm 5.4990.$$

Thus we claim that 95% of all houses 2200 square feet in size will sell for between

$$37.0796 - 5.4990 = 31.5806 \approx \$31,581$$

and

$$37.0796 + 5.4990 = 42.5786 \approx \$42,579.$$

Referring to our sample, we note that houses 5 and 19 are both 2200 square feet in size and that they sold for $35,000 and $37,000, respectively. Since both of these prices lie in the interval from $31,581 to $42,579, we see that neither house is priced unusually high—or low—for its size (at least, they are not among the $2\frac{1}{2}$% lowest or highest priced houses).

EXERCISES   *Section 6.8*

p. 521   6.10 a. Referring to the TEACHERS data, give a 95% confidence interval on the number of teachers in a particular state paying average annual salary of $3900. Interpret its meaning. Compare to the interval obtained in Problem 6.9a and explain the difference.

p. 522   b. Referring to the PRECIPITATION data, give a 99% confidence interval for the temperature of a particular city if it has 55 inches of precipitation. Interpret the meaning of this interval, in terms specific to the problem. The city "Ne" had 55 inches of precipitation and temperature 56.3 degrees; would it be considered unusual in temperature? Explain.

p. 524   c. Give a 95% prediction interval for the sales of a particular store which employs 50 people, using the SALES data. Interpret the meaning of this interval.

p. 525   d. Referring to the HEATING data, suppose that on a given day the temperature was 72 degrees. Form a 99% prediction interval on the heating oil bill and interpret it, with reference to the range of temperatures represented in the sample.

6.9 FINDING THE STANDARD ERROR OF ESTIMATE ON COMPUTER PRINTOUT

All of the simple regression computer programs we discussed in Chapter 4 give the standard error of estimate, and it is usually not very difficult to locate.

BMD PROGRAMS

For the BMDP1R program for our house size-price problem, refer to Figure 4.3, p.116. It is hard to miss that STD. ERROR OF EST. = 2.5506, since this figure is printed at the top of the analysis. For the BMDP2R, the standard error of estimate is shown in Step 1, in the third row from the top, Figure 4.7, p. 120.

SPSS PROGRAMS

To find the standard error of estimate when using the SPSS package, refer to Figure 4.15, p. 136, and the list of simple statistics given at the bottom of the scatter diagram. The second statistic given is

STD ERR OF EST = 2.55059,

so there is no trouble finding it here. In addition, the standard error can be found in Figure 4.16 in Step 1 of the regression routine, printed out under the label STANDARD ERROR.

SAS PROGRAMS

Using the SAS printout, refer to Figure 4.22, p. 149. Under ERROR MEAN SQUARE we find $s_{y \cdot x}^2 = 6.50553097$, and below and to the right of this figure we find

STD DEV

2.55059424

for the standard error of estimate. Note that under a column labeled STD ERROR OF ESTIMATE, we find two figures, but these have to do with how much a and b vary from sample to sample from the same population and are not equal to $s_{y \cdot x}$. We will see in the next chapter that these are important measures.

The SAS and the BMD calculate a coefficient of variation, C.V. = 7.3930, and the coefficient of variation can be calculated by hand from the SPSS once $\overline{Y}$ and $s_{y \cdot x}$ have been located.

EXERCISES   *Section 6.9*

6.11   Using the output of either the BMD, the SPSS, or the SAS programs run in Chapter 4, find the standard error of estimate for
p. 521   a. the TEACHERS data,
p. 522   b. the PRECIPITATION data,
p. 524   c. the SALES data,
p. 525   d. the HEATING data.
Compare these values to those obtained in your previous analyses.

6.12   Using the output of either the BMD, the SPSS, or the SAS programs run in Chapter 4, find the coefficient of variation for
a. the TEACHERS data,
b. the PRECIPITATION data,
c. the SALES data,
d. the HEATING data.
Compare these values to those obtained in your previous analyses.

6.13   Using the BMD, the SPSS, or the SAS programs run in Chapter 4, compare the standard error found for the data set you are analyzing when X was corrected for its mean to the standard error when X was not corrected for its mean. Does correcting X for its mean affect the standard error of estimate?

6.10 SUMMARY

We now know how far we can expect our estimates of Y based on X to differ from the actual observed Y-values, and can use this information to obtain interval estimates on both the typical Y-value and an individual Y-value at a given value of X.

The strength of a relationship is determined not only by the slope of the regression line, but also by the amount of scatter (error) around the line. In order to see how strong the relationship really is, and especially to see if it is strong enough to rule out the possibility that it is spurious—that is, coincidental—we turn to the study of tests of significance in Chapter 7.

Chapter 7

INFERENCE IN SIMPLE LINEAR REGRESSION

7.1 INTRODUCTION

Central to the study of regression are hypothesis tests to determine whether or not we can be reasonably confident that our two variables are, in fact, related. We realize the possibility that we could have drawn a sample which indicates a relationship that does not exist in the population. We will review statistical tests of hypotheses, illustrating each step with reference to the house size-price example. The confidence intervals we consider this time have to do with the parameters of the model rather than with predicted values.

7.2 CAN WE BE CONFIDENT THAT A RELATIONSHIP EXISTS?

STRENGTH OF THE LINEAR RELATIONSHIP

In obtaining the regression equation relating prices to sizes of houses, we found $b = 12.9$. This was interpreted to mean that if two houses differed in size by 1000 square feet, then the larger house would cost on the average \$12,900 more than the smaller house. Our standard error of estimate, $s_{y \cdot x} = \$2550$, gave the average amount of error we could expect in estimating price from size alone.

Now, a difference of 1000 square feet in the floor space of two houses is quite a considerable difference. While a price difference of \$12,900 also seems large, is it *really* a large difference in price for two houses that are 1000 square feet different in size? Is a difference of \$12.90 per square foot really large when we consider that the average variation around the mean price is \$2550, regardless of the size of the house? In other words, *taking the size of the error into account, is b = 12.9 large enough for us to be reasonably sure that a linear relationship between size and price really exists?*

We know that it is possible to draw a sample indicating a relationship when, in fact, there is no relationship in the population. With a simple random sample, there is always the chance, however small, that the sample relationship is a spurious one. How can we tell in a particular problem if the value obtained for b is large enough (in absolute value) for us to be reasonably sure that our sample did not come from a population in which $\beta = 0$? That is, in a given problem, is b significantly different from zero? Are the chances of observing $b = 12.9$ when there is no relationship so small that we can be quite confident that the relationship is real? In order to answer this question, we turn to the problem of hypothesis, or *significance*, tests on β.

Significant
Difference

Hypothesis
Tests

Actually, this question should be the first one answered in any regression study. If it is found that a relationship actually does exist, then it makes sense to use the regression equation to make estimates and to measure the size of the error involved in estimation. However, without reasonable confidence that the two variables are related, it hardly makes sense to predict one from the other or to measure error. We have delayed this central question until now only because it was necessary to calculate b, $s_{y \cdot x}$, and several other quantities in order to answer the question.

ADEQUACY OF THE MODEL

No less central a question is whether or not the data fit the proposed model. For our house size-price data, we assumed a simple linear regression model

$$Y = \alpha + \beta X + \varepsilon$$

and found the regression equation and standard error of estimate corresponding to this model. We need to make reasonably certain that this model is appropriate to describe the relationship. From our inspection of the scatter diagram of the data, the relationship appeared to be reasonably linear; but we noted that price appears to increase with size at a decreasing rate—a leveling-off of prices can be seen from the largest houses in the sample. We opted for the simple linear model, but we should make sure our choice was appropriate.

We first address the question of whether β is significantly different from zero; later we use similar techniques to test the adequacy of the model.

7.3 HYPOTHESIS TESTS ON THE REGRESSION COEFFICIENT

Some review of the procedure for testing hypotheses may be in order. Generally, there are five steps to any hypothesis test, as outlined below.

STEP 1   *Set Up Null and Alternative Hypotheses*

Null Hypothesis

The null hypothesis is the hypothesis of no effect, formulated for the express purpose of being *rejected*, or disproved. In simple regression problems, the null hypothesis will typically be

$$H_0: \beta = 0.$$

This states that there is no relationship in the population, and we hope that this is a false statement.

Alternative
Hypothesis
    The alternative hypothesis is a statement of the experimental hypothesis, or what we suspect is true and want to test. We hope our sample will provide sufficient evidence that we can *accept* the alternative hypothesis. For simple regression studies, the alternative may be any of the following, depending on the particular problem:

$$H_1: \beta > 0$$

or $H_1: \beta < 0$

or $H_1: \beta \neq 0.$

If we hope to establish a *direct* relationship, then our alternative hypothesis is $H_1: \beta > 0$; but if we seek an inverse relationship, the alternative $H_1: \beta < 0$ is appropriate. In some instances we have no prior suspicion, or are indifferent to whether the relationship is direct or inverse. We simply want to know if any relationship exists. Then we can use the alternative $H_1: \beta \neq 0$.

    By setting up the hypotheses, we state what the problem is. In the house size-price example, we probably want to know if $b = 12.9$ indicates a true direct relationship. Thus the null and alternative hypotheses are

$$H_0: \beta = 0 \quad \text{(no relationship)},$$

$$H_1: \beta > 0 \quad \text{(direct relationship)}.$$

Most computer programs use the alternative $H_1: \beta \neq 0$ so that the same program can be used whether the problem involves an inverse or a direct relationship.

STEP 2   *Choose a Significance Level*

We can never be 100% certain of our conclusion unless we have the entire population at hand, so we must specify what level of certainty is required in a particular problem in order to go on and use the regression equation to make predictions. A certainty of 95% or 99% is often required, although a minimum certainty of 90% is sometimes acceptable. It is not the problem of the statistician to specify the required degree of certainty; rather, it is up to the researcher. Any one researcher can employ a variety of degrees of certainty for various problems, and any two researchers can require different degrees of confidence for the same problem, as well. The researcher's personality and the nature of the particular problem and of its consequences determine how certain one must be of the results: life-and-death consequences require higher degrees of confidence than do preliminary pilot studies, and researchers with cautious personalities tend to require the higher degrees of certainty.

Significance
Level

Once the degree of confidence has been specified for a particular problem, the significance level is the difference between this percentage and 100%: for example, if 95% certainty is required, then the significance level, denoted by α (not to be confused with the α in our model $Y = \alpha + \beta X + \varepsilon$), is 0.05, or 5%. This is the probability of rejecting H_0 (and thus accepting H_1) when, in fact H_0 is true. That is, α is the probability of concluding that a relationship exists when there is no relationship in the population. If we *can* reject H_0, then we can be at least $100(1 - \alpha)$% (e.g., 95%) sure that a relationship does exist. However, if we cannot reject H_0, this does not mean that we are, say, 95% sure that there is no relationship. *If we cannot reject H_0, then we cannot be at least 95% sure that a relationship does exist.* We might be only 93% sure of a relationship, but if α is specified to be 0.05% then being 93% sure is not enough. The significance level is specified prior to collection and analysis of data and, as stated before, is the *largest* probability of falsely rejecting a null hypothesis deemed acceptable in the problem at hand. If 95% certainty is specified, the researcher cannot hedge and say, "Oh well, I guess 93% is okay." If 93% certainty is acceptable, this should have been specified at the outset, and *not* after analysis of data. Otherwise one commits the scientific sin of forcing the data to show what is wanted, and objectivity is lost.

STEP 3   *Determine the Critical Region*

Critical Region
Rejection
Region

The critical region, or rejection region, consists of a set of values that constitute sufficient evidence for the researcher to be $100(1 - \alpha)$% sure that H_1 is true. Usually, and always for our purposes, the critical region must be determined by using statistical tables. The statement in the alternative hypothesis and the level of significance must both be taken into account in determining the rejection region. For regression problems, we shall use tables of the Student's t-distribution and Fisher's

F-Distribution F-distribution.*

CRITICAL REGIONS USING THE t-DISTRIBUTION

In Chapter 6 we used the t-distribution to form confidence intervals. It can also be used to test hypotheses concerning the values of α and β in the model

$$Y = \alpha + \beta X + \varepsilon.$$

The t-statistic essentially compares the sample regression coefficient b to an hypothesized value of β, usually zero. A difference $b - \beta$ close to zero is evidence that H_0 is true. Values far away from zero, either positive or negative, give evidence that H_0 is false and thus H_1 is true. If $H_1: \beta > 0$, we need to see that t is much *larger* than zero; if $H_1: \beta < 0$, we need to see that t is considerably *less* than zero: and $H_1: \beta \neq 0$ can be established as true if t is *either* a positive number sufficiently larger than zero *or* a negative number much less than zero. To determine how large (in absolute value) t must be in order to be sufficiently different from zero, we refer to the significance level, α.

* The F-distribution was named for its inventor, Sir Ronald A. Fisher, one of the founders of statistics as a tool for science.

We defined $t_{v,\alpha}$ to be that value of t with v degrees of freedom that cuts off the largest $100\alpha\%$ of all t-values. (See diagram at top of Table 1.) That is, only $100\alpha\%$ of all t_v-values are greater than $t_{v,\alpha}$. Then if we observe a value greater than or equal to $t_{v,\alpha}$, chances are at least $100(1 - \alpha)\%$ that β really is greater than zero; i.e., that $H_1: \beta > 0$ is true. The critical region corresponding to the hypotheses

$$H_0: \beta = 0$$

$$H_1: \beta > 0$$

is thus:

reject H_0 if $t > t_{v,\alpha}$.

Recall that regardless of the number of degrees of freedom, the t-distribution is symmetric around the point $t = 0$. Thus, if $t_{v,\alpha}$ cuts off the largest $100\alpha\%$ of all t_v-values, $-t_{v,\alpha}$ cuts off the smallest $100\alpha\%$. Since chances are small that we would observe one of the $100\alpha\%$ smallest possible t-values if $H_0: \beta = 0$ is true, we take such an observation as evidence that H_0 is false and that $H_1: \beta < 0$ is true. That is, if

$$H_0: \beta = 0$$

$$H_1: \beta < 0,$$

then the critical region is:

reject H_0 if $t < -t_{\alpha,v}$.

For the two-sided test, either very large *or* very small t-values give evidence that $H_0: \beta = 0$ is false. Looking for the *most extreme $100\alpha\%$* of all t-values, we look at the same time for the *$100(\alpha/2)\%$ largest* and the *$100(\alpha/2)\%$ smallest*. Thus, if

$$H_0: \beta = 0$$

$$H_1: \beta \neq 0,$$

we reject H_0 if $t > t_{v,\alpha/2}$ or $t < -t_{v,\alpha/2}$.
For example, if $v = 15$, $\alpha = 0.05$, and

$$H_0: \beta = 0$$

$$H_1: \beta > 0,$$

then the critical region is:

reject H_0 if $t > t_{15,.05} = 1.753$.

If

$$H_0 : \beta = 0$$
$$H_1 : \beta < 0,$$

then reject H_0 if $t < -t_{15,.05} = -1.753$.

If

$$H_0 : \beta = 0$$
$$H_1 : \beta \neq 0,$$

then, since $0.05/2 = 0.025$, reject H_0 if

$$t > t_{15,.025} = 2.131 \quad \text{or} \quad t < -t_{15,.025} = -2.131.$$

Note carefully that the two-sided t-test has a two-sided critical region.

CRITICAL REGIONS USING THE *F*-DISTRIBUTION

Sometimes the hypotheses $H_0 : \beta = 0$, $H_1 : \beta \neq 0$ are tested using an F-statistic instead of a t-statistic. In simple regression, an F-statistic is the square of a t-statistic. Thus, F's are always positive, and large positive F-values are evidence that H_1 is true. Since F is never negative, we cannot test for the direction of the relationship but can use F only when the alternative is $H_1 : \beta \neq 0$. Standard computer programs usually test using the two-sided alternative and the F-test.

An F-statistic has two sets of degrees of freedom, denoted v_1 and v_2, and usually called degrees of freedom for the numerator and degrees of freedom for the denominator, respectively. (F is a ratio of variances, as will be seen in Step 4.) In simple regression, v_1 is always 1 and v_2 is always $n - 2$. We recognize $n - 2$ as the degrees of freedom for $s_{y \cdot x}^2$ and when we study multiple regression, v_1 will be the number of independent variables, k, and v_2 will be $n - p = n - (k + 1)$.

Table II of the Appendix, p. 000, gives critical values of the F-distribution. Note that there is one table for $\alpha = 0.01$ and another for $\alpha = 0.05$. More extensive F-tables are available elsewhere. Both tables list v_1 values across the top and v_2 values down the left column. If $F_{v_1, v_2, \alpha}$ is that value of F with v_1 and v_2 degrees of freedom which cuts off the largest $100\alpha \%$ of all F_{v_1, v_2}-values, then to test

$$H_0 : \beta = 0$$
$$H_1 : \beta \neq 0,$$

we reject H_0 if $F > F_{v_1, v_2, \alpha}$.

For example, if $\alpha = 0.05$ and $v_1 = 1$, $v_2 = 15$, then reject H_0 if $F > F_{1,15,.05} = 4.54$. Note that $t_{15,.025}$ was found to be 2.131, and that $(2.131)^2 = 4.54$; that is, $t_{v,\alpha/2}^2 = F_{1,v,\alpha}$ and the two-sided t- and F-tests are equivalent. Note carefully that the two-sided F-test has a *one-sided* critical region.

STEP 4   *Take a Sample and Calculate the Value of a Test Statistic*

Note that the hypotheses, significance level, and critical region have all been determined prior to the actual collection of data. That is, Steps 1–3 define the problem and the evidence necessary to determine if a relationship exists. In Step 4 we actually gather evidence and boil the entire sample down into one value, a t- or an F-statistic.

THE t-STATISTIC

t-statistic   If $H_1: \beta > 0$ or if $H_1: \beta < 0$ we will test with a t-statistic. The alternative $H_1: \beta \neq 0$ can be tested with either statistic, as noted in Step 3. The t-statistic is

$$t = \frac{b - \beta}{\sqrt{s_{y \cdot x}^2 / S_{XX}}}$$

The numerator, $b - \beta$, compares the sample evidence b to an hypothesized value for β, the *value stated in the null hypothesis*. If we are testing for the existence of a relationship, then

$$H_0: \beta = 0$$

and β will be zero in the t-statistic.

(It is also possible to test $H_0: \beta = \beta_0$, where β_0 is some value other than zero, using the t-statistic; in such cases one substitutes the value β_0 for β in the t-statistic. For example, suppose that for a given city an equation

$$\hat{Y} = 1.25 + 0.75X$$

is found relating income (X) in hundreds of dollars and weekly amount spent on beer (Y) in dollars. Then $b = 0.75$ is a *marginal propensity to consume* (MPC), and the equation is called a *consumption function*. Suppose further that in a nationwide study, the MPC for beer is found to be 0.50, and the question arises as to whether the MPC for this city is significantly greater than that for the nation. Then

$$H_0: \beta = 0.50$$
$$H_1: \beta > 0.50$$

and the numerator of t would be $b - \beta = 0.75 - 0.50$. While such tests are easily done using the t-statistic, our primary concern in regression is to show that $H_0: \beta = 0$ is false. Note that F can test H_0 only for $\beta = 0$.)

The numerator of t is in units Y/X, since b and β both measure differences in Y for differences in X. Thus the size of the difference $b - \beta$ depends upon the units in which X and Y are measured. It is thus necessary to take units of measurement into account when deciding if the difference between b and β is large enough to be attributable to something other than chance. Note that the denominator of t is also in units Y/X: $s_{y \cdot x}^2$ is in Y^2-units, S_{XX} is in X^2-units, and then the square root is taken. Thus the denominator of t takes units of measurements into account and corrects for them, making the comparison of b to $\beta = 0$ independent of the units of measurement of either X or Y.

For example, if X is an income variable measured in dollars, then b might be 0.02. But if the same data are used with X measured in thousands of dollars, so that an income of \$16,000 is called just 16, then b would be 20. Both values of b measure the same relationship although they seem different because of the difference in units. The t-statistic eliminates reflection of units in b and enables a unitless comparison of b to zero to be made.

Standard Error of the Regression Coefficient

The denominator of t, $[s_{y \cdot x}^2/S_{XX}]^{1/2}$, called the standard error of the regression coefficient, estimates how much b will vary from sample to sample of the same size from the same population. Imagine several different people all sampling 20 randomly chosen houses from this same city. Because each sample is different, each person will obtain a slightly different value for b. If the standard deviation of all these b-values were calculated, then we would have an estimate of how much b varies from sample to sample, and this would be the standard error of b. Using this standard error in the denominator of the t-statistic gives us a statistic that estimates how many standard deviations b is away from $\beta = 0$. The tabulated values of t tell how many (estimated) standard deviations difference is considered too extreme to be attributed solely to chance.

Further study of the denominator of t shows where the degrees of freedom, v, for the t-statistic come from. In the denominator of t is $s_{y \cdot x}^2$, whose own denominator (which we called degrees of freedom in Chapter 6) is $n - p$; thus t has $v = n - 2$ degrees of freedom in simple regression problems.

In our example concerning prices of houses of varying sizes, we found

$$b = 12.8982,$$

$$S_{XX} = 13.56,$$

$$s_{y \cdot x}^2 = 6.5058.$$

The value $b = 12.9$ appears to be larger than zero, but we must remove units to make the comparison. Calculating the t-statistic,

$$t = \frac{12.8982 - 0}{\sqrt{\dfrac{6.5058}{13.56}}} = 12.8982 \sqrt{\frac{13.56}{6.5058}}$$

$$= 12.8982 \sqrt{2.0843} = 12.8982(1.4437)$$

$$= 18.6212.$$

This value represents a condensation of all the sample data into one figure which we may compare in size to zero. Note that we estimate that b is 18.6212 standard deviations away from $\beta = 0$.

THE F-STATISTIC

F-Statistic   The F-statistic can only be used to test $H_0: \beta = 0$, and it accomplishes the same "deunitizing" as does the t-statistic. Since an F is the square of a t in simple regression, consider

$$t^2 = \left[\frac{b - 0}{\sqrt{s_{y \cdot x}^2 / S_{XX}}} \right]^2 = \frac{b^2 S_{XX}}{s_{y \cdot x}^2} = \frac{(S_{XY}^2 / S_{XX}^2) S_{XX}}{s_{y \cdot x}^2}$$

$$= \frac{S_{XY}^2 / S_{XX}}{s_{y \cdot x}^2} = \frac{b S_{XY}}{s_{y \cdot x}^2} = F.$$

The numerator of F measures the relationship between X and Y, while the denominator measures the variation in Y due to error. Thus F compares variation in Y due to its relationship with X to variation in Y due to error. If the relationship is strong enough to overcome error variation, then F will be large enough for us to reject $H_0: \beta = 0$.

In the denominator of F is $s_{y \cdot x}^2$, with degrees of freedom $n - p$, or $n - 2$ in simple regression. This accounts for the value of v_2, the denominator degrees of freedom for the F-ratio. The term in the numerator represents the relationship of Y to *one* predictor variable X. Thus $v_1 = 1$. We shall see in multiple regression that v_1 will always equal the number of predictor variables, and that the relationship of Y with all k predictors will be measured in the numerator of F.

SUMS OF SQUARES AND MEAN SQUARES

Regression   Often, the quantity $b S_{XY}$ is referred to as the mean square due to regression, or
Mean Square   regression mean square. The quantity $s_{y \cdot x}^2$ in the denominator is called the mean square due to error, or error mean square. F then is the ratio of the regression mean square to

Error Mean   the error mean square, or
Square
$$F = \frac{MS(R)}{MS(E)}.$$

The reasons for this terminology are fairly obvious. Consider

$$s_{y \cdot x}^2 = \frac{\sum (Y_i - \hat{Y}_i)^2}{n - 2}.$$

The numerator, $\sum (Y_i - \hat{Y}_i)^2$, is a *sum of squares*. When a sum is divided by the number of terms, a mean is obtained; thus $s_{y \cdot x}^2$ is a *mean* of squares, or mean square. Since the squared quantities are errors, $s^2_{y \cdot x}$ is an error mean square. Similarly, $bS_{XY} = b^2 S_{XX}$, as was shown in the derivation of F from the square of a t, and $S_{XX} = \sum(X_i - \bar{X})^2$ is also a sum of squares. Since S_{XX} gets multiplied by b^2, the numerator measures relationship of Y with X(regression). The degrees of freedom are 1, so that the sum of squares gets divided by 1 to become a mean square. Another name for mean square is simply variance. Thus, the F-statistic compares variation due to regression to variation due to error.

ANALYSIS OF VARIANCE

All of these pieces can be shown nicely in their respective roles if one calculates the F-statistic by means of an *analysis of variance table.* As the name implies, we are analyzing the variation in Y into its component parts: one part due to relationship with X and one

ANOVA

part due to error. The general form of the analysis of variance (ANOVA) table for a simple regression is:

ANOVA

Source of Variation	Sum of Squares	Degrees of Freedom	Mean Square	F
Regression	$SS(R) = bS_{XY}$	1	$MS(R) = \dfrac{bS_{XY}}{1} = bS_{XY}$	$\dfrac{MS(R)}{MS(E)} = \dfrac{bS_{XY}}{s^2_{y \cdot x}}$
Error	$SS(E) = S_{YY} - bS_{XY}$	$n - 2$	$MS(E) = \dfrac{S_{YY} - bS_{XY}}{n - 2} = s^2_{y \cdot x}$	
Total	$SS(T) = S_{YY}$	$n - 1$		

Note that total variation in Y is the variance of Y, or $s^2 y = S_{YY}/(n - 1) = SS(T)/(n - 1)$.

Our example about house sizes and prices produces the following table.

ANOVA

Source	SS	df	MS	F
Regression	12.8982(174.9) = 2255.8952	1	2255.8952	346.75
Error	117.1048 (by subtraction)	18	6.5058	
Total	2373	19		

Here, $F = 346.75$ means that the variation due to regression is an estimated 346.75 times as great as the variation due to error. Note that we found $t = 18.6212$, and that $t^2 = 346.75 = F$.

STEP 5   *Compare Test Statistic to the Critical Region and Make Appropriate Conclusion*

TESTING FOR DIRECT RELATIONSHIPS

We might want to test whether or not a significant *direct* relationship exists between prices and sizes of houses. The hypotheses are

$$H_0: \beta = 0$$

$$H_1: \beta > 0.$$

If α is chosen to be 0.05, then we will reject H_0 if $t > t_{18,.05} = 1.734$. In Step 4, we found $t = 18.6212$; since $t > 1.734$, we reject H_0 and conclude we are more than 95% sure that there is a direct relationship between the size of a house and its price. Thus, the larger the house, generally the greater its price. Thus we may use the regression equation to estimate price from size, and can know the average error involved in the estimates.

Suppose, on the other hand, that t had been found to be some small number, say $t = 1.234$. Then since $t \not> 1.734$, we *could not* reject H_0 and we *could not* be 95% sure that house prices are related to size. In this case, we would not be sufficiently certain that a relationship exists to be willing to use the regression equation to estimate price from size alone.

Note that if t is not greater than the critical value, we *do not* conclude that we are 95% sure that no relationship exists. This is not at all the case. There is *some* evidence of a relationship, but not enough to be 95% confident that it is real.

TESTING FOR INVERSE RELATIONSHIPS

In the house size-price problem, one would probably not hypothesize the existence of an inverse relationship, but should one wish to test

$$H_0: \beta = 0$$

$$H_1: \beta < 0,$$

then H_0 will be rejected if $t < -t_{18,.05} = -1.734$. We found $t = 18.6212$, and t is *not* less than -1.734. Thus, we cannot be 95% sure that an inverse relationship exists.

Let us consider an instance in which it would be reasonable to test for an inverse relationship. Suppose $X = $ price charged for an item in various locations and $Y = $ number of items sold per week. Suppose further that the regression equation, based on 32 observations, was found to be

$$\hat{Y} = 178.4 - 4.96X,$$

with $s_{y \cdot x}^2 = 4$ and $S_{XX} = 16$. We want to see if we can be 95% sure that fewer items are sold in locations charging higher prices. Then

$$H_0: \beta = 0$$

$$H_1: \beta < 0,$$

and if $\alpha = 0.05$, we will reject H_0 if $t < -t_{.05,30} = -1.697$. The test statistic is

$$t = \frac{-4.96 - 0}{\sqrt{4/16}} = \frac{-4.96}{\sqrt{1/4}} = \frac{-4.96}{1/2} = 2(-4.96) = -9.92.$$

Since $-9.92 < -1.697$, we reject H_0 and conclude we are more than 95% sure that locations charging higher prices sell fewer items than those charging less.

TESTING TWO-SIDED ALTERNATIVES

If the objective is to determine whether or not *any* relationship is present, either direct or inverse, then the hypotheses are

$$H_0: \beta = 0$$

$$H_1: \beta \neq 0;$$

and either a t- or an F-test can be used. For the t-test in the house size-price example, H_0 is rejected if $t > t_{18,.025} = 2.101$ or if $t < -t_{18,.025} = -2.101$. Then since $t = 18.6212 > 2.101$, we reject H_0 and conclude we are more than 95% sure that the price of a house is related to its size. (Note that in the conclusion we do not specify the direction of the relationship, although we can tell the direction by noting the sign of t.)

If the F-test is used, then we must be testing the two-sided alternative to $H_0: \beta = 0$ and we reject H_0 if $F > F_{1,18,.05} = 4.41$. Since $F = 346.75 > 4.41$, reject H_0 and conclude we are more than 95% sure that the sizes and the prices of houses are related. (Recall that we cannot tell, just by looking at the value of F, whether the relationship is direct or inverse.)

Obviously the first question to be answered in a regression problem is whether or not a relationship exists. Unless one can be reasonably confident that the two variables are related, it certainly does not make sense to make estimates of one from the other, measure the error involved, and interpret the regression constant and the regression coefficient. The reader can see, however, that the question of significance of the relationship had to be postponed until after the study of error, because so many intermediate quantities had to be calculated in order to find the t- and F-statistics.

Section 7.3

p. 521   *7.1*   Refer to the TEACHERS data.
  a.   Use a *t*-test and $\alpha = 0.05$ to test for a significant direct relationship. Show all five steps of the hypothesis test and state conclusion in terms specific to the problem.
  b.   Repeat part a, but test for the presence of a relationship.
  c.   Repeat part b, but use an *F*-test. Calculate the *F*-statistic in an analysis of variance table.

p. 522   *7.2*   Refer to the PRECIPITATION data and repeat Exercise 7.1.

p. 524   *7.3*   Refer to the SALES data and repeat Exercise 7.1, but use $\alpha = 0.01$.

p. 525   *7.4*   Refer to the HEATING data. Repeat Exercise 7.1, but in part a, test for a significant inverse relationship.

p. 526   *7.5*   Test to see if you can be 99% sure that the FERTILIZER data show a direct relationship. Use *Z*-values.

7.4 TESTS ON THE REGRESSION CONSTANT

It is also possible to test hypotheses concerning α, the regression constant. However, at best α gives very specific, as opposed to general, information about the relationship—it gives the average *Y*-value at a specified *X*-value, $X = 0$; but often α doesn't even have any physical meaning to the problem. There is, however, an instance in which one is interested in testing an hypothesis concerning the regression constant. For some pairs of variables, if $X = 0$, it seems reasonable that *Y* should also be zero. For example, if X = number of items sold and Y = revenue from sales of the item, then it stands to reason that if no items are sold, then no revenue will be accrued. However, because of sampling variability, the fitted regression line could show an intercept different from zero. If the nature of the problem is such that it seems reasonable that the line should pass through the point $(0,0)$, then we might want to test

$$H_0: \alpha = 0$$

against some alternative ($H_1: \alpha < 0$ or $H_1: \alpha > 0$ or $H_1: \alpha \neq 0$). If H_0 cannot be rejected, then we have no evidence that the line should not be fitted through the origin, and thus failure to reject H_0 can provide some justification for adopting the model

$$Y = \beta^* X + \varepsilon$$

and fitting an equation of the form

$$\hat{Y} = b^* X,$$

where $b^* = \sum X_i / Y_i / \sum X_i^2$ (refer to Chapter 5 and Appendix D).

The statistic for testing an hypothesis concerning α is

$$t = \frac{a - \alpha}{\sqrt{s_{y \cdot x}^2 (1/n + \bar{X}^2/S_{XX})}}$$

with $n - 2$ degrees of freedom. The mechanics of the test are exactly the same as for the t-test on β. It is not always necessary to test $H_0: \alpha = 0$. The test statistic allows for an hypothesized value other than zero—but it is seldom of interest to test any other null hypothesis. The denominator of the t-statistic above is the <u>standard error of the regression constant</u>. It estimates how much the value of a varies from sample to sample.

Standard Error of the Regression Constant

In our house size-price example, it might seem reasonable to suppose that a house of no size should sell for zero dollars. Let us test to see whether we might want to fit the regression line through the origin:

$$H_0: \alpha = 0$$

$$H_1: \alpha > 0.$$

If $\alpha = 0.05$, reject H_0 if $t > t_{.05, 18} = 1.734$.

$$t = \frac{8.7036 - 0}{\sqrt{6.5058[1/20 + (2)^2/13.56]}} = \frac{8.7036}{1.4981} = 5.8098.$$

Since $5.8098 > 1.734$, we can be more than 95% sure that the intercept is above zero. Thus we will not adopt the model $Y = \beta^* X + \varepsilon$. Recall that in Chapter 5 we fitted a regression line through the origin for these data and saw that this line did not follow the data points very well at all. Perhaps zero thousand square feet is not the actual lower bound on size.

EXERCISES   *Section 7.4*

p. 521   7.6   a.   Referring to TEACHERS data, test to see whether or not you can be 95% sure that the intercept $a = -4.1414$ is below the origin. Conclude whether you would use the regression equation

$$\hat{Y} = -4.1414 + 7.3102X; \text{ or would you use}$$

$$\hat{Y} = 6.2908X?$$

p. 522   b.   Referring to PRECIPITATION data, test to see whether or not you can be 95% sure that the intercept $a = 56.4928$ is positive. Conclude whether you would use the regression equation

$$\hat{Y} = 56.4928 + 0.0058X; \text{ or would you use}$$

$$\hat{Y} = 0.924X?$$

    c.   Referring to SALES data, test to see whether or not you can be 95% sure that the intercept $a = -0.3256$ is different from zero. Would you use the regression equation

$$\hat{Y} = -0.3256 + 0.3842X, \text{ or would you use}$$

$$\hat{Y} = 0.3749X ?$$

7.7   a.   Referring to the TEACHERS data and using the regression equation fitted through the origin, $\hat{Y} = 6.2908X$, test to see if you can be 95% sure that there is a direct relationship.

       b.   Repeat part a using the SALES data.

7.5 MATRIX APPROACH

THE ANALYSIS OF VARIANCE TABLE

When the regression problem is formulated in terms of matrices and vectors, we should be able to do the hypothesis tests in terms of matrices and vectors, as well. This will be helpful to us when the need arises to test hypotheses concerning regression parameters in more complicated models.

Recall that we could write and solve normal equations using

$$\mathbf{y} = \begin{bmatrix} Y_1 \\ Y_2 \\ \vdots \\ Y_n \end{bmatrix}, \quad \hat{\boldsymbol{\beta}} = \begin{bmatrix} a \\ b \end{bmatrix}, \quad \text{and} \quad \mathbf{X} = \begin{bmatrix} 1 & X_1 \\ 1 & X_2 \\ \vdots & \vdots \\ 1 & X_n \end{bmatrix}.$$

Consider the following quantities:

$$\mathbf{y'y} = [Y_1 \quad Y_2 \quad \cdots \quad Y_n] \begin{bmatrix} Y_1 \\ Y_2 \\ \vdots \\ Y_n \end{bmatrix} = Y_1^2 + Y_2^2 + \cdots + Y_n^2 = \sum_{i=1}^{n} Y_i^2$$

and

$$\hat{\boldsymbol{\beta}}' \mathbf{X'y} = [a \quad b] \begin{bmatrix} \sum Y_i \\ \sum X_i Y_i \end{bmatrix} = a \sum Y_i + b \sum X_i Y_i.$$

We will show that $\mathbf{y'y}$ can be used as a total sum of squares and $\hat{\boldsymbol{\beta}}'\mathbf{X'y}$ can be used as a regression sum of squares. Then the analysis of variance table in terms of matrices and vectors can be written

ANOVA

Source	SS	df
Regression	$\hat{\boldsymbol{\beta}}'\mathbf{X'y}$	2
Error	$\mathbf{y'y} - \hat{\boldsymbol{\beta}}'\mathbf{X'y}$	$n-2$
Total	$\mathbf{y'y}$	n

This is not quite the same as our original analysis of variance table: for one thing, the regression sum of squares has two degrees of freedom instead of one, and the total sum of squares has n degrees of freedom instead of $n-1$. In addition, the total sum of squares is $\sum Y_i^2$ instead of S_{YY}. We can make a few alterations in this table to show how it is not really very different from the one we previously used.

Consider again the quantity

$$\hat{\boldsymbol{\beta}}'\mathbf{X'y} = a \sum Y_i + b \sum X_i Y_i.$$

Recall that

$$a = \bar{Y} - b\bar{X},$$

so that

$$\hat{\boldsymbol{\beta}}'\mathbf{X'y} = (\bar{Y} - b\bar{X})(\sum Y_i) + b \sum X_i Y_i$$

$$= \frac{(\sum Y_i)^2}{n} - \frac{b(\sum X_i)(\sum Y_i)}{n} + b \sum X_i Y_i$$

$$= \frac{(\sum Y_i)^2}{n} + bS_{XY}.$$

In our previous ANOVA table, the regression sum of squares involved only bS_{XY}, containing only the regression coefficient, and not the regression constant. That regression sum of squares had one degree of freedom associated with the slope; $\hat{\boldsymbol{\beta}}'\mathbf{X'y}$ contains an intercept term as well as the slope and thus has two degrees of freedom.

That is, we can break down $\hat{\boldsymbol{\beta}}'\mathbf{X}'\mathbf{y}$ like this:

ANOVA

Source	SS	df
Regression	$\hat{\boldsymbol{\beta}}'\mathbf{X}'\mathbf{y}$	2
$\begin{cases} \text{Intercept} \\ \text{Slope} \end{cases}$	$\begin{cases} (\sum Y_i)^2/n \\ bS_{XY} \end{cases}$	$\begin{cases} 1 \\ 1 \end{cases}$

To make the two analysis of variance tables alike, if we rewrite the regression sum of squares as

$$SS(R) = \hat{\boldsymbol{\beta}}'\mathbf{X}'\mathbf{y} - \frac{(\sum Y_i)^2}{n},$$

then it will have one degree of freedom. But then

$$SS(R) + SS(E) = SS(T),$$

so

$$\left[\hat{\boldsymbol{\beta}}'\mathbf{X}'\mathbf{y} - \frac{(\sum Y_i)^2}{n} \right] + [\mathbf{y}'\mathbf{y} - \hat{\boldsymbol{\beta}}'\mathbf{X}'\mathbf{y}] = \mathbf{y}'\mathbf{y} - \frac{(\sum Y_i)^2}{n} = \sum Y_i^2 - \frac{(\sum Y_i)^2}{n} = S_{YY}$$

and $SS(T) = S_{YY}$ will have $n - 1$ degrees of freedom. In summary, we can write the analysis of variance table using vectors and matrices as

ANOVA

Source	SS	df
Regression	$\hat{\boldsymbol{\beta}}'\mathbf{X}'\mathbf{y} - \dfrac{(\sum Y_i)^2}{n}$	1
Error	$SS(T) - SS(R)$	$n - 2$
Total	$\mathbf{y}'\mathbf{y} - \dfrac{(\sum Y_i)^2}{n}$	$n - 1$

and have a table equivalent to the one originally presented.

A NUMERICAL EXAMPLE

For our house size-price example,

$$\mathbf{y} = \begin{bmatrix} 32 \\ 24 \\ \vdots \\ 28 \end{bmatrix}, \qquad \mathbf{X} = \begin{bmatrix} 1 & 1.8 \\ 1 & 1.0 \\ \vdots & \vdots \\ 1 & 1.5 \end{bmatrix}, \qquad \mathbf{X'y} = \begin{bmatrix} 690 \\ 1554.9 \end{bmatrix}$$

and $\hat{\beta}$ was found to be

$$\hat{\beta} = \begin{bmatrix} a \\ b \end{bmatrix} = \begin{bmatrix} 8.7036 \\ 12.8982 \end{bmatrix}.$$

Now,

$$\mathbf{y'y} = \begin{bmatrix} 32 & 24 & \cdots & 28 \end{bmatrix} \begin{bmatrix} 32 \\ 24 \\ \vdots \\ 28 \end{bmatrix} = 26{,}178,$$

and

$$\hat{\beta}'\mathbf{X'y} = \begin{bmatrix} 8.7036 & 12.8982 \end{bmatrix} \begin{bmatrix} 690 \\ 1554.9 \end{bmatrix} = 26{,}060.8952$$

The analysis of variance table then is

ANOVA

Source	SS	df
Regression	$26{,}060.8952 - \dfrac{(690)^2}{20} = 2255.8952$	1
Error	(by subtraction) $= 117.1048$	18
Total	$26{,}178 - \dfrac{(690)^2}{20} = 2373$	19

and this is the same as we obtained in Section 7.3.

When we turn to the consideration of more complicated models, such as polynomial or multiple regressions, the most convenient way to set up an analysis of variance table will be by using matrices and vectors. The algebraic expressions for the regression sum of squares in these more complicated models will be so involved that we will be glad for a simple expression like $\boldsymbol{\beta}'\mathbf{X}'\mathbf{y} - (\sum Y_i)^2/n$.

CORRECTING X FOR ITS MEAN

Of course this approach is not affected if the X-values are corrected for their mean. If X is corrected for its mean, then

$$\hat{\boldsymbol{\beta}} = \begin{bmatrix} a' \\ b \end{bmatrix}, \qquad \mathbf{X} = \begin{bmatrix} 1 & X_1 - \bar{X} \\ 1 & X_2 - \bar{X} \\ \vdots & \vdots \\ 1 & X_n - \bar{X} \end{bmatrix}, \qquad \mathbf{X}'\mathbf{y} = \begin{bmatrix} \sum Y_i \\ S_{XY} \end{bmatrix},$$

so that

$$\hat{\boldsymbol{\beta}}'\mathbf{X}'\mathbf{y} = \begin{bmatrix} a' & b \end{bmatrix} \begin{bmatrix} \sum Y_i \\ S_{XY} \end{bmatrix} = a' \sum Y_i + b S_{XY}.$$

Since $a' = \bar{Y}$,

$$\hat{\boldsymbol{\beta}}'\mathbf{X}'\mathbf{y} = \frac{(\sum Y_i)^2}{n} + b S_{XY},$$

as before; thus the analysis of variance table and the tests of significance are unaffected if X is corrected for its mean.

7.6 CONFIDENCE INTERVALS ON THE PARAMETERS IN THE MODEL

CONFIDENCE INTERVALS FOR β

Interval Estimate of β

Often it is desired to have an interval estimate of β, rather than just a point estimate, since we know that b, being based only on sample information, gives only an estimate of the true relationship between X and Y.

Recall that

$$t = \frac{b - \beta}{\sqrt{s_{y \cdot x}^2 / S_{XX}}}$$

follows the student t-distribution with $v = n - 2$ degrees of freedom. Then since $100(1 - \alpha)\%$ of all t_v-values lie between $-t_{v,\alpha/2}$ and $+t_{v,\alpha/2}$, we can be $100(1 - \alpha)\%$ confident that

$$-t_{v,\alpha/2} < \frac{b - \beta}{\sqrt{s_{y \cdot x}^2 / S_{XX}}} < t_{v,\alpha/2}.$$

Playing around with the above inequality to obtain β by itself in the middle,

$$-t_{v,\alpha/2} \sqrt{\frac{s_{y \cdot x}^2}{S_{XX}}} < b - \beta < t_{v,\alpha/2} \sqrt{\frac{s_{y \cdot x}^2}{S_{XX}}},$$

$$-b - t_{v,\alpha/2} \sqrt{\frac{s_{y \cdot x}^2}{S_{XX}}} < -\beta < b + t_{v,\alpha/2} \sqrt{\frac{s_{y \cdot x}^2}{S_{XX}}},$$

$$b + t_{v,\alpha/2} \sqrt{\frac{s_{y \cdot x}^2}{S_{XX}}} > \beta > b - t_{v,\alpha/2} \sqrt{\frac{s_{y \cdot x}^2}{S_{XX}}},$$

or

$$b - t_{v,\alpha/2} \sqrt{\frac{s_{y \cdot x}^2}{S_{XX}}} < \beta < b + t_{v,\alpha/2} \sqrt{\frac{s_{y \cdot x}^2}{S_{XX}}}.$$

If we require a 95% confidence interval on β, in the house size-price example then

$$t_{v,\alpha/2} = t_{18,.025} = 2.101$$

and

$$12.8982 - 2.101 \sqrt{\frac{6.5058}{13.56}} < \beta < 12.8982 + 2.101 \sqrt{\frac{6.5058}{13.56}}$$

$$12.8982 - 1.4553 < \beta < 12.8982 + 1.4553$$

$$11.4429 < \beta < 14.3535,$$

or

$$11.44 < \beta < 14.35.$$

Thus, we can say with 95% confidence that if one house is 1000 square feet larger than another, then it will sell for between \$11,440 and \$14,350 more than the smaller house. This is a very strong statement of the relationship between the sizes and prices of houses.

The reader might note that if no relationship exists—if $H_0: \beta = 0$ cannot be rejected—then the confidence interval on β will straddle zero. For example, suppose we had found $b = 1.23$ in our example, with $s_{y \cdot x}^2$ and S_{XX} remaining the same. Then the

95% confidence interval would be 1.23 ± 1.45, or $-0.22 < \beta < 2.68$, which would say with 95% confidence that a house 1000 square feet larger in size than another will cost between \$220 *less* and \$2680 *more* than the smaller house. This kind of statement is equivalent to saying that no relationship exists.

CONFIDENCE INTERVALS FOR α

Confidence
Interval on α

Recognizing that the regression constant can vary from sample to sample, one might want to form a <u>confidence interval on α</u> as well, especially when the intercept has meaning. As with the confidence interval on β, we begin with

$$P\left(-t_{\alpha/2,n-2} < \frac{a - \alpha}{\sqrt{s_{y \cdot x}^2(1/n + \bar{x}^2/S_{XX})}} < t_{\alpha/2,n-2}\right) = 1 - \alpha,$$

and then manipulate the inequality to obtain α by itself in the middle of the interval. The $100(1 - \alpha)\%$ confidence interval on α is found to be

$$a - t_{\alpha/2,n-2}\sqrt{s_{y \cdot x}^2\left(\frac{1}{n} + \frac{\bar{x}^2}{S_{XX}}\right)} < \alpha < a + t_{\alpha/2,n-2}\sqrt{s_{y \cdot x}^2\left(\frac{1}{n} + \frac{\bar{x}^2}{S_{XX}}\right)}.$$

Although it is artificial to find a confidence interval on α for the house size-price problem, since it has no physical meaning, we will construct one for illustration. For a 95% confidence interval on α, $t_{.025,18} = 2.101$; so we have

$$8.7036 \pm 2.101\sqrt{6.5058\frac{1}{20} + \frac{(2)^2}{13.56}},$$

or

$$8.7036 \pm 3.1476,$$

or

$$5.5560 < \alpha < 11.8512.$$

We are 95% sure that the intercept of the regression equation lies between 5.5560 and 11.8512. (There is no interpretation in terms of sizes and prices of houses, since it makes no sense to talk about houses zero thousand square feet in size.)

EXERCISES   *Section 7.6*

7.8   Form and interpret the following confidence intervals on β:

p. 521   a. A 95% confidence interval, using the TEACHERS data.

p. 522
p. 524
p. 525
    b. A 95% confidence interval, using the PRECIPITATION data.
    c. A 99% confidence interval, using the SALES data.
    d. A 95% confidence interval, using the HEATING data.

7.9   Form and interpret the following confidence intervals on α:
    a. A 95% confidence interval, using the TEACHERS data.
    b. A 99% confidence interval, using the PRECIPITATION data.
    c. A 90% confidence interval, using the SALES data.
    d. A 98% confidence interval, using the HEATING data.

7.7 TESTING FOR SIGNIFICANCE USING THE COMPUTER

Most of the output from all three computer programs we have studied—the BMD, the SPSS, and the SAS—is concerned with tests of the significance of the regression. Let us look at the three packages in turn and examine what kinds of tests are performed by each.

THE BMDP1R AND BMDP2R

Refer to the printout for the house size-price problem shown in Figure 4.3, p. 116. We have already identified a, b, and $s_{y \cdot x}^2$ from this printout. We noted that the standard error of the regression coefficient is

$$s_b = \sqrt{\frac{s_{y \cdot x}^2}{S_{XX}}},$$

which estimates how much b varies from sample to sample. This goes in the denominator of the t-statistic. From the printout in Figure 4.3 we find $s_{y \cdot x}^2 = 6.506$ as the mean square for deviation about regression. S_{XX} was found to be 13.56. Thus

$$s_b = \sqrt{\frac{6.506}{13.56}} = 0.693$$

is the standard error of the regression coefficient.

To calculate the t-statistic for testing $H_0: \beta = 0$, one then needs only to divide b, the regression coefficient, by its standard error:

$$t = \frac{12.898}{0.693} = 18.62,$$

which is very close to what we obtained in Section 7.2. Note that t should not be calculated as the square root of the F-statistic given in the analysis of variance table, because this method will not yield a sign on the t-statistic.

The entries in the analysis of variance table should now be familiar to the reader. Comparing the ANOVA table in Figure 4.3 to the one obtained in Section 7.3, we see very little difference between them. What differences do exist can be attributed to rounding error and the number of decimal places carried.

The BMDP1R actually looks up critical values for us and compares the values of the F- and t-statistics to these critical values. Note the 0.00000 listed under P(TAIL) in Figure 4.3. P(TAIL) is the probability that our calculated F will exceed the critical F-value if $H_0:\beta = 0$ is true. To five decimal places, this probability is zero. If we had chosen α to be 0.00002, we would still reject H_0. Similarly, under P(2TAIL) we have the probability that $t > t_{\alpha/2.n-2}$ or $t < -t_{\alpha/2.n-2}$. Thus, as long as the two-sided alternative is being tested, one need only look at the probabilities associated with the t- or F-values, and if they are smaller than the α specified for the problem, H_0 can be rejected.

Figure 4.7, p. 120, showed the BMDP2R analysis for the house size-price example. Look at Step 1 there. The analysis of variance table is just like that from the BMDP1R, except that significance levels are not given. Below the table, under VARIABLES IN EQUATION, we find a and b and s_b. Thus, while the BMDP2R does not calculate t-statistics, it gives us all the information necessary to do so.

THE SPSS

The REGRESSION routine of the SPSS gives essentially the same output as does the BMD, as can be seen from Figure 4.16, p. 138. The analysis of variance table in Step Number 1 is essentially the same, except for rounding error, as was found in Section 7.2 and in Figure 4.3. Note that no row for total sum of squares or total degrees of freedom is given, however. Under VARIABLES IN THE EQUATION we find a, b, the beta coefficient (as explained in Chapter 4), and the standard error of the regression coefficient (explained under the BMD output).

The table of statistics following the SCATTERGRAM routine, as shown in Figure 4.15, actually performs the test of significance for the simple linear regression. Note that in the third column of the STATISTICS table, the first entry is

SIGNIFICANCE – 0.00001.

This says that the F-statistic for testing $H_0:\beta = 0$ is larger than $F_{.00001,1,18}$. (Equivalently, the t-statistic, in absolute value, is larger than $t_{.000005,18}$.) Thus, one actually needs only the SCATTERGRAM routine to find a and b and test for the significance of a simple linear regression.

THE SAS

Figure 4.22, p. 149, shows the analysis of the simple regression problem using SAS. We first note that the analysis of variance table is essentially the same as we found in the

BMD and SPSS programs. The source of variation labeled MODEL is what we have previously called REGRESSION. The SAS gives almost every other piece of information one could want, as well. The actual test of significance is made:

PRØB > F

0.0001

means that the regression is significant at the $\alpha = 0.0001$ level of significance. In addition, in the last row of the table are given values of the t-statistics for testing $H_0: \alpha = 0$ and $H_0: \beta = 0$, and the significance levels for both tests are also given. The SAS is the only one of these three programs that provides any test to see if $\alpha = 0$ (if the line should be fitted through the origin).

COMPARISON

All three programs will test for the significance of the simple linear regression using an F-test in an analysis of variance table. As before, the choice of which program to use is determined by what is available, what results are printed out, and the personal preference of the user. For example, the SAS is the only program that calculates a t-statistic for testing $H_0: \alpha = 0$, so the user who suspects that the equation should be fitted through the origin will find the SAS most useful. On the other hand, very often the intercept is of little interest in a given problem and one might opt instead for the BMD because of its compact format.

EXERCISES   *Section 7.7*

7.10   Using either the BMD, SPSS, or SAS simple regression programs run in Chapter 4, test for the significance of the relationship in

p. 521   a.   the TEACHERS data,
p. 522   b.   the PRECIPITATION data,
p. 524   c.   the SALES data,
p. 525   d.   the HEATING data.
                 Use $\alpha = 0.05$.

7.11   Using either the BMD, SPSS, or SAS simple regression programs run in Chapter 4, test for
         a.   a significant direct relationship in the TEACHERS data.
         b.   a significant direct relationship in the PRECIPITATION data.
         c.   a significant direct relationship in the SALES data.
         d.   a significant inverse relationship in the HEATING data.
         Use $\alpha = 0.05$.

7.12   Using the SAS programs run in Chapter 4,
         a.   Can you be 95% sure that the regression equation $\hat{Y} = -4.1414 + 7.3102X$, which was fitted to the TEACHERS data, should have a negative intercept? Would you use the above regression equation or would you use $\hat{Y} = 6.2908X$?

b.   We found $\hat{Y} = 56.4928 + 0.0058X$ for the PRECIPITATION data. Can you be 95% sure that the intercept should be positive, or would you use $\hat{Y} = 0.924X$?

c.   For the SALES data, we found $\hat{Y} = -0.3256 + 0.3842X$. Can you be 95% sure that an intercept should be included in this equation, or should you use the equation $\hat{Y} = 0.3749X$?

7.13   Does correcting X for its mean have any effect on the tests of significance? Compare the F-statistics for simple regression programs run in Chapter 4 for the TEACHERS, PRECIPITATION, SALES, or HEATING data when X was corrected for its mean to when X was not corrected. Make the comparison for either the BMD, the SPSS, or the SAS.

7.8   SUMMARY

We test an hypothesis concerning the regression coefficient in a simple regression problem to decide whether or not we can be reasonably sure that two variables are related. If a relationship is found, then it makes sense to use one variable to predict the other, to measure the expected error in prediction, and to explain the data by interpreting the regression equation. By testing hypotheses or forming confidence intervals on the regression constant we can decide whether or not the regression equation should be fitted through the origin. These concerns are usually central to the study of the simple linear model.

Often, however, our objective is not really to estimate the value of one variable from the value of another. It might be sufficient just to know the extent to which two variables are related. In the next chapter we shall see that relationships can be studied without even finding a regression equation.

The house size-price problem that we have been using for illustration is such a well-behaved little problem that so far it has afforded us small opportunity to see what can go wrong when we try to analyze actual data. Our study of the simple linear model will not be complete, then, until we tackle this problem—in Chapter 9.

Chapter 8 CORRELATION

8.1 INTRODUCTION

Correlation

Although the simple regression problem has already been fully analyzed, we might profit by looking at the question of a relationship from a slightly different viewpoint. As long as both X and Y are random variables, we may look at the <u>correlation</u> between them. By correlation we mean the strength, or degree, of linear association or relationship; that is, to what extent the two variables behave alike or vary together. Since we are interested in the simultaneous variation of two variables, it makes no sense to talk about correlation if X is a controllable variable.

To a large extent, measures of correlation tell us the same thing as measures we obtained in the study of regression. Their value lies partly in their ability to measure the same thing in a slightly different way. For example, $s_{y \cdot x}$ gives a measure of the strength of the relationship since it gives the estimated average amount that the actual Y-values differ from the estimated Y-values. To determine if $s_{y \cdot x}$ in a particular problem indicates a large or small error, however, we must consider the units used to measure Y. It sometimes helps to have a measure of strength that does not depend upon the units of measurement; and while the coefficient of variation may be used, its interpretation is rather subjective.

In addition, a correlation analysis is sometimes preferable to a regression analysis because we are mainly interested in determining the degree to which the two variables are related, and not in predicting one from the other. Finally, in multiple regression, the concepts of correlation are central to the understanding of how several variables act together to predict another.

8.2 THE COEFFICIENT OF DETERMINATION

THE PROPORTION OF EXPLAINED VARIATION

When developing the calculating formula for $s_{y\cdot x}^2$ and when discussing the F-statistic used to test $H_0: \beta = 0$, we noted that the total variation in Y could be broken down into the sum of two parts: variation in Y due to its relationship with X and variation in Y due to error. If

$$\sigma_y^2 = \text{population variance of } Y,$$

$$\sigma_{y\cdot x}^2 = \text{population of variance of } Y, \text{ given } X \text{ (i.e., variance of points around the regression line)},$$

then

$$\sigma_y^2 = \sigma_{Y\,\text{due to}\,X}^2 + \sigma_{y\cdot x}^2$$

is the mathematical statement of the component parts of the variance of Y. If the objective is to measure the strength of the relationship, then we want to measure the *proportion*, or percentage, *of variation in Y which can be explained by its relationship with X*,

$$\frac{\sigma_{Y\,\text{due to}\,X}^2}{\sigma_Y^2} = \frac{\text{amount of variation in } Y \text{ because of } X}{\text{total amount of variation in } Y}.$$

Population
Coefficient of
Determination

We call this measure the <u>population coefficient of determination</u> and denote it by ρ^2 (ρ is the Greek letter rho).
Then

$$\rho^2 = \frac{\sigma_{Y\,\text{due to}\,X}^2}{\sigma_Y^2} = \frac{\sigma_Y^2 - \sigma_{y\cdot x}^2}{\sigma_Y^2}$$

$$= 1 - \frac{\sigma_{y\cdot x}^2}{\sigma_Y^2}.$$

This last form for ρ^2 indicates that the percentage of variation in Y due to X is 100% of variation in Y minus the percentage due to error.

Suppose that X and Y are perfectly related. Then all points lie on the straight line $\mu_{y\cdot x} = \alpha + \beta X$, and $\sigma_{y\cdot x}^2 = 0$. Thus,

$$\rho^2 = 1 - \frac{0}{\sigma_Y^2} = 1;$$

in other words, variation in X explains 100% of the variation in Y, which varies only because X varies. At the other extreme, if X and Y are not related, then the only reason Y

varies is because of what we have called error, so that

$$\sigma^2_{y \cdot x} = \sigma^2_y$$

and

$$\frac{\sigma^2_{y \cdot x}}{\sigma^2_y} = 1$$

and

$$\rho^2 = 1 - 1 = 0.$$

In this case, variation in X explains 0% of the variation in Y. Thus,

$$0 \leq \rho^2 \leq 1,$$

regardless of the units on X or Y, and the closer ρ^2 is to 1, the stronger the relationship between the two variables.

SAMPLE COEFFICIENT OF DETERMINATION

Sample Coefficient of Determination

In order to estimate ρ^2 from a sample, we define the sample coefficient of determination

$$r^2 = 1 - \frac{SS(E)}{SS(T)} = \frac{SS(T) - SS(E)}{SS(T)} = \frac{SS(R)}{SS(T)},$$

where $SS(E)$ is the error sum of squares, $SS(R)$ is the regression sum of squares, and $SS(T)$ is the total sum of squares:

$$SS(T) = S_{YY},$$
$$SS(R) = bS_{XY},$$
$$SS(E) = S_{YY} - bS_{XY}.$$

(Sums of squares are used instead of mean squares because we are not interested in making estimates at this point; thus there is no need to adjust for degrees of freedom. That is, we may understand the denominator of n in both $MS(E)$ and $MS(T)$, and these n's will cancel out in the ratio.) Then

$$r^2 = \frac{SS(R)}{SS(T)} = \frac{bS_{XY}}{S_{YY}} = \frac{(S_{XY}/S_{XX})S_{XY}}{S_{YY}}$$

$$= \frac{(S_{XY})^2}{S_{XX}S_{YY}}$$

gives a quick calculating formula for r^2. Note that this formula does not require a knowledge of b. In a correlation study the value of b may never be calculated if the researcher is not interested in making estimates of Y from X. Note also that since S_{XY} is the only quantity in r^2 that could be negative, and since S_{XY} is squared, r^2 must always be positive. It cannot detect the direction of the relationship.

A NUMERICAL EXAMPLE

In our example about house sizes and prices, the following quantities were found:

$$S_{XY} = 174.9, \quad S_{XX} = 13.56, \quad S_{YY} = 2373.$$

Then

$$r^2 = \frac{(174.9)^2}{(13.56)(2373)} = 0.9507 \approx 95\%.$$

Thus, differences in the sizes of houses account for about 95% of the variation in prices. Obviously this indicates a very strong relationship. All other variables, including chance, put together can account for at most an additional 5% of the variation in price. (This conclusion of a strong relationship is supported by the sizes of the t- and F-statistics calculated in the last chapter.)

HOW LARGE MUST r^2 BE?

What if we had found $r^2 = 0.33$? Is this interpreted as strong, weak, moderate, or what? Since $0 \leq r^2 \leq 1$, and since $0.33 < 0.50$, one might say that an r^2 of 0.33 indicates a moderately weak relationship. However, it means that X by itself explains 33% of the total variability in Y, leaving only two-thirds of the variability to be explained by *all other* possible predictors taken together. While the interpretation of how strong is strong is rather subjective, it can be argued that $r^2 = 0.33$ indicates a moderately strong relationship.

It takes a bit of exposure for one to be able to appreciate how strong relationships are. Let us look at some graphs. In the following series of figures, all X-values are between 1 and 5, and Y-values range from 4 to 17. Figure 8.1 shows three situations in which the coefficients of determination are zero. Data points on each graph let one verify this. In part (a), we see that for a given X-value, Y is sometimes relatively large and sometimes relatively small, but *on the average Y* is constant regardless of the value of X. In this instance, X does not really explain anything about Y, so $r^2 = 0$.

In part (b), we see that Y very definitely is related to X: as X increases, Y first increases and then decreases. It would be possible to fit a parabola exactly through these points. However, there is *no* linear relationship between the two variables: regardless of the value of X, the average value of Y is constant. It is important to keep in mind that

$$r^2 = \frac{(S_{XY})^2}{S_{XX}S_{YY}}$$

measures the strength of a *linear* association. This is not to say, however, that r^2 cannot be used to measure the strength of a curvilinear relationship. If we break the total sum of squares into a sum of squares due to a curvilinear regression and a sum of squares due to error, then

$$r^2 = 1 - \frac{SS(E)}{SS(T)} = \frac{SS(R)}{SS(T)}$$

will measure the strength of the curvilinear relationship. However, in this case, $SS(R)/SS(T)$ will not equal $(S_{XY})^2/S_{XX}S_{YY}$.

FIGURE 8.1   *Examples of Coefficients of Determination Equal to Zero*

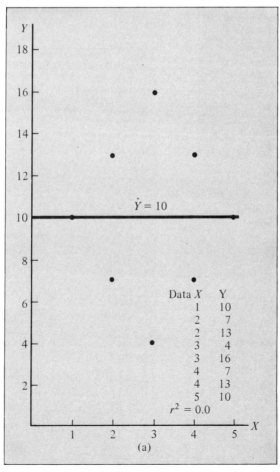

FIGURE 8.1   *(continued)*

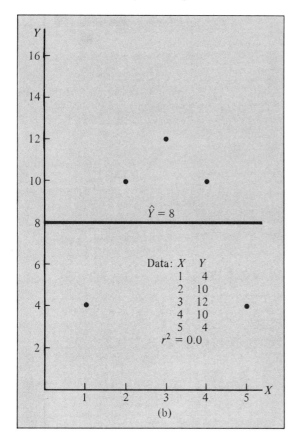

Data: X   Y
       1   4
       2   10
       3   12
       4   10
       5   4
$r^2 = 0.0$

$\hat{Y} = 8$

(b)

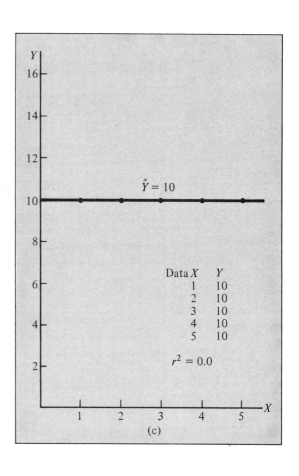

$\hat{Y} = 10$

Data X   Y
      1   10
      2   10
      3   10
      4   10
      5   10

$r^2 = 0.0$

(c)

In part (c) of Figure 8.1, all points lie on the line. So we might be tempted to say that we have a perfect linear relationship and the coefficient of determination should be 1 instead of 0. However, since $Y = 10$ everywhere, X carries no information about Y, and $r^2 = 0$.

At the other end of the scale are situations in which $r^2 = 1$. Two of these are shown in Figure 8.2. In part (a) the variables are perfectly directly related, and in part (b) the variables are perfectly inversely related. Whether Y increases or decreases as X increases is irrelevant. What matters is that all points lie on a straight nonhorizontal line so that if X is known, Y is known without error. Thus X tells 100% of what there is to know about Y.

Intermediate values of r^2 are shown in Figure 8.3. Since all three graphs are drawn to the *same scale*, we can compare the scatter of the points around the lines. While all three graphs show direct relationships, the direction makes no difference to r^2.

FIGURE 8.2   *Examples of Data for Which $r^2 = 1$*

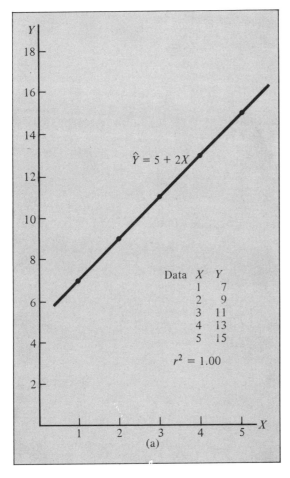

FIGURE 8.2   *(continued)*

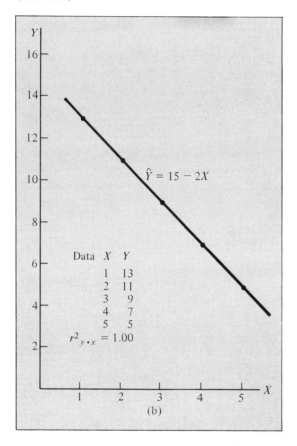

Data X Y
      1   13
      2   11
      3   9
      4   7
      5   5
$r^2_{y \cdot x} = 1.00$

$\hat{Y} = 15 - 2X$

(b)

In part (a) of Figure 8.3 some points are about six units away from the line. While there is a definite tendency for Y to increase with X, one could get only a very rough approximation to the value of Y by knowing the value of X. You can verify that $r^2 \approx \frac{1}{4}$, so 26% of the variation in Y can be explained by X.

The coefficient of determination doubles in part (b) of Figure 8.3. Here the farthest point from the line is about 4 units away. Predictions of Y from X can be made much more accurately in this case, since variation in X explains about 52% of the variation in Y.

Finally, in part (c) $r^2 = 75\%$. No point is more than about 3 units from the line, and we are a lot closer to knowing Y if we know X than we were in either of the two previous cases.

While the relationship in part (c) of Figure 8.3 is certainly stronger than the one in (a), it is arguable whether the one in (a) can really be considered weak. It should, of course, be kept in mind that only five data points were used in the example for the sake of simplicity. Still, in (a), one variable alone can explain 26% of the variation in another, leaving over 74% to be explained by all other predictors, how numerous or few they may be. Further support of the contention that $r^2 = 0.26$ is not necessarily weak is presented in the next section.

FIGURE 8.3   Data for Which r^2 Is Approximately $\frac{1}{4}$, $\frac{1}{2}$, and $\frac{3}{4}$

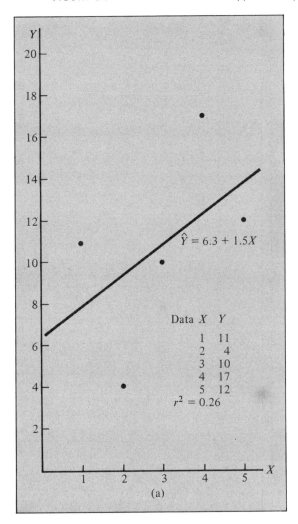

$\hat{Y} = 6.3 + 1.5X$

Data X   Y

1   11
2   4
3   10
4   17
5   12

$r^2 = 0.26$

(a)

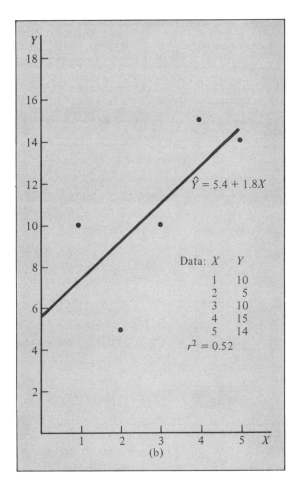

$\hat{Y} = 5.4 + 1.8X$

Data: X   Y

1   10
2   5
3   10
4   15
5   14

$r^2 = 0.52$

(b)

FIGURE 8.3   *(continued)*

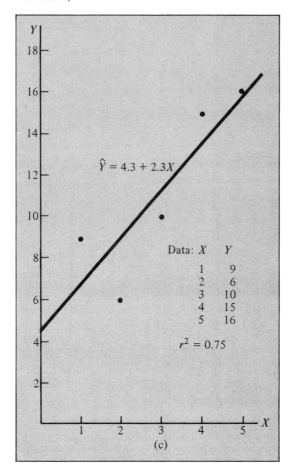

$\hat{Y} = 4.3 + 2.3X$

Data:
X	Y
1	9
2	6
3	10
4	15
5	16

$r^2 = 0.75$

(c)

EXERCISES   *Section 8.2*

p. 521   *8.1*   a. For the TEACHERS data, calculate the coefficient of determination and interpret its meaning.

p. 522   b. Calculate and interpret the coefficient of determination for the PRECIPITATION data.

p. 524   c. Calculate and interpret the coefficient of determination for the SALES data.

p. 525   d. Referring to the HEATING data, calculate the coefficient of determination and interpret its meaning.

8.3   THE CORRELATION COEFFICIENT

Since the sample coefficient of determination, r^2, is always nonnegative, it cannot indicate whether the relationship is direct or inverse. When we need to know the direction of the relationship as well as its strength, another measure of correlation is used.

The population correlation coefficient, ρ, is defined as

$$\rho = \pm\sqrt{\rho^2},$$

so that

$$-1 \leq \rho \leq 1.$$

When $\rho = -1$ a perfect inverse relationship exists; $\rho = 0$ indicates no relationship; $\rho = 1$ indicates a perfect direct relationship; and values of r close to $+1$ or -1 indicate strong relationships. Note that while ρ^2 is a proportion, or percentage, ρ is *not*. It is a "pure" number whose sole function is to be compared to ± 1.

The sample correlation coefficient, r, is calculated:

$$r = \frac{S_{XY}}{\sqrt{S_{XX}S_{YY}}},$$

and the sign on S_{XY} determines the sign on r.

In our example,

$$r = \frac{174.9}{\sqrt{(13.56)(2373)}} = 0.975,$$

or

$$r = \pm\sqrt{r^2} = \pm\sqrt{0.9507} = \pm 0.975$$

and the plus sign is retained upon inspection of the sign of S_{XY}. Since r is very close to $+1$, a very strong direct relationship is indicated.

To interpret the strength of the relationship indicated by the value of r, one commonly thinks of that segment of the real-number line from -1 to 1:

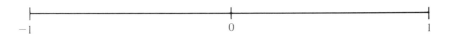

Those values between -1 and 0 indicate inverse relationships and those between 0 and 1 indicate direct relationships. At the ends of the segment, ± 1 indicate perfect relationships while in the middle, at zero, there is no relationship. If we define "moderate" to be halfway between none and perfect, then moderate would be located at

$\pm\frac{1}{2}$. Then perhaps $\pm\frac{3}{4}$ would stand for moderately strong and $\pm\frac{1}{4}$ would denote moderately weak:

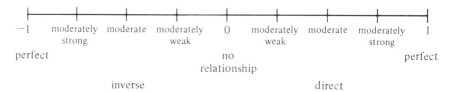

Suppose $r^2 = 0.33$ and $r = -0.5744$. The negative sign indicates an inverse relationship, and since r is over halfway from 0 to -1, we would interpret $r = -0.5744$ as a moderately strong inverse relationship. This backs up our interpretation of $r^2 = 0.33$ indicating a moderately strong relationship.

Referring back to Figure 8.3, in part (a), we found $r^2 = 0.26$. Since the relationship is direct, $r = +\sqrt{0.26} = 0.51$, which we see as a moderate relationship, rather than a weak one. Similarly, in part (b), $r = +\sqrt{0.52} = 0.72$ indicates a moderately strong direct relationship. Finally, the $r^2 = 0.75$ from part (c) yields $r = 0.87$ which probably translates as strong.

Note that r and r^2 are two different methods for estimating the strength of the relationship between two variables, and different standards are used to interpret them. Moreover, while r^2 can be interpreted in terms of the problem, r has no physical meaning.

8.4 TEST OF SIGNIFICANCE

Since sampling error can cause a spurious relationship, how large does r have to be (in absolute value) for us to be reasonably certain that a relationship exists? While this question has already been answered by the hypothesis tests on β, we can run an equivalent test using r instead of b.

In order to test for a significant correlation, we test

$$H_0: \rho = 0$$

against the appropriate alternative:

$$H_1: \rho > 0$$

if we seek a direct relationship,

$$H_1: \rho < 0$$

for an inverse relationship, or

$$H_1: \rho \neq 0$$

for either a positive or negative correlation. The test statistic is a t:

$$t = \frac{r}{\sqrt{(1 - r^2)/(n - 2)}} = r\sqrt{\frac{n - 2}{1 - r^2}},$$

where t obviously has $n - 2$ degrees of freedom. The critical region corresponding to the significance level and the alternative hypothesis is determined in the same way as for the t-test on β.

In our example,

$$H_0: \rho = 0$$

$$H_1: \rho > 0$$

seem to be the reasonable hypotheses to test, so if $\alpha = 0.05$, we will reject H_0 if $t > t_{18,.05} = 1.734$. Then

$$t = 0.975\sqrt{\frac{20 - 2}{1 - 0.9507}} = 18.6301.$$

Since $18.63 > 1.734$, reject H_0 and conclude you are more than 95% sure that there is a positive correlation between house size and price—generally the larger the house, the higher the price.

COMPARISON TO TEST ON β

Notice that this test on ρ is almost exactly the same as the test for $\beta > 0$. The critical region is the same, the conclusion is the same, and the value of the t-statistic is almost the same. Only the approaches are different. A researcher who is interested only in the strength of the relationship will utilize the correlation approach. Working just with r and r^2, and never even calculating a, b, or $s_{y \cdot x}$, one can tell if the relationship is significant, determine its direction, and measure its strength.

LIMITATIONS OF THE TEST STATISTIC

The foregoing test statistic can only be used to test a null hypothesis that the correlation is *zero*. If we want to test an hypothesis such as

$$H_0: \rho = 0.5,$$

we must use something other than the t-statistic given here. There are other statistics available for testing such hypotheses. Since these can be found in introductory texts, they will be omitted here, because our primary interest is detecting the existence of a relationship, and thus we will typically want to test $H_0: \rho = 0$.

ASSUMPTIONS

The hypothesis test given here depends for its validity on a very important assumption. Recall that when we studied confidence intervals on $\mu_{y \cdot x}$ and Y_i^* we made the assumption that for each X-value in the population, the subpopulation of associated Y-values was a *normal* subpopulation. When conducting a correlation analysis, we said that X, as well as Y, must be a random variable. In order for the correlation coefficient and its test of significance to be valid, we must make the very important assumption that the *X-values, as well as the Y-values, come from a normal population.*

In many studies of relationships this assumption of normality is not too difficult to justify (although it should be checked out by methods presented in the next chapter) if the data collected represent *measurements* (or sometimes *counts*) of the values of a random variable. In other instances, the data may consist of rankings or classifications instead of measurements. An example of data consisting of rankings might be students' ranks in graduating class; examples of data consisting of classifications might be $5 =$ strongly agree; $4 =$ agree; ..., $1 =$ strongly disagree; or $1 =$ male, $0 =$ female. For ranked data or categorical data special measures of correlation have been developed, and the reader is referred to texts dealing with the study of nonparametric statistics.

EFFECT OF SAMPLE SIZE

We found a very large positive correlation for our house size-price problem, but it must be remembered that these were artificial data. Seldom in practice does one find such a high correlation (positive or negative) between two variables. For large samples, r can be relatively small and still be significantly different from zero. For example, if $n = 200$ and we are testing the alternative $H_1 : \rho > 0$ with $\alpha = 0.05$, then $H_0 : \rho = 0$ will be rejected if

$$t = r \sqrt{\frac{200 - 2}{1 - r^2}} > t_{.05, 198} \approx 1.658$$

or if

$$r^2 \left(\frac{198}{1 - r^2} \right) > (1.658)^2 = 2.749$$

or if

$$198 r^2 > 2.749(1 - r^2) = 2.749 - 2.749 r^2$$

or if

$$(198 + 2.749) r^2 > 2.749$$

or if

$$r^2 = \frac{2.749}{200.749} > 0.0137$$

or if

$$r > 0.117.$$

Thus, even though $r = 0.12$ is not very large, if $n = 200$ we can be 95% sure that r is really greater than zero, and did not occur just by chance. Another sample of size 200 would be expected to produce a correlation coefficient approximately this same size. If, however, a sample of only ten pairs of observations had resulted in a correlation of 0.12, it would be much more likely that a nonzero correlation was observed just by chance and, in fact, the true correlation is zero. Another sample of size 10 might not reproduce a correlation coefficient as large as this one. The point is that, a low correlation found to be significant in a large sample only means that we are reasonably sure that a weak relationship is most likely present, one not due just to chance.

EXERCISES   *Sections 8.3 and 8.4*

p. 521   *8.2*   a. For the TEACHERS data, calculate the correlation coefficient and interpret its meaning.
                   b. Using $\alpha = 0.05$, test for a significant positive correlation and state your conclusion.

p. 522   *8.3*   a. Calculate the correlation coefficient for the PRECIPITATION data and interpret its meaning.
                   b. Using $\alpha = 0.05$, test for the existence of a significant correlation and state your conclusion.

p. 524   *8.4*   a. Calculate the correlation coefficient for the SALES data and interpret its meaning.
                   b. Using $\alpha = 0.01$, test for a significant positive correlation and state your conclusion.

p. 525   *8.5*   a. For the HEATING data, calculate and interpret the correlation coefficient.
                   b. Using $\alpha = 0.05$, test for a significant negative correlation and state your conclusion.

8.5 THE COMPUTER PRINTOUT

It is easy to locate the coefficient of determination and the correlation coefficient in the three programs we studied in Chapter 4.

BMD PROGRAMS

For the BMDP1R program for the house size-price example, refer back to Figure 4.3, p. 116. The correlation coefficient is obviously given by

MULTIPLE R ... 0.9750,

although the adjective multiple is not applicable when we have only one predictor. Note carefully that r is calculated as the *positive* square root of r^2 by this program, regardless of the direction of the relationship. One must look at the sign of b to determine the sign of r. A correlation coefficient is also given in Figures 4.10 and 4.11, in which second- and third-degree polynomials, respectively, were fitted. These measures are not strictly appropriate because the correlation coefficient measures degree of linear association only.

The "multiple" coefficients of determination are also prominently displayed in Figures 4.3, 4.10, and 4.11. Even in a polynomial regression, r^2 can be interpreted as the percentage of variation in Y explained by the model, so that this is a meaningful figure.

The correlation coefficient and coefficient of determination are indicated in the same way in the BMDP2R output as in the P1R printout. See, for example, Step 1 of Figure 4.7.

SPSS PROGRAMS

The SPSS SCATTERGRAM printout for this problem is shown in Figure 4.15. At the bottom of the diagram both r and r^2 are given in the table of statistics. We find

CØRRELATIØN (R) — 0.97501

and

R SQUARED — 0.95065.

These values are found also in the REGRESSIØN routine in Figure 4.16. MULTIPLE R in this case is just plain old r; since this is actually a multiple regression program, some of the terminology is not strictly appropriate to the simple regression problem. The ADJUSTED R SQUARE is not really of much interest to us. It is calculated as

$$1 - \frac{s_{y.x}^2}{s_y^2},$$

using the error variance (mean square) and the total variance (mean square) instead of a ratio of sums of squares. Thus, the sample size relative to the number of independent variables is taken into account, giving a more conservative estimate of the amount of variation in Y explained by X.

When higher-order polynomial equations are fitted by the SPSS, as in Figures 4.19 and 4.20, values for r and r^2 are again given. Only in Step 1, in which the simple linear equation is found, however, does the correlation coefficient have meaning. Since this REGRESSIØN routine is intended for use in a multiple linear regression, not all the measures calculated when fitting a curvilinear regression are appropriate.

SAS PROGRAMS

Figure 4.22 shows the simple regression analysis for the house size-price problem using the SAS. One can easily find

R-SQUARE

0.950653.

In order to find the correlation coefficient, the square root of r^2 must be taken, and one must look at the regression coefficient to determine the sign of the root.

EXERCISES   *Section 8.5*

8.6   Using either the BMD, the SPSS, or the SAS computer programs run in Chapter 4, find r and r^2 for

p. 521   a.   TEACHERS data,
p. 522   b.   PRECIPITATION data,
p. 524   c.   SALES data,
p. 525   d.   HEATING data.

8.7   Does correcting X for its mean affect the value for r (or r^2)? Refer to the computer programs run in Chapter 4 and compare r- or r^2-values for the TEACHERS, PRECIPITATION, SALES, or HEATING data for when X was not corrected for its mean and when X was corrected for its mean.

8.6 SUMMARY

Methods of correlation enable us to study the strength of the relationship between variables and test for the significance of the relationship without calculating a regression equation.

Correlation analysis is more than an alternative to regression. It is essential to the understanding of the interrelationships among the variables in a multiple regression.

We shall turn to the study of multiple regression after examining some problems that can crop up when using simple regression to analyze actual data.

Chapter 9

APPROPRIATENESS OF THE MODEL

9.1 INTRODUCTION

Our house size-price problem as well as the problems in the exercises are artificially constructed. Seldom would one encounter real data that are so well behaved. Even with the house size-price problem, however, we noted that some of our assumptions really should be checked out. For example, should a simple linear model have been used, or is there a leveling-off of prices for higher-priced houses? Is the scatter of the points around the line truly uniform? Can we justify the assumption that for each size house, the prices are normally distributed around the line?

It is now time to check out our assumptions for the house size-price problem and see what corrective measures can be taken if they are not met. Of course, these assumptions normally would be verified before fitting the regression equation. Some new examples will be introduced so as to illustrate the kinds of problems that often arise with real data.

9.2 THE ASSUMPTION OF UNIFORM SCATTER

In Chapter 6 we assumed that regardless of the value of X, the variability in the associated Y-values remained constant. Otherwise we could not have estimated the error in estimation. Without uniform scatter, the accuracy of our estimates would depend on the value of X for which we were trying to estimate Y.

LOOKING AT THE SCATTER DIAGRAM

In order to determine whether or not the scatter of the points is uniform around the regression line, we can look at the scatter diagram with the regression equation graphed, as in Figure 2.8. The data points fit snugly within two lines parallel to and equidistant from the regression line. So it appears that the scatter is uniform for our house size-price example. Had the scatter been greater for some size houses than for others, nonhomogeneous scatter would be indicated.

PLOTTING THE RESIDUALS

If a curvilinear equation had been fitted to the data, we could still inspect the scatter diagram for uniform scatter by seeing if the dots lay snugly within curved lines following the regression equation. However, this inspection might not be as easy as it sounds. Indeed, the fact that the regression line in Figure 2.8 rises might somewhat obscure the pattern of variation we are looking for. Inspection of a graph for uniform scatter can be made easier if we adopt a method whereby we need not take into account any rise, fall, or curve in the regression equation. Toward this purpose, we plot the

Residuals   residuals, $Y_i - \hat{Y}_i$ against the X-values. The residuals can be read off one of the computer printouts obtained in Chapter 4.

For our house size-price data, refer to the list of residuals in the SAS analysis in Figure 4.22, p. 149. The plot of the residuals was obtained from SAS in Figure 4.24, which is reproduced in Figure 9.1. Note that the residuals are deviations about the horizontal line $Y - \hat{Y} = 0$. By plotting residuals against X, then, we have rotated the

FIGURE 9.1   *SAS Plot of Residuals from Figure 4.24*

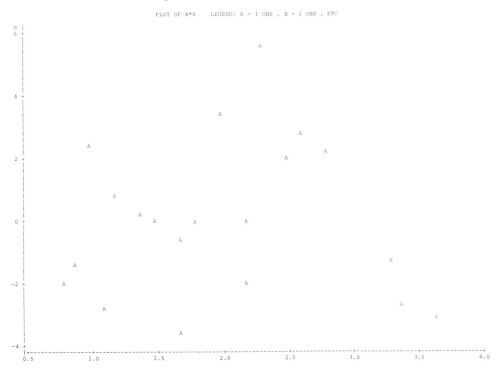

regression line in Figure 2.8 until it is horizontal, carrying the dots along in their same relative positions. This way, we can look just at the vertical scatter of the dots. The brain has less information to process at one time.

OUTLIERS

Outlier

From the plot of residuals, we see something that was not readily apparent from the scatter diagram: all but one of the residuals lies between -4 and $+4$. If *all* of the residuals were within 4 units of the regression line (represented as the line $Y_i - \hat{Y}_i = 0$ in the plot of residuals), we could be quite confident that the scatter is uniform. If several points were outside the 4-unit limits, then we would know that the scatter is not uniform. With just one point lying outside these limits, we conclude that the scatter is generally uniform *except* for one unusual case. Such an unusual case is referred to as an outlier—it is not typical of the rest of the data.

An outlier in the data warns us to stop and investigate. How could this have come about? One would first make sure that an outlier did not result from an error in the data. Human frailty sometimes results in data being misrecorded; digits can be transposed, the measuring instrument can malfunction, the observer may be distracted. If we find that the outlier resulted from a mistake, then our only course of action is to throw the observation away. This is a drastic action because each piece of information bears a cost. But misinformation distorts a relationship.

If one value is extremely different from all the others, its effect can be drastic. As an example, consider

X	Y
0	3
1	4
2	5
3	6
4	2

A quick inspection shows that the first four values all lie on the line

$$\hat{Y} = 3 + X,$$

but that the point $(4, 2)$ lies below this line. (The reader might want to graph these points.) Suppose that X and Y are, in fact, perfectly related as in the above equation, and that the observation $(4, 2)$ really should have been recorded as $(4, 7)$. If we fit a straight line to these data, our calculations are

X	Y	X^2	XY
0	3	0	0
1	4	1	4
2	5	4	10
3	6	9	18
4	2	16	8

Sum: 10 20 30 40

$$S_{XY} = \frac{5(40) - (10)(20)}{5} = 0,$$

$$S_{XX} = \frac{5(30) - (10)^2}{5} = 10,$$

$$\overline{Y} = \frac{20}{5} = 4, \ \overline{X} = \frac{10}{5} = 2.$$

Then

$$b = 0/10 = 0 \quad \text{and} \quad a = 4 - (0)(2) = 4.$$

The regression equation is found to be

$$\hat{Y} = 4 - X,$$

which is quite a different relationship from the one that actually exists!

This is not to say that we automatically discard all outliers. On the contrary, the more information we have, the more accurate our description of the relationship. If it cannot be established that an outlier came about as the result of an error in data collection, then we have to live with this extreme observation and try to discover what it is telling us.

Why would this particular house be atypically high-priced for its size? There are many possible reasons: perhaps it has a terrific view, or a pool, tennis courts, and stable, or all new carpet and paint. An outlier (or several outliers) indicates that we should take other variables into account. For accurate predictions, we should choose a multiple regression model in preference to the simple linear model in order to take these other factors into account.

On the other hand, this one particular house might not really be an outlier. Usually a true outlier has a residual four standard deviations or more from zero. Recall our assumption that at each X-value, the Y-values are normally distributed around the regression line. In a normal distribution, more than 99% of all values lie within three standard deviations of their mean. Thus, it is unusual to see an observation three standard deviations greater or less than the mean, and it is quite rare to find a value four or more standard deviations different from the mean. In our example, House 11 has a residual of about 5.63 (see Figure 9.1, for example). Since the standard error of estimate is about 2.5, this residual is only a little over two standard deviations from zero. How can we tell whether or not to consider it an outlier? One method is to fit a regression equation to the other $n - 1 = 19$ observations, holding out the questionable one. Using this new regression equation, find a $100(1 - \alpha)\%$ confidence interval for the prices of houses this size. If the actual price lies outside the confidence limits, then we can conclude that this observation did not come from the same population as the rest of our data and it is an outlier.

IS HOUSE NO. 11 AN OUTLIER?

In our example, if we hold out House 11 and redo the analysis, we find, for $n = 19$

$$\sum X_i = 37.7, \qquad \sum Y_i = 646, \qquad \sum X_i Y_i = 1{,}453.7,$$
$$\sum X_i^2 = 88.27, \qquad \sum Y_i^2 = 24{,}242.$$

Then

$$S_{XX} = \frac{19(88.27) - (37.7)^2}{19} = 13.4653,$$

$$S_{YY} = \frac{19(24{,}242) - (646)^2}{19} = 2278,$$

$$S_{XY} = \frac{19(1{,}453.7) - (37.7)(646)}{19} = 171.9,$$

and

$$b = 12.7661,$$
$$a = 34 - 12.7661(1.9842) = 8.6694.$$

Our new regression equation is

$$\hat{Y} = 8.67 + 12.77X.$$

The new standard error will be the square root of

$$s_{y \cdot x}^2 = \frac{2278 - 12.7661(171.9)}{17} = 4.9122,$$

or

$$s_{y \cdot x} = 2.2163.$$

For a 99% confidence interval on the price of a house 2300 square feet in area,

$$\hat{Y}_{2 \cdot 3} = 8.6694 + 12.7761(2.3) = 38.0544,$$

$$t_{.005,17} = 2.898,$$

$$38.0544 \pm 2.898(2.2163)\sqrt{1 + \frac{1}{19} + \frac{(2.3 - 1.9842)^2}{13.4653}}$$

$$38.0544 \pm 6.6128,$$

or

$$31.3316 < \hat{Y}_{2 \cdot 3} < 44.6672.$$

The actual price of House 11 is $44,000. Since this is within the range of $31,442 to $44,667, we conclude that House 11 is not an outlier. (At least, we cannot be 99% sure that it is; we used a 99% confidence interval instead of a 95% confidence interval because we want to be quite certain that we really have an outlier before we throw an observation away.) Thus we will retain House 11 in our data, using the original analysis based on all 20 observations. We are aware, however, that to improve the accuracy of our estimates of prices, we need to include variables other than size in our model. The next chapter and all those following will address the problem of multiple linear regression.

There are other methods for detecting outliers, among them, a quick small-sample test, given in Section 15.2.

A TEST FOR HOMOGENEOUS SCATTER

Having decided that House 11 is not an outlier, we return to the problem of whether or not the scatter can be considered homogeneous. Besides a visual interpretation of the plot of residuals, there are statistical tests to help us decide whether or not the variances of the Y-values remain constant regardless of the value of X. One of these tests involves

Split-Sample Analysis

a split-sample analysis.* Split the sample in half, calculate $s_{y \cdot x}^2$ for each of the halves, and compare the two values for $s_{y \cdot x}^2$ using an F-test. How is the sample split in half? This depends upon where you suspect the differences in scatter might be.

Suppose the plot of residuals showed a pattern as in Figure 9.2. The scatter appears to be larger for large X-values than for smaller X-values. Then as one split of the sample, we would choose the $n/2$ observations with the smallest X-values, and put the $n/2$ observations with the largest X-values in the other split. (If n is an odd number, then one split will have one more observation than the other, but it doesn't matter which split is the larger.) A similar split would be used if the plot of residuals showed larger scatter at small X-values than at large X-values.

FIGURE 9.2   *Plot of Residuals Showing Scatter Increasing with X*

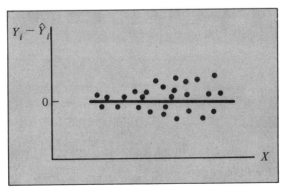

───────────
* The term "split sample" here has a different connotation than in discriminant analysis, in which part of the sample is used to verify results found for the other part.

TABLE 9.1   *Splitting the Sample to Test for Uniform Scatter*

		Split 1				Split 2	
	House No.	*Size (X)*	*Price (Y)*	*House No.*	*Size (X)*	*Price (Y)*	
small	6	0.8	17	17	1.4	27	
	12	0.9	19	20	1.5	28	
	2	1.0	24	3	1.7	27	
	8	1.1	20	15	1.7	30	
	13	1.2	25	1	1.8	32	
large	7	3.6	52	9	2.0	38	
	14	3.4	50	5	2.2	35	
	18	3.3	50	19	2.2	37	
	4	2.8	47	11	2.3	44	
	10	2.6	45	16	2.5	43	
Sum		20.7	349		19.3	341	
Sum of Squares	55.11		14,189		38.45	11,989	
Sum of Products			877.5			677.4	
S_{XX}			12.261			1.201	
S_{YY}			2008.9			360.9	
S_{XY}			155.07			19.27	
$s_{y \cdot x}^2$			5.9581			6.4642	

Referring to Figure 9.1, the plot of residuals for the house size-price data, the greatest scatter seems to occur at middle-sized values of X. Thus as one split of the sample we could choose the middle 10 observations, and put the 5 largest and 5 smallest houses together in the other split. Let us try it as shown in Table 9.1. Are the two variances in the table, 5.9581 and 6.4642, different enough to indicate a real difference in scatter, or is the difference just due to sampling error? To answer this question, let s_1^2 and s_2^2 denote the two variances. Form an F-test statistic,

$$F = \frac{s_1^2}{s_2^2},$$

in order to test

$$H_0 : \sigma_1^2 = \sigma_2^2$$
$$H_1 : \sigma_1^2 \neq \sigma_2^2,$$

where σ_1^2 and σ_2^2 denote the true variances of the points around the regression lines in the two subsamples. Let n_1 denote the number of observations in subsample number 1, and n_2 the observations in the other subsample. Then s_1^2 has $n_1 - 2$ degrees of freedom and s_2^2 has $n_2 - 2$ degrees of freedom, and the F-statistic will have $v_1 = n_1 - 2$ degrees of freedom in the numerator and $v_2 = n_2 - 2$ degrees of freedom in the denominator.

FIGURE 9.3   *Finding the Two-Sided Critical Region for an F-Test*

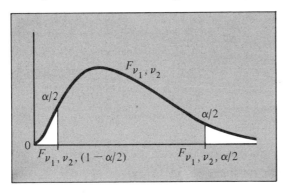

We have chosen a two-sided alternative hypothesis so that we can detect if $\sigma_1^2 > \sigma_2^2$ as well as if $\sigma_2^2 > \sigma_1^2$. We thus need a two-sided critical region, for values of F either considerably less than 1 *or* considerably greater than 1 can be taken as evidence that $\sigma_1^2 \neq \sigma_2^2$. Thus, we need to find F-values that cut off the upper $100(\alpha/2)\%$ of the area under the F-curve and the lower $100(\alpha/2)\%$, as illustrated in Figure 9.3.

Our tables give only *upper-tail* percentage points, but we can still use them to find the lower-tail critical values. Suppose $\alpha = 0.10$. Then we want to cut off $\alpha/2 = 0.05$ of the area under the upper tail and 5% of the area under the lower tail. From the F-tables, we can easily find $F_{\nu_1,\nu_2,0.05}$. The problem is to cut off 5% of the area under the lower tail, which is equivalent to cutting off 95% of the area under the upper tail. Thus we can denote this percentage point as $F_{\nu_1,\nu_2,0.95}$. The upper-tail tables do not include $\alpha = 0.95$. However, we can find $F_{\nu_1,\nu_2,0.95}$ by means of the relation

$$F_{\nu_1,\nu_2,0.95} = \frac{1}{F_{\nu_2,\nu_1,0.05}}.$$

Note carefully that numerator and denominator degrees of freedom are switched.

TEST FOR HOMOGENEOUS SCATTER IN HOUSE SIZE-PRICE EXAMPLE

In our example, $\nu_1 = \nu_2 = 8$, so

$$F_{8,8,0.95} = \frac{1}{F_{8,8,0.05}} = \frac{1}{3.44} = 0.29.$$

Thus, if α is chosen to be 0.10, we will reject H_0 if

$$F > F_{8,8,0.05} = 3.44$$

or if

$$F < F_{8,8,0.95} = 0.29.$$

The test statistic can be formed either as

$$F = \frac{5.9581}{6.4642} = 0.9217$$

or as

$$F = \frac{6.4642}{5.9581} = 1.0849,$$

depending upon which split sample was denoted as sample number one. Whichever way F is formed, we cannot reject

$$H_0: \sigma_1^2 = \sigma_2^2.$$

Thus, we cannot be 90 % sure that the scatter is nonuniform. In the absence of sufficient evidence to conclude nonuniform scatter, we will assume that the scatter is close enough to being homogeneous.

TRANSFORMATIONS TO OBTAIN HOMOGENEOUS SCATTER

What if we had a problem in which $H_0: \sigma_1^2 = \sigma_2^2$ was rejected, and we were forced to conclude that the scatter is not homogeneous? In this case, $s_{y \cdot x}$ is meaningless because the error in estimation depends on which X-value you are using to make the estimate. Then everything that depends on $s_{y \cdot x}$, such as significance tests, is also invalid. Unfortunately, heteroscedastic data is common in economics, especially when X is an income variable; it often happens that $\sigma_{y \cdot x}$ is proportional to X, that is,

$$\sigma_{y \cdot x_i} = cX_i,$$

Transformation of Data

for some constant c. In this case, we can use a transformation on the data to make the scatter homogeneous.

The first problem is to determine if the standard error is proportional to the X-value, and if so, to estimate the constant c. Again a split-sample type analysis can be employed. Divide the data into several subsamples, depending on the sizes of the X-values, and calculate $s_{y \cdot x}$ for each subsample. By comparing the $s_{y \cdot x}$-values, one can determine whether or not the standard error is proportional to X and obtain an estimate of c. For example, suppose we divide a set of data into four subsamples and find the following values of $s_{y \cdot x}$.

Subsample	X-values	$s_{y \cdot x}$
1	0–9	1
2	10–19	3
3	20–29	5
4	30–39	7

We see that as X increases by 10, $s_{y \cdot x}$ increases by 2. (Of course, real data would seldom show such a clear-cut relationship.) That is, $s_{y \cdot x} \approx 0.2 X_i$.

Once we have determined that the scatter around the regression line is proportional to the value of X, we can transform the model so that in the transformed model the scatter is uniform. Begin with the original model

$$Y = \alpha + \beta X + \varepsilon,$$

in which the variance of the errors, or residuals, is $\sigma^2_{y \cdot x} = c^2 X^2$. Instead of regressing Y on X, regress the ratio of Y to X on the reciprocal of X (regress Y/X on $1/X$). Of course, this will yield different values for the slope and intercept and for the errors. Denoting these by β', α', and ε', respectively, we can write the model as

$$\frac{Y}{X} = \alpha' + \beta' \frac{1}{X} + \varepsilon',$$

or

$$Y' = \alpha' + \beta' X' + \varepsilon'.$$

How does this model relate to the original one? Multiplying through by X, in the transformed model,

$$Y = \alpha' X + \beta' + \varepsilon' X,$$

so that α' corresponds to the slope and β' corresponds to the intercept of a simple linear model. That is,

$$Y' = \frac{Y}{X}, \quad X' = \frac{1}{X}, \quad \alpha' = \beta, \quad \beta' = \frac{\alpha}{X}, \quad \text{and} \quad \varepsilon' = \frac{\varepsilon}{X}.$$

In this transformed model the variance of the new errors is $1/X^2$ times the variance of the original errors, or

$$\frac{1}{X^2}(c^2 X^2) = c^2.$$

Thus the variance of the errors in the transformed model is constant—it does not depend on the value of X—and the scatter will be uniform in the transformed model.*

AN EXAMPLE OF TRANSFORMING TO OBTAIN HOMOGENEITY

Let us work an example. Suppose we have the following data relating $Y =$ weekly beer expenditure in dollars to $X =$ weekly income in dollars. In Table 9.2, the symbol $\Sigma(\cdot)$ indicates that a sum is taken, either of X-values or of Y-values. $\Sigma(\cdot)^2$, of course, indicates a sum of squares.

TABLE 9.2   *Hypothetical Data Relating Beer Expenditure to Income*

Family	Average Weekly Beer Expenditure ($)	Weekly Income ($00)
1	2.40	0.9
2	3.00	1.0
3	2.50	1.1
4	3.10	1.2
5	3.20	1.3
6	2.40	1.3
7	3.20	1.5
8	2.40	1.8
9	3.60	2.0
10	2.60	2.1
11	2.40	2.6
12	4.10	2.8
13	1.80	3.0
14	3.00	3.2
15	5.00	3.2
$\Sigma(\cdot)$	44.7	29.0
$\Sigma(\cdot)^2$	142.15	65.82
$\Sigma X_i Y_i$	89.41	

$$S_{YY} = 8.944 \quad S_{XX} = 9.7533 \quad S_{XY} = 2.99$$

* We used the theorem that if each value in a set of data is multiplied by a constant, then the variance of the new set will be the square of the constant times the variance of the original set of data. For example, suppose that X_i, $i = 1, 2, \ldots, n$ are all multiplied by a constant k, yielding a new set of values $kX_1, kX_2, \ldots, kX_n$. The mean of this new set will be

$$\frac{1}{n}\sum_{i=1}^{n} kX_i = k\left(\frac{1}{n}\sum_{i=1}^{n} X_i\right) = k\bar{X},$$

and the variance will be

$$\frac{1}{n-1}\sum_{i=1}^{n}(kX_i - k\bar{X})^2 = \frac{1}{n-1}\sum_{i=1}^{n}[k(X_i - \bar{X})]^2$$

$$= \frac{1}{n-1}\sum_{i=1}^{n}[k^2(X_i - \bar{X})^2]$$

$$= k^2\frac{1}{n-1}\sum_{i=1}^{n}(X_i - \bar{X})^2 = k^2 s^2.$$

From the table, we find

$$b = \frac{2.99}{9.7533} = 0.3066$$

$$a = 2.98 - (0.3066)(1.9333) = 2.3973,$$

so that the regression equation is

$$\hat{Y} = 2.3873 + 0.3066X.$$

The data and regression line are plotted in Figure 9.4. There we can see that the scatter of the points around the line increases with X. Thus, the error involved in estimating beer consumption depends upon what income we are talking about. To see if the error is proportional to the value of X, let us divide the data into 5 groups of 3 families each, according to ascending values of income, and calculate $s_{y\cdot x}^2$ for each group. Calculations are shown in Table 9.3, where we see that the ratio of the standard error to the mean X-value is about 0.44 for each group. Thus, $\sigma_{y\cdot x} \approx 0.44X$, or $\sigma_{y\cdot x}^2 \approx 0.2X^2$. The transformation

$$Y' = \frac{Y}{X}, \qquad X' = \frac{1}{X}$$

FIGURE 9.4   *Scatter Diagram and Regression Equation for Beer Consumption Example*

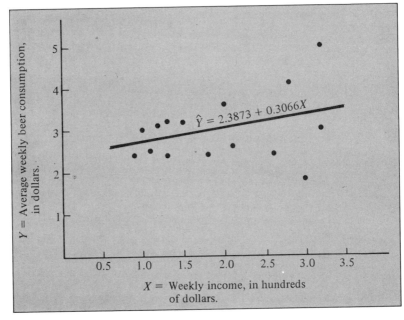

is thus appropriate for these data. For the transformed values in Table 9.4, we find the regression equation

$$b' = \frac{2.3944}{0.9966} = 2.4026$$

$$a' = 1.7882 - (2.4026)(0.6194) = 0.3$$

$$\hat{Y}' = 0.3 + 2.4026X'.$$

Figure 9.5 shows the transformed data and their corresponding regression equation plotted. We note that the scatter of the transformed data around their regression line is reasonably uniform.

We now have an equation relating the ratio of consumption to income to the reciprocal of income. How can this equation be used to estimate consumption from income? Suppose a family has weekly income of $200. Then, by using the original equation,

$$\hat{Y} = 2.3873 + 0.3066(2) = 3.0005.$$

TABLE 9.3   *Beer Consumption Data Divided into 5 Groups.*

Families	Calculations		$s_{y \cdot x}^2$	$\overline{X}$	$s_{y \cdot x}/\overline{X}$
1–3	$\sum Y_i = 7.90$ $\sum Y_i^2 = 21.01$ $\sum X_i = 3.00$ $\sum X_i^2 = 3.02$	$\sum X_i Y_i = 7.91$ $S_{YY} = 0.2067$ $S_{XX} = 0.0200$ $S_{XY} = 0.0100$	0.2017	1.0000	0.4491
4–6	$\sum Y_i = 8.7$ $\sum Y_i^2 = 25.61$ $\sum X_i = 3.80$ $\sum X_i^2 = 4.82$	$\sum X_i Y_i = 11.00$ $S_{YY} = 0.3800$ $S_{XX} = 0.0067$ $S_{XY} = -0.0200$	0.3203	1.2667	0.4468
7–9	$\sum Y_i = 9.20$ $\sum Y_i^2 = 28.96$ $\sum X_i = 5.30$ $\sum X_i^2 = 9.49$	$\sum X_i Y_i = 16.32$ $S_{YY} = 0.7467$ $S_{XX} = 0.1267$ $S_{XY} = 0.0667$	0.7120	1.7667	0.4776
10–12	$\sum Y_i = 9.1$ $\sum Y_i^2 = 29.33$ $\sum X_i = 7.50$ $\sum X_i^2 = 19.01$	$\sum X_i Y_i = 23.18$ $S_{YY} = 1.7267$ $S_{XX} = 0.2600$ $S_{XY} = 0.4300$	1.0155	2.5000	0.4031
13–15	$\sum Y_i = 9.80$ $\sum Y_i^2 = 37.24$ $\sum X_i = 9.40$ $\sum X_i^2 = 29.48$	$\sum X_i Y_i = 31.00$ $S_{YY} = 5.2267$ $S_{XX} = 0.0267$ $S_{XY} = 0.2933$	2.0057	3.2000	0.4426

TABLE 9.4   *Beer Consumption Data Transformed to Give Uniform Scatter.*

Family	$Y' = \dfrac{Y}{X}$	$X' = \dfrac{1}{X}$
1	2.6667	1.1111
2	3.0000	1.0000
3	2.2727	0.9091
4	2.5833	0.8333
5	2.4615	0.7692
6	1.8462	0.7692
7	2.1333	0.6667
8	1.3333	0.5556
9	1.8000	0.5000
10	1.2381	0.4762
11	0.9231	0.3846
12	1.4643	0.3571
13	0.6000	0.3333
14	0.9375	0.3125
15	1.5625	0.3125
$\sum(\cdot)$	26.8225	9.2904
$\sum(\cdot)^2$	55.1956	6.7507
$\sum X'Y'$	19.0072	

$S_{Y'Y'} = 7.2325$
$S_{X'X'} = 0.9966$
$S_{X'Y'} = 2.3944$

If we had not performed an analysis of the untransformed data, we could transform $X = 2$ to $X' = 1/2 = 0.5$ and use the regression equation for the transformed data

$$\hat{Y}' = 0.3 + 2.4026(0.5) = 1.5013.$$

Then from $Y' = Y/X$, we have $Y'X = Y$; and $\hat{Y} = 2(1.5013) = 3.0026$. An easier way, however, is to interchange a' and b' in the equation for the transformed data:

$$\hat{Y} = 2.4026 + 0.3X,$$

so that for $X = 2$,

$$\hat{Y} = 2.4026 + 0.3(2) = 3.0026.$$

This latter method requires much less calculation.

FIGURE 9.5   *Scatter Diagram and Regression Line for Transformed Beer Consumption Data*

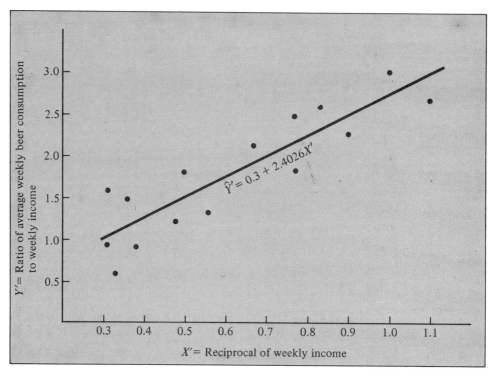

EXERCISES   *Section 9.2*

9.1   In order to see if its secretaries were being paid according to their abilities, a large company sampled 20 of its secretaries, obtaining an efficiency rating and monthly salary for each. The ratings, on a scale from 1.0 to 5.0, were obtained from input from supervisors on various attributes. The results were

Employee	Rating	Salary ($)	Employee	Rating	Salary ($)
Mrs. Adams	3.9	790	Mrs. Irving	4.4	770
Ms. Bright	3.6	740	Mrs. Jacobs	4.8	780
Mrs. Collins	3.1	750	Miss Knight	2.3	700
Miss Dale	4.9	820	Ms. Lawley	2.5	730
Ms. Evans	3.1	720	Ms. Moore	3.6	850
Mrs. Fisher	2.4	700	Mrs. Owen	2.5	690
Mrs. Foster	4.9	790	Miss Paul	3.0	720
Miss Greene	3.6	770	Mrs. Priestly	2.3	720
Miss Hill	4.4	800	Ms. Williams	3.2	730
Mrs. Hustings	2.2	720	Mrs. Wilson	4.0	750

a. Fit a simple linear regression equation to these data, using rating to predict salary.
b. Plot the residuals from the equation obtained in part a.

    c. Office gossip has it that Ms. Moore has been dating a senior executive of the company. From examination of the plot of residuals, does Ms. Moore's salary appear to be out of line?

    d. Form a 99 % confidence interval to see if Ms. Moore's salary represents an outlier observation.

p. 521 9.2 Refer to the TEACHERS data.

    a. Plot the residuals. (See computer printout from Chapter 4 for residuals.)

    b. Test for homogeneous variances by dividing the data into two groups, one group consisting of the states paying lowest salaries and the other of the states paying highest salaries.

p. 524 9.3 Refer to the VALUE ADDED data.

    a. Plot the residuals resulting from fitting a straight line to the data. (See computer printout from Chapter 4.)

    b. Test for homogeneous variances by dividing the data into two groups and calculating $s_{y \cdot x}^2$ for each. In one group, use the seven industries with the smallest number of employees; in the other, use the eight industries employing the largest number of workers.

    c. Transform the data by $Y' = Y/X$ and $X' = 1/X$ and fit a straight line to the transformed data. Plot the transformed data and regression line. Do the transformed data appear to exhibit uniform scatter around their regression line?

    d. Test to see if the scatter for the transformed data is homogeneous. Use the same groups as in part b.

9.3 THE ASSUMPTION OF THE SIMPLE LINEAR MODEL

We fit a simple linear regression equation to a set of data because we have hypothesized the model

$$Y = \alpha + \beta X + \varepsilon,$$

meaning that we think that Y should be linearly related to X, except for error. Just because we hypothesize this model and can fit a simple linear regression line to any set of data does not mean that we have done the correct thing. In the last section we noted that outliers can suggest that a multiple-predictor model might be more appropriate. Plots of residuals can reveal omitted predictors in more ways than one; and even if we use only one predictor, we should check to see that the relationship is truly linear, not curvilinear. In this section, we shall investigate ways to detect when the simple linear model is appropriate. We first examine the data for clues pointing to a curvilinear model and then turn to the question of how to spot the need for more than one predictor.

SHOULD A CURVILINEAR MODEL BE USED?

Plotting the data in a scatter diagram is a first and invaluable step in determining whether a linear or a curvilinear model should be adopted. A pronounced curvature often shows up in the plot of data points. In other cases, the curvature may be more subtle or may be partially obscured by the choice of scales on the horizontal and vertical axes. One can always fit more closely and thus obtain more accurate predictions with a curvilinear equation, but this does not mean that a curved regression line is to be preferred to a simple linear one. We are generally interested in more than accurate predictions. We also want to describe the underlying relationship and interpret the meaning of the regression equation. Thus the simpler description is usually the preferred one.

We chose a simple linear model for the house size-price example because our original plot of the data indicated that the ratio of size to price was relatively constant. It could be argued, however, that there appears to be a leveling-off of the prices for the larger houses in the sample; price increases with size, but at a decreasing rate. If this is indeed the case, then we should have adopted a curvilinear model. We should examine the data more closely to see if there is significant curvature. We realize that this apparent leveling-off of prices could be just a function of the particular sample we happened to draw.

Examining Figure 2.8, in which the regression line was plotted on the scatter diagram for the house size-price data, we see that for houses between about 800 and 2200 square feet in size, the associated prices appear to vary reasonably randomly around the line. However, for houses larger than 2200 square feet, we see first one group of prices above the line and then another group of prices below the line. Such a pattern of variation could indicate systematic variation rather than random error. It might be a clue that the wrong model has been used. For example, suppose we have a definitely curvilinear relationship, as illustrated in Figure 9.6, but fit a straight-line regression

FIGURE 9.6   *A Straight Line Fitted to Curvilinear Data*

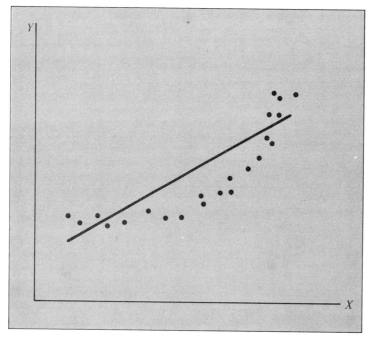

equation. Here we see a very definite pattern to the scatter of the points around the line: at first there is a group of points all above the line, then a group all below the line, and then another group all above the line.

A systematic pattern often shows up better in a plot of residuals than in the scatter diagram. The plot of residuals from Figure 9.6 would first show a group of positive residuals, then a group of negative residuals, and then another group of positive residuals. Looking at the plot of residuals from the house size-price data, we see for the larger houses first a group of positive residuals and then a group of negative residuals. Perhaps this is evidence that a curvilinear equation should have been fitted. We now need to test for significant curvature in the data. If we find it, we then need to decide on the kind of curvilinear model to adopt; or perhaps we can make a transformation on the data to achieve linearity.

LACK-OF-FIT TESTS

Pure Error

Lack of Fit

When we have two or more Y-values corresponding to at least one X-value, we can test to see if the linear model is appropriate. With a designed experiment, it is relatively easy to take enough observations to have *replications* in the data. In a survey, you may or may not observe two items with the same X-value but differing Y-values. The objective is to break the error sum of squares down into two pieces, one due to pure error, and one due to lack of fit. Without at least one replication, we cannot get a measure of pure error, and thus cannot separate it from lack of fit. In our example, both Houses 5 and 19 have 2200 square feet, and both 3 and 15 have 1700 square feet. The differences in prices of Houses 5 and 19 and of Houses 3 and 15 are due to (1) omitted predictors and (2) random variation. They are *not* due to failure of the simple *linear* model. Thus we will be able to break out from the sum of squares for error a piece measuring just pure error.

To do this, we want to measure the variance in prices of houses the same size. Consider Houses 5 and 19, whose prices are 35 and 37 thousand dollars, respectively. We can calculate the variance of these two prices by finding first their mean and then the average squared deviation of the prices from their mean. Call this mean $\bar{Y}_{2.2}$. Then

$$\bar{Y}_{2.2} = \frac{35 + 37}{2} = 36;$$

and their variance will be

$$s_{2.2}^2 = \frac{(35 - 36)^2 + (37 - 36)^2}{2 - 1} = 2.$$

We can do a similar thing for the prices of Houses 3 and 15, which are 27 and 30 thousand dollars, respectively:

$$\bar{Y}_{1.7} = \frac{27 + 30}{2} = 28.5,$$

$$s_{1.7}^2 = \frac{(27 - 28.5)^2 + (30 - 28.5)^2}{2 - 1} = 4.5.$$

The assumption of homogeneous variances becomes quite important now, because we are going to *pool* these two variances into a single estimate of the pure-error variance. Unless we can justify the assumption of homoscedasticity, we cannot put these two sample variances together; they would not be estimates of the same quantity.

It is actually easier in larger examples to pool sums of squares instead of variances. (In our case, it makes no difference because the denominators are both equal to 1.) To pool the sums of squares, simply add up the numerators of $s_{2.2}$ and $s_{1.7}^2$, and call this sum $SS(pure\ error)$:

$$SS(pure\ error) = 2 + 4.5 = 6.5.$$

Of course, had we replications at other X-values, we could have simply calculated s^2 for those cases and added them in, too. The degrees of freedom associated with this sum of squares is n minus the number of *different* X-values in the sample. In our example, there are 18 different values of X, so our degrees of freedom are $20 - 18 = 2$ for pure error.

The sum of squares for lack of fit, $SS\ (lack\ of\ fit)$, is the difference between the error sum of squares and the sum of squares for pure error:

$$SS(lack\ of\ fit) = SSE - SS(pure\ error).$$

Its degrees of freedom will be the number of different X-values minus 2, since the degrees of freedom for lack of fit and the degrees of freedom for pure error must sum to $n - 2$. In our example, degrees of freedom for lack of fit are $18 - 2 = 16$.

Our analysis of variance table for the house size-price problem now looks like this:

ANOVA

Source	SS	df	MS
Regression	2255.8952	1	
Error	117.1048	18	
⎰ Lack of Fit	⎰ 110.6048	⎰16	6.9128
⎱ Pure	⎱ 6.5	⎱ 2	3.250
Total	2373	19	

We can now use an F-test to see whether or not the linear model is appropriate; that is, we test

$$H_0: Y = \alpha + \beta X + \varepsilon$$

against

$$H_1: Y \neq \alpha + \beta X + \varepsilon.$$

If we form an F-test statistic

$$F = \frac{MS(lack\ of\ fit)}{MS(pure\ error)},$$

with 16 and 2 degrees of freedom, then large values of F will indicate that more error is coming from lack of fit than from random variability and that the simple linear model is not appropriate. In our example,

$$F = \frac{6.9128}{3.250} = 2.127.$$

Comparing to the critical value $F_{16,2,0.05} \approx 19.13$, we are unable to reject

$$H_0: Y = \alpha + \beta X + \varepsilon.$$

We thus conclude that the slight curvature apparent from the scatter diagram is most probably due only to sampling variability, and not to any real leveling-off effect in the population.

A caution is in order: note that we had repeated observations at only two values of X. This is not much information on which to base an estimate of the pure-error variance. It is quite possible that we have over-estimated the variance due to pure error. This lack-of-fit test is more trustworthy when many observations are repeated. Even though the foregoing test does not indicate a significant lack of fit of the data to a simple linear model, we shall shortly see a surprising result that may seem to contradict this conclusion.

This same test can be used to decide if a multiple regression model or a curvilinear model is appropriate. Note also that a rejection of $H_0: Y = \alpha + \beta X + \varepsilon$ only tells you that the simple linear model is not appropriate. It does not tell you which model you should use. If H_0 is rejected, then, you must go back to the scatter diagram or plot of residuals in order to get an idea of what model should be used. The new model can be tested out just as this simple linear model was. On the other hand, failure to conclude that the model $Y = \alpha + \beta X + \varepsilon$ does not fit the data does not necessarily imply that this is THE model. In our example, we found that the simple linear model fits the data reasonably well, and this is all we found.

This test for lack of fit can, we have said, be used only when there are repeated observations at at least one value of X. Without these, one must rely on examination of the scatter diagram or of the plot of residuals in determining the appropriateness of the proposed model.

ALTERNATIVES TO THE STRAIGHT-LINE MODEL: POLYNOMIAL MODELS

Suppose we conclude that a simple linear model is not appropriate, and that there is curvature in the data. We then have two choices: (1) find another model, or (2) transform the data so that the transformed data are linear. We will consider first some alternative models.

Perhaps the simplest curvilinear model to fit is a <u>polynomial model</u>

$$Y = \alpha + \beta_1 X + \beta_2 X^2 + \cdots + \beta_k X^k + \varepsilon.$$

We have already discussed some polynomial models and used the computer to fit polynomial equations in Chapter 4. Let us examine them a bit more closely now. Figure 4.25 showed the results of fitting a second-degree polynomial equation

$$Y = a + b_1 X + b_2 X^2$$

by SAS76 to the house size-price data. It is reproduced below, in Figure 9.7, even though we have determined that a simple linear model is adequate for this example. The equation was found to be

$$Y = 1.77590 + 20.46853X - 1.75566X^2,$$

and $a = 1.77590$ would have been the average price of houses zero thousand square feet in size, had there been any such thing; $b_1 = 20.46853$ is the slope of the tangent to the curve at $X = 0$; and $b_2 = -1.75566$ is half the amount that the slope of the tangent changes for each one-unit increase in X. These values for a, b_1, and b_2 are found by solving normal equations. How are the entries in the analysis of variance table obtained?

It is now that we are glad that the regression problem can be formulated in terms of matrices and vectors. Recall from Section 7.5 that the analysis of variance table can be written as

ANOVA

Source	SS	df
Regression	$\hat{\beta}'X'y - \dfrac{(\sum Y_i)^2}{n}$	k
Error	by subtraction	$n - (k + 1)$
Total	$y'y - \dfrac{(\sum Y_i)^2}{n}$	$n - 1$

where k is the number of parameters in the model, *excluding* the intercept. In our present example, since we are fitting the model

$$Y = \alpha + \beta_1 X + \beta_2 X_2^2 + \varepsilon,$$

FIGURE 9.7   SAS Quadratic Regression from Figure 4.25

QUADRATIC REGRESSION
GENERAL LINEAR MODELS PROCEDURE

DEPENDENT VARIABLE: Y

SOURCE	DF	SUM OF SQUARES	MEAN SQUARE	F VALUE	PR > F	R-SQUARE	C.V.
MODEL	2	2284.14810434	1142.07405217	218.51	0.0001	0.962557	6.6266
ERROR	17	88.85189566	5.22658210			STD DEV	Y MEAN
CORRECTED TOTAL	19	2372.00000000				2.28617193	34.50000000

SOURCE	DF	TYPE I SS	F VALUE	PR > F	DF	TYPE IV SS	F VALUE	PR > F
X	1	2255.90044248	431.62	0.0001	1	199.26157592	38.12	0.0001
X*X	1	28.24766186	5.40	0.0327	1	28.24766186	5.40	0.0327

PARAMETER	ESTIMATE	T FOR H0: PARAMETER=0	PR > |T|	STD ERROR OF ESTIMATE
INTERCEPT	1.77590491	0.54	0.5939	3.26847611
X	20.46852858	6.17	0.0001	3.31500152
X*X	-1.75566671	-2.32	0.0327	0.75519175

OBSERVATION	OBSERVED VALUE	PREDICTED VALUE	RESIDUAL
1	32.00000000	32.93092863	-0.93092863
2	24.00000000	20.48877678	3.51122322
3	27.00000000	31.49855561	-4.49855561
4	47.00000000	45.32343636	1.67656364
5	35.00000000	38.30928933	-3.30928933
6	17.00000000	17.02710748	-0.02710748
7	52.00000000	52.70929690	-0.70929690
8	20.00000000	22.16694173	-2.16694173
9	38.00000000	35.69033525	2.30966475
10	45.00000000	43.12583989	1.87416011
11	44.00000000	39.56609667	4.43390333
12	19.00000000	18.77549870	0.22450130
13	25.00000000	23.80999355	1.19000645
14	50.00000000	51.07351057	-1.07351057
15	30.00000000	31.49855561	-1.49855561
16	43.00000000	41.97437195	1.0256805
17	27.00000000	26.99075778	0.00924222
18	50.00000000	50.20294770	-0.20294770
19	37.00000000	38.30928933	-1.30928933
20	28.00000000	28.52847019	-0.52847019

SUM OF RESIDUALS   -0.00000000
SUM OF SQUARED RESIDUALS   88.85189566
SUM OF SQUARED RESIDUALS - ERROR SS   -0.00000000
FIRST ORDER AUTOCORRELATION   -0.24754441
DURBIN-WATSON D   2.47899672

we have $k = 2$. Now,

$$\mathbf{y} = \begin{bmatrix} 32 \\ 24 \\ \vdots \\ 28 \end{bmatrix}, \qquad \mathbf{X} = \begin{bmatrix} 1 & 1.8 & (1.8)^2 \\ 1 & 1.0 & (1.0)^2 \\ \vdots & \vdots & \vdots \\ 1 & 1.5 & (1.5)^2 \end{bmatrix} = \begin{bmatrix} 1 & 1.8 & 3.24 \\ 1 & 1.0 & 1.00 \\ \vdots & \vdots & \vdots \\ 1 & 1.5 & 2.25 \end{bmatrix},$$

$$\mathbf{X'y} = \begin{bmatrix} 690.0 \\ 1554.9 \\ 3965.89 \end{bmatrix}, \qquad \mathbf{y'y} = 26{,}178;$$

and from the printout we find

$$\hat{\boldsymbol{\beta}} = \begin{bmatrix} a \\ b_1 \\ b_2 \end{bmatrix} = \begin{bmatrix} 1.77590 \\ 20.46853 \\ -1.75566 \end{bmatrix}.$$

Then

$$SST = 26{,}178 - \frac{(690)^2}{20} = 2373,$$

$$\hat{\boldsymbol{\beta}}'\mathbf{X'y} = [1.77590 \quad 20.46853 \quad -1.75566] \begin{bmatrix} 690.0 \\ 1554.9 \\ 3965.89 \end{bmatrix} = 26{,}089.1338,$$

$$SSR = 26{,}089.1338 - \frac{(690)^2}{20} = 2284.1338$$

and

$$SSE = SST - SSR = 88.8662.$$

Comparing these figures to those for the analysis of variance table in Figure 9.7, we see that they are the same, except for rounding error. From Figure 9.7 we also see an F-value of 218.51. Comparing this F to an F with 2 and 17 degrees of freedom, using $\alpha = 0.05$, we find

$$218.51 > F_{2,\,17,\,0.05} = 3.59,$$

so that the regression is significant. The question now is, do we need the X^2-term?

Let us refer first to Figure 9.7 and the analysis presented by the SAS program. Note that below the general analysis of variance table are lines containing degrees of freedom, sums of squares, F-ratios, and probabilities associated with Type I and Type IV sums of squares. Let us first consider the Type I sums and their associated F-values.

SEQUENTIAL SUMS OF SQUARES

Sequential
Sums of
Squares

We first note that the Type I sum of squares associated with X is 2255.90044348, and (except for rounding error) this is the same as the regression sum of squares when the simple linear model was used. These Type I sums of squares are referred to as sequential sums of squares. First, a simple regression is run using the first variable mentioned in the MODEL statement in the program, X (refer to Chapter 4), and the regression sum of squares for the model

$$Y = \alpha + \beta_1 X + \varepsilon$$

is given. Then, since X^2 is the next variable mentioned in

MODEL Y $= X X * X / P$;

a two-variable regression is run using X and X^2. Then the difference in the regression sum of squares using X and X^2 and the regression sum of squares using X alone is calculated. That is,

$$SSR_1(X) = SSR(X),$$

$$SSR_1(X^2) = SSR(X, X^2) - SSR(X) = SSR(X^2 \mid X),$$

where SSR_1 indicates a Type I regression sum of squares, $SSR(X, X^2)$ indicates the regression (or model) sum of squares with 2 degrees of freedom and $SSR(X^2 \mid X)$ indicates a conditional sum of squares for X^2, given X. That is,

$$SSR_1(X) = SSR(X) = 2255.90044248$$

and

$$SSR_1(X^2) = 2284.14810434 - 2255.90044248$$

$$= 28.24766186.$$

Thus, the Type I sums of squares for X^2 measures how much the regression sum of squares increases (or equivalently, how much the error sum of squares decreases) when the quadratic term is added to the model. It is a "marginal" kind of sum of squares, measuring additional reduction of error achieved by adding on a quadratic term, over and above (under and below?) error achieved by using the linear term alone. Note that had the model statement read

MODEL Y $= X * X$ X;

then we would first see $SSR_1(X^2)$, obtained by fitting the model

$$Y = \alpha + \beta_2 X^2 + \varepsilon,$$

and then $SSR_1(X)$ would be

$$SSR_1(X) = SSR(X \mid X^2) = SSR(X, X^2) - SSR(X^2).$$

That is, these sequential sums of squares are calculated according to the order in which the predictors are given in the model statement.

The question now is, do we achieve a significant reduction in error by using a quadratic regression rather than a simple linear regression? In order to answer this question, we form a test statistic for the quadratic term. This test statistic is an F, formed by

$$F = \frac{MSR}{MSE},$$

as before. The error mean square is found in the table for the overall test:

$$MSE = 5.22658210$$

with 17 degrees of freedom. For the regression sum of squares, we use the Type I sum of squares divided by its degrees of freedom. From the printout, we see that the linear term and the quadratic term each have one degree of freedom. Thus, to test

H_0: no significant linear regression

H_1: significant linear regression,

we would use

$$F = \frac{SSR_1(X)/1}{MSE} = \frac{2255.90044248}{5.22658210} = 431.62.$$

Note that even though $SSR_1(X)$ was obtained using only the simple linear model, the error mean square used in this test is the one from the *polynomial* model. From the printout we see that this F is greater than $F_{.0001,1,17}$, so we conclude that there is a significant linear regression. To test

H_0: quadratic regression does not significantly improve over linear regression

H_1: quadratic regression significantly improves over linear regression,

the test statistic is

$$F = \frac{SSR_1(X^2)}{MSE} = \frac{28.24766186/1}{5.22658210} = 5.40.$$

We want to be quite sure that we need a quadratic regression before we will adopt one, so $\alpha = 0.01$ is chosen, and we will reject H_0: no significant improvement if $F > F_{.01,1,17} = 8.40$. Since $5.40 \not> 8.40$, we cannot be 99% sure that a quadratic regression should be used. Note from the printout that this F is significant at $\alpha = 0.0327$. If the significance level had been chosen to be $\alpha = 0.05$, we would have concluded that a quadratic regression should be used. This would have contradicted the results of our lack-of-fit test, which indicated no significant lack of fit using the simple linear model. If $\alpha = 0.05$ had been used for testing the significance of the quadratic regression, then we would have to conclude that the lack-of-fit test was not really trustworthy, since it had only two degrees of freedom for pure error. Recall that not finding significant lack of fit does not imply that the correct model was used, but only indicates that the model being tested is not too inconsistent with the data.

PARTIAL SUMS OF SQUARES

Turning to the Type IV sums of squares and their associated F-values, we see that

$$SSR_{IV}(X^2) = SSR_I(X^2) = 28.24766186$$

with one degree of freedom. Thus,

$$SSR_{IV}(X^2) = SSR(X^2 \mid X) = SSR(X, X^2) - SSR(X).$$

Partial Sums
of Squares

However, $SSR_{IV}(X)$ is not the simple sum of squares for regression on X. These Type IV sums are called partial sums of squares, and are conditional on all other variables in the model being fitted. That is,

$$SSR_{IV}(X) = SSR(X \mid X^2) = SSR(X, X^2) - SSR(X^2).$$

These partial sums of squares are also marginal, in that they measure the increase in the regression sum of squares attained by one variable, given that the other variable was already in the model. The order in which the variables are entered in the MODEL statement has no effect on these Type IV sums of squares. The F-statistics are formed as before. To test

H_0: addition of linear term to quadratic term achieves no significant reduction in error

H_1: addition of linear term to quadratic term significantly reduces the error mean square,

we use

$$F = \frac{SSR_{IV}(X)/1}{MSE} = \frac{199.26157592}{5.22658210} = 38.12,$$

and this F is significant at $\alpha = 0.0001$. We conclude that MSE would be significantly reduced by using the linear term in addition to the quadratic term over what could be achieved by using the quadratic term alone. To test

H_0: addition of quadratic term to linear term does not significantly increase the regression sum of squares

H_1: regression sum of squares is significantly reduced by addition of quadratic term to linear term,

we use

$$F = \frac{SSR_{IV}(X^2)/1}{MSE} = \frac{28.24766186}{5.22658210} = 5.40.$$

As before, if $\alpha = 0.01$, we conclude that we cannot be 99 % sure that it is really necessary to use a quadratic equation to describe the relationship.

Still referring to the SAS analysis of the problem, a quick check shows us that the t-statistics presented are Type IV, or partial t-statistics. For example, the t corresponding to the linear term is 6.17, and $(6.17)^2 = 38.0689$, which is approximately equal to the Type IV F, 38.12. As before, we form t-statistics by dividing the regression coefficients by their standard errors; for example,

$$t_X = \frac{b_1}{\text{standard error of } b_1} = \frac{20.46852858}{3.31500152} = 6.17.$$

We can use these t-statistics to test one- or two-sided alternatives to the hypotheses

$H_{01}: \alpha = 0$

$H_{02}: \beta_1 = 0$

and

$H_{03}: \beta_2 = 0.$

The probabilities given on the printout correspond to the two-sided alternatives, as can be shown by comparing these probabilities to those given for the F-statistics.

BMD AND SPSS PROGRAMS

We discussed the SAS printout first because it gives both sequential and partial sums of squares and F-tests. The BMD and the SPSS analyses each give only one test. Examination of Figure 4.10, in which the BMDP1R was used to fit a second-degree polynomial, shows that partial t-statistics only are given by this analysis. This is because the P1R performs the analysis with both predictors, X and X^2, in the model. It does not enter them one at a time. The P2R stepwise analysis also shows partial F-values, even though the steps are in some sense sequential, in that the order of entry of the linear, quadratic, and cubic terms was determined by specifying levels 1, 2, and 3, respectively. (See Figure 4.13, pp. 130–131.)

Both the BMD and SPSS programs are designed more for use with a multiple regression than with a polynomial regression, so their individual F-tests are based on Type IV, or partial, sums of squares. The SPSS fit of a second-degree polynomial is shown in Figure 4.19. There, we see under Step 2 the F-values 38.125 and 5.405, which are the same as the Type IV F-values given by SAS. (Discussion of values labeled BETA IN, PARTIAL, and TOLERANCE are postponed until our study of multiple regression.)

DIFFICULTIES WITH POLYNOMIAL MODELS

Polynomial models are not the only curved models that can be fitted to data. In fact, polynomial models can be poor models to fit, especially if one is forced to use the regression equation to predict Y-values for X-values outside the range of the data, or if the theory behind the data indicates that a polynomial model is not appropriate. For example, Figure 9.8 shows a second-degree polynomial fitted through some points. For the range of data in the sample, the quadratic regression provides a very nice description. But if one is forced to extrapolate beyond that range, the estimates are likely to be quite far off. The reason is that by fitting an equation

$$\hat{Y} = a + b_1 X + b_2 X^2,$$

we have fitted a portion of a parabola. Once the parabola reaches its minimum (or maximum) value, it will turn and proceed to rise (or fall).

FIGURE 9.8   *A Second-Degree Polynomial Equation Fitted to Some Data, with Regression Curve Extended beyond the Range of the Data*

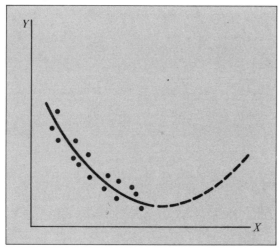

EXPONENTIAL MODELS

In situations like that depicted in Figure 9.8, it is often more likely (and the theory behind the data may indicate) that instead of beginning to rise again, the response may tend to level off at or around some minimum value, as shown in Figure 9.9. The regression equation in this case should asymptotically approach the horizontal axis; it gets slowly closer and closer to the horizontal axis without ever actually touching it. Of course, curves which increase at a decreasing rate can asymptotically approach some upper limit, too. In these cases, we would use some kind of exponential model. For example, models such as

Exponential Model

$$Y = \alpha \beta^X \varepsilon$$

are said to be *intrinsically linear* because logarithms of both sides of the equation give us

$$\log Y = \log \alpha + X \log \beta + \log \varepsilon$$

or

$$Y' = \alpha' + \beta'X + \varepsilon',$$

where

$$Y' = \log Y, \quad \alpha' = \log \alpha, \quad \beta' = \log \beta, \quad \varepsilon' = \log \varepsilon.$$

This is of the same form as a simple linear model. In order to fit the equation, one would proceed as usual for a simple linear regression, but use X and log Y instead of X and Y. From this analysis would result a', which estimates log α, and b', which estimates log β. Taking antilogs of a' and b' we then have estimates of α and β in the model

$$Y = \alpha \beta^X \varepsilon$$

that we can use to draw the curve.

FIGURE 9.9   *Data Leveling Off around a Minimum Value*

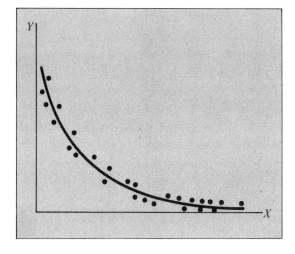

Other exponential-type regression models that can be fitted are

$$Y = \alpha + \beta \log X + \varepsilon,$$

commonly used by economists in relating demand (Y) to price (X), and

$$\log Y = \alpha + \beta \log X + \varepsilon.$$

Those working in specialized areas will recognize the type of function often used for their particular kinds of data.

These exponential models give us an idea of what transformations can be used on curvilinear data so that the transformed data will be linear. These are called, appropriately enough, *logarithmic* transformations.

SHOULD A MULTIPLE REGRESSION MODEL BE USED?

We have seen that outliers in the data can indicate that an important predictor, or several important predictors, have not been considered. While Chapter 10 and those following will deal with multiple regression, we will briefly consider here some ways to spot the omission of predictors.

What would you think if you saw a plot like the one in Figure 9.10? Perhaps one ought to take into account that most of the state is rural, with a few large metropolitan areas.

Sometimes one can identify possible sources of variation which have not been taken into account. For example, suppose Figure 9.11 shows the relationship between age and earnings for a group of people. At first glance, all seems to be in order. But what if the dots represent males and the circles females? In that case, the sex variable should be taken into account if one's objective is to make accurate estimates of earnings.

TIME DEPENDENCIES

Sometimes data cannot be collected all at once, but must be obtained in a time sequence that might affect the response. For example, suppose a gymnastics coach is interested in relating the performance of one of his gymnasts to the number of hours sleep on the previous night. Suppose data are collected over several months, and a plot of performance rating by hours sleep shows a well-behaved linear regression. We are aware, however, that performance is quite related to total number of hours practice, or learning, so that it is quite likely that performance ratings will improve over time, regardless of the number of hours sleep. If this is the case, then if the residuals are plotted in time order, a picture such as that in Figure 9.12 might well appear. In this case, time should be included as a predictor. How this can be done will be discussed in Chapters 10 through 14.

FIGURE 9.10   *Plot of Hypothetical Data and Regression Equation Relating Newspaper Circulation to Number of Registered Voters by County in a State*

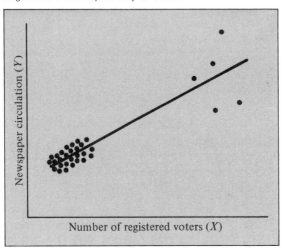

FIGURE 9.11   *Scatter Diagram and Regression Line Relating Earnings to Age*

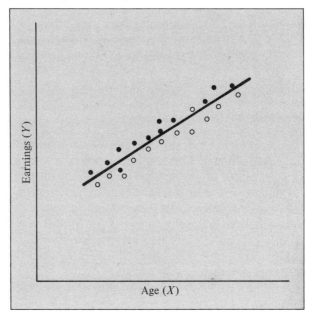

FIGURE 9.12   *Residuals Plotted in Time Order, Showing a Practice Effect*

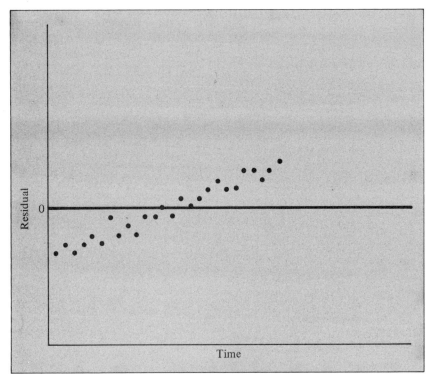

EXERCISES   *Section 9.3*

p. 527   *9.4*   Refer to the DEPRIVATION data and analysis in the Analyses of Data Sets section.
   a.   Fit a simple linear regression equation to the data. Test for the significance of the regression.
   b.   Plot the residuals from the simple linear equation and notice any pattern that indicates that the simple linear model is not appropriate.
   c.   Test for lack of fit to the simple linear model.
   d.   Show the vectors $\mathbf{y}$ and $\hat{\boldsymbol{\beta}}$ and the matrix $\mathbf{X}$ necessary to fit a second-degree polynomial equation to these data. Set up and solve the normal equations and state the regression equation.
   e.   Using an analysis of variance table, test for the significance of the regression using the second-degree polynomial model.
   f.   Using sequential F-tests, test for the significance of the linear and quadratic terms, entering the linear term first.
   g.   Use the SAS or the BMDP2R and fit a second-degree polynomial equation to the data. Compare the results to those in part f.
   h.   Test for lack of fit to the second-degree polynomial equation. Are your results surprising? (Refer to the plot of the regression equation in your solution to Exercise 2.16.) Compare the F-statistic for this problem to the one obtained in part c and comment.

9.5   Suppose that a forester needs a quick way to estimate the volume of a stand of trees, so that when they are harvested he can know approximately how many board-feet the stand will yield. The forester decides to measure the circumferences at 24 inches above the ground of a sample of 25 trees, and then records the volume for each tree when it is harvested. He hopes to be able to use circumference as an indicator of volume. His data are as follows:

Tree	Circumference (ft)	Volume (cu ft)
1	4.7	114
2	2.7	96
3	3.5	100
4	2.5	110
5	4.0	112
6	3.6	108
7	2.7	100
8	5.5	132
9	5.0	118
10	2.6	100
11	2.5	94
12	6.0	144
13	3.8	110
14	5.8	134
15	3.9	106
16	2.8	102
17	5.8	138
18	3.5	102
19	3.6	102
20	2.5	100
21	3.5	106
22	4.7	118
23	3.6	106
24	2.5	98
25	5.8	132

a.   Fit a straight line to the data and plot the residuals. Is there any pattern in the residuals which indicates that a simple linear model is not appropriate?
b.   Test for lack of fit. Comment on the effectiveness of the lack-of-fit test for detecting an inappropriate model.
c.   Fit a second-degree polynomial equation to these data.
d.   Test for the significance of the polynomial regression.
e.   Using sequential F-tests, test for the significance of the linear term and of the quadratic term, given the linear term.

p. 523   9.6   For the POLICE data, referring to the second-degree polynomial equation $\hat{Y} = 2010.2296 - 35.3156X + 0.2096X^2$:
a.   Test for the significance of the polynomial regression.
b.   Use a sequential F-test to test for the significance of first the linear term and then the quadratic term, given the linear term.
c.   Could a lack-of-fit test be performed on these data? Would it be advisable?

9.7   In an evaluation of a typist, a recent manuscript was examined for the number of errors per page. The number of words per page were to be used as the indicator of the number of errors. In the 23-page manuscript, the following numbers of errors were recorded.

Page	Number of Words	Number of Errors
1	228	10
2	230	11
3	221	9
4	208	5
5	222	8
6	197	4
7	180	1
8	229	10
9	212	7
10	190	3
11	209	6
12	207	6
13	185	4
14	209	9
15	193	6
16	209	8
17	207	8
18	208	9
19	229	13
20	206	8
21	208	9
22	207	7
23	185	5

a. Fit a simple linear regression equation to these data. Plot the data and the regression equation.

b. The pages of the manuscript were typed in order and at one sitting. Those evaluating the study were also interested in any fatigue factor which might be present. Calculate the residuals and plot them in time order. Is there any effect of fatigue on the number of errors? Should it be taken into account in order to improve predictions of number of errors from number of words per page?

9.8 In a study to relate IQ to performance on a standardized test, 26 grade-school students were randomly selected from two classes at the same school. Each student was given the same IQ test and was taught from the same textbook. All students took the standardized test at the same time. The results were as follows:

Student	IQ (X)	Score (Y)	Teacher
1	114	90	A
2	118	90	B
3	110	84	B
4	109	88	A
5	94	78	A
6	113	90	B
7	114	88	B
8	110	88	B
9	107	80	B
10	106	82	B
11	106	84	B
12	109	84	B
13	109	90	A
14	109	86	B

Student	IQ (X)	Score (Y)	Teacher
15	96	80	A
16	130	98	A
17	123	96	A
18	108	82	B
19	106	86	B
20	113	92	A
21	125	94	A
22	124	96	A
23	111	86	B
24	120	94	A
25	104	82	A
26	117	92	A

a. Plot the data, indicating whether each student was taught by teacher A or teacher B.
b. Fit and plot a simple linear regression equation.
c. Do you think that the teacher should be taken into account when trying to predict score on the standardized test? Explain.

9.4 INDEPENDENCE OF ERRORS

Every item in our sample is a unique individual and thus each of our measurements contains variations attributable to just that individual. Provided our model

$$Y = \alpha + \beta X + \varepsilon$$

is correct, in that the relationship is truly linear and no important predictors have been omitted, then ε represents the effect of only these individual differences. Recall that we could detail the model as

$$Y_1 = \alpha + \beta X_1 + \varepsilon_1$$
$$Y_2 = \alpha + \beta X_2 + \varepsilon_2$$
$$\vdots$$
$$Y_N = \alpha + \beta X_N + \varepsilon_N.$$

Then ε_1 is attributable to the individuality of the first item in the sample, ε_2 is attributable to the uniqueness of the second item, and so on. Since each error term is attributable only to the individual, the size of any error ε_i should have no effect on the size of any other error ε_j. That is, the error terms are independent, provided that the model is correct. All our analyses of the regression problem are based on the assumption that the errors are independent; obviously, we must check to see if this assumption holds. If not, we must find out why, and take remedial measures, if possible. If the errors are not independent, the model is usually incorrect; perhaps some predictors are omitted, or perhaps a curvilinear model should have been used.

We estimate the ε_i's by $e_i = Y_i - \hat{Y}_i$, $i = 1, 2, \ldots, n$ found by the regression equation. Now, these estimates will *not* be independent, because all n observations were used to find the regression equation and thus to find $\hat{Y}_i$; so the value of e_i is dependent upon all observations in the sample, and not just the ith one. Thus, even if the model is correct, there will be some correlation among the residuals, but this will be small if the sample size is reasonably large.

In most of our examples so far, we have no reason to doubt that the error terms are independent. For example, in our house size-price data, why would the individual variability in one house affect that in another? The 20 houses form a random sample chosen from a large population of houses, and the fact that the sample was random says that the choosing of one house has no effect on whether or not any other house will be chosen.

SERIAL CORRELATION

Serial
Correlation

When the data were collected over time, however, it is important, as we saw in the last section, to see if any time effect is incorporated in the residuals. Such time effects could be growth, decay, practice or boredom effects, change in conditions over time, and so on. Correlation over time, called serial correlation, or *autocorrelation*, means that the level of response in one time period affects the level of response in the next period. Thus the residual in one time period is related to the residual in the next. In order to see if any serial correlation among the errors is present, take the residuals in pairs and find their correlation coefficient. For example, suppose we have the residuals shown in Table 9.5. We can see that they get larger at each successive time period, with a few exceptions. In other words, the residuals are *positively serially* correlated. In order to measure this serial correlation, pair up the residuals as shown in Table 9.6. Then calculate the serial correlation using

$$r_{t,t-1} = \frac{\sum_{i=2}^{n} e_i e_{i-1}}{\sum_{i=2}^{n} e_{i-1}^2} = \frac{26.22}{30.8} = 0.8513,$$

where $r_{t,t-1}$ denotes the correlation between the residual at time t and the residual at time $t - 1$.

TABLE 9.5   *Residuals per Time Period.*

Time Period	ε_i
1	−2.6
2	−2.1
3	−2.6
4	−1.3
5	0.6
6	−0.1
7	0.8
8	2.4
9	2.1
10	2.8

TABLE 9.6   *Calculation of Serial Correlation.*

Time Period i	Residual e_i	Residual for Preceding Period e_{i-1}	$e_i e_{i-1}$
1	−2.6		—
2	−2.1	−2.6	5.46
3	−2.6	−2.1	5.46
4	−1.3	−2.6	3.38
5	0.6	−1.3	−0.78
6	−0.1	0.6	−0.06
7	0.8	−0.1	−0.08
8	2.4	0.8	1.92
9	2.1	2.4	5.04
10	2.8	2.1	5.88
Sum			26.22
Sum of Squares		30.8	

TEST FOR SERIAL CORRELATION

As we noted, there will be some correlation among residuals whenever the data are used to fit a regression equation. For a given set of data, if we suspect that an effect of time is present, we can calculate the serial correlation coefficient and then test it to see whether it is large enough (in absolute value) to be attributable to an effect of time. For this test, we use a special statistic called the <u>Durbin-Watson d-statistic</u>. Its form depends upon whether one wishes to test for a significant positive serial correlation or a negative one. Let $\rho_{t,t-1}$ denote the true serial correlation. To test

<u>Durbin-Watson d-Statistic</u>

$$H_0: \rho_{t,t-1} = 0$$

$$H_1: \rho_{t,t-1} > 0,$$

use

$$d = \frac{\displaystyle\sum_{i=2}^{n} (e_i - e_{i-1})^2}{\displaystyle\sum_{i=1}^{n} e_i^2}.$$

To test

$$H_0 : \rho_{t,t-1} = 0$$
$$H_1 : \rho_{t,t-1} < 0,$$

use the test statistic

$$d^- = 4 - d.$$

Refer to the Tables of the Durbin-Watson statistic, Table III of the Appendix. Note that there are separate tables for $\alpha = 0.05$ and $\alpha = 0.01$. Note also that the tables are subdivided into parts for different numbers of parameters in the model, p. The sample size begins at $n = 15$; otherwise the serial correlation would not be based on enough information to be trustworthy. Under each value of p are listed two values, d_L and d_U, denoting lower and upper limits, respectively, on the value of d. To test the alternative $H_1 : \rho_{t,t-1} > 0$, the procedure is

reject H_0 if $d < d_L$,

accept H_0 if $d > d_U$;

the test is inconclusive   if $d_L \le d \le d_U$.

The procedure for testing the alternative $H_1 : \rho_{t,t-1} < 0$ is

reject H_0 if $d^- < d_L$,

accept H_0 if $d^- > d_U$,

and the test is inconclusive   if $d_L \le d^- \le d_U$.

Some of the rationale behind this test is based on the fact that d has a range from 0 to 4. Values of d around 2 usually indicate that no positive serial correlation is present. (Note that the tabulated values of d_U are generally less than 2.) To test for a negative serial correlation, $d^- = 4 - d$ will convert the d-value based on data with negative serial correlation to a frame of reference comparable to the value of d for positively serially correlated data.

TESTING FOR SERIAL CORRELATION: AN EXAMPLE

Let us consider an example. Suppose that a certain department store tried out a television advertising campaign in order to see what effect, if any, television advertising had on its sales. Data on the number of minutes' advertising per week and the sales in the store were recorded for twenty successive weeks, beginning a week ahead of the advertising. Hypothetical data are given in Table 9.7. Since there could very easily be a cumulative effect of advertising over time, it is necessary to examine the residuals for serial correlation. Calculation of the serial correlation coefficient and of the Durbin-Watson statistic are shown in Table 9.8.

TABLE 9.7   *Data and Calculation for Television Advertising and Sales Example*

Week	Minutes of TV Advertising (X)	Weekly Sales (Y) ($000)	Residual
1	0	8.8	−0.5363
2	5	9.0	−0.9553
3	9	9.2	−1.2505
4	4	10.0	0.1685
5	6	9.6	−0.4791
6	13	10.4	−0.5457
7	18	10.8	−0.7647
8	5	9.8	−0.1553
9	4	9.4	−0.4315
10	9	10.2	−0.2505
11	17	12.0	0.5591
12	13	11.0	0.0543
13	7	10.8	0.5971
14	9	10.8	0.3495
15	6	10.6	0.5209
16	6	11.0	0.9209
17	9	11.2	0.7495
18	7	11.0	0.7971
19	5	10.4	0.4447
20	15	11.4	0.2067
$\sum(\cdot)$	167	207.4	−0.0006
$\sum(\cdot)^2$	1813	2164.68	
$\sum XY$		1783.6	

$S_{XX} = 418.55 \quad S_{XY} = 51.81 \quad S_{YY} = 13.942$
$b = 0.1238 \quad a = 9.3363 \quad \hat{Y} = 9.3363 + 0.1238X$

Since we fitted a simple linear regression equation, $p = 2$. Recall that $n = 20$. If $\alpha = 0.05$, we will reject

$$H_0 : \rho_{t,t-1} = 0$$

in favor of

$$H_1 : \rho_{t,t-1} > 0$$

if $d < d_L = 1.20$.

TABLE 9.8   *Calculation of Serial Correlation Coefficient and Durbin-Watson Statistics*

Week	e_i	e_{i-1}	$e_i e_{i-1}$	$e_i - e_{i-1}$
1	−0.5363	—	—	—
2	−0.9553	−0.5363	0.5123	−0.4190
3	−1.2505	−0.9553	1.1946	−0.2952
4	0.1685	−1.2505	−0.2107	1.4190
5	−0.4791	0.1685	−0.0807	−0.6476
6	−0.5457	−0.4791	0.2614	−0.0666
7	−0.7647	−0.5457	0.4173	−0.2190
8	−0.1553	−0.7647	0.1188	0.6094
9	−0.4315	−0.1553	0.0670	−0.2762
10	−0.2505	−0.4315	0.1081	0.1810
11	0.5591	−0.2505	−0.1401	0.8096
12	0.0543	0.5591	0.0304	−0.5048
13	0.5971	0.0543	0.0324	0.5428
14	0.3495	0.5971	0.2087	−0.2476
15	0.5209	0.3495	0.1821	0.1714
16	0.9209	0.5209	0.4797	0.4000
17	0.7495	0.9209	0.6902	−0.1714
18	0.7971	0.7495	0.5974	0.0476
19	0.4447	0.7971	0.3545	−0.3524
20	0.2067	0.4447	0.0919	−0.2380
$\sum(\cdot)$			4.9153	
$\sum(\cdot)^2$	7.5289	7.4762		4.8967

$$r_{t,t-1} = \frac{4.9153}{7.4762} = 0.6575 \qquad d = \frac{5.8967}{7.5289} = 0.6504$$

Since $d = 0.6504 < 1.20$, we conclude that we are more than 95% sure that there is positive serial correlation in the data. This means that sales have increased over time, as well as with the amount spent on television advertising.

ADJUSTING FOR SERIAL CORRELATION

Regression
on First
Differences

If we want to use only television advertising to estimate sales, we must adjust the data so as to eliminate the time effect. This way we will not violate the assumption of independent error terms. One way is to perform a regression on first differences. A first difference is the difference between a value in one time period and the value in the preceding time period. That is, if we let X' and Y' denote the first differences for the X- and Y-values, respectively, then

$$X_i' = X_i - X_{i-1}$$

$$Y_i' = Y_i - Y_{i-1}.$$

These first differences should take the time effect into account if the value in one time period is affected by the value in the preceding time period. If the interdependence extends back beyond the previous time period, then first differences may not be effective in removing the time dependency. Thus, when transforming the data into first differences, one should make sure that the residuals resulting from the transformed variables are independent. If so, the transformation has effectively eliminated the time dependency; but if not, some other transformation should be made on the data (a second-difference transformation, perhaps). It is also possible for first differences to overcorrect. Then, if the original data show positive serial correlation, the transformed data show negative serial correlation, and vice versa. This is what happened in our example, as can be seen from the calculations in Table 9.9. Since $d^- = 1.13 < 1.20 = d$, we see a significant *negative* serial correlation in the transformed data. Thus, the first-difference transformation was ineffective in this case.

Another method for removing serial correlation uses the transformation

$$Y_i' = Y_i - \hat{\rho} Y_{i-1},$$
$$X_i' = X_i - \hat{\rho} X_{i-1},$$

where $\hat{\rho}$ is an estimate of the serial correlation $\rho_{t,t-1}$. Note that the first-difference transformation can be considered a special case of this transformation, with $\rho_{t,t-1}$ estimated to be 1. We can obtain an estimate of $\rho_{t,t-1}$ from the serial correlation coefficient calculated in Table 9.7 for the original (untransformed) data. There, $r_{t,t-1} = 0.6575$. Table 9.10 shows the residuals, serial correlation, and Durbin-Watson test for the data transformed by

$$Y_i' = Y_i - 0.6575 Y_{i-1}$$
$$X_i' = X_i - 0.6575 X_{i-1}.$$

The symbol $\sum(\cdot)(\cdot)$ denotes a cross-product. From Table 9.10 we see that $d^- = 1.5939 > 1.4 = d_U$, so we can conclude that there is no negative serial correlation in these transformed data. If this transformation had also failed to remove serial correlation, then it would have been necessary to revise our estimate of $\rho_{t,t-1}$ and employ the transformation $Y_i' = Y_i - \hat{\rho} Y_{i-1}, X_i' = X_i - \hat{\rho} X_{i-1}$ using this new value of $\hat{\rho}$. Usually this new value should be slightly larger than $r_{t,t-1}$ in absolute value, because $r_{t,t-1}$ tends to under-estimate ρ. You can work yourself to death, but the idea is to keep repeating the process until a value for $\hat{\rho}$ is found so that the transformed data do not exhibit any serial correlation. This is called an *iterative process*. It is iterated, or repeated, increasing $|\hat{\rho}|$ by small amounts at each step, until a satisfactory result is obtained.

COMPUTER PRINTOUT

The computer programs we have studied all give the value of the Durbin-Watson statistic, either automatically or as an option. However, this statistic does not always have relevance to the problem. For example, in the house size-price example, the data

TABLE 9.9   *First Differences and Durbin-Watson Test for Television Advertising and Sales Example*

Week	X_i	X_{i-1}	Y_i	Y_{i-1}	$X_i' = X_i - X_{i-1}$	$Y_i' = Y_i - Y_{i-1}$	e_i	e_{i-1}	$e_i - e_{i-1}$
1	0	—	8.8	—	—	—	—0.3212	—	—
2	5	0	9.0	8.8	5	0.2	—0.2299	—0.3212	0.0913
3	9	5	9.2	9.0	4	0.2	1.1918	—0.2299	1.4217
4	4	9	10.0	9.2	—5	0.8	—0.6473	1.1918	—1.8391
5	6	4	9.6	10.0	2	—0.4	0.0962	—0.6473	0.7435
6	13	6	10.4	9.6	7	0.8	—0.1212	0.0962	—0.2174
7	18	13	10.8	10.4	5	0.4	0.1222	—0.1212	0.2434
8	5	18	9.8	10.8	—13	—1.0	—0.3734	0.1222	—0.4956
9	4	5	9.4	9.8	—1	—0.4	0.2788	—0.3734	0.6522
10	9	4	10.2	9.4	5	0.8	1.0049	0.2788	0.7261
11	17	9	12.0	10.2	8	1.8	—0.6995	1.0049	—1.7044
12	13	17	11.0	12.0	—4	—1.0	0.2831	—0.6995	0.9826
13	7	13	10.8	11.0	—6	—0.2	—0.2473	0.2831	—0.5304
14	9	7	10.8	10.8	2	0.0	0.0092	—0.2473	0.2565
15	6	9	10.6	10.8	—3	—0.2	0.3353	0.0092	0.3261
16	6	6	11.0	10.6	0	0.4	—0.1386	0.3353	—0.4739
17	9	6	11.2	11.0	3	0.2	—0.0821	—0.1386	0.0565
18	7	9	11.0	11.2	—2	—0.2	—0.4821	—0.0821	—0.4000
19	5	7	10.4	11.0	—2	—0.6	0.0223	—0.4821	0.5044
20	15	5	11.4	10.4	10	1.0			
		$\sum(\cdot)$			15	2.6			
		$\sum(\cdot)^2$			585	9.4	4.2632	4.2627	12.2354
		$\sum X_i Y_i$			54.4				

$$\hat{Y}' = 0.0647 + 0.0913 X'$$

$$r_{t,t-1} = \frac{-1.9064}{4.2627} = -0.4472$$

$$d = \frac{12.2354}{4.2632} = 2.87$$

$$d^- = 4 - 2.87 = 1.13$$

were *not* collected over time. The numbering of the houses in the sample was completely arbitrary. Nevertheless, the computer calculated Durbin-Watson statistics using the residuals in the order given. A different numbering of the houses would have resulted in a different value for the Durbin-Watson statistic, although no other results from the printout (other than the order of values in the list of residuals) would have been affected. It is only when data are entered in a time order that the test for serial correlation has any meaning.

Note that the SPSS computer program plots (standardized) residuals in the order given. This is a very handy feature when the data are collected in a time order.

EXERCISES   *Section 9.4*

9.9   a.   For Exercise 9.7 relating the number of typing errors to the number of words per page, calculate the serial correlation coefficient and test for significant serial correlation.

TABLE 9.10   *Transformed Data and Durbin-Watson Test for Television Advertising and Sales Data.*

Week	$Y_i' = Y_i - 0.6575Y_{i-1}$	X_i'	e_i	e_{i-1}	$e_i - e_{i-1}$
1	—	—	—	—	—
2	3.2140	5.0000	−0.5897	—	—
3	3.2825	5.7125	−0.5860	−0.5897	0.0037
4	3.9510	−1.9175	0.7761	−0.5860	1.3621
5	3.0250	3.3700	−0.6305	0.7761	−1.4066
6	4.0880	9.0550	−0.0843	−0.6305	0.5462
7	3.9620	9.4525	−0.2464	−0.0843	−0.1621
8	2.6990	−6.8350	−0.0289	−0.2464	0.2175
9	2.9565	0.7125	−0.4575	−0.0289	−0.4286
10	4.0195	6.3700	0.0913	−0.4575	0.5488
11	5.2935	11.0825	0.9369	0.0913	0.8456
12	3.1100	1.8225	−0.4049	0.9369	−1.3418
13	3.5675	−1.5475	0.3590	−0.4049	0.7639
14	3.6990	4.3975	−0.0499	0.3590	−0.4089
15	3.4990	0.0825	0.1423	−0.0499	0.1922
16	4.0305	2.0550	0.4945	0.1423	0.3522
17	3.9675	5.0550	0.1588	0.4945	−0.3357
18	4.6360	1.0825	0.1884	0.1588	0.0296
19	3.1675	0.3975	−0.2178	0.1884	−0.4062
20	4.5620	11.7125	0.1481	−0.2178	0.3659
$\sum(\cdot)$	69.7300	67.0600	−0.0005		
$\sum(\cdot)^2$	262.8539	647.9824	3.5453	3.5234	8.5303
$\sum(\cdot)(\cdot)$	283.5037		−0.9047		

$$\hat{Y}' = 3.3492 + 0.0909X'$$

$$r_{t,t-1} = \frac{-0.9047}{3.5234} = -0.2568$$

$$d = \frac{8.5303}{3.5453} = 2.4061$$

$$d^- = 4 - 2.4061 = 1.5939$$

    b.   Try to remove the serial correlation by using first differences. Test to see if this method has been effective.

p. 525   9.10   Refer to the HEATING data. Suppose that the temperature and oil bills were recorded for 15 consecutive days. Test for serial correlation.

9.5 THE ASSUMPTION OF NORMALITY

THE IMPORTANCE OF THE ASSUMPTION

In order for F- or t-tests and confidence intervals to be valid, the values of the test statistics must be calculated from data that are normally distributed. This is why we made the assumption in Chapter 6 that in the population, the possible Y-values for each X-value are normally distributed around the line $\mu_{y \cdot x} = \alpha + \beta X$. Usually this is not an unreasonable assumption to make. For example, for a given size house we would expect a greater concentration of prices near the mean price than far away from it, and we would expect about as many prices above the mean as below. This kind of structure of prices for a given size gives at least a reasonably symmetric, mound-shaped distribution. While the distribution of prices for a given size might not be exactly normal, as long as the distribution is reasonably symmetric, the F- and t-statistics can

Robust

be trusted. This is because F and t are what is called robust to the assumption of normality. This means that as long as the data are not too far from normal, the F-statistic will still follow the F-distribution reasonably well, and the probabilities found from the t-tables will be reasonably applicable to the calculated value of the t-statistic. If we cannot assume that the data come from a normal population, then we can still look for "large" versus "small" t- or F-values in order to see if a variable is a strong predictor. But we cannot answer the question, "How large is *large*?" The probability values from the t- and F-tables will no longer be applicable, so we cannot use the tables to find critical values beyond which "large" (in absolute value) t- or F-values lie.

A TEST FOR NORMALITY

To check out the normality assumption, one quick and easy test is based on the properties of the normal distribution: about 68 % of all normal values lie within one standard deviation of their mean, and so on. In a *standard normal* distribution, $N(0, 1)$, this means that about 68 % of all standard normal values lie between -1 and $+1$, about 95 % lie between -2 and $+2$, and more than 99 % lie between -3 and $+3$.

Now if the Y-values are normally distributed at each X-value, then the errors $\varepsilon_i = Y_i - \mu_{y \cdot x_i}$ will also be normal. Since $\mu_{y \cdot x_i}$ is the true mean of the Y-values at $X = X_i$, the errors will have mean zero. The variance of the Y-values, and thus the variance of the errors, is $\sigma_{y \cdot x_i}^2$. That is, $\varepsilon_i \sim N(0, \sigma_{y \cdot x_i}^2)$. If we standardize the errors by dividing them by their standard deviation, then $\varepsilon_i / \sigma_{y \cdot x_i} \sim N(0, 1)$. In the sample, we can

Standardized
Residuals

estimate these standardized errors by standardized residuals, $e_i / s_{y \cdot x}$, and these residuals can be used to give a quick check on the normality assumption. Approximately 68 % of the standardized residuals should lie between plus and minus 1, and so on.

For our house size-price data, the standardized residuals are shown in Table 9.11. We see that 13 of the standardized residuals, or $13/20 = 0.65 = 65$ % are between -1 and $+1$. There are 17 standardized residuals, 95 %, between -2 and $+2$. One hundred percent of the standardized residuals lie between -3 and $+3$. From this close association between our percentages and those expected for normal data, we have little reason to doubt that our data are reasonably close to normal. Had we found poor agreement with the normal percentages, we would need to transform the data so that the normality assumption would hold. Fortunately, the violation of the normality assumption often occurs along with violation of the homoscedasticity assumption. Thus, the reciprocal transformation described in Section 9.2 for achieving equal variances will almost always give transformed data for which the normality assumption holds.

This method for checking the normality assumption is probably one of the quickest and easiest methods available. It is also the roughest. Other methods, such as plots on normal probability paper or goodness-of-fit tests are also often used to check this assumption.

EXERCISES   *Section 9.5*

p. 521
p. 525

9.11   Form standardized residuals and check for normality in
   a.   the TEACHERS data,
   b.   the HEATING data.
         (Computer printouts in Chapter 4 will show residuals.)

TABLE 9.11   *Standardized Residuals for House Size-Price Example (Residuals Taken from Figure 4.24)*

House	e_i	$e_{i/2.5506}$
1	0.0796	0.0312
2	2.3982	0.9403
3	−3.6305	−1.4235
4	2.1814	0.8553
5	−2.0796	−0.8154
6	−2.0221	−0.7929
7	−3.1372	−1.2301
8	−2.8916	−1.1338
9	3.5000	1.3724
10	2.7611	1.0826
11	5.6305	2.2077
12	−1.3119	−0.5144
13	0.8186	0.3210
14	−2.5575	−1.0028
15	−0.6305	−0.2472
16	2.0509	0.8042
17	0.2389	0.0937
18	−1.2677	−0.4971
19	−0.0796	−0.0312
20	−0.0509	−0.0200

9.6   SUMMARY

Whether a simple linear model or a more complicated one is proposed for a given set of data, it is important to check to see that the model is appropriate—or at least not too inappropriate—and that the assumptions underlying the model have been met. In this chapter we discussed several of the assumptions, ways in which they could be checked out, and measures to be taken if they were not met.

In subsequent chapters we will look at multiple regression models, which are used if one predictor is found to be insufficient. It should be kept in mind that if a multiple regression model is used, the same kind of examination of it is necessary: are the variances constant? are the data normal? does the model fit the data? is there a time effect? Mathematically, almost any model can be used with almost any set of data—with regression equations, correlation coefficients, F- and t-values, and so on, being calculated. Just because such calculations can be made, however, does not say that they are appropriate. Whether a simple or multiple regression model, linear or curvilinear, is adopted, one should be careful to investigate whether or not the theory (model) is supported by the evidence (data). Only then can the inquiry being conducted have meaningful scientific value.

Chapter 10 MULTIPLE REGRESSION

10.1 INTRODUCTION

Very seldom would one perform a simple regression study if the objective is to be able to predict fairly accurately the value of some variable. In most situations, the value of a dependent variable is affected by several related variables. Obviously, the more information we have, the better our predictions should be.

In addition, the researcher often can think of several independent variables that *might* be related to a dependent variable. Without running an experiment and collecting data, however, one does not know which variables will indeed be strong predictors of the response variable. Thus, data will be collected on all k predictors and the response. Then a multiple regression analysis is used to eliminate those predictors that are not effective.

10.2 THE MULTIPLE LINEAR REGRESSION MODEL

Multiple
Linear
Regression

In multiple linear regression, then, our model is

$$Y = \alpha + \beta_1 X_1 + \beta_2 X_2 + \cdots + \beta_k X_k + \varepsilon.$$

That is, Y is a linear function of k predictor variables, $X_1, X_2, \ldots, X_k$, but an imperfect one. That the function is linear can be seen because no regression coefficient (β_i) is raised to a power greater than one (refer to the normal equations presented in the next section). Note that the error term ε is still present. Regardless of the number of factors we take into account, individual variation will prevent us from ever predicting Y exactly accurately.

GEOMETRIC INTERPRETATION

What kind of geometric interpretation can we make for a multiple regression problem? Consider the case in which we have two predictors, X_1 and X_2, of Y. This will lead to a regression equation of the form

$$\hat{Y} = a + b_1 X_1 + b_2 X_2,$$

which is the equation of a two-dimensional plane in three-dimensional space.

Look at the corner made by two walls and the floor of the room you are sitting in. The corner which runs from floor to ceiling, formed by the joining of the left-hand and the right-hand walls, is the Y-axis. The corner made by the right-hand wall and the floor is the X_1-axis, and the corner made by the left-hand wall and the floor is the X_2-axis. The point $(x_1 = 0, x_2 = 0, y = 0)$ is the corner made by the two walls and the floor. Below the floor, Y is negative; on the other side of the left-hand wall, X_1 is negative; and on the other side of the right-hand wall, X_2 is negative. Now, imagine a thin board floating in space near the corner. The height of the board from the floor is the Y-intercept, a. The slope of the board with respect to the right-hand wall is b_1, and the slope of the board with respect to the left-hand wall is b_2.

When the number of predictors is greater than two, the real-world geometrical interpretation breaks down. For those with good abstract imaginations,

$$\hat{Y} = a + b_1 X_1 + b_2 X_2 + \cdots + b_k X_k$$

is the equation of a k-dimensional hyperplane in $(k + 1)$-dimensional space. Because of the difficulty in interpreting the general regression problem in geometric terms, we emphasize interpreting the regression constant and regression coefficients in terms specific to the problem.

OBJECTIVES OF THE ANALYSIS

The objectives in a multiple regression problem are essentially the same as for a simple regression. We want to estimate a and $b_1, \ldots, b_k$; test $b_1, \ldots, b_k$ for the significance of the associated predictors; use the regression equation to estimate Y from $X_1, X_2, \ldots, X_k$; and measure the error involved in estimation. While the objectives remain the same, the calculations and interpretations become more complicated the more predictors we have.

We will use as an illustrative example the problem of estimating the price (Y) of a house from knowing its size (X_1) and also its age (X_2). The sizes and prices will be the same as in the simple regression problem. What we have done is add ages of houses to the existing data. Note carefully that in real life, one would *not* first go out and collect data on sizes and prices, analyze the simple regression problem, and then go out again and find the ages of the twenty houses sampled. Rather, one would collect all data which might be pertinent on all twenty houses at the outset. Then the analysis performed would throw out predictors which turn out not to be needed.

10.3 NORMAL EQUATIONS

NORMAL EQUATIONS FOR THE TWO-PREDICTOR MODEL

When using age and size to predict prices of houses, our model is

$$Y = \alpha + \beta_1 X_1 + \beta_2 X_2 + \varepsilon.$$

We shall use the data of Table 10.1 to find a, b_1, and b_2 in the regression equation

$$\hat{Y} = a + b_1 X_1 + b_2 X_2.$$

In order to find a, b_1, and b_2, we will need to solve three normal equations in three unknowns. Recalling the trick for writing down normal equations in Chapter 2, we obtain the normal equations

$$\sum Y = na + b_1 \sum X_1 + b_2 \sum X_2$$
$$\sum X_1 Y = a \sum X_1 + b_1 \sum X_1^2 + b_2 \sum X_1 X_2$$
$$\sum X_2 Y = a \sum X_2 + b_1 \sum X_1 X_2 + b_2 \sum X_2^2.$$

TABLE 10.1   *Sizes, Ages, and Prices of Twenty Houses*

House No.	$X_1 = Size$ (000 sq ft)	$X_2 = Age$ (years)	Y ($000)
1	1.8	30	32
2	1.0	33	24
3	1.7	25	27
4	2.8	12	47
5	2.2	26	35
6	0.8	25	17
7	3.6	28	52
8	1.1	29	20
9	2.0	25	38
10	2.6	2	45
11	2.3	30	44
12	0.9	23	19
13	1.2	12	25
14	3.4	33	50
15	1.7	1	30
16	2.5·	12	43
17	1.4	17	27
18	3.3	16	50
19	2.2	22	37
20	1.5	29	28

(Note that the indices of summation have been deleted; for example, $\sum X_1$ really means

$$\sum_{i=1}^{n} X_{1i} = X_{11} + X_{12} + \cdots + X_{1n},$$

or the sum of all the X_1-values (sizes), of which there are n.) Of course, if we had three predictors, then we would have to solve four equations in four unknowns, and so on.

NORMAL EQUATIONS FOR HOUSE SIZE–PRICE–AGE EXAMPLE

The normal equations indicate that the following calculations need to be performed on the data:

$$\sum X_1 = 40, \qquad \sum X_1^2 = 93.56, \qquad \sum X_1 Y = 1554.9;$$
$$\sum X_2 = 430, \qquad \sum X_2^2 = 10{,}990, \qquad \sum X_2 Y = 14{,}540;$$
$$\sum Y = 690, \qquad \sum Y^2 = 26{,}178, \qquad \sum X_1 X_2 = 844.7.$$

Since what we are calling Y now is the same as in simple regression, $\sum Y$ and $\sum Y^2$ have not changed from the earlier example. What was just X in simple regression is now called X_1, so that what was $\sum X$ before is now $\sum X_1$; what was $\sum X^2$ is now $\sum X_1^2$; and what was $\sum X Y$ is now $\sum X_1 Y$. The same sorts of calculations are performed using X_2 and, in addition, $\sum X_1 X_2$ is also calculated.

For this example, then, the normal equations are

$$690 = 20a + 40b_1 + 430b_2$$
$$1554.9 = 40a + 93.56b_1 + 844.7b_2$$
$$14{,}540 = 430a + 844.7b_1 + 10{,}990b_2.$$

Now the problem is to solve these three equations in three unknowns. While several equivalent alternative methods can be used, one reason the matrix approach of Chapter 3 was formulated was so that we could use matrix methods to obtain a solution.

NORMAL EQUATIONS IN MATRIX FORM

Define

$$\mathbf{X} = \begin{bmatrix} 1 & X_{11} & X_{21} \\ 1 & X_{12} & X_{22} \\ \vdots & \vdots & \vdots \\ 1 & X_{1n} & X_{2n} \end{bmatrix} = \begin{bmatrix} 1 & 1.8 & 30 \\ 1 & 1.0 & 33 \\ \vdots & \vdots & \vdots \\ 1 & 1.5 & 29 \end{bmatrix}$$

Then

$$\mathbf{X'X} = \begin{bmatrix} n & \sum X_1 & \sum X_2 \\ \sum X_1 & \sum X_1^2 & \sum X_1 X_2 \\ \sum X_2 & \sum X_1 X_2 & \sum X_2 \end{bmatrix} = \begin{bmatrix} 20 & 40 & 430 \\ 40 & 93.56 & 844.7 \\ 430 & 844.7 & 10{,}990 \end{bmatrix}.$$

Note that $\mathbf{X'X}$ is symmetric. If we had more than 2 predictors, it is a simple extension to define $\mathbf{X}$ and $\mathbf{X'X}$. If there are k predictors, $\mathbf{X}$ will be $n \times (k + 1)$. The first column of $\mathbf{X}$ will contain ones, the second column will contain X_1-values, the third will contain X_2 values,..., and the $(k + 1)$st column will contain values of X_k. $\mathbf{X'X}$ will then be $(k + 1) \times (k + 1)$ and be of the form

$$\begin{bmatrix} n & \sum X_1 & \sum X_2 & \cdots & \sum X_k \\ & \sum X_1^2 & \sum X_1 X_2 & \cdots & \sum X_1 X_k \\ & & \sum X_2^2 & \cdots & \sum X_2 X_k \\ & & & & \vdots \\ \text{sym.} & & & & \sum X_k^2 \end{bmatrix}.$$

As always,

$$\mathbf{y} = \begin{bmatrix} Y_1 \\ Y_2 \\ \vdots \\ Y_n \end{bmatrix} = \begin{bmatrix} 32 \\ 24 \\ \vdots \\ 28 \end{bmatrix}.$$

so that

$$\mathbf{X'y} = \begin{bmatrix} \sum Y \\ \sum X_1 Y \\ \sum X_2 Y \end{bmatrix} = \begin{bmatrix} 690 \\ 1554.9 \\ 14{,}540 \end{bmatrix}.$$

$\mathbf{X'y}$ also extends directly for $k > 2$ and is of the form

$$\begin{bmatrix} \sum Y \\ \sum X_1 Y \\ \sum X_2 Y \\ \vdots \\ \sum X_k Y \end{bmatrix}.$$

Finally,

$$\hat{\beta} = \begin{bmatrix} a \\ b_1 \\ b_2 \end{bmatrix},$$

and this also extends directly to

$$[a \quad b_1 \quad b_2 \quad \cdots \quad b_k]'$$

if more than two predictors are used. Then the normal equations are

$$(\mathbf{X}'\mathbf{X})\hat{\beta} = \mathbf{X}'\mathbf{y},$$

or

$$\begin{bmatrix} 20 & 40 & 430 \\ 40 & 93.56 & 844.7 \\ 430 & 844.7 & 10{,}990 \end{bmatrix} \begin{bmatrix} a \\ b_1 \\ b_2 \end{bmatrix} = \begin{bmatrix} 690 \\ 1554.9 \\ 14{,}540 \end{bmatrix}.$$

SOLVING THE NORMAL EQUATIONS

In this example, as in simple regression, the normal equations are solved by

$$\hat{\beta} = (\mathbf{X}'\mathbf{X})^{-1}\mathbf{X}'\mathbf{y},$$

so all we have to do is to invert $\mathbf{X}'\mathbf{X}$ and perform a multiplication. The beauty of the matrix approach is that once $\mathbf{X}$, $\mathbf{y}$, and $\hat{\beta}$ are appropriately defined, the normal equations are solved by $\hat{\beta} = (\mathbf{X}'\mathbf{X})^{-1}\mathbf{X}'\mathbf{y}$, regardless of the number of predictors we have in any problem. The tedious arithmetic involved in inverting $\mathbf{X}'\mathbf{X}$ is best done by a computer. (Note that Exercise 4.19 used the computer to find this inverse.) We find

$$(\mathbf{X}'\mathbf{X})^{-1} = \begin{bmatrix} 0.67164255 & -0.16300717 & -0.01375015 \\ -0.16300717 & 0.07448317 & 0.00065306 \\ -0.01375015 & 0.00065306 & 0.00057879 \end{bmatrix}.$$

(Compare this to the answer to Exercise 4.19.) Notice that eight decimal places are presented in $(\mathbf{X}'\mathbf{X})^{-1}$. It is necessary to carry many decimal places because of the considerable rounding error incurred in inversion. A partial check of the calculations is given by

$$(\mathbf{X'X})^{-1}\mathbf{X'X} = \begin{bmatrix} 0.99999970 & -0.00000054 & -0.00000850 \\ -0.00000054 & 0.99999837 & -0.00002000 \\ -0.00000850 & -0.00002000 & 0.99997738 \end{bmatrix}.$$

$$\approx \mathbf{I}.$$

Carrying fewer decimal places results in $(\mathbf{X'X})^{-1}\mathbf{X'X}$ deviating further from $\mathbf{I}$; carrying more decimal places will bring the product closer to $\mathbf{I}$.

Proceeding with the solution, we obtain

$$\hat{\boldsymbol{\beta}} = \begin{bmatrix} 0.67164255 & -0.16300717 & -0.01375015 \\ -0.16300717 & 0.07448317 & 0.00065306 \\ -0.01375015 & 0.00065306 & 0.00057879 \end{bmatrix} \begin{bmatrix} 690 \\ 1554.9 \\ 14,540 \end{bmatrix}$$

$$= \begin{bmatrix} 10.04632987 \\ 12.83442613 \\ -0.05655391 \end{bmatrix} = \begin{bmatrix} a \\ b_1 \\ b_2 \end{bmatrix}.$$

Substituting back into the normal equations, we see that the check is pretty good:

$$20(10.04632987) + 40(12.83442613) + 430(-0.05655391) = 689.98546130$$

$$40(10.04632987) + 93.56(12.83442613) + 844.7(-0.05655391) = 1554.87101574$$

$$430(10.0463298) + 844.7(12.83442613) + 10,990(-0.05655391) = 14,539.63499511.$$

If even more decimal places were carried in $(\mathbf{X'X})^{-1}$, the check would come out even better. The question as to why so many decimal places need be carried will become apparent later. Suffice it to say for now that any element of $(\mathbf{X'X})^{-1}$ is determined in part at least, by *every* element of the original matrix, $\mathbf{X'X}$. It is the relationships among all the elements of $\mathbf{X'X}$ that cause the difficulty in $(\mathbf{X'X})^{-1}$. Had only six decimal places been carried, the first equation would have given 691.3173, the second 1557.474, and the third 14,573.631, showing only a very rough check.

CORRECTING *X* FOR ITS MEAN

Some of this rounding error can be reduced if we correct X_1 and X_2 for their means. That is, instead of using the original X_1- and X_2-values, use

$$U_1 = X_1 - \bar{X}_1$$

and

$$U_2 = X_2 - \bar{X}_2.$$

The vector **y** will be unaffected, but

$$\hat{\beta} = \begin{bmatrix} a' \\ b_1 \\ b_2 \end{bmatrix}$$

and

$$\mathbf{X} = \begin{bmatrix} 1 & X_{11} - \bar{X}_1 & X_{21} - \bar{X}_2 \\ 1 & X_{12} - \bar{X}_1 & X_{22} - \bar{X}_2 \\ \vdots & \vdots & \vdots \\ 1 & X_{1n} - \bar{X}_1 & X_{2n} - \bar{X}_2 \end{bmatrix},$$

where $\bar{X}_1$ is the mean of the X_1-values and $\bar{X}_2$ is the mean of the X_2-values. Then

$$\mathbf{X'X} = \begin{bmatrix} n & \sum(X_{1i} - \bar{X}_1) & \sum(X_{2i} - \bar{X}_2) \\ \sum(X_{1i} - \bar{X}_1) & \sum(X_{1i} - \bar{X}_1)^2 & \sum(X_{1i} - \bar{X}_1)(X_{2i} - \bar{X}_2) \\ \sum(X_{2i} - \bar{X}_2) & \sum(X_{1i} - \bar{X}_1)(X_{2i} - \bar{X}_2) & \sum(X_{2i} - \bar{X}_2)^2 \end{bmatrix}$$

$$= \begin{bmatrix} n & 0 & 0 \\ 0 & S_{11} & S_{12} \\ 0 & S_{12} & S_{22} \end{bmatrix},$$

since

$$\sum(X_{1i} - \bar{X}_1) = \sum(X_{2i} - \bar{X}_2) = 0$$

$$\sum(X_{1i} - \bar{X}_1)^2 = S_{X_1 X_1}, \quad \text{or } S_{11} \text{ for convenience,}$$

$$\sum(X_{2i} - \bar{X}_2)^2 = S_{X_2 X_2}, \quad \text{or } S_{22},$$

and

$$\sum(X_{1i} - \bar{X}_1)(X_{2i} - \bar{X}_2) = S_{X_1 X_2}, \quad \text{or } S_{12}.$$

That is, S_{11} is an S_{XX} using X_1-values; S_{22} is an S_{XX} using X_2-values; and S_{12} is like an S_{XY}, except instead of using an X and a Y, we use an X_1 and an X_2. Calculating formulas for these three quantities are

$$S_{11} = \frac{n \sum X_1^2 - (\sum X_1)^2}{n} = \frac{20(93.56) - (40)^2}{20} = 13.56$$

(which is what S_{XX} was in the simple regression problem),

$$S_{22} = \frac{n \sum X_2^2 - (\sum X_2)^2}{n} = \frac{20(10{,}990) - (430)^2}{20} = 1745,$$

$$S_{12} = \frac{n \sum X_{12} - (\sum X_1)(\sum X_2)}{n} = \frac{20(844.7) - (40)(430)}{20} = -15.3.$$

So in our example, if X_1 and X_2 are corrected for their means,

$$\mathbf{X'X} = \begin{bmatrix} 20 & 0 & 0 \\ 0 & 13.56 & -15.3 \\ 0 & -15.3 & 1745 \end{bmatrix},$$

which is close to being diagonal and thus is relatively simple to invert. Think of $\mathbf{X'X}$ partitioned as

$$\mathbf{X'X} = \left[\begin{array}{c|cc} 20 & 0 & 0 \\ \hline 0 & 13.56 & -15.3 \\ 0 & -15.3 & 1745 \end{array} \right].$$

Then

$$(\mathbf{X'X})^{-1} = \left[\begin{array}{c|cc} \frac{1}{20} & 0 & 0 \\ \hline 0 & & \\ & \mathbf{S}^{-1} & \\ 0 & & \end{array} \right],$$

where

$$\mathbf{S} = \begin{bmatrix} 13.56 & -15.3 \\ -15.3 & 1745 \end{bmatrix}.$$

Then

$$\mathbf{S}^{-1} = \frac{1}{(13.56)(1745) - (-15.3)^2} \begin{bmatrix} 1745 & 15.3 \\ 15.3 & 13.56 \end{bmatrix}$$

$$= \frac{1}{23,428.11} \begin{bmatrix} 1745 & 15.3 \\ 15.3 & 13.56 \end{bmatrix}$$

$$= \begin{bmatrix} 0.074483 & 0.000653 \\ 0.000653 & 0.000579 \end{bmatrix},$$

and

$$(\mathbf{X}'\mathbf{X})^{-1} = \begin{bmatrix} 0.05 & 0 & 0 \\ 0 & 0.074483 & 0.000653 \\ 0 & 0.000653 & 0.000579 \end{bmatrix}.$$

Partitioning $\mathbf{X}'\mathbf{X}$ makes the inversion relatively easy.

Proceeding to solve the normal equations, we note that if X_1 and X_2 are corrected for their means, then

$$\mathbf{X}'\mathbf{y} = \begin{bmatrix} \sum Y_i \\ \sum(X_{1i} - \bar{X}_1)Y_i \\ \sum(X_{2i} - \bar{X}_2)Y_i \end{bmatrix} = \begin{bmatrix} \sum Y_i \\ S_{1Y} \\ S_{2Y} \end{bmatrix},$$

where

$$S_{1Y} = S_{X_1Y} = \sum(X_{1i} - \bar{X}_1)(Y_i)$$

$$= \sum(X_{1i} - \bar{X}_1)(Y_i - \bar{Y}) = \frac{n\sum X_1Y - (\sum X_1)(\sum Y)}{n}$$

$$= \frac{20(1554.9) - (40)(690)}{20} = 174.9$$

(which is what S_{XY} was when our only predictor was $X = X_1 =$ size), and

$$S_{2Y} = S_{X_2Y} = \sum(X_{2i} - \bar{X}_2)(Y_i) = \sum(X_{2i} - \bar{X}_2)(Y_i - \bar{Y})$$

$$= \frac{n\sum X_2Y - (\sum X_2)(\sum Y)}{n}$$

$$= \frac{20(14,540) - (430)(690)}{20} = -295.$$

Then if

$$\mathbf{X'y} = \begin{bmatrix} 690 \\ 174.9 \\ -295 \end{bmatrix},$$

the solution to the normal equations is

$$\hat{\beta} = \begin{bmatrix} a' \\ b_1 \\ b_2 \end{bmatrix} = \begin{bmatrix} 0.05 & 0 & 0 \\ 0 & 0.074483 & 0.000653 \\ 0 & 0.000653 & 0.000579 \end{bmatrix} \begin{bmatrix} 690 \\ 174.9 \\ -295 \end{bmatrix}$$

$$= \begin{bmatrix} 34.5 \\ 12.834442 \\ -0.056595 \end{bmatrix}.$$

Note that $a' = \bar{Y} = 34.5$, and b_1 and b_2 have changed from before only slightly. Using these values to check the normal equations we find

$$20(34.5) + 0(12.834442) + 0(-0.056595) = 690$$

$$0(34.5) + 13.56(12.834442) - 15.3(-0.056595) = 174.900938$$

$$0(34.5) - 15.3(12.834442) + 1745(-0.056595) = -295.125238.$$

When X_1 and X_2 are corrected for their means, we only need to carry six decimal places to obtain as good a check on the solution to the normal equations as we got by carrying eight decimal places when the predictors were not corrected for their means. Thus a considerable savings in effort and rounding error can be accomplished in multiple regression if the predictors are corrected for their means. As you can imagine, the savings become even more pronounced when more than two predictors are involved, and the calculations must be done by hand.

For the sake of simplicity, we shall use the following values for a, b_1, and b_2, except when they are needed in other calculations:

$a = 10.05,$

$b_1 = 12.83,$

$b_2 = -0.06,$

so that our regression equation is

$$\hat{Y} = 10.05 + 12.83X_1 - 0.06X_2.$$

In the next section, we shall examine the interpretation and use of this equation.

EXERCISES   *Section 10.3*

SALES II Data

Referring to the SALES data, which gives data about the weekly retail sales of stores and their numbers of employees, suppose that we had also the following information on the sizes of the stores, in thousands of square feet.

$Y = $ Average Weekly Retail Sales ($000)	$X_1 = $ Number of Employees	$X_2 = $ Size (000 sq ft)
7	17	7
17	39	9
10	32	8
5	17	4
7	25	5
15	43	9
11	25	8
13	32	10
19	48	12
3	10	5
17	48	12
15	42	10
14	36	10
12	30	10
8	19	8

10.1 a. Find the simple linear regression equation relating Y to X_2.
  b. Find the multiple linear regression equation

$$\hat{Y} = a + b_1 X_1 + b_2 X_2.$$

(You may want to use the computer program given in Chapter 4, or one similar to it, to invert $\mathbf{X'X}$.)

RESTAURANT Data

In an effort to decide whether or not to apply for a liquor license, a restaurant owner sampled ten comparable restaurants in his town and recorded the total profits (Y), the food sales (X_1), and the liquor sales (X_2) for the past month. His sample yielded the following results. (All figures are in thousands of dollars.)

Restaurant	Y	X_1	X_2
1	4.5	3.3	4.0
2	2.0	1.7	1.5
3	1.7	1.5	2.0
4	2.3	1.7	3.0
5	4.0	3.0	3.0
6	4.1	3.2	3.7
7	2.5	1.6	2.0
8	3.7	3.5	4.0
9	3.6	3.0	3.5
10	3.0	3.0	3.5

10.2 For the RESTAURANT data, find the multiple linear regression equation relating profits to sales of food and liquor.

DRUG Data

A pharmaceutical company testing a new pain-killing drug tests the drug on twenty people suffering from arthritis. The time elapsed, in minutes, from taking the drug until a noticeable relief in pain is detected, is to be predicted from the dosage (in grams) and the age of the patient (in years). The results are given below.

Patient	Time	Dosage	Age
1	11	2	59
2	3	2	57
3	20	2	22
4	25	2	12
5	27	2	18
6	15	5	40
7	10	5	64
8	34	5	27
9	14	5	54
10	34	5	22
11	35	7	33
12	28	7	49
13	23	7	29
14	21	7	32
15	33	7	20
16	27	10	43
17	8	10	61
18	3	10	69
19	12	10	62
20	14	10	61

10.3 For the DRUG data, find the regression equation relating time to relief to dosage and age.

AUTO Data

A company's motor pool collected the following information concerning the cost of maintenance and repairs of the vehicles during the last year.

Vehicle No.	Age	Mileage during the year (thousands)	Cost ($00)
1	3	12	1.2
2	3	36	1.5
3	2	10	0.9
4	4	14	1.5
5	3	16	1.0
6	5	22	4.0
7	2	14	0.8
8	2	10	0.8
9	2	24	1.0
10	5	14	2.0
11	5	18	2.0
12	5	21	3.0
13	5	28	3.2
14	0*	27	0.5
15	5	14	1.9

* Age 0 denotes a car less than 1 year old.

10.4 Using the AUTO data, find the multiple linear regression equation relating maintenance and repair costs to age and mileage of the vehicles.

ADS Data

A large discount department store chain advertises on television (X_1), on the radio (X_2), and in newspapers (X_3). A sample of 12 of its stores in a certain area showed the following advertising expenditures and revenues during a given month. (All figures are in thousands of dollars.)

Store	Revenues	TV	Radio	Paper
1	84	13	5	2
2	84	13	7	1
3	80	8	6	3
4	50	9	5	3
5	20	9	3	1
6	68	13	5	1
7	34	12	7	2
8	30	10	3	2
9	54	8	5	2
10	40	10	5	3
11	57	5	6	2
12	46	5	7	2

10.5 From the ADS data, find the regression equation relating revenues to the three kinds of advertising expenditures.

CHILDREN Data

In a study of grade school children, ages, heights (in inches), weights and scores on a physical fitness test were recorded, as follows.

Child	Score	Age	Height	Weight
1	58	7	47.5	53
2	54	7	45.0	50
3	55	9	52.5	85
4	74	7	48.0	52
5	86	9	55.0	76
6	98	8	51.0	64
7	96	9	53.0	75
8	70	7	46.0	75
9	40	7	48.0	68
10	67	9	50.5	74
11	41	6	45.0	40
12	41	7	48.5	66
13	47	8	50.5	65
14	45	8	49.0	70
15	92	9	51.5	70
16	50	7	46.5	60
17	98	9	53.5	77
18	42	8	45.0	65
19	64	8	52.5	65
20	70	8	51.5	67

10.6 Find the multiple linear regression equation relating the scores to the ages, heights, and weights of the children.

GLUE Data

An experiment was run to test the adhesive strength of a new glue (in pounds) under different conditions of temperature (°F) and humidity (percent).

Strength	Temperature (°F)	Humidity (%)
190	80	40
189	80	40
192	90	40
190	90	40
196	80	60
193	80	60
195	90	60
196	90	60
201	80	80
200	80	80
203	90	80
205	90	80

10.7 For the GLUE data, find the multiple linear regression equation relating strength to temperature and humidity. (*Hint*: code the X-values.)

BOTTLES Data

Returnable soft drink bottles must be sterilized before they are refilled. An experiment was run by a bottler to study the effects of three different temperature levels (coded low, -1; medium, 0; and high, $+1$) and four different sterilizing times (coded short, -2; medium short, -1; medium long, $+1$; and long, $+2$) on the number of impurities remaining. The results are given below.

Time	Temperature	Impurities
-2	-1	4
-2	-1	5
-2	0	6
-2	0	4
-2	1	5
-2	1	5
-1	-1	3
-1	-1	5
-1	0	4
-1	0	4
-1	1	2
-1	1	3
1	-1	4
1	-1	3
1	0	3
1	0	3
1	1	2
1	1	1
2	-1	2
2	-1	1
2	0	1
2	0	0
2	1	0
2	1	0

10.8 Find the multiple linear regression equation relating the number of impurities remaining to sterilization time and temperature, for the BOTTLES data.

YIELD Data

To determine the effects of nitrogen, potassium, and phosphorus fertilizers on a certain crop, high, medium, and low concentrations (coded $-1, 0, 1$) of the three elements were tested. The yields were as follows.

Yield	Nitrogen	Potassium	Phosphorus
539	−1	−1	−1
319	−1	−1	0
164	−1	−1	1
228	−1	0	−1
207	−1	0	0
178	−1	0	1
180	−1	1	−1
250	−1	1	0
340	−1	1	1
491	0	−1	−1
305	0	−1	0
514	0	−1	1
380	0	0	−1
364	0	0	0
366	0	0	1
192	0	1	−1
171	0	1	0
187	0	1	1
354	1	−1	−1
415	1	−1	0
326	1	−1	1
239	1	0	−1
292	1	0	0
446	1	0	1
250	1	1	−1
527	1	1	0
477	1	1	1

10.9 a. Find the multiple linear regression equation relating yield (in pounds per acre) to concentrations of the three elements.

b. Find each of the three simple linear regression equations.

c. Find all three two-predictor linear regression equations.

d. Compare the results in parts a, b, and c.

10.4 INTERPRETATION AND USE OF THE MULTIPLE REGRESSION EQUATION

ESTIMATING AN AVERAGE

The same considerations for the use of the regression equation with a single predictor apply for each predictor in multiple regression. First, as before, the regression equation is intended to give only an estimated *average* Y-value for the corresponding X-values. For example, we could estimate that the price of a 26-year-old house having 2200 square feet would be

$$Y = 10.04632987 + 12.83442613(2.2) - 0.05655391(26)$$

$$= 36.81166570,$$

or about $36,812. The interpretation is that the *average* price of houses 2200 square feet in size and 26 years old is estimated at $36,812. House No. 5 in our sample, of this size and age, sells for $35,000, which is close to, but not equal to the predicted price. Factors other than age and size, as well as random variation, lowered the price of House No. 5 below the typical price of houses that age and size. After we calculate the standard error of estimate for this problem, we will be able to give a confidence interval for these prices and can then tell whether or not House No. 5 is unusually low-priced.

RELEVANT RANGE

As in simple regression, it is valid to make estimates of price only for houses in the range of sizes *and* the range of ages sampled. For example, since we have information only on houses between one and 33 years old, we cannot use the regression equation to predict the price of a 50-year-old house, even if it is in our size range. We do not know whether prices will be more or less expensive for older houses or if they will behave in the same way as prices of houses between one and 33 years old. In addition, if any or all of the predictors are controlled experimentally, then valid inferences can be made only at the experimental levels chosen for those variables.

THE REGRESSION CONSTANT

In order for the regression constant, a, to have meaning, the two conditions specified in Chapter 5 must hold for *all* predictors. Thus, in order for a to be the average Y-value when X_1 and X_2 are zero, then

1. it must be physically possible for *both* X_1 and X_2 to be zero, and
2. one must have data around $X_1 = 0$ *and* around $X_2 = 0$.

Naturally, if we have $k > 2$ predictors, the two conditions must hold for *all* k of them. For the house size-age-price example, even though a brand-new house would probably be considered to have age zero, it is impossible for a house to have no size. Thus, $a = 10.05$ does *not* mean that the average house of age zero and size zero will sell for $10,050. Rather, some other factors besides age and size determine this constant, which is needed in order to predict price accurately from size and age.

NET REGRESSION COEFFICIENTS

Net Regression
Coefficient

When there is more than one predictor present, $b_1, b_2, \ldots, b_k$ are called <u>net regression co-efficients</u>. Their interpretation is similar to that in simple regression. For instance, b_1 measures the effect of X_1 on Y, by "netting out," or controlling for, the effect of X_2 on Y. Thus in our example, $b_1 = 12.83$ means that for two houses the same age (holding X_2 constant), if one house is 1000 square feet larger than the other, then on the average it will cost \$12,830 more than the smaller house. And $b_2 = -0.06$ means that if two houses are the same size but one is one year older than the other, then on the average the older house will sell for \$60 less than the newer house.

EXERCISES   *Section 10.4*

10.10 Interpret the meaning of a, b_1, and b_2 in

p. 524   a.   the SALES data,
p. 531   b.   the RESTAURANT data,
p. 532   c.   the DRUG data,
p. 532   d.   the AUTO data,
p. 535   e.   the GLUE data,
p. 535   f.   the BOTTLES data.

10.11 Interpret the meaning of a, b_1, b_2, and b_3 in

p. 533   a.   the ADS data,
p. 533   b.   the CHILDREN data,
p. 536   c.   the YIELD data.

p. 531   *10.12* a.   Using the regression equation found for the SALES II data, estimate the sales of a store 7000 square feet in size employing 17 people. Compare your answer to the actual sales in the first store listed in the sample which had sales of \$7000, and explain any differences.

          b.   Using the regression equation found for the RESTAURANT data, estimate the total profits in a restaurant with \$1700 food sales and \$3000 liquor sales. Compare this estimate to the actual profits of Restaurant No. 4 which had revenues of \$2300, and explain any differences.

          c.   Can you use the regression equation found for the DRUG data to estimate how long it will take a 29-year-old patient to experience relief from pain if he is given three grams of drug? Explain why or why not.

          d.   Referring to the AUTO data, suppose another car was one-year old and was driven 15,000 miles. Estimate the maintenance and repair costs and interpret the meaning of the estimate.

          e.   Referring to the ADS data, can you use the regression equation to estimate the sales of a store spending \$10,000 on television advertising, \$5000 on radio ads, and \$6000 on newspaper ads? Explain why or why not.

          f.   Use the regression equation found for the CHILDREN data to estimate the physical fitness score of a nine-year-old child who is 50.5 inches tall and weighs 74 pounds. Compare this estimate to Child No. 10, who scored 67, and explain any differences.

g. Use the regression equation from the GLUE data to estimate the strength of the glue at 80 degrees temperature and 60 % humidity. Compare this estimate to the actual strengths found in the sample, 193 and 196 pounds, and explain any differences.

h. Referring to the BOTTLES data, what is the average number of impurities you would expect to find in bottles sterilized at high temperatures for a medium short time? Compare this estimate to comparable cases in the experiment in which there were 3 and 4 impurities left, and explain any differences.

i. What is the estimated average yield of the crop when fertilized with high concentrations of all three elements in the YIELD data?

10.5 MULTICOLLINEARITY

THE COEFFICIENTS HAVE CHANGED

In the foregoing interpretation of the regression coefficients, the value of b_1 is different from the value obtained for b in simple regression, even though the same data were used and b and b_1 were both supposed to measure the effect of size on price. In this case $b_1 = 12.83$, while in the simple regression problem, $b = 12.9$. Granted, this is not much of a difference, but it is a real one; it is not caused by rounding error. Why has this value changed? The obvious reason is that b_1 is trying to "net out" or control for, the effect of age on price and measure only the effect of size, while b in the simple regression problem ignored the fact that age as well as size might affect price.

One could readily imagine that if a simple regression were run relating price to age alone, the regression coefficient would not exactly equal -0.06. Further, if a third variable were included, we expect b_1 and b_2 to be different from 12.83 and -0.06, since both b_1 and b_2 would be holding the effect of X_3 constant, which they are not doing at present.

CAN YOU HOLD ONE VARIABLE CONSTANT WHILE VARYING THE OTHER?

This is all fine, *provided* it is possible to hold one variable constant while allowing the other to vary. Of course it can be done *mathematically*, but can it be done in the real world? For example, suppose that in the last 40 years, the trend in housing has been to build smaller and smaller houses. In that case, generally the older the house, the larger it is, and newer houses tend to be smaller. In that case, there is a direct relationship between the two predictors size and age. Then it will be impossible to consider houses the same size but of varying ages, or houses the same age with varying sizes. Instead, if you look at new houses, you will see small sizes, and if you look at old houses, you will see large sizes.

If the predictors themselves are related, then, *can* b_1 actually net out the effect of X_2, and *can* b_2 really control for X_1? Obviously not. Even though b_1 is trying to measure only the effect of X_1 on Y and none of the effect of X_2, if both predictors are related, then b_1 will be contaminated to some extent. It will include some of the effect of X_2 on Y. A similar statement can be made for b_2. We see in our example that b_1 must be contaminated with some of the effect of X_2, because its value in the multiple regression equation has changed from what it was in the simple regression equation. Since the X_1-values and Y-values are the same in each instance, the change in b_1 must be caused by some X_2-values entering into its calculation.

CALCULATION OF REGRESSION COEFFICIENTS

To see how this comes about, let us suppose that X_1 and X_2 are corrected for their means so, that

$$\mathbf{X'X} = \begin{bmatrix} n & 0 & 0 \\ 0 & S_{11} & S_{12} \\ 0 & S_{12} & S_{22} \end{bmatrix}, \quad \mathbf{X'y} = \begin{bmatrix} \sum y \\ S_{1Y} \\ S_{2Y} \end{bmatrix},$$

and

$$\mathbf{b} = \begin{bmatrix} a' \\ b_1 \\ b_2 \end{bmatrix}.$$

Then

$$\mathbf{b} = (\mathbf{X'X})^{-1}\mathbf{X'y}$$

$$= \begin{bmatrix} \dfrac{1}{n} & 0 & 0 \\ 0 & \dfrac{S_{22}}{S_{11}S_{22} - S_{12}^2} & \dfrac{-S_{12}}{S_{11}S_{22} - S_{12}^2} \\ 0 & \dfrac{S_{12}}{S_{11}S_{22} - S_{12}^2} & \dfrac{S_{11}}{S_{11}S_{22} - S_{12}^2} \end{bmatrix} \begin{bmatrix} \sum Y \\ S_{1Y} \\ S_{2Y} \end{bmatrix}$$

$$= \begin{bmatrix} \dfrac{1}{n}\sum Y \\ \dfrac{S_{22}}{S_{11}S_{22} - S_{12}^2}S_{1Y} - \dfrac{S_{12}}{S_{11}S_{22} - S_{12}^2}S_{2Y} \\ \dfrac{-S_{12}}{S_{11}S_{22} - S_{12}^2}S_{1Y} + \dfrac{S_{11}}{S_{11}S_{22} - S_{12}^2}S_{2Y} \end{bmatrix} = \begin{bmatrix} a \\ b_1 \\ b_2 \end{bmatrix}$$

Since $b_1 = (S_{22}S_{1Y} - S_{12}S_{2Y})/(S_{11}S_{22} - S_{12}^2)$, we see that X_2-values are entering into the calculation of b_1 (and a similar statement can, of course, be made about b_2). Only if the covariance S_{12} between the predictors is zero will b_1 measure only the effect of X_1 on Y: if $S_{12} = 0$, then

$$b_1 = \frac{S_{22}S_{1Y} - 0S_{2Y}}{S_{11}S_{22} - 0^2} = \frac{S_{1Y}}{S_{11}},$$

which involves only values of X_1 and Y. The stronger the relationship between X_1 and X_2, the greater the contamination of the regression coefficient by predictors whose effect it is not trying to measure. This holds true, of course, regardless of the number of predictors involved; for example, if X_7 is related to X_4, X_5, and X_{10}, then b_7 will try to measure *only* the effect of X_7, but will in fact also measure some of the effects of X_4, X_5, and X_{10}.

Multi-
collinearity
  The situation of having correlated predictor variables is called multicollinearity. This is a nicely descriptive name: "collinear" means related linearly, and the "multi-" indicates that any or all predictors may be related with any or all others. When multicollinearity is present, the net regression coefficients are said to be *unreliable* measures of the effects of their associated predictor variables; they not only measure the effect of the related predictor, but are confounded with the effects of other predictors related to it.

MULTICOLLINEARITY IN HOUSE SIZE-PRICE-AGE EXAMPLE

Let us see to what extent X_1 and X_2 are related in our problem by calculating the correlation coefficient between X_1 and X_2. The formula, similar to that presented in Chapter 8, uses X_1 and X_2 instead of X and Y. We subscript r to indicate that we are calculating the correlation between the two predictors:

$$r_{12} = r_{X_1 X_2} = \frac{S_{12}}{\sqrt{S_{11}S_{22}}}$$

$$= \frac{-15.3}{\sqrt{(13.56)(1745)}} = -0.0995.$$

From this we see that our sample gives some evidence of an inverse relationship between the size of a house and its age; older houses tend to be smaller than newer houses. Since $r_{12}^2 = (-0.0995)^2 = 0.0099 \approx 0.01$, we can say that about one percent of the variation in sizes can be explained by differences in ages, or equivalently, that about one percent of the variation in age can be accounted for by differences in size. This is obviously not a very strong relationship, and an hypothesis test will show that r_{12} is not significantly different from zero. However, since r_{12} is not zero, our sample shows some correlation between predictors. Thus *in the sample* it is not possible to hold X_1 or X_2 completely constant while allowing the other to vary. Therefore b_1 and b_2 are not completely reliable measures of the effects of the associated predictors.

THE VARIANCES OF THE REGRESSION COEFFICIENTS

If we have serious multicollinearity between predictor variables, the net regression coefficients are unreliable not only because they don't measure what they are supposed to, but also because their magnitudes and even their signs may be quite far off from what they should be. Refer back to $(\mathbf{X'X})^{-1}$ from the case in which we are using X_1 and X_2 as predictors; $(\mathbf{X'X})^{-1}$ is called a variance-covariance matrix.

Variance-
Covariance
Matrix

Note the diagonal elements of the submatrix of $(\mathbf{X'X})^{-1}$ obtained by deleting the first row and column of $(\mathbf{X'X})^{-1}$. These diagonal elements when multiplied by $\sigma^2_{y \cdot x}$ are the *variances of the net regression coefficients*. That is, the variance of b_1 is

$$\sigma^2_{y \cdot x} \frac{S_{22}}{S_{11}S_{22} - S_{12}^2}$$

and the variance of b_2 is

$$\sigma^2_{y \cdot x} \frac{S_{11}}{S_{11}S_{22} - S_{12}^2}.$$

(The off-diagonal elements, $-S_{12}/(S_{11}S_{22} - S_{12}^2)$ are used to find the covariances between the regression coefficients.) Consider the denominators

$$S_{11}S_{22} - S_{12}^2$$

of these variances. Note that they can be written as

$$1 - \frac{S_{12}^2}{S_{11}S_{22}} = 1 - r_{12}^2$$

if the numerators are also divided by $S_{11}S_{22}$. Now suppose that X_1 and X_2 are very highly correlated, so that r_{12}^2 is close to 1 and $1 - r_{12}^2$ is close to zero. This will make the ratios $S_{11}/(S_{11}S_{22} - S_{12}^2)$ and $S_{22}/(S_{11}S_{22} - S_{12}^2)$ very large, regardless of the values of S_{11} and S_{22}.

What does a large variance for a regression coefficient mean? It means that from sample to sample drawn from this same population the value of the regression coefficient can fluctuate wildly. However, in the long run, if many, many samples are drawn from this same population, the various values of the regression coefficient will *average* out to be correct. But in any one sample the value obtained can be so far off as to even have the wrong sign. Thus, if multicollinearity is severe, not only does b_1 measure a lot of the effect of X_2 as well as the effect of X_1, but you cannot trust b_1 to be anywhere near the correct magnitude or even have the correct sign.

REDUNDANCY

Suppose we had calculated $r_{12} = \pm 1$. Then, in the sample at least, X_1 and X_2 would be perfectly related, meaning that they are, in essence, the same predictor. For example, if X_1 were size in thousands of square feet, and X_2 were size in hundreds of square feet, then X_1 and X_2 are actually the same predictor variable. The two would give redundant information—what one explains about Y is exactly what the other explains. In this case, b_1 and b_2 would be completely unreliable, since b_1 would measure the effect of both X_1 and X_2, and the same goes for b_2. Furthermore, $(\mathbf{X'X})^{-1}$ would not exist, because the denominators $1 - r_{12}^2$ would be zero: we would not even be able to find values for b_1 and b_2. For example, suppose that $X_2 = kX_1$, for k some constant. (X_1 might be a measurement in feet and X_2 a measurement in inches.) Then

$$
\begin{aligned}
S_{12} &= \sum (X_1 - \bar{X}_1)(X_2 - \bar{X}_2) \\
&= \sum (X_1 - \bar{X}_1)(kX_1 - k\bar{X}_1) \\
&= kS_{11}, \\
S_{22} &= \sum (X_2 - \bar{X}_2)^2 = \sum (kX_1 - k\bar{X}_1)^2 = k^2 S_{11},
\end{aligned}
$$

and

$$
\mathbf{X'X} = \begin{bmatrix} n & 0 & 0 \\ 0 & S_{11} & kS_{11} \\ 0 & kS_{11} & k^2 S_{11} \end{bmatrix},
$$

where the third row (column) of $\mathbf{X'X}$ is k times the second row (column). Recall from Chapters 3 and 4 that if one row or column of a matrix is a linear combination of other rows or columns, then the inverse cannot be found—the system of equations cannot be solved. Note that the determinant of the above matrix is

$$
n(k^2 S_{11}^2 - k^2 S_{11}^2) = 0.
$$

Even if X_1 and X_2 are not perfectly correlated, but are very highly correlated, the rows/columns of $\mathbf{X'X}$ will be very nearly linear combinations of other rows/columns and the determinant will be close to zero. Then the elements of the inverse matrix become quite large and coefficients become quite large. Recall what happened in Section 4.5 when we used the SAS matrix inversion routine to investigate what happened when the matrix to be inverted was singular or nearly so. The matrix

$$
\begin{bmatrix}
1 & 2 & 3 & 4 \\
2.001 & 4 & 6 & 7.999 \\
4 & 7.999 & 12 & 16 \\
8 & 16 & 24.001 & 32
\end{bmatrix}
$$

gave rise to the diagonal elements

$$1600.2, \ -0.0000000000889039, \ -0.000000000071039, \text{ and } -600.$$

Note, however, that $\mathbf{X'X}$ always will be symmetric, will never have negative diagonal elements, will never have a negative determinant; and the effect of ill-conditioning will be to make the elements all very *large*, as contrasted to the very small elements obtained in this example. The matrix above is not an $\mathbf{X'X}$ matrix: it is simply a matrix constructed for purposes of showing what happens when a nearly singular matrix is inverted.

A word of caution is in order. Since we are illustrating the effects of multicollinearity in the two-predictor case, it is possible to examine r_{12}^2 to see the extent of the multicollinearity: if there is any present, it is a linear combination between X_1 and X_2. When we have more than two predictors, however, life is not so easy. Suppose we have predictors X_1, X_2, X_3, and X_4. If X_1 and X_3 are highly correlated, this will show up in r_{13}^2. But suppose that X_3 is a linear combination, or nearly so, of X_1 and X_4. This might not show up in r_{13}^2 or r_{34}^2, and it might be quite a chore to determine which variables the multicollinearity involves. More advanced techniques, such as *eigenvalue* or *characteristic root* analysis are needed in order to spot such relationships among predictors.

In summary, the stronger the relationship between predictors, the more unreliable b_1 and b_2 become as measures of the influences of their associated predictors. Only if $r_{12} = 0$ in the sample are b_1 and b_2 completely reliable; only then can we know the separate effects of the various independent variables on Y.

DESIGNED EXPERIMENTS

In social science studies multicollinearity almost always fouls up the works. Thus the social science researcher can hardly ever know the individual effects of his predictors on the dependent variable. This is an unavoidable fact of life. Almost any variables of interest to a social scientist will be bound up with influences from other variables of interest. One just cannot completely divorce from each other such variables as age, attitudes, incomes, spending habits, amount of education, marital status, and so on. The laboratory scientist who is familiar with the techniques of experimental design, however, is more fortunate. This scientist often has control over the values of the independent variables and can choose levels of $X_1 X_2, \ldots, X_k$ in such a way that all correlations among predictors are zero in the sample. For example, suppose a chemist is studying the effect of temperature (X_1) and pressure (X_2) on reaction time (Y). He (she) chooses three evenly spaced temperature levels, which he calls high, medium, and low and codes to $+1$, 0, and -1, respectively. Similarly, two pressure levels, high and low, are specified and coded to $+1$, and -1. The experiment is replicated three times under each combination of temperature level and pressure level. Table 10.2 shows the experimental design.

TABLE 10.2   *Experimental Design for Three Temperature*
Levels and Two Pressure Levels

X_1 Temperature	X_2 Pressure	Y Reaction Time
-1	-1	Y_1
-1	-1	Y_2
-1	-1	Y_3
-1	1	Y_4
-1	1	Y_5
-1	1	Y_6
0	-1	Y_7
0	-1	Y_8
0	-1	Y_9
0	1	Y_{10}
0	1	Y_{11}
0	1	Y_{12}
1	-1	Y_{13}
1	-1	Y_{14}
1	-1	Y_{15}
1	1	Y_{16}
1	1	Y_{17}
1	1	Y_{18}

Note here that Y_1, Y_2, and Y_3 are reaction times for trials run at low temperature and low pressure; Y_4, Y_5, and Y_6 are reaction times for trials run at low temperature and high pressure, and so on. We note that $\sum X_1 = \sum X_2 = 0$, since temperature and pressure levels are coded. Also, $\sum X_1^2 = 12$, $\sum X_2^2 = 18$, and $\sum X_1 X_2 = 0$. Then

$$S_{12} = \frac{18(0) - (0)(0)}{18} = 0$$

and

$$r_{12} = \frac{0}{\sqrt{S_{11}S_{22}}} = 0,$$

so that X_1 and X_2 are uncorrelated in the sample. Thus the chemist can know from his regression analysis what the effect of temperature is on reaction time, and independently what the effect of pressure is. In his equation, b_1 will measure *only* the effect of temperature, uncontaminated by the effect of pressure, and b_2 will measure only the effect of pressure and none of the effect of temperature. (Because in this example neither X_1 nor X_2 is a random variable, it is not strictly correct to calculate a correlation coefficient for them. You are encouraged to satisfy yourself, in the design represented in Table 10.2, that there is no tendency for any given temperature level to occur in conjunction with any particular level of pressure.)

Note also that the calculations involved in solving the normal equations are also considerably simplified if the predictors are not correlated. If X_1 and X_2 are not correlated, S_{12} will be zero and if X_1 and X_2 are corrected for their means,

$$\mathbf{X'X} = \begin{bmatrix} n & 0 & 0 \\ 0 & S_{11} & 0 \\ 0 & 0 & S_{22} \end{bmatrix},$$

which is diagonal. Thus $(\mathbf{X'X})^{-1}$ is easily found to be

$$(\mathbf{X'X})^{-1} = \begin{bmatrix} \dfrac{1}{n} & 0 & 0 \\ 0 & \dfrac{1}{S_{11}} & 0 \\ 0 & 0 & \dfrac{1}{S_{22}} \end{bmatrix}$$

and

$$\hat{\boldsymbol{\beta}} = \begin{bmatrix} \dfrac{1}{n} & 0 & 0 \\ 0 & \dfrac{1}{S_{11}} & 0 \\ 0 & 0 & \dfrac{1}{S_{22}} \end{bmatrix} \begin{bmatrix} \sum Y \\ S_{1Y} \\ S_{2Y} \end{bmatrix} = \begin{bmatrix} \sum \dfrac{Y}{n} \\ \dfrac{S_{1Y}}{S_{11}} \\ \dfrac{S_{2Y}}{S_{22}} \end{bmatrix}.$$

This extends easily to $k > 2$, of course. Note that in this case, $b_1 = S_{1Y}/S_{11}$ and $b_2 = S_{2Y}/S_{22}$, which are both of the form $b = S_{XY}/S_{XX}$. Thus, when predictors are uncorrelated, the multiple regression is just the combination of two simple regressions. That is, a simple regression on the same data using X_1 and Y will yield the same value for b as found by $b_1 = S_{1Y}/S_{11}$. If the value of the regression coefficient is unaffected by the other predictors, the predictors are uncorrelated. When the regression coefficient changes in the presence of other predictors, multicollinearity is indicated.

EXERCISES   *Section 10.5*

10.13 a. In Exercise 2.8 we found $\hat{Y} = -0.3256 + 0.3842X_1$, where $X_1 = $ number of employees and $Y = $ sales. In Exercise 10.1a, we found $\hat{Y} = -3.6183 + 1.7896X_2$, where $X_2 = $ size of store and $Y = $ sales. In Exercise 10.1b we found $Y = -2.447907 + 0.269825X_1 + 0.667393X_2$. Is there multicollinearity present in the data? Explain how you can tell.

b. Recall the following results:
Exercise 10.9a: $\hat{Y} = 322.259259 + 51.166667X_1 - 47.388889X_2$
$+ 8.055556X_3$;
Exercise 10.9b: $\hat{Y} = 322.259259 + 51.166667X_1$
$\hat{Y} = 322.259259 + 47.388889X_2$
$\hat{Y} = 322.259259 + 8.055556X_3$;
Exercise 10.9c: $\hat{Y} = 322.259259 + 51.166667X_1 - 47.388889X_2$
$\hat{Y} = 322.259259 + 51.166667X_1 + 8.055556X_3$
$\hat{Y} = 322.259259 - 47.388889X_2 + 8.055556X_3$.
Is there multicollinearity present in the data? Explain how you can tell.

p. 531 *10.14* a. For the RESTAURANT data, calculate the correlation coefficient between food and liquor sales. Interpret its meaning, with reference to implications in multicollinearity.
p. 532 b. For the DRUG data, calculate the correlation coefficient between drug dosage and patient age. Interpret its meaning, with reference to implications to multicollinearity.
p. 532 c. Calculate the correlation coefficient between age and mileage in the AUTO data. Interpret its meaning, with reference to implications to multicollinearity.
p. 535 d. Calculate the correlation coefficient between temperature and humidity in the GLUE data. Interpret its meaning, with reference to implications to multicollinearity. Is it really appropriate to calculate a correlation coefficient in this problem?

p. 533 *10.15* a. Calculate the correlation coefficients between TV and radio advertising, between TV and newspaper advertising, and between radio and newspaper advertising in the ADS data. Interpret their meanings with reference to implications to multicollinearity.
p. 533 b. Refer to the CHILDREN data and calculate correlations between age and height, age and weight, and height and weight. Interpret their meanings with reference to implications to multicollinearity.

10.6 THE STANDARD ERROR OF ESTIMATE

Just as in simple regression, we want to find a measure of the variability of the actual values around the average predicted by the regression equation. Recall that in simple regression, we started with the total sum of squares S_{YY} and subtracted off the sum of squares which measured variation in Y due to X. We follow the same procedure here, except that from S_{YY} we must subtract off variability due to both X_1 and X_2. For the case of two independent variables,

$$s_{y \cdot x}^2 = \frac{S_{YY} - (b_1 S_{1Y} + b_2 S_{2Y})}{n - 3} = \frac{S_{YY} - b_1 S_{1Y} - b_2 S_{2Y}}{n - 3}$$

is the calculating formula for the error variance. We are still measuring the average squared amount that the actual Y-values differ from the predicted:

$$s_{y \cdot x}^2 = \frac{\sum_{i=1}^{n} (Y_i - \hat{Y}_i)^2}{n - 3},$$

where now

$$\hat{Y}_i = a + b_1 X_{1i} + b_2 X_{2i}.$$

Note that we used a boldface $\mathbf{x}$ in the subscript on s^2, to indicate that we have more than one independent variable. Alternative notation is

$$s^2_{y \cdot x_1, x_2},$$

but this becomes more cumbersome as the number of predictor variables increases.

In the denominator of $s^2_{y \cdot x}$ the $n - 3$ conforms to the general rule that the denominator is $n - (k + 1) = n - p$. The calculating formula extends easily to any number of predictors: the general formula is

$$s^2_{y \cdot x} = s^2_{y \cdot x_1, x_2, \ldots, x_k} = \frac{S_{YY} - b_1 S_{1Y} - b_2 S_{2Y} - \cdots - b_k S_{kY}}{n - (k + 1)}$$

$$= \frac{S_{YY} - \sum_{i=1}^{k} b_i S_{iY}}{n - (k + 1)} = \frac{SS(E)}{n - (k + 1)} = MS(E).$$

The standard error of estimate, $s_{y \cdot x}$, is the square root of $s^2_{y \cdot x}$. It measures the average amount that the actual values differ from the estimated values. For our house size-price-age example,

$$s^2_{y \cdot x} = \frac{2373 - 12.83442613(174.9) - (-0.05655391)(-295)}{20 - 3}$$

$$= \frac{111.57546641}{17} = 6.56326273;$$

$$s_{y \cdot x} = 2.56188655.$$

We interpret the standard error of estimate to mean that when we use the regression equation to estimate the price of a house from its size and age, our estimates will be off from the actual prices by \$2600, on the average.

CONFIDENCE INTERVALS ON MEAN RESPONSE

The standard error of estimate can be used to form confidence intervals on the mean price of houses of a given size and age and also confidence or prediction intervals on the price of a particular house of a given size and age. Let $\mu_{y \cdot x_0}$ denote the true mean Y-value at $X_1 = X_{10}, X_2 = X_{20}, \ldots, X_k = X_{k0}$, where $X_{10}, X_{20}, \ldots, X_{k0}$ denote specified values of $X_1, X_2, \ldots, X_k$, respectively. We can denote these specified X-values as

$$\mathbf{x}_0 = \begin{bmatrix} 1 \\ X_{10} \\ X_{20} \\ \vdots \\ X_{k0} \end{bmatrix}$$

The $\mu_{y \cdot x_0}$ is estimated by $\hat{Y}_0 = a + b_1 X_{10} + b_2 X_{20} + \cdots + b_k X_{k0}$. Then the general form of a $100(1 - \alpha)\%$ confidence interval on $\mu_{y \cdot x_0}$ is

$$\hat{Y}_0 + t_{\alpha/2, n-p} s_{y \cdot x} \sqrt{\mathbf{x}_0' (\mathbf{X}'\mathbf{X})^{-1} \mathbf{x}_0}.$$

In our example, we estimated the price of a house 2200 square feet in size and 26 years old to be $\hat{Y}_0 = 36.81166570$. Then

$$\mathbf{x}_0 = \begin{bmatrix} 1 \\ 2.2 \\ 26 \end{bmatrix}$$

and

$$\mathbf{x}_0'(\mathbf{X}'\mathbf{X})^{-1}\mathbf{x}_0 =$$

$$\begin{bmatrix} 1 & 2.2 & 26 \end{bmatrix} \begin{bmatrix} 0.67164255 & -0.16300717 & -0.01375015 \\ -0.16300717 & 0.07448317 & 0.00065306 \\ -0.01375015 & 0.00065306 & 0.00057879 \end{bmatrix} \begin{bmatrix} 1 \\ 2.2 \\ 26 \end{bmatrix}$$

$$= \begin{bmatrix} -0.04447712 & 0.017065140 & 0.00273512 \end{bmatrix} \begin{bmatrix} 1 \\ 2.2 \\ 26 \end{bmatrix} = 0.06417931.$$

Then a 95 % confidence interval on the mean price of houses 2200 square feet in size and 26 years old is

$$36.81166570 \pm 2.110(2.5742)\sqrt{0.06417931},$$

or

$$36.81166570 \pm 1.37601209,$$

or

$$35.43565361 < \mu_{y \cdot x0} > 38.18767779.$$

We are 95 % sure that the mean price of houses 2200 square feet in size and 26 years old is between \$35,436 and \$38,188.

CONFIDENCE INTERVALS ON INDIVIDUAL RESPONSE

The confidence interval for an individual house is formed similarly. Let Y_0^* denote the actual price of a house of size

$$\mathbf{x}_0 = \begin{bmatrix} 1 \\ X_{10} \\ X_{20} \\ \vdots \\ X_{k0} \end{bmatrix}.$$

Then a $100(1 - \alpha)$ % confidence interval on Y_0^* is given by

$$\hat{Y}_0 - t_{\alpha/2, n-p} s_{y \cdot x} \sqrt{1 + \mathbf{x}_0'(\mathbf{X}'\mathbf{X})^{-1}\mathbf{x}_0} < Y_0^* < \hat{Y}_0 + t_{\alpha/2, n-p} s_{y \cdot x} \sqrt{1 + \mathbf{x}'(\mathbf{X}'\mathbf{X})^{-1}\mathbf{x}}.$$

Thus a 95 % prediction interval for the price of a house 2200 square feet in size and 26 years old is

$$36.81166570 \pm 2.110(2.5742)\sqrt{1 + 0.06417931},$$

or

$$36.81166570 \pm 5.60314863,$$

or

$$31.20851707 < Y_0^* < 42.41481433.$$

The interpretation is that 95 % of all houses 26 years old and 2200 square feet in size will cost between \$31,208 and \$42,415. We note that the price of \$35,000 for House 5, which is this size and age, lies in this interval. Thus, House 5 is not considered unusual in price for houses its size and age.

COEFFICIENT OF VARIATION

A coefficient of variation can be formed in exactly the same way as for simple regression studies:

$$CV = \frac{s_{y \cdot x}}{\bar{Y}}.$$

As before, CV measures the size of the error relative to the size of the response variable being measured. Is the error large because the numbers we are working with are large or because the estimates are inaccurate? In our example,

$$CV = \frac{2.56188655}{34.5} = 0.0742 = 7.42\%$$

says that our estimates will be off, on the average, 7.42%.

EFFECT OF MULTICOLLINEARITY ON STANDARD ERROR OF ESTIMATE

Let us now pause and examine the calculating formula for $s_{y \cdot x}^2$ a bit more closely. One objective in a multiple regression study is to make reasonably accurate estimates, and one feels that more predictors mean more accurate estimates. Thus, if X_1 and X_2 are both effective predictors of Y, then the estimates made using

$$\hat{Y} = a + b_1 X_1 + b_2 X_2$$

should be closer to the actual values than those made by using only one predictor. To say that the estimates are more accurate means that the error in estimation should be smaller using two predictors than when using only one, or that

$$s_{y \cdot x_1, x_2}^2 < s_{y \cdot x}^2.$$

This should also be apparent from examination of the numerator of the calculating formula for $s_{y \cdot x}^2$:

$$S_{YY} - b_1 S_{1Y} - b_2 S_{2Y}.$$

In simple regression, we would subtract off $b S_{XY}$ from S_{YY}, but with two predictors we subtract off two terms. Now $b S_{XY}$ in simple regression will not equal $b_1 S_{1Y}$ in multiple regression if multicollinearity is present because b will not equal b_1. However, it seems reasonable that even if X_1 and X_2 are correlated, $b_1 S_{1Y} + b_2 S_{2Y}$ should be larger than $b S_{XY}$, so that the numerator of $s_{y \cdot x}^2$ becomes smaller the more predictors we have. Thus, we expect that if X_1 and X_2 are both significant predictors of Y, the standard error should be smaller when two predictors are used than when only one is considered. Similarly, the coefficient of variation should be smaller in a multiple regression.

You may have noticed that the house size-price-age example seems to contradict what we have just said. There, the standard error of estimate using two predictors is *greater* than the standard error using only one: recall that $s_{y \cdot x} = \$2550$ when we were predicting price from size alone, but $s_{y \cdot x} = \$2562$ using age and size. Let us examine what happened.

In Chapter 6, we found

$$s_{y \cdot x}^2 = \frac{117.0996}{18},$$

while for the multiple regression problem, we got

$$s_{y \cdot x}^2 = \frac{111.57546641}{17}.$$

We note that, as advertised, the numerator of $s_{y \cdot x}^2$ is smaller than the numerator of $s_{y \cdot x}^2$. However, the denominator of $s_{y \cdot x}^2$ is *also* smaller, and this is what accounts for $s_{y \cdot x}^2$ being larger than $s_{y \cdot x}^2$. Thus, the reduction in total variability (S_{YY}) obtained by including age (X_2) as a predictor of price is not great enough to overcome the concurrent reduction in degrees of freedom. The question immediately arises: is X_2 an effective predictor of Y? Obviously not. We cannot predict price more accurately using age and size than we could by using size alone. Only if the effect of X_2 is so strong that $b_2 S_{2Y}$ reduces the numerator of $s_{y \cdot x}^2$ enough to overcome the reduction in degrees of freedom will it be worthwhile to retain X_2 as a predictor. One might suspect that X_2 is not a significant predictor by simply noting the size of $b_2 = -0.06$. This is not a very large value, especially when compared to the size of the error ($\$60$ vs $\$2600$). We will confirm that X_2 is not a significant predictor later on when we perform the multiple regression hypothesis test for the significance of the predictor.

Other strange things can happen to the standard error in pathological problems which arise with startling frequency in many actual studies. Especially when the degree of multicollinearity among the predictors is great, many seemingly contradictory results can arise. The most frequent of these is for b_i and S_{iY} to have different signs. In such cases, the quantity

$$- b_i S_{iY}$$

will be positive, so that addition of the predictor X_i may increase the error mean square, $MS(E) = (S_{YY} - \sum_{i=1}^{k} b_i S_{iY})/(M - 2)$. Thus, it is a pretty sure bet that if inclusion of a predictor results in an increase in $s_{y \cdot x}^2$, that predictor has no business being in the regression equation. Note that the converse does not hold, however: just because inclusion of a predictor results in a reduction in error does not necessarily mean that that predictor is significant. The final determinant of whether or not a predictor belongs in the regression equation is the hypothesis test, which we shall get to in due time.

EXERCISES   *Section 10.6*

10.16 Calculate the standard error of estimate for the following data sets. State its units and interpret its meaning.

p. 531   a. SALES II,
p. 531   b. RESTAURANT,
p. 532   c. DRUG,
p. 532   d. AUTO,
p. 533   e. ADS,
p. 533   f. CHILDREN,
p. 535   g. GLUE,
p. 535   h. BOTTLES,
p. 536   i. YIELD.

10.17 a. Calculate the standard error of estimate for Exercise 10.1a, p. 306, in which X_2, size of store, alone was used as a predictor of sales.

b. Recall that when sales were predicted from number of employees alone, the standard error of estimate was found to be $1439.70. Does the size of the store, or the number of employees, when taken alone, provide the better estimates of sales?

c. Compare the two standard errors for the two simple regressions to the standard error from the multiple regression. Are predictions of sales better using two predictors than either one alone?

10.7 SUMMARY

In this chapter we extended techniques applied in simple regression to the analysis of a multiple regression problem. While most of the earlier results extend to the present problem in a relatively straightforward manner, we have already seen that life becomes considerably more complicated when we are dealing with several predictors than when we only had to worry about one.

In order to examine more carefully the effect and implications of multicollinearity in a multiple regression problem, we will turn to the study of how variables can be correlated in multiple regression.

Chapter 11

CORRELATION IN MULTIPLE REGRESSION

11.1 INTRODUCTION

In the last chapter, we saw how the simple correlation coefficient between two predictor variables can be used as a measure of the extent of multicollinearity in a multiple regression problem. Predictors, however, are not the only variables that might be correlated. One might ask, for instance, if one predictor is correlated with the response, or what percentage of the variation in the response is explained by both predictors together. These questions are addressed here.

11.2 CORRELATIONS BETWEEN PAIRS OF VARIABLES

We saw in Chapter 10 that we could calculate the correlation coefficient or coefficient of determination between the two predictor variables, X_1 and X_2, in our two-predictor multiple regression. This correlation coefficient,

$$r_{12} = \frac{S_{12}}{\sqrt{S_{11}S_{22}}} = -0.0995,$$

is a measure of the degree of multicollinearity among the predictors. In multiple regression studies with $k > 2$, a similar correlation coefficient can be calculated for all pairs of independent variables, to measure the degree of multicollinearity for each pair.

What is perhaps of greater interest is to calculate the correlation between each predictor variable and the response variable, Y. This allows us to measure the strength of each predictor separately. The correlation coefficient between X_1 and Y is

$$r_{1Y} = r_{X_1Y} = \frac{S_{1Y}}{\sqrt{S_{11}S_{YY}}} = \frac{174.9}{\sqrt{(13.56)(2373)}} = 0.975.$$

From this we see (as in the simple regression problem) that X_1, size, is a very strong predictor of price, Y, since r_{1Y} is very close to 1. Since r_{1Y} is positive, we also know that larger houses tend to cost more than smaller houses. The coefficient of determination, $r_{1Y}^2 = 0.9507$ says that approximately 95 % of the variation in prices can be explained entirely by differences in the sizes of the houses. Similarly,

$$r_{2Y} = \frac{S_{2Y}}{\sqrt{S_{22}S_{YY}}} = \frac{-295}{\sqrt{(1745)(2373)}} = -0.1449$$

indicates a weak inverse relationship—older houses tend to cost less than newer houses, but not with a very pronounced effect. Only $r_{2Y}^2 = 0.0210$, or 2 % of the variation in prices, can be accounted for by the differences in ages of the houses. This result supports our suspicion that age is not a significant predictor of price.

EXERCISES   *Section 11.2*

11.1   Calculate the simple coefficients of determination between Y and each of the predictors, and explain their meanings, for the

11.3 MULTIPLE COEFFICIENT OF DETERMINATION

The question arises as to what percentage of variation in price is explained by both predictors, together. Since the predictors themselves are correlated, we cannot just add r_{1Y}^2 and r_{2Y}^2 to get the percentage of variation in Y explained by X_1 and X_2 together. Since $r_{12} \neq 0$, X_1 and X_2 act alike to some extent; some of the variation in Y explained by X_1 is also explained by X_2, and vice versa, so that X_1 and X_2 overlap in what they tell about Y. The amoeba-like shape in Figure 11.1 represents total amount of variation in Y. Percentages of variation explained by X_1 alone and X_2 alone are indicated, and the total amount of variation explained by both X_1 and X_2 together is crosshatched. Since the variation explained by X_1 alone and that explained by X_2 alone are not disjoint, their sum is greater than the variation explained by both predictors together. Of course, if X_1 and X_2 are not correlated, these pieces will be disjoint and can be simply added together to determine the total amount of variation explained.

FIGURE 11.1   *Percentage of Variation in Y Explained by X_1 and X_2, where X_1 and X_2 are correlated*

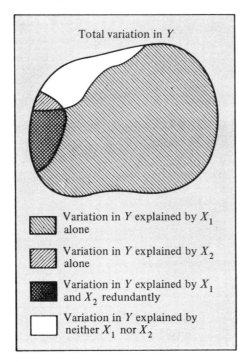

Note in the figure that the variation in Y explained by X_1 and X_2 at the same time (redundantly) is *not* r_{12}^2, since r_{12}^2 does not measure variation in Y.

How then, can we measure the percentage of variation in Y explained by the regression—by both X_1 and X_2? Recall from Chapter 8 that the coefficient of determination is defined as the ratio of variation in Y explained by X to total variation. If we have two predictors, then, the <u>multiple coefficient of determination</u> will be the ratio of the variation in Y explained by X_1 and X_2 to the total variation in Y:

<u>Multiple Coefficient of Determination</u>

$$r_{xy}^2 = \frac{\text{variation in } Y \text{ explained by } X_1 \text{ and } X_2}{\text{total variation in } Y}.$$

(The notation r_{xy}^2 is adopted to denote that we are now considering more than one X.) Then

$$r_{xy}^2 = \frac{SS(R)}{SS(T)} = \frac{SS(\text{Regression on } X_1) + SS(\text{Regression on } X_2)}{SS(T)}$$

$$= \frac{b_1 S_{1Y} + b_2 S_{2Y}}{S_{YY}},$$

or in general, for k predictors,

$$r_{xy}^2 = \frac{b_1 S_{1Y} + b_2 S_{2Y} + \cdots + b_k S_{kY}}{S_{YY}} = \frac{\sum_{i=1}^{k} b_i S_{iY}}{S_{YY}}.$$

Recalling that we partition variation in Y into two parts, one part due to regression and the other due to error, we can equivalently write r_{xy}^2 as

$$r_{xy}^2 = \frac{SS(T) - SS(E)}{SS(T)} = 1 - \frac{SS(E)}{SS(T)}.$$

This latter form is convenient, because we earlier found $SS(E)$ when calculating the standard error. For our example,

$$r_{xy}^2 = 1 - \frac{111.57546641}{2373} = 1 - 0.0470 = 0.9530.$$

Differences in the sizes and ages together explain 95.30% of the variation in house prices. Note that size alone explained 95.07% of the variation in price, so addition of age as a predictor increased the percentage of explained variation by only 0.23%. This constitutes further evidence that age is not a significant predictor of price.

Note that it does not make sense to define a "multiple correlation coefficient" by taking the square root of r_{xy}^2, primarily because no meaningful sign could be attached indicating the direction of the multiple relationships. In our example, Y is directly related to X_1 but inversely related to X_2, so that one would not know which sign to attach to the square root. This statistic is often reported, even though it is a meaningless measure.

In addition, no easy calculating formula can be given for r_{xy}^2 as was given for r_{1Y}^2, r_{2Y}^2, or r_{12}^2, since r_{xy}^2 measures correlation between one group of k independent variables and a single dependent variable, instead of between one variable and one other.

EXERCISES   *Section 11.3*

11.2   Calculate the multiple coefficient of determination and interpret its meaning with reference to the simple coefficients of determination found in Exercise 11.1, for the

11.3 Explain why $r_{1Y}^2 + r_{2Y}^2 \neq r_{xy}^2$ in the
   a. SALES II data,
   b. RESTAURANT data,
   c. DRUG data,
   d. AUTO data.

11.4 Explain why $r_{1Y}^2 + r_{2Y}^2 + r_{3Y}^2 \neq r_{xy}^2$ in the
   a. ADS data,
   b. CHILDREN data.

11.5 Explain why $r_{1Y}^2 + r_{2Y}^2 = r_{xy}^2$ in the
   a. GLUE data,
   b. BOTTLES data,
   c. and why $r_{1Y}^2 + r_{2Y}^2 + r_{3Y}^2 = r_{xy}^2$ in the YIELD data.

11.4 PARTIAL CORRELATION

The correlation coefficient r_{1Y} does not take into account the fact that the predictor X_2 is included in the regression, and similarly, r_{2Y} does not take the presence of X_1 into account. Since r_{12} is nonzero, we know that some of the variation in Y explained by one variable is also explained by the other. We might ask the question: Once X_1 has explained all it can about Y, what percentage of the left-over variation is explained by X_2? To answer this, we calculate a partial correlation coefficient.

Partial
Correlation
Coefficient

DERIVATION OF THE FORMULA

Suppose we want to correlate X_1 and Y, controlling for X_2. If we run a simple regression using X_2 to predict Y, we would obtain a regression equation of the form

$$\hat{Y} = \bar{Y} + b_1(X_2 - \bar{X}_2),$$

if X_2 is corrected for its mean. The residuals,

$$Y - \hat{Y} = Y - \bar{Y} - b_1(X_2 - \bar{X}_2),$$

then do not depend on X_2, but rather on variables other than X_2 affecting Y. Denote these residuals as W_1:

$$W_1 = Y - \hat{Y}.$$

Similarly, if we were to run a simple regression to predict X_1 from X_2, we would obtain an equation of the form

$$\hat{X}_1 = \bar{X}_1 + b_2(X_2 - \bar{X}_2),$$

and the residuals

$$W_2 = X_1 - \hat{X}_1 = X_1 - \overline{X}_1 - b_2(X_2 - \overline{X}_2)$$

would not depend on X_2.

Since neither W_1 nor W_2 depends on X_2, the correlation between W_1 and W_2 will be the correlation between X_1 and Y, controlling for X_2, or having taken out the effect of X_2 on both X_1 and Y. To calculate the correlation between W_1 and W_2, we first note that

$$\sum W_1 = \sum W_2 = 0.$$

This can be seen by

$$\begin{aligned}
\sum W_1 &= \sum [Y_i - \overline{Y} - b_1(X_{2i} - \overline{X}_2)] \\
&= \sum [(Y_i - \overline{Y}) - b_1(X_{2i} - \overline{X}_2)] \\
&= \sum (Y_i - \overline{Y}) - b_1 \sum (X_{2i} - \overline{X}_2) \\
&= 0 - b_1(0) = 0,
\end{aligned}$$

and similarly for $\sum W_2$. Therefore,

$$r_{W_1 W_2} = \frac{S_{W_1 W_2}}{\sqrt{S_{W_1 W_1} S_{W_2 W_2}}} = \frac{\sum W_1 W_2}{\sqrt{\sum W_1^2 \sum W_2^2}}.$$

To find $\sum W_1^2$,

$$\begin{aligned}
\sum W_1^2 &= \sum [(Y_i - \overline{Y}) - b_1(X_{2i} - \overline{X}_2)]^2 \\
&= \sum [(Y_i - \overline{Y})^2 - 2b_1(Y_i - \overline{Y})(X_{2i} - \overline{X}_2) + b_1^2(X_{2i} - \overline{X}_2)^2] \\
&= S_{YY} - 2b_1 S_{2Y} + b_1^2 S_{22} \\
&= S_{YY} - 2\left(\frac{S_{2Y}}{S_{22}}\right) S_{2Y} + \left(\frac{S_{2Y}^2}{S_{22}^2}\right) S_{22} \\
&= S_{YY} - \frac{S_{2Y}^2}{S_{22}}.
\end{aligned}$$

Similarly,

$$\sum W_2^2 = S_{11} - \frac{S_{12}^2}{S_{22}}.$$

(You can verify this for yourself, following the steps used to find $\sum W_1^2$.) Finally,

$$\sum W_1 W_2 = \sum [(Y_i - \overline{Y}) - b_1(X_{2i} - \overline{X}_2)] \, [(X_{1i} - \overline{X}_1) - b_2(X_{2i} - \overline{X}_2)]$$

$$= S_{1Y} - \frac{S_{12}S_{2Y}}{S_{22}}.$$

Then

$$\begin{aligned}
r_{W_1 W_2} &= \frac{S_{1Y} - (S_{12}S_{2Y}/S_{22})}{\sqrt{S_{YY} - (S_{2Y}^2/S_{22})}\sqrt{S_{11} - (S_{12}^2/S_{22})}} \\[2mm]
&= \frac{S_{22}}{S_{22}} \frac{S_{1Y} - (S_{12}S_{2Y}/S_{22})}{\sqrt{(S_{YY}/S_{YY})[S_{YY} - (S_{2Y}^2/S_{22})]}\sqrt{(S_{11}/S_{11})[S_{11} - (S_{12}^2/S_{22})]}} \\[2mm]
&= \frac{S_{1Y}S_{22} - S_{12}S_{2Y}}{S_{22}\sqrt{S_{YY}(1 - r_{2Y}^2)}\sqrt{S_{11}(1 - r_{12}^2)}} \\[2mm]
&= \frac{S_{1Y}S_{22} - S_{12}S_{2Y}}{\sqrt{S_{YY}S_{11}S_{22}^2}\sqrt{(1 - r_{2Y}^2)(1 - r_{12}^2)}} \\[2mm]
&= \frac{\dfrac{S_{1Y}S_{22}}{\sqrt{S_{YY}S_{11}S_{22}^2}} - \dfrac{S_{12}S_{2Y}}{\sqrt{S_{YY}S_{11}S_{22}^2}}}{\sqrt{(1 - r_{2Y}^2)(1 - r_{12}^2)}} \\[2mm]
&= \frac{r_{1Y} - r_{12}r_{2Y}}{\sqrt{(1 - r_{12}^2)(1 - r_{2Y}^2)}}.
\end{aligned}$$

Thus, the partial correlation between X_1 and Y, controlling for X_2, is calculated by using the formula

$$r_{1Y.2} = \frac{r_{1Y} - r_{12}r_{2Y}}{\sqrt{1 - r_{12}^2}\sqrt{1 - r_{2Y}^2}}.$$

Note that this partial correlation begins with the simple correlation r_{1Y} and then subtracts off a piece measuring the correlation between the two predictors, r_{12}, and the correlation, r_{2Y}, between the other predictor and Y. Then a scaling is performed in the denominator. In a similar fashion, the partial correlation between X_2 and Y, controlling for X_1, is

$$r_{2Y.1} = \frac{r_{2Y} - r_{12}r_{1Y}}{\sqrt{1 - r_{12}^2}\sqrt{1 - r_{1Y}^2}}.$$

PARTIAL CORRELATIONS FOR HOUSE SIZE-PRICE-AGE EXAMPLE

In our example,

$$r_{1Y.2} = \frac{0.975 - (-0.0995)(-0.1449)}{\sqrt{1 - 0.0099}\sqrt{1 - 0.0210}} = 0.9757$$

and

$$r_{2Y.1} = \frac{-0.1449 - (-0.0995)(0.975)}{\sqrt{1 - 0.0099}\sqrt{1 - 0.9507}} = -0.2168.$$

Notice that taking X_2 into account, the relationship between X_1 and Y is direct, and taking X_1 into account, the relationship between X_2 and Y is inverse. Since $r_{1Y.2}^2 = (0.9757)^2 = 0.9520$, once X_2(age) has explained all the variation in Y(price) that it can, then of the unexplained variation X_1(size) explains 95.2%. And since $r_{2Y.1}^2 = (-0.2168)^2 = 0.0470$, once X_1(size) has explained all it can about Y, X_2 explains 4.7% of what is left. These results again support our suspicion that X_2 is not a significant predictor.

Note that it may be possible for both r_{1Y} and r_{2Y} to be relatively large, but if r_{12} is also large, then X_1 will not improve much over what X_2 can do, and vice versa, due to the large degree of redundancy. In this case, the less important predictor will be discarded: even though by itself it is a strong predictor, in conjunction with the other independent variable with which it is highly correlated it will not add very much information about Y. Neither $r_{1Y.2}$ nor $r_{2Y.1}$ will be very large, even though r_{1Y} and r_{2Y} are reasonably close to ± 1.

An alternative way of interpreting a partial correlation coefficient can be illustrated by the following example. Suppose X_1 = age of child, X_2 = number of hours spent watching television, and Y = score on a manual dexterity test. We might expect to see the following relationships among these three variables: X_1 and Y will probably be directly related, X_1 and X_2 will probably also be directly related (the older the child, the later his bedtime), and perhaps because of the relationship between X_1 and X_2, we will also see X_2 and Y directly related. Suppose, however, that we narrowed down the list and looked at a subset of the children in the sample who are all the same age. When we look at the correlation between X_2 and Y for just these children, we would not be surprised to see a negative correlation: for children the same age, the more time spent in a passive activity such as television viewing, the lower the score on a manual dexterity test. This latter correlation is a partial correlation—it holds X_1 constant and looks at the correlation between X_2 and Y; as such it measures something quite different from that which is measured by the simple correlation between X_2 and Y.

HIGHER-ORDER PARTIAL CORRELATIONS

When only one variable is controlled for, as in the above example, we have a *first-order* partial correlation. Given more than two predictors, all but one must be controlled for when correlating one predictor and Y. The *order* of the partial correlation is the number of controlled variables. For example, if X_1, X_2, and X_3 predict Y, then the second-order partial correlations are $r_{1Y \cdot 23}$, $r_{2Y \cdot 13}$, and $r_{3Y \cdot 12}$. The formulas for second-order and higher partial correlations appear complicated and require calculation of all lower-order partial correlations. However, their general form is not hard to grasp. For example,

$$r_{1Y \cdot 23} = \frac{r_{1Y \cdot 2} - (r_{13 \cdot 2})(r_{3Y \cdot 2})}{\sqrt{1 - r_{13 \cdot 2}^2}\sqrt{1 - r_{3Y \cdot 2}^2}},$$

or

$$r_{1Y \cdot 23} = \frac{r_{1Y \cdot 3} - (r_{12 \cdot 3})(r_{2Y \cdot 3})}{\sqrt{1 - r_{12 \cdot 3}^2}\sqrt{1 - r_{2Y \cdot 3}^2}},$$

is the formula for one possible second-order partial correlation—the correlation between X_1 and Y, correcting for X_2 and X_3. Note that the two formulas are equivalent; you can start off with $r_{1Y \cdot 2}$ or with $r_{1Y \cdot 3}$ and you condition on the same variable in each term; both give the same result. One third-order partial correlation, between X_1 and Y controlling for X_2, X_3, and X_4, is given by the formula

$$r_{1Y \cdot 234} = \frac{r_{1Y \cdot 23} - (r_{14 \cdot 23})(r_{4Y \cdot 23})}{\sqrt{1 - r_{14 \cdot 23}^2}\sqrt{1 - r_{4Y \cdot 23}^2}}.$$

There are two formulas equivalent to this, one starting with $r_{1Y \cdot 24}$ and another beginning with $r_{1Y \cdot 34}$. Other second- and third-order as well as higher-order partial correlations are calculated similarly.

EXERCISES   *Section 11.4*

11.6   Calculate the partial correlation coefficients $r_{1Y \cdot 2}$ and $r_{2Y \cdot 1}$ and interpret their meanings in the

p. 531   a.   SALES II data,
p. 531   b.   RESTAURANT data,
p. 532   c.   DRUG data,
p. 532   d.   AUTO data.

11.7   Find the partial correlations $r_{1Y \cdot 23}$, $r_{2Y \cdot 13}$, and $r_{3Y \cdot 12}$ and interpret their meanings for the
p. 533   a.   ADS data,
p. 533   b.   CHILDREN data.

11.8   a.   One might expect that in reference to the CHILDREN data, overweight children would tend to score low on a physical fitness test. If we consider the partial correlation $r_{3Y \cdot 12}$ to be the correlation between weight and score, holding age and height constant, can we conclude that for children the same age and height, the score is lower for heavier children than for lighter-weight children? Recall that $r_{3Y \cdot 12} = -0.2544$.

    b. Why does the inverse relationship between weight and score for children the same age and height not show up on the simple correlation, $r_{3Y} = 0.3681$?

11.9 Write the two formulas equivalent to the formula for $r_{1Y.234}$ given in the text.

11.10 Using the ADS data, show that the two formulas for $r_{3Y.12}$ give the same result.

11.5 SUMMARY

There are many correlations that can be calculated in multiple regression problems. Simple correlations between pairs of predictors measure pairwise multicollinearity in the problem. Simple correlations between each predictor and the response tell how strong each predictor is by itself. Partial correlations between a predictor and the response measure the strength of the predictor in the presence of other predictors. Finally, a multiple coefficient of determination measures what percentage of the variation in the response is explained by the entire set of predictors.

    We will see that all three kinds of correlations—simple, partial, and multiple— are important in testing the significance of the various predictors and are essential to the understanding of the entire multiple regression problem. Chapter 12 takes up the study of significance tests in multiple regression.

Chapter 12

TESTS OF SIGNIFICANCE IN MULTIPLE REGRESSION

12.1 INTRODUCTION

The researcher, having collected data on several predictor variables and one response variable, has found a regression equation, a standard error of estimate, and several measures of correlation. Just because such measures can be calculated, however, does not necessarily mean that they are appropriate. Before including a predictor in the regression equation, one should be reasonably certain that it is an effective predictor, and not redundant with any other predictors.

We are now ready to test our predictors for significance. From the sizes of the net regression coefficients, the standard error of estimate, and the various correlations in our house size-price-age example, we suspect that size, but not age, is an important predictor of price. These suspicions will be supported by the tests of hypotheses.

12.2 TEST OF THE SIGNIFICANCE OF THE REGRESSION

Before testing each predictor separately, we will test for the *significance of the regression*. This test will tell us if *any* of the predictors is significantly different from zero. If we find that at least one predictor is significant, then we turn to the question of which one or ones. If the test of the significance of the regression cannot detect any significant predictors, however, then further testing is generally unnecessary. You can usually forget the whole question. If your problem is at all reasonable, the overall test for significant predictors can be expected to reject the null hypothesis that no predictors are significant. (Of course, there are exceptions.)

THE HYPOTHESIS

We want to test

$$H_0: \beta_i = 0 \quad \text{for all } \beta_i, \, i = 1, 2, \ldots, k$$

$$H_1: \text{some } \beta_i \neq 0.$$

Note that the alternative hypothesis is two-sided. We are simultaneously testing all k predictors in order to detect those inversely related to Y as well as those directly related. The null and alternative hypotheses can alternatively be written in vector notation as

$$H_0: \boldsymbol{\beta^*} = \mathbf{0}$$

$$H_1: \boldsymbol{\beta^*} \neq \mathbf{0},$$

where $\boldsymbol{\beta^*}$ is the $k \times 1$ subvector of $\boldsymbol{\beta}$ obtained by deleting the first element, α. That is,

$$\boldsymbol{\beta} = [\alpha \quad \beta_1 \quad \beta_2 \quad \cdots \quad \beta_k]', \qquad \boldsymbol{\beta^*} = [\beta_1 \quad \beta_2 \quad \cdots \quad \beta_k]',$$

and

$$\mathbf{0}_k = [0 \quad 0 \quad \cdots \quad 0]'.$$

THE TEST

The test is run using the following analysis of variance, where F has k and $n - (k + 1)$ degrees of freedom.

ANOVA

General Regression Significance Test

Source	SS	df	MS	F
Regression $(X_1, X_2, \ldots, X_k)$	$SS(R) = \sum_{i=1}^{k} b_i S_{iY}$	k	$MS(R) = \dfrac{\sum_{i=1}^{k} b_i S_{iY}}{k}$	$\dfrac{MS(R)}{MS(E)}$
Error	$SS(E) = S_{YY} - \sum_{i=1}^{k} b_i S_{iY}$	$n - (k + 1)$	$MS(E) = \dfrac{S_{YY} - \sum_{i=1}^{k} b_i S_{iY}}{n - (k + 1)} = s_{y \cdot x}^2$	
Total	$SS(T) = S_{YY}$	$n - 1$		

We compare F to $F_{k, n-(k+1), \alpha}$ and reject H_0 if F is greater than this critical value. If we can reject H_0, we are at least $100(1 - \alpha)\%$ sure that at least one of the k predictors is significant, and then we might proceed to look for it. If H_0 cannot be rejected, then usually none of the predictors is strong enough to bother with.

GENERAL REGRESSION SIGNIFICANCE TEST FOR HOUSE SIZE-PRICE-AGE EXAMPLE

In our example,

ANOVA

Source	SS	df	MS	F
Regression	2261.4245	2	1130.7122	172.2780
Error	111.5755	$20 - 3 = 17$	6.5633	
Total	2373	$20 - 1 = 19$		

If $\alpha = 0.05$, we will reject $H_0 : \boldsymbol{\beta} = 0$ in favor of $H_1 : \boldsymbol{\beta} \neq \mathbf{0}$ if $F > F_{2, 17, 0.05} = 3.59$. Since $172.2780 > 3.59$, we can be more than 95% sure that either size or age, or both, is a significant predictor of price.

GENERAL TEST IN TERMS OF MATRICES

Before proceeding to find which predictors are significant, it will help to formulate the general test in terms of matrices and vectors. Recall from Chapter 7 that the test for significance can be formulated as

ANOVA

Source	SS	df
Regression	$\hat{\boldsymbol{\beta}}' \mathbf{X}' \mathbf{y}$	$k + 1$
Error	$\mathbf{y}' \mathbf{y} - \hat{\boldsymbol{\beta}}' \mathbf{X}' \mathbf{y}$	$n - k - 1$
Total	$\mathbf{y}' \mathbf{y}$	n

or equivalently as

ANOVA

Source	SS	df
Regression	$\hat{\beta}'\mathbf{X}'\mathbf{y} - \dfrac{(\sum Y)^2}{n}$	k
Error	$\mathbf{y}'\mathbf{y} - \hat{\beta}'\mathbf{X}'\mathbf{y}$	$n - k - 1$
Total	$\mathbf{y}'\mathbf{y} - \dfrac{(\sum Y_i)^2}{n} = S_{YY}$	$n - 1$

We now want to show that this latter form is completely equivalent to what we have already done. We do this by showing that

$$\hat{\beta}'\mathbf{X}'\mathbf{y} - \frac{(\sum Y)^2}{n} = b_1 S_{1Y} + b_2 S_{2Y}$$

in the two-predictor case. First,

$$\hat{\beta}'\mathbf{X}'\mathbf{y} = \begin{bmatrix} a & b_1 & b_2 \end{bmatrix} \begin{bmatrix} \sum Y \\ \sum X_1 Y \\ \sum X_2 Y \end{bmatrix}$$

$$= a\sum Y + b_1 \sum X_1 Y + b_2 \sum X_2 Y$$

$$= a\sum Y + \frac{nb_1 \sum X_1 Y}{n} - \frac{b_1(\sum X_1)(\sum Y)}{n} + \frac{b_1(\sum X_1)(\sum Y)}{n}$$

$$+ \frac{nb_2 \sum X_2 Y}{n} - \frac{b_2(\sum X_2)(\sum Y)}{n} + \frac{b_2(\sum X_2)(\sum Y)}{n}$$

$$= a\sum Y + b_1 \left[\frac{n\sum X_1 Y - (\sum X_1)(\sum Y)}{n} \right] + \frac{b_1(\sum X_1)(\sum Y)}{n}$$

$$+ b_2 \left[\frac{n\sum X_2 Y - (\sum X_2)(\sum Y)}{n} \right] + \frac{b_2(\sum X_2)(\sum Y)}{n}$$

$$= a\sum Y + b_1 S_{1Y} + b_2 S_{2Y} + \frac{b_1(\sum X_1)(\sum Y)}{n} + \frac{b_2(\sum X_2)(\sum Y)}{n}.$$

Then

$$\hat{\beta}'X'y - \frac{(\sum Y)^2}{n} = b_1 S_{1Y} + b_2 S_{2Y}$$

$$+ \left[a\sum Y - \frac{(\sum Y)^2}{n} + \frac{b_1(\sum X_1)(\sum Y)}{n} + \frac{b_2(\sum X_2)(\sum Y)}{n} \right]$$

$$= b_1 S_{1Y} + b_2 S_{2Y} + (\sum Y) \left[a - \frac{\sum Y}{n} + \frac{b_1(\sum X_1)}{n} + \frac{b_2(\sum X_2)}{n} \right].$$

Recall that the first normal equation for the two-predictor model is

$$\sum Y = na + b_1 \sum X_1 + b_2 \sum X_2,$$

or

$$a = \frac{\sum Y}{n} - \frac{b_1 \sum X_1}{n} - \frac{b_2 \sum X_2}{n},$$

so that

$$a - \left[\frac{\sum Y}{n} - \frac{b_1 \sum X_1}{n} - \frac{b_2 \sum X_2}{n} \right] = a - \frac{\sum Y}{n} + \frac{b_1 \sum X_1}{n} + \frac{b_2 \sum X_2}{n} = 0.$$

Thus,

$$\hat{\beta}'X'y - \frac{(\sum Y)^2}{n} = b_1 S_{1Y} + b_2 S_{2Y}.$$

To use the matrix formulation in our house size-price-age problem,

$$\hat{\beta}'X'y = [10.04632987 \quad 12.83442613 \quad -0.05655391] \begin{bmatrix} 690 \\ 1554.9 \\ 14,540 \end{bmatrix}$$

$$= 26,065.9229;$$

$$\hat{\beta}'X'y - \frac{(\sum Y)^2}{n} = 26,065.9229 - \frac{(690)^2}{20}$$

$$= 2260.9229.$$

Except for rounding error this is the same as we found before.

EXERCISES   *Section 12.2*

12.3 INDIVIDUAL TESTS: GENERAL

We now turn to the question of which predictors are significant. That is, we want to test the set of k hypotheses

$$H_{01}: \beta_1 = 0, \quad H_{02}: \beta_2 = 0, \ldots, H_{0k}: \beta_k = 0$$

$$H_{11}: \beta_1 \neq 0, \quad H_{12}: \beta_2 \neq 0, \ldots, H_{1k}: \beta_k \neq 0.$$

We have indicated two-sided alternatives here in order to consider the most general case, although one might also be interested in one-sided alternatives. Provided that all alternatives are two-sided, $H_{0i}: \beta_i = 0, i = 1, 2, \ldots, k$ can all be tested using

$$F_i = \frac{b_1^2}{c_{ii} s_{y \cdot x}^2}$$

with 1 and $n - (k + 1)$ degrees of freedom. The c_{ii} in the denominator of F_i is the (i, i) element of the inverse of the matrix $\mathbf{S}$ defined in Chapter 10. Recall that $\mathbf{S}$ is the $k \times k$ submatrix obtained from $\mathbf{X'X}$ by deleting the first row and column of $\mathbf{X'X}$. (In Chapter 10, $\mathbf{S}$ was defined when X_1 and X_2 had been corrected for their means. While it can make life easier if the X's are corrected for their means, it is not absolutely necessary.) That is,

$$\mathbf{X'X} = \begin{bmatrix} n & \sum X_1 & \sum X_2 & \cdots & \sum X_k \\ \sum X_1 & \sum X_1^2 & \sum X_1 X_2 & \cdots & \sum X_1 X_k \\ \sum X_2 & \sum X_1 X_2 & \sum X_2^2 & \cdots & \sum X_2 X_k \\ \vdots & \vdots & \vdots & & \vdots \\ \sum X_k & \sum X_1 X_k & \sum X_2 X_k & \cdots & \sum X_k^2 \end{bmatrix};$$

so

$$
\mathbf{S} = \begin{bmatrix} \sum X_1^2 & \sum X_1 X_2 & \cdots & \sum X_1 X_k \\ \sum X_1 X_2 & \sum X_2^2 & \cdots & \sum X_2 X_k \\ \vdots & \vdots & & \vdots \\ \sum X_1 X_k & \sum X_2 X_k & \cdots & \sum X_k^2 \end{bmatrix}.
$$

and

$$
\mathbf{S}^{-1} = \begin{bmatrix} c_{11} & c_{12} & \cdots & c_{1k} \\ c_{21} & c_{22} & \cdots & c_{2k} \\ \vdots & \vdots & & \vdots \\ c_{k1} & c_{k2} & \cdots & c_{kk} \end{bmatrix}.
$$

Individual
Significance
Tests

The individual significance tests are much easier and neater when the predictors are uncorrelated. Thus, we will consider separately the two cases:

1. predictors uncorrelated,
2. predictors correlated.

12.4 INDIVIDUAL TESTS: UNCORRELATED PREDICTORS

THE VARIANCE-COVARIANCE MATRIX

We will consider the simpler case first: that in which $X_1, X_2, \ldots, X_k$ are all mutually uncorrelated. If all predictors are uncorrelated, then let us assume that they have all been corrected for their means, so that

$$
\mathbf{X}'\mathbf{X} = \begin{bmatrix} n & 0 & 0 & \cdots & 0 \\ 0 & S_{11} & S_{12} & \cdots & S_{1k} \\ 0 & S_{21} & S_{22} & \cdots & S_{2k} \\ \vdots & \vdots & \vdots & & \vdots \\ 0 & S_{k1} & S_{k2} & \cdots & S_{kk} \end{bmatrix}
$$

and

$$
\mathbf{S} = \begin{bmatrix}
S_{11} & S_{12} & \cdots & S_{1k} \\
S_{21} & S_{22} & \cdots & S_{2k} \\
\vdots & \vdots & & \vdots \\
S_{k1} & S_{k2} & \cdots & S_{kk}
\end{bmatrix}.
$$

$\mathbf{S}$ is called the variance-covariance matrix for the independent variables. But if all predictors are uncorrelated, then

$$
S_{12} = S_{13} = \cdots = S_{1k} = S_{23} = S_{24} = \cdots = S_{2k} = \cdots = S_{(k-1)k} = 0;
$$

that is, the only nonzero elements of $\mathbf{S}$ are $S_{11}, S_{22}, \ldots, S_{kk}$ and

$$
\mathbf{X'X} = \mathrm{diag}(n, S_{11}, S_{22}, \ldots, S_{kk}).
$$

Thus,

$$
\hat{\boldsymbol{\beta}} = (\mathbf{X'X})^{-1}\mathbf{X'y} = \begin{bmatrix}
1/n & 0 & 0 & \cdots & 0 \\
0 & 1/S_{11} & 0 & \cdots & 0 \\
0 & 0 & 1/S_{22} & \cdots & 0 \\
\vdots & \vdots & \vdots & & \vdots \\
0 & 0 & 0 & \cdots & 1/S_{kk}
\end{bmatrix}
\begin{bmatrix}
\sum y \\
S_{1Y} \\
S_{2Y} \\
\vdots \\
S_{kY}
\end{bmatrix}
$$

$$
= \begin{bmatrix}
\sum y / n \\
S_{1Y}/S_{11} \\
S_{2Y}/S_{22} \\
\vdots \\
S_{kY}/S_{kk}
\end{bmatrix}
$$

so that $b_i = S_{iY}/S_{ii}$ for $i = 1, 2, .., k$, and b_i depends only on X_i. Further, $c_{ii} = 1/S_{ii}$ for $i = 1, 2, \ldots, k$, since $\mathbf{S}^{-1} = \mathrm{diag}(1/S_{11}, 1/S_{22}, \ldots, 1/S_{kk})$.

THE F-TEST

Then the F-statistic for testing $H_{0i}: \beta_i = 0$ is

$$
F_i = \frac{b_i^2}{c_{ii}s_{y \cdot x}^2} = \frac{b_i^2}{(1/S_{ii})s_{y \cdot x}^2} = \frac{b_i^2 S_{ii}}{s_{y \cdot x}^2}.
$$

Since

$$b_i = S_{iY}/S_{ii},$$

$$b_i^2 S_{ii} = \frac{S_{iY}^2}{S_{ii}^2} S_{ii} = \frac{S_{iY}}{S_{ii}} S_{iY} = b_i S_{iY}.$$

Thus

$$F_i = \frac{b_i S_{iY}}{s_{y\cdot x}^2}.$$

In the analysis of variance table that follows, $F_i = b_i S_{iY}/s_{y\cdot x}^2$ has 1 and $n - (k + 1)$ degrees of freedom.

ANOVA

Source	SS	df	MS	F
Regression	$\displaystyle\sum_{i=1}^{k} b_i S_{iY}$	k	$\displaystyle\sum_{i=1}^{k} b_i S_{iY}/k$	
$\quad$ on X_1	$b_1 S_{1Y}$	1	$b_1 S_{1Y}$	F_1
$\quad$ on X_2	$b_2 S_{2Y}$	1	$b_2 S_{2Y}$	F_2
$\quad\;\;\vdots$	$\vdots$	$\vdots$	$\vdots$	$\vdots$
$\quad$ on X_k	$b_k S_{kY}$	1	$b_k S_{kY}$	F_k
Error	$S_{YY} - \displaystyle\sum_{i=1}^{k} b_i S_{iY}$	$n - (k + 1)$	$s_{y\cdot x}^2$	
Total	S_{YY}	$n - 1$		

We compare each F to $F_{1,n-(k+1),\alpha}$ and reject H_{0i} for those F_i's which are greater than this critical value, retaining the associated predictors in the regression equation. Those predictors which are not significant are dropped from the model.

Note that the form of F_i is the same as the form of F in a simple regression problem and that the regression sum of squares for the entire regression, $SS(R)$, breaks down into k independent pieces, each measuring the effect of only one predictor. That is,

$$SS(R) = \sum_{i=1}^{k} b_i S_{iY} = b_1 S_{1Y} + b_2 S_{2Y} + \cdots + b_k S_{kY}$$

$$= SSR(X_1) + SSR(X_2) + \cdots + SSR(X_k).$$

Since predictors are uncorrelated, we know the effect of each X_i on Y separately, as measured by b_i, and this measure is not affected by the inclusion or exclusion of any other predictors. The only way in which the multiple regression with k significant predictors differs from k-significant simple regressions is in the size of the error mean square: $s^2_{y \cdot x}$ will typically be smaller than any of the $s^2_{y \cdot x}$-values in the simple regressions, which means simply that we can predict more accurately using k variables than using any one of them alone.

Any of the k variables found not to be significant are dropped from the regression equation. This will not affect the values for a' and the remaining b_i's, since $a' = \bar{y}$ and $b_i = S_{iY}/S_{ii}$. However, the standard error of estimate should be recalculated using only those X_i's that remain in the model.

THE t-TEST

In case any one-sided hypotheses are to be tested, a t-test must be employed, since F can be used only for two-sided alternatives. Since $t^2_v = F_{1,v}$, we can see that

$$t_i = \frac{b_i}{\sqrt{c_{ii}s^2_{y \cdot x}}}$$

with $n - (k + 1)$ degrees of freedom. If predictors are uncorrelated, $c_{ii} = 1/S_{ii}$; so

$$t_i = \frac{b_i}{\sqrt{s^2_{y \cdot x}/S_{ii}}}.$$

Note that this is exactly the same form as the t-test in simple regression, and that t_i depends only on X_i. All t-tests, just like all F-tests, will be independent when predictors are uncorrelated.

A NUMERICAL EXAMPLE

As an example, suppose we have the following data from an experiment to determine the effects of altitude and cooking temperature on the baking time required for a new cake mix which is to be marketed. Altitudes are below sea level (coded to -1), normal (0), and high ($+1$), at equally spaced intervals. Temperatures are 350°, 400°, and 450°, also coded to $-1, 0$, and 1, respectively. Two cakes were baked at each combination of altitude and temperature, yielding the results shown in Table 12.1. From the calculations in the table, we have

$$\mathbf{X'X} = \begin{bmatrix} 18 & 0 & 0 \\ 0 & 12 & 0 \\ 0 & 0 & 12 \end{bmatrix}, \quad (\mathbf{X'X})^{-1} = \begin{bmatrix} \frac{1}{18} & 0 & 0 \\ 0 & \frac{1}{12} & 0 \\ 0 & 0 & \frac{1}{12} \end{bmatrix},$$

$$\mathbf{X'y} = \begin{bmatrix} 1196 \\ 138 \\ -70 \end{bmatrix}, \quad \text{and} \quad \hat{\boldsymbol{\beta}} = \begin{bmatrix} 66.4444 \\ 11.5000 \\ -5.8333 \end{bmatrix}.$$

To test

$$H_0 : \beta_1 = \beta_2 = 0$$

$$H_1 : \text{some } \beta_i \neq 0, \quad i = 1, 2,$$

we will reject H_0 if $F > F_{0.05, 2, 15} = 3.68$.

TABLE 12.1   *Baking Times at Three Altitudes and Three Temperatures*

Altitude	Temperature	Baking Time (Minutes)
−1	−1	60
−1	−1	58
−1	0	56
−1	0	50
−1	1	50
−1	1	45
0	−1	78
0	−1	76
0	0	70
0	0	68
0	1	65
0	1	63
1	−1	83
1	−1	80
1	0	77
1	0	75
1	1	70
1	1	72
$\sum (\cdot) \quad 0$	0	1196
$\sum (\cdot)^2 \quad 12$	12	81,630
$\sum X_1 Y = 138$ $S_{YY} = 2162.4444$	$S_{12} = 0$	
$\sum X_2 Y = -70$ $S_{1Y} = 138$	$S_{11} = 12$	
$\sum X_1 X_2 = 0$ $S_{2Y} = -70$	$S_{22} = 12$	
$n = 18$		

ANOVA

Source	SS	df	MS	F
Regression	1995.3310	2	997.6655	89.5498
Error	167.1134	15	11.1409	
Total	2162.4444	17		

Since $89.5498 > 3.68$, conclude that either altitude or temperature, or both, are significant predictors of baking time. In order to determine which predictor(s) is significant, if we want to test the two-sided alternatives then we break the regression sum of squares,

$$SSR = 1995.3310,$$

into

$$SSR(X_1) = b_1 S_{1Y} = 11.5(138) = 1587$$
$$= b_1^2 S_{11} = (11.5)^2(12) = 1587$$

and

$$SSR(X_2) = b_2 S_{2Y} = (-5.8333)(-70) = 408.3310$$
$$= b_2^2 S_{22} = (-5.8333)^2(12) = 480.3287.$$

Note that $1587 + 408.3310 = 1995.3310$, so that (except for possible rounding error)

$$SSR(X_1) + SSR(X_2) = SSR.$$

Then to test

$$\begin{matrix} H_{01}: \beta_1 = 0 \\ H_{11}: \beta_1 \neq 0 \end{matrix} \quad \text{and} \quad \begin{matrix} H_{02}: \beta_2 = 0 \\ H_{12}: \beta_2 \neq 0, \end{matrix}$$

reject H_0 if $F_i > F_{0.05,1,15} = 4.54$, for $i = 1,2$.

ANOVA

Source	SS	df	MS	F
Regression	1995.3310	2		
$\begin{cases} \text{on } X_1 \\ \text{on } X_2 \end{cases}$	$\begin{cases} 1587.0000 \\ 408.3310 \end{cases}$	$\begin{cases} 1 \\ 1 \end{cases}$	1587 408.3310	142.4481 36.6515
Error	167.1134	15	11.1409	
Total	2162.4444	17		

Since both F_1 and F_2 are greater than 4.54, conclude that both temperature and altitude are significant predictors of baking time.

Had we wanted to test

$$H_{01}: \beta_1 = 0 \qquad H_{02}: \beta_2 = 0$$
$$\text{and}$$
$$H_{11}: \beta_1 > 0 \qquad H_{12}: \beta_2 < 0,$$

we would reject H_{01} if $t_1 > t_{0.05,15} = 1.753$ and reject H_{02} if $t_2 < -t_{0.05,15} = -1.753$. We obtain

$$t_1 = \frac{11.5}{\sqrt{11.1409/12}} = 11.5\sqrt{12/11.1409} = 11.9352,$$

$$t_2 = \frac{-5.8333}{\sqrt{11.1409/12}} = -6.0540.$$

Since $t_1 > 1.753$ and $t_2 < -1.753$, reject H_{01} and H_{02} and conclude that the higher the altitude, the longer the baking time, and the higher the temperature, the shorter the baking time. The point is that since the data are from a carefully designed experiment, it is possible to know the effect of temperature completely separate from the effect of altitude. Had two separate simple regressions been run, $SSR(X_1)$ and $SSR(X_2)$ would not have changed at all. The values of the F-statistics would be a bit smaller, however, since $s_{y \cdot x}^2$ for either simple regression would be larger than $s_{y \cdot x}^2$ in the multiple regression.

EXERCISES   *Section 12.4*

12.2 Regardless of the results of the overall regression test, test for the significance of the individual predictors in the
p. 535   a. GLUE data,
p. 535   b. BOTTLES data,
p. 536   c. YIELD data.
Comment on any unusual results. Show the regression equation you would use in each problem.

12.5 INDIVIDUAL TESTS: CORRELATED PREDICTORS

THE VARIANCE-COVARIANCE MATRIX

If $X_1, X_2, \ldots, X_k$ are correlated, as is usually the case in a social science study, then life becomes more confusing. In this case, $\mathbf{X}'\mathbf{X}$ will not be diagonal, even if $X_1, X_2, \ldots, X_k$ are all corrected for their means. Then $\mathbf{S}$ is not diagonal, and neither is $\mathbf{S}^{-1}$. Let us first consider the $k = 2$ case when X_1 and X_2 are corrected for their means, and then see how matters become even more complicated if $k > 2$. It was shown earlier that

$$\mathbf{X'X} = \begin{bmatrix} n & 0 & 0 \\ 0 & S_{11} & S_{12} \\ 0 & S_{21} & S_{22} \end{bmatrix} = \left[\begin{array}{c|cc} n & 0 & 0 \\ \hline 0 & & \\ 0 & & \mathbf{S} \end{array} \right]$$

and that

$$(\mathbf{X'X})^{-1} = \left[\begin{array}{c|cc} \dfrac{1}{n} & 0 & 0 \\ \hline 0 & & \\ 0 & & \mathbf{S}^{-1} \end{array} \right] = \begin{bmatrix} \dfrac{1}{n} & 0 & 0 \\ 0 & \dfrac{S_{22}}{S_{11}S_{22} - S_{12}^2} & \dfrac{-S_{12}}{S_{11}S_{22} - S_{12}^2} \\ 0 & \dfrac{-S_{12}}{S_{11}S_{22} - S_{12}^2} & \dfrac{S_{11}}{S_{11}S_{22} - S_{12}^2} \end{bmatrix}.$$

Then

$$\hat{\boldsymbol{\beta}} = (\mathbf{X'X})^{-1}\mathbf{X'y} = \begin{bmatrix} \dfrac{1}{n} & 0 & 0 \\ 0 & \dfrac{S_{22}}{|\mathbf{S}|} & \dfrac{-S_{12}}{|\mathbf{S}|} \\ 0 & \dfrac{-S_{12}}{|\mathbf{S}|} & \dfrac{S_{11}}{|\mathbf{S}|} \end{bmatrix} \begin{bmatrix} \sum y \\ S_{1Y} \\ S_{2Y} \end{bmatrix}$$

$$= \begin{bmatrix} \sum y/n \\ \dfrac{S_{22}S_{1Y} - S_{12}S_{2Y}}{S_{11}S_{22} - S_{12}^2} \\ \dfrac{S_{11}S_{2Y} - S_{12}S_{1Y}}{S_{11}S_{22} - S_{12}^2} \end{bmatrix} = \begin{bmatrix} a' \\ b_1 \\ b_2 \end{bmatrix}.$$

We see that b_1 involves not only the effect of X_1 on Y, as measured by S_{1Y} and S_{11}, but also the effect of X_2 on X_1 (measured by S_{12}) and the effect of X_2 on Y (measured by S_{2Y} and S_{22}). Of course, this is the effect of multicollinearity—that S_{12} is nonzero brings into b_1 terms involving X_2 and Y. Obviously, if predictors are correlated,

$$b_i \neq \frac{S_{iY}}{S_{ii}},$$

as was the case when predictors were not correlated. In addition, c_{ii} will not involve just X_i: for example,

$$c_{11} = \frac{S_{22}}{S_{11}S_{22} - S_{12}^2}$$

involves something other than just S_{11}. Dividing numerator and denominator of c_{11} by $S_{11}S_{22}$,

$$c_{11} = \frac{S_{22}/S_{11}S_{22}}{(S_{11}S_{22} - S_{12}^2)/S_{11}S_{22}} = \frac{1/S_{11}}{1 - r_{12}^2} = \frac{1}{S_{11}(1 - r_{12}^2)},$$

and similarly,

$$c_{22} = \frac{1}{S_{22}(1 - r_{12}^2)}.$$

Thus, c_{ii} involves the correlation between the two predictor variables, as well as S_{ii}, for $i = 1, 2$.

THE *F*- AND *t*-TESTS

The *F*-statistic for testing $H_{0i}:\beta_i = 0$ for $i = 1, 2$ is then

$$F_i = \frac{b_i^2}{c_{ii}s_{y\cdot x}^2} = \frac{b_i^2 S_{ii}(1 - r_{12}^2)}{s_{y\cdot x}^2}.$$

The *t*-statistic is, of course,

$$t_i = \frac{b_i}{\sqrt{s_{y\cdot x}^2/S_{ii}(1 - r_{12}^2)}}.$$

THE EFFECT OF MULTICOLLINEARITY

Let us examine the effect of multicollinearity on the tests of significance. The greater the degree of correlation between the predictors, the closer r_{12}^2 will be to 1. In the numerator of F_i, then, if X_1 and X_2 are very highly correlated, the quantity $(1 - r_{12}^2)$ will be close to zero, thus making the *F*-statistic close to zero. Also recall that $c_{ii}s_{y\cdot x}^2$ is an estimate of the variance of b_i. If X_1 and X_2 are highly correlated, c_{ii} will be quite large, making the *F*-statistic very small. The same thing will happen to the *t*-statistic. Now, it is only for *F*-values considerably larger than zero (or *t*-values larger than zero in absolute value) that we can reject $H_{0i}.\beta_i = 0$. Thus, if predictors are correlated, the hypothesis tests become very conservative, in that they will not allow H_{0i} to be rejected unless a predictor affects *Y* strongly enough to overcome its correlation with the other predictor. That is, b_i must be large enough, taking units of measurement into account, to overcome the effect of r_{12}^2 on the *F*- or the *t*-statistic. Thus it is possible that one predictor which is quite strong by itself will not be significant when used in conjunction with another strong predictor with which it is highly correlated. The fact that two predictors are highly correlated means that they are redundant—they are both doing the same thing—so that one is superfluous when used in conjunction with the other and should be discarded.

The denominator of the t-statistic,

$$\left[\frac{s_{y \cdot x}^2}{S_{ii}(1 - r_{12}^2)} \right]^{1/2},$$

is called the standard error of the regression coefficient. It measures how much b_i is expected to vary from sample to sample of the same size from the same population. Recall that if this denominator is small, the value of b_i can be trusted pretty well. On the other hand, if the standard error of the regression coefficient is large, then we cannot believe that β_i is close to the value we have obtained for b_i. Severe multicollinearity, which is manifested by a large value of r_{12}^2, will result in a large standard error. The same argument holds when we have more than two predictors. If multicollinearity is severe, the determinant of $\mathbf{X'X}$ will be small, resulting in the elements of $(\mathbf{X'X})^{-1}$ being quite large. These large elements c_{ii} appear in the standard errors of the regression coefficients and make these standard errors large.

Thus, when multicollinearity is present in the data, one should look carefully at the sizes of the standard errors of the regression coefficients. If these errors are too large, the whole regression equation is untrustworthy, and another sample would give quite different values for b_i, $i = 1, 2, \ldots, k$. (Techniques called *ridge regression* have been recently introduced in attempts to handle this difficulty. Ridge regression is beyond the scope of this text.)

Further, if predictors are correlated, the regression sum of squares does not break down into k independent tests on the individual predictors, as it did in the previous case. The analysis of variance for correlated predictors, where $F_i = b_i^2 / c_{ii} s_{y \cdot x}^2$ with 1 and $n - (k + 1)$ degrees of freedom, is given below.

ANOVA

Source	SS	df	MS	F
Regression	$\sum\limits_{i=1}^{k} b_i S_{iY}$	k	$\sum\limits_{i=1}^{k} b_i S_{iY}/k$	
on X_1	b_1^2/c_{11}	1	b_1^2/c_{11}	F_1
on X_2	b_2^2/c_{22}	1	b_2^2/c_{22}	F_2
$\vdots$	$\vdots$	$\vdots$	$\vdots$	$\vdots$
on X_k	b_k^2/c_{kk}	1	b_k^2/c_{kk}	F_k
Error	$S_{YY} - \sum\limits_{i=1}^{k} b_i S_{iY}$	$n - (k + 1)$	$s_{y \cdot x}^2$	
Total	S_{YY}	$n - 1$		

INDIVIDUAL TESTS FOR THE HOUSE SIZE-PRICE-AGE EXAMPLE

In the house size-price-age example, recall that

$$c_{11} = 0.07448317 \quad \text{and} \quad c_{22} = 0.0005789.$$

Thus

$$SSR(X_1) = b_1^2/c_{11} = \frac{(12.83442613)^2}{0.07448317} = 2211.5398$$

$$= \frac{b_1^2}{S_{11}(1 - r_{12}^2)} = (12.83442613)^2(13.56)(1 - 0.0099)$$

$$= 2211.5240$$

and

$$SSR(X_2) = b_2^2/c_{22} = \frac{(-0.05655391)^2}{0.00057879} = 5.5253$$

$$= b_2^2 S_{22}(1 - r_{12}^2) = (-0.05655391)^2(1745)(1 - 0.0099)$$

$$= 5.5253.$$

Thus

ANOVA

Source	SS	df	MS	F
Regression	2261.4245	2		
$\begin{cases} \text{on } X_1 \\ \text{on } X_2 \end{cases}$	$\begin{cases} 2211.5398 \\ 5.5253 \end{cases}$	$\begin{cases} 1 \\ 1 \end{cases}$	2211.5398 5.5253	336.9555 0.8418
Error	111.5755	17	6.5633	
Total	2373	19		

We will reject $H_{01}: \beta_1 = 0$ and $H_{02}: \beta_2 = 0$ in favor of $H_{11}: \beta_1 \neq 0$ and $H_{12}: \beta_2 \neq 0$ if $F_i > F_{0.05,1,17} = 4.45$. Since $F_1 > 4.45$, conclude that size is a significant predictor of price. But since $F_2 \ngtr 4.45$, we cannot be 95 % sure that the age of a house is related to its price. Note that because of multicollinearity, $SSR(X_1) + SSR(X_2) \neq SSR$:

$$2211.5398 + 5.5253 = 2217.0651 \neq 2261.4245.$$

Thus, when the predictors are correlated, the regression sum of squares cannot be partitioned into two disjoint pieces which sum to *SSR*.

If one-sided alternatives are to be tested, then t-tests can be used. For example, if

$$H_{01}:\beta_1 = 0 \qquad H_{02}:\beta_2 = 0$$
$$\text{and}$$
$$H_{11}:\beta_1 > 0 \qquad H_{12}:\beta_2 < 0,$$

we will reject H_{01} if $t_1 > t_{0.05,17} = 1.740$, and we will reject H_{02} if $t_2 < -t_{0.05,17} = -1.740$. The test statistics give

$$t_1 = \frac{12.83442613}{\sqrt{6.56326273/13.56(1 - 0.0099)}} = 18.3563$$

and

$$t_2 = \frac{-0.05655391}{\sqrt{6.56326273/1745(1 - 0.0099)}} = -0.9176.$$

Then since $t_1 > 1.740$, we are more than 95 % sure that the larger a house, generally the higher its price, but since $t_2 \nless -1.740$, we cannot be 95 % sure that older houses cost less than newer houses.

THE REDUCED MODEL

Since we have only very slight evidence that age is related to price, we will drop $X_2 = $ age from the model $Y = \alpha + \beta_1 X_1 + \beta_2 X_2 + \varepsilon$ and work with the *reduced model*

$$Y = \alpha + \beta X + \varepsilon,$$

where $Y = $ price and $X = $ size. Since we cannot be reasonably certain that the slight relationship we saw between age and size is not just due to chance, there is no reason to retain age as a predictor. Another sample of size 20 from the same population would not be expected to show any relationship between age and price. The regression equation we will use to predict price from size is the equation found in the simple regression problem:

$$\hat{Y} = 8.7 + 12.9X,$$

and the standard error is $s_{y \cdot x} = 2.55$ thousand dollars. Note that had we not already performed the simple regression of price on size, but had only run the multiple regression, it would now be necessary to rework the problem using only $X = $ size and Y. Since predictors are correlated, the term involving X_2 in the multiple regression equation cannot simply be dropped, as it was in the case of independent predictors. The regression coefficient on size changes depending upon whether or not X_2 is included in the equation.

EXTENSION TO LARGER PROBLEMS

The reader can appreciate how the situation becomes even more complicated when $k > 2$. For example, if $k = 3$, then

$$
S = \begin{bmatrix} S_{11} & S_{12} & S_{13} \\ S_{12} & S_{22} & S_{23} \\ S_{13} & S_{23} & S_{33} \end{bmatrix},
$$

and it can be shown that

$$
c_{11} = \frac{1}{S_{11}} \cdot \frac{1 - r_{23}^2}{1 - r_{23}^2 - r_{13}^2 - r_{12}^2 + 2S_{12}S_{13}S_{23}/S_{11}S_{22}S_{23}},
$$

and similarly for c_{22} and c_{33}. Note that c_{11} involves more than X_1. It involves all pairwise correlations among all three predictor variables and a term involving all three predictor variables at once. Thus a test on β_1 will depend not only on the effect of X_1 on Y, but also on the effects of X_2 and X_3 on Y, and on how X_1 is correlated with X_2 and X_3 and how X_2 is correlated with X_3. Obviously, the test on β_1 will not be independent of the test on β_2 and β_3. Thus, the analysis of the multiple regression problem with correlated predictors is best done with the aid of a computer, using one of the several programs available. The interpretation of the output of some representative programs is discussed in the next chapter.

The reader should note that we have considered here a reasonably well-behaved problem. Strange things can happen when several predictors are highly correlated, so it will be very instructive for the student to carefully study the results of the sample problems in the exercises for Chapters 10–12.

PARTIAL SUMS OF SQUARES AND t- AND F-STATISTICS

Partial Tests

The t- and F-tests presented in this chapter are partial tests. They test for the significance of the predictor, conditional on all the other predictors in the model. These partial tests let us see if a predictor adds something to what has already been explained by the other predictors, or whether the predictor is redundant once the other predictors have been included. When a partial t- or F-test tells us to reject $H_0 : \beta_i = 0$, we conclude that, given that all other predictors are being used, X_i can still yield significant amount of information about Y. When H_0 cannot be rejected, we conclude that either X_i is a worthless predictor or it is giving the same information as some other predictor. In our house size-price-age example, since the partial F-test rejected $H_{01} : \beta_1 = 0$, the conclusion is that even if $X_2 =$ age has explained all it can about $Y =$ price, $X_1 =$ size can still improve the percentage of explained variation. In other words, even if age is used as a predictor, we still need size. Since $H_{02} : \beta_2 = 0$ was not rejected, either age is a worthless predictor, or it adds nothing to what size can tell alone. If size is used as a predictor, age is not needed.

We discussed partial sums of squares and their associated t- and F-tests briefly in Chapter 9 in connection with polynomial regressions. But there they were less appropriate than were sequential sums of squares and tests. One often sees these partial sums of squares calculated (for the two-predictor case) as

$$SSR(X_2 \mid X_1) = SSR(X_1, X_2) - SSR(X_1)$$

and

$$SSR(X_1 \mid X_2) = SSR(X_1, X_2) - SSR(X_2).$$

In our home size-price-age example,

$$SSR(X_2 \mid X_1) = 2261.42453359 - 2255.90044248 = 5.52409111$$

and

$$SSR(X_1 \mid X_2) = 2261.42453359 - 49.87106017 = 2211.55347342.$$

Except for rounding error, these are the same values for the individual regression sums of squares as were found previously.

Suppose we had a three-predictor model. Then the partial sums of squares could be calculated as

$$SSR(X_1 \mid X_2, X_3) = SSR(X_1, X_2, X_3) - SSR(X_2, X_3),$$
$$SSR(X_2 \mid X_1, X_3) = SSR(X_1, X_2, X_3) - SSR(X_1, X_3),$$
$$SSR(X_3 \mid X_1, X_2) = SSR(X_1, X_2, X_3) - SSR(X_1, X_2).$$

The extension to $k = 4$ and greater is obvious. All these partial sums of squares measure the increase in the regression sum of squares or, equivalently, the reduction in error, acheived by adding a predictor to those already in the model. They are sometimes referred to as extra sums of squares, or sums of squares if last; i.e., if this variable were the last one added to the model.

Extra Sums of
Squares

Sums of
Squares If Last

SEQUENTIAL SUMS OF SQUARES

In some instances sequential sums of squares are useful in multiple regression analysis. Suppose the researcher has an order of preference for the variables in a problem. For example, X_4 seems intuitively important as a predictor of Y. Next in importance comes X_2, because it is very convenient to measure. Then comes X_3 because it is inexpensive.

Last in importance is X_1. The researcher is then interested first in what X_4 can do by itself, then in what X_2 adds to X_4, then what X_3 adds to X_2 and X_4, and finally in what X_1 can explain, given X_2, X_3, and X_4. He (she) would then enter the variables in this predetermined order of importance and would calculate sequential sums of squares:

$$SSR(X_4),$$

$$SSR(X_2 \mid X_4) = SSR(X_2, X_4) - SSR(X_4),$$

$$SSR(X_3 \mid X_2, X_4) = SSR(X_2, X_3, X_4) - SSR(X_2, X_4),$$

$$SSR(X_1 \mid X_2, X_3, X_4) = SSR(X_1, X_2, X_3, X_4) - SSR(X_2, X_3, X_4).$$

Note that

$$SSR(X_4) + SSR(X_2 \mid X_4) + SSR(X_3 \mid X_2, X_4) + SSR(X_1 \mid X_2, X_3, X_4)$$
$$= SSR(X_1, X_2, X_3, X_4),$$

since each sequential sum of squares represents an additional piece of the regression sum of squares.

The question is, in a set of k interrelated predictors, how can we find a subset of predictors that are all effective, given the others? We might like to find a subset of predictors such that they explain the largest possible percentage of variation in Y. Or we might want to first find the single strongest predictor, then add the next strongest predictor, given the first, then the third strongest predictor, given the first two, and so on, until the next strongest predictor is no longer needed, given the others. There are many different ways to define the "best" multiple regression model that can be formed using a set of k possible predictors. In the next chapter we shall investigate some of these techniques. The work required is so involved as to be almost prohibitive without the aid of the computer.

EXERCISES   *Section 12.5*

12.3 Regardless of the results of the overall regression significance test, test for the significance of the individual predictors using partial F-tests in the

p. 531   a. SALES II data,
p. 531   b. RESTAURANT data,
p. 532   c. DRUG data,
p. 532   d. AUTO data,
p. 533   e. ADS data,
p. 533   f. CHILDREN data.

Comment on any unusual results. Show the regression equation you would use in each problem.

12.6 CONFIDENCE INTERVALS ON NET REGRESSION COEFFICIENTS

Confidence intervals on $\beta_1, \beta_2, \ldots, \beta_k$ can be formed in just the same way as we found a confidence interval on β in simple regression. Recall that the general form of a $100(1 - \alpha)\%$ confidence interval on any parameter is

point estimate $\pm t_{v,\alpha/2}$ (standard error of the point estimate).

The point estimate of $\beta_1, \beta_2, \ldots, \beta_k$ are $b_1, b_2, \ldots, b_k$, respectively. The standard error of any b_i is

$$\sqrt{c_{ii}s^2_{y\cdot x}}.$$

Thus, a $100(1 - \alpha)\%$ confidence interval on β_i will look like

$$b_i - t_{\alpha/2, n-p}s_{y\cdot x}\sqrt{c_{ii}} < \beta_i < b_i + t_{\alpha/2, n-p}s_{y\cdot x}\sqrt{c_{ii}}.$$

In our house size-price-age example, suppose we wanted to construct a 95% confidence interval on β_2. Then

$$b_2 = -0.05655391,$$

$$s^2_{y\cdot x} = 6.5633,$$

$$c_{22} = 0.0005789,$$

$$t_{0.925, 17} = 1.740;$$

and the 95% confidence interval on β_2 is

$$-0.05655391 \pm 1.74\sqrt{6.5633(0.0005789)},$$

or

$$-0.05655391 \pm 0.107247,$$

or

$$-0.1638 < \beta_2 < 0.0507.$$

Thus we can be 95% sure that for houses the same size, if one house is one year older than another, the older house will sell for between \$163.80 less and \$50.7 more than the newer house. Obviously, this does not give us very much information, and can be taken as evidence equivalent to an hypothesis test that age is not a predictor of price. A similar confidence interval could be constructed for β_1.

Of course, once a predictor is dropped from the model, both $s_{y \cdot x}^2$ and c_{ii} will change for the predictors left in the model. Thus, unless one is using confidence intervals to determine which predictors to delete and which to retain, it is probably best to wait until the final model is found before constructing confidence intervals on the net regression coefficients. If we used the reduced model from the house size-price-age problem, our confidence interval on β_1 would of course be the same one as we found in simple regression.

EXERCISES   *Section 12.6*

12.4 Give 95 % confidence intervals on the regression coefficients in the
- p. 531   a. SALES II data,
- p. 532   b. DRUG data,
- p. 535   c. GLUE data,
- p. 535   d. BOTTLES data.
           Interpret the meaning of these confidence intervals.

12.7 SUMMARY

The first test of significance to be performed in a multiple regression problem is intended to detect the presence of any significant predictors in the set under consideration. Provided we have evidence of a relationship for the response with at least one of the predictors, we then proceed to perform individual tests to tell which predictors are significant. The procedure is quite straightforward if there is no multicollinearity in the problem; but the results can be confusing if predictors are correlated. For example, the general regression significance test can indicate the presence of at least one significant predictor but each individual test might indicate that its predictor is not effective.

Since we need the computer in order to fully investigate a multiple regression problem of any size, the next chapter introduces the use of the computer in multiple regression problems.

Chapter 13 USE OF THE COMPUTER IN MULTIPLE REGRESSION

13.1 INTRODUCTION

Obviously if any multicollinearity is present in a multiple regression problem, only a masochist would choose to analyze the problem with only the aid of a calculator. Even when $k = 2$, the calculations are quite laborious. Rounding error is always a problem, and in large problems human error can creep in, too. When we turn to the search for the *best* set of predictors, the labor involved is almost prohibitive without the computer. However, several computer programs can perform the multiple regression analysis quite quickly and accurately. We will discuss the output of three commonly used packages, the BMD, SPSS, and SAS. In this chapter, we shall see how to use the computer to perform the general regression significance test of

$$H_0 : \beta_1 = \beta_2 = \cdots = \beta_k = 0.$$

The next chapter deals with various ways to find the best set of predictors.

13.2 THE BMDP1R

The first regression program listed in the *BMDP Biomedical Computer Programs* manual is called "Multiple Linear Regression." This is the program we used in Chapter 4 to analyze simple regression problems. It can perform the general regression test as well and will also give the regression coefficients, multiple correlation, and other measures of interest.

FIGURE 13.1   *Multiple Regression for House Size-Price-Age Data Using BMDP1R: Worksheet*

	1-10										11-20										21-30										31-40										
1	2	3	4	5	6	7	8	9	10	11	12	13	14	15	16	17	18	19	20	21	22	23	24	25	26	27	28	29	30	31	32	33	34	35	36	37	38	39	40	41
{	S	Y	S	T	E	M			C	A	R	D	S																											
P	R	Ø	B	L	E	M					T	I	T	L	E	=	'	H	Ø	U	S	E		S	I	Z	E		P	R	I	C	E		A	G	E	'	.	/
I	N	P	U	T							V	A	R	I	A	B	L	E	S	=	3	.																		
											F	Ø	R	M	A	T	=	'	(	F	3	.	1	,	F	3	.	0	,	F	3	.	0	)	'	.			
											C	A	S	E	S	=	2	0	.	/																			
V	A	R	I	A	B	L	E				N	A	M	E	S	=	S	I	Z	E	,	A	G	E	,	P	R	I	C	E	.	/								
P	R	I	N	T							C	Ø	V	A	R	.																								
											C	Ø	R	R	.																									
											D	A	T	A	.	/																							
P	L	Ø	T								R	E	S	I	D	U	A	L	S	.	/																			
R	E	G	R	E	S						D	E	P	E	N	D	E	N	T	=	P	R	I	C	E	.	/													
E	N	D	/																																					
1	.	8		3	0		3	2																																
1	.	0		3	3		2	4																																
		:																																						
1	.	5		2	9		2	8																																
F	I	N	I	S	H	/																																		

RUNNING THE PROGRAM

Figure 13.1 shows how to run the house size-price-age data through the BMDP1R. The program shown is very much like that of Figure 4.1, in which we used the BMDP1R (or BMDP2R) for simple regression. There are just a few statements in it that we have not seen before. One is in the PLØT paragraph, in which we request a plot of RESIDUALS. This requests that the residuals and squared residuals be plotted against the predicted values. This is usually the plot of residuals of interest in multiple regression. Recall that in simple regression, we plotted the residuals against X, but we now have more than one X in the model. We could request a plot of residuals and predicted and observed values against either predictor by using the VARIABLE statement. For example, adding

VARIABLE = SIZE, AGE.

would request that one set of plots be made against $X_1 =$ SIZE and that another set of plots be made against $X_2 =$ AGE.

In the PRINT paragraph, in addition to requesting that the data be printed, we ask for the covariance matrix (COVAR) and the correlation matrix (CORR).

PRELIMINARY STATISTICS

Figure 13.2 shows the simple statistics and the covariance and correlation matrices from the BMDP1R analysis of the house size-price-age problem. Note that in both

FIGURE 13.2   *Preliminary Calculations from BMDP1R*

VARIABLE	MEAN	STANDARD DEVIATION	ST.DEV/MEAN	MINIMUM	MAXIMUM
1 SIZE	2.00000	0.84480	0.42240	0.80000	3.60000
2 AGE	21.49991	9.58341	0.44574	1.00000	33.00000
3 PRICE	34.49994	11.17563	0.32393	17.00000	52.00000

COVARIANCE MATRIX

	SIZE 1	AGE 2	PRICE 3
SIZE 1	0.7137		
AGE 2	-0.8053	91.8418	
PRICE 3	9.2053	-15.5263	124.8947

CORRELATION MATRIX

	SIZE 1	AGE 2	PRICE 3
SIZE 1	1.0000		
AGE 2	-0.0995	1.0000	
PRICE 3	0.9750	-0.1450	1.0000

matrices, only the entries on and below the diagonal are printed. This is done for ease of reading since both matrices are symmetric. The entries should look familiar. The covariance matrix contains entries $S_{ij}/(n-1)$. For example, in the $(2, 1)$ position of the covariance matrix we have

$$-0.8053 = \frac{S_{21}}{n-1} = \frac{S_{12}}{n-1} = \frac{-15.3}{19}.$$

Other entries are obtained similarly. Note that all entries on the diagonal must be positive because they are the variances of the variables.

Entries in the correlation matrix can be obtained from entries in the covariance matrix. For example, the entry in the $(3, 2)$ position of the correlation matrix is

$$-0.145 = r_{2Y} = \frac{S_{2Y}}{(S_{22}S_{YY})^{1/2}} = \frac{S_{2Y}/(n-1)}{[(S_{22}/n-1)(S_{YY}/n-1)]^{1/2}}$$

$$= \frac{-15.5263}{[(91.8418)(124.8947)]^{1/2}}.$$

Of course, since any variable is perfectly and directly correlated with itself, all entries in the diagonal of the correlation matrix are $+1$.

THE GENERAL REGRESSION SIGNIFICANCE TEST

Figure 13.3 shows the general regression significance test and associated statistics. We see some familiar values, such as

$$s_{y \cdot x} = 2.5619, \quad a = 10.046, \quad b_1 = 12.834, \quad b_2 = -0.057, \quad r_{xy}^2 = 0.9530.$$

FIGURE 13.3   *General Regression Significance Test from BMDP1R (Part 1)*

MULTIPLE R	0.9762	STD. ERROR OF EST.	2.5619
MULTIPLE R-SQUARE	0.9530		

ANALYSIS OF VARIANCE

	SUM OF SQUARES	DF	MEAN SQUARE	F RATIO	P(TAIL)
REGRESSION	2261.419	2	1130.709	172.271	0.00000
RESIDUAL	111.580	17	6.564		

VARIABLE	COEFFICIENT	STD. ERROR	STD. REG COEFF	T	P(2 TAIL)
INTERCEPT	10.046				
SIZE 1	12.834	0.699	0.970	18.356	0.0
AGE 2	-0.057	0.062	-0.048	-0.917	0.372

FIGURE 13.3   *(Part 2)*

		PREDICTED	VARIABLES		
NO.	RESIDUAL	VALUE	1 SIZE	2 AGE	3 PRICE
1	0.5474	31.4526	1.8000	30.0000	32.0000
2	2.9845 *	21.0155	1.0000*	33.0000*	24.0000
3	-3.4518 *	30.4518	1.7000	25.0000	27.0000
4	1.6955	45.3045	2.8000	12.0000	47.0000*
5	-1.8125	36.8125	2.2000	26.0000	35.0000
6	-1.9008	18.9008	0.8000*	25.0000	17.0000*
7	-2.6677*	54.6677	3.6000*	28.0000	52.0000*
8	-2.5251	22.5251	1.1000*	29.0000	20.0000*
9	3.6978*	34.3022	2.0000	25.0000	38.0000
10	1.6972	43.3028	2.6000	2.0000**	45.0000
11	6.1301 **	37.8699	2.3000	30.0000	44.0000
12	-1.2973	20.2973	0.9000*	23.0000	19.0000*
13	0.2306	24.7694	1.2000	12.0000	25.0000
14	-1.8182	51.8182	3.4000*	33.0000*	50.0000*
15	-1.8084	31.8084	1.7000	1.0000**	30.0000
16	1.5458	41.4542	2.5000	12.0000	43.0000
17	-0.0537	27.0537	1.4000	17.0000	27.0000
18	-1.4956	51.4956	3.3000*	16.0000	50.0000*
19	-0.0386	37.0386	2.2000	22.0000	37.0000
20	0.3412	27.6588	1.5000	29.0000	28.0000

NOTE – NEGATIVE CASE NUMBER DENOTES A CASE WITH MISSING VALUES.
     THE NUMBER OF STANDARD DEVIATIONS FROM THE MEAN IS DENOTED BY UP TO
       3 ASTERISKS TO THE RIGHT OF EACH RESIDUAL OR VARIABLE.
     MISSING VALUES ARE DENOTED BY MORE THAN THREE ASTERISKS.

SERIAL CORRELATION OF RESIDUALS = −0.0282

FIGURE 13.3   *(Part 3)*

ESTIMATE

FIGURE 13.3   *(Part 4)*

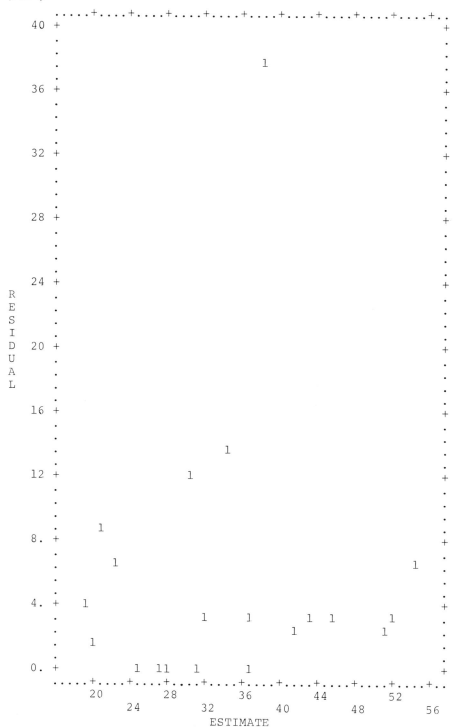

(The square root of this last figure is also taken, but you will recall that it is not interpretable.) The analysis of variance table is the same as was obtained in Chapter 12 (except for rounding error). Note that we do not even have to carry F-tables around with us: P(TAIL) gives the probability of finding an F-value as large as or larger than this one to be zero (to five decimal places). In addition, we see the partial t-values for testing $H_{01}:\beta_1 = 0$ and $H_{02}:\beta_2 = 0$. Note that the standard error of the regression coefficient is $\sqrt{c_{ii}s_{y\cdot x}^2}$ so that, for example,

$$t_2 = \frac{-0.057}{0.062} = -0.917.$$

P(2 TAIL) says that unless our significance level is $\alpha = 0.372$ or greater, AGE is not a significant predictor, given SIZE.

STANDARDIZED REGRESSION COEFFICIENTS

Standardized Regression (Beta) Coefficient

The quantities labeled STD. REG CØEFF are standardized regression coefficients, also called beta coefficients, or beta weights. These are computed so that the regression coefficients may be made comparable in terms of their sizes, even though the predictor variables are measured in different units. That is, one cannot compare $b_1 = 12.834$ to $b_2 = -0.057$ in order to see whether X_1 or X_2 is the stronger predictor, because b_1 measures the effect on price of a one-thousand-square-foot difference in size but b_2 measures the effect of a one-year difference in age. Not only is difference of 1000 square feet not comparable to a difference of one year, but we also see that b_1 and b_2 will be affected by the sizes of the units on X_1 and X_2. For example, if age were measured in units of ten years, then b_2 would be -0.57 instead of -0.057. Of course, all one need do to determine which predictor is stronger is to look at the correlation coefficients. However, some researchers, notably sociologists and psychologists, prefer a regression equation in which the regression coefficients themselves indicate the relative strengths of their associated variables. Besides, although correlations measure the strengths and directions of the predictor variables, they do not tell *how much* Y changes when X_i changes. Thus these investigators want to remove the units on the regression coefficients in order to make them directly comparable.

Another instance in which beta coefficients are considered quite useful is that in which an arbitrary measurement scale has been assigned to one of the predictor variables. While an engineer, for instance, has ready-made units for measuring height, weight, pressure, or strength, the social scientist must often make up a scale. For example, a sociologist studying the effect on social class of income and degree of political conservatism has the problem of quantifying the response *social class* and the predictor *degree of conservatism*. A scale from 0 to 4 might do for political conservatism, with 0 meaning not at all conservative, 2 meaning middle of the road and 4 meaning very conservative, but this is not a natural scale in the sense that length or height or income measurements are. Is a person who rates 4 on the conservatism scale twice as conservative as a person who rates 2? Even though the scale is completely arbitrary, the social scientist's results are very much influenced by the choice: the regression coefficients using a 0–4 scale are not at all the same values he would have obtained with a 0–10 scale, for example. Thus, to remove the influence of this arbitrary scaling, the social scientist often employs a standardized regression coefficient, which provides a unitless measure of the effects of the independent variable on Y. How is this done?

Consider $b_2 = -0.057$ and recall that its units are Y-units over X_2-units (price per age). If the standard deviation of the Y-values were calculated,

$$s_y = \left[\frac{S_{YY}}{n-1} \right]^{1/2},$$

its units would be Y-units (thousands of dollars). Similarly, one could calculate the standard deviation of the X_2-values,

$$s_2 = s_{x_2} = \left[\frac{S_{22}}{n-1} \right]^{1/2},$$

and its units would be the units of X_2 (years). Then multiplying b_2 by the ratio s_2/s_y would remove the units on b_2. The resulting quantity is what is referred to as a beta coefficient or beta weight or standardized regression coefficient. The terminology is unfortunate because the word beta commonly refers to a population parameter β, estimated by a regression coefficient b, which can never be known from sample information only. A beta weight is, of course, something quite different from a beta. To avoid confusion, we will denote the beta weight by BW. Then for the ith beta weight, $i = 1, 2, \ldots, k$,

$$BW_i = b_i \frac{s_i}{s_y},$$

where s_i is the standard deviation of the ith predictor variable. In our example, we have from Figure 13.5 that

$$s_1 = 0.8448,$$
$$s_2 = 9.5834,$$
$$s_y = 11.1756.$$

Then

$$BW_1 = 12.834 \left(\frac{0.8448}{11.1756} \right) = 0.9702$$

and

$$BW_2 = -0.057 \left(\frac{9.5834}{11.1756} \right) = -0.0485.$$

Since BW_1 is greater in absolute value, we see that X_1 (size) is a stronger predictor of price than is X_2 (age).

Note that if multicollinearity is present in the data, a beta weight is not equal to a correlation coefficient. While this might seem obvious, some people nevertheless believe that a beta weight is a correlation coefficient. Only when there is no multicollinearity present in the data are correlation coefficients and beta coefficients equal: if there is no multicollinearity, then for example, $b_1 = S_{1Y}/S_{11}$, and

$$BW_1 = \frac{S_{1Y}}{S_{11}} \cdot \frac{[S_{11}/(n-1)]^{1/2}}{[S_{YY}/(n-1)]^{1/2}}$$

$$= \frac{S_{1Y}}{(S_{11}S_{YY})^{1/2}} = r_{1Y}.$$

If multicollinearity is present, then $b_1 \neq S_{1Y}/S_{11}$, and BW_1 is not a correlation coefficient.

The interpretation of exactly what a beta weight measures is a bit awkward. Recall that b_1, for example, measures the number of units and direction that Y changes for a one-unit increase in X_1. BW_1, since it has been standardized by the standard deviations of the variables involved, measures the direction and number of *standard deviations Y* changes for each one *standard deviation* increase in X_1. The direction and number of standard deviations Y changes is measured with reference to s_Y, and a one-standard-deviation increase in X_1 is measured with reference to s_1. That is, since $s_1 = 0.8448$ and $s_Y = 11.1756$, then $BW_1 = 0.9702$ says that if X_1 increases 0.8448 units, then the amount of Y will increase 0.9702(11.1756) = 10.8426 units. (Note that 0.8448(12.83446) = 10.8426, also.)

One final comment about beta weights. Although beta weights are hardly ever used when X is a controlled variable, there is no reason they could not be. If X is controlled, however, the proper standardization is

$$BW_i = b_i \frac{s_i}{s_{y \cdot x}}.$$

This will not be calculated by the computer, however.

LIST OF RESIDUALS AND PLOTS

Continuing with the examination of Figure 13.3, we next come to the table listing the observations, predicted values, residuals, and the serial correlation coefficient. The residuals are calculated based on *both* predictors, even though AGE is not significant. Finally, plots of the residuals and the squared residuals are made against $\hat{Y}$. These should be examined for the presence of outliers or patterns indicating failure of the model to hold. More on this in Chapter 15.

EXERCISES *Section 13.2*

13.1 Perform the general regression significance test using the BMDP1R for the

p. 531 a. SALES II data,
p. 531 b. RESTAURANT data,
p. 532 c. DRUG data,
p. 532 d. AUTO data,
p. 533 e. ADS data,
p. 533 f. CHILDREN data,
p. 535 g. GLUE data,
p. 535 h. BOTTLES data,
p. 536 i. YIELD data.

Compare the results to the results of the previous analyses for the overall test and the individual tests.

13.3 THE SPSS REGRESSION ROUTINE

RUNNING THE PROGRAM

There are very few changes necessary in the program given in Figure 4.12 in order to run a multiple regression. The worksheet needed to analyze the house price-size-age problem using the SPSS REGRESSIØN routine is shown in Figure 13.4. The PAGESIZE and RUN NAME cards are the same as for a simple regression. The DATA LIST card is the same type as before, but we now have more than two variables.

FIGURE 13.4   *Worksheet for SPSS Multiple Regression*

```
{SYSTEM CARDS
PAGESIZE                NØEJECT
RUN NAME                HØUSES
DATA LIST               FIXED/1 PRICE 1-2 SIZE 3-6 AGE 7-9
INPUT MEDIUM            CARD
N ØF CASES              20
PRINT FØRMATS           PRICE(0),SIZE(1),AGE(0)
LIST CASES              CASES=20/VARIABLES=ALL
READ INPUT DATA
32 1.8 30
24 1.0 33
        ⋮
28 1.5 29
REGRESSIØN              VARIABLES=PRICE,SIZE,AGE/
                        REGRESSIØN=PRICE WITH SIZE,AGE(2) RESID=0/
STATISTICS              ALL
FINISH
```

Also we chose to use their names rather than calling them Y, X_1, and X_2. We omit the SCATTERGRAM and its associated STATISTICS cards, although they could be used if desired. All other cards are exactly the same as before, except for the two REGRESSIØN cards. In the variable list we now need three variables, instead of two. For consistency, we are using variable names PRICE, SIZE, and AGE.

The REGRESSIØN card which begins in column 16 says that price is to be predicted from size and age, and the 2 in parentheses indicates that the overall test is to be run. (Actually, to obtain the overall analysis from SPSS, any even number between 2 and 998 can be in the parentheses. We will see in the next chapter that an odd number calls for a different kind of analysis.)

Finally, we ask for all the available statistics, which will include tables of means and standard deviations, correlations, analysis of residuals, and so on. You may want to refer back to Chapter 4 to refresh your memory on the details of this program, or to the SPSS manual.

PRELIMINARY INFORMATION

We are now ready to examine the printout from the SPSS REGRESSIØN routine for a multiple regression. The program in Figure 13.4 first prints the data, as a check of the accuracy of keypunching. It also lets one have the entire problem at hand so that it is not necessary to refer to a separate listing of the data. Then follow tables of means and standard deviations, and correlation coefficients. Since these tables are now quite familiar to the reader, they are not shown here.

GENERAL REGRESSION SIGNIFICANCE TEST

ADJUSTED
R-SQUARE

Figure 13.5 shows the general regression significance test and a summary table. First we see some correlations and the standard error of estimate. These figures should also be quite familiar, with the exception of the ADJUSTED R-SQUARE value. Recall from Chapter 4 that this takes degrees of freedom into account. That is, while one usually calculates

$$r_{xy}^2 = 1 - \frac{SSE}{SST},$$

the adjusted coefficient of determination is calculated as

$$1 - \frac{MSE}{SST/(n-1)} = 1 - \frac{s_{y \cdot x}^2}{s_y^2}.$$

The analysis of variance below the list of correlations is essentially the same as we saw before. Under VARIABLES IN THE EQUATION we can find the regression constant and the regression coefficients and their standard errors, and the partial F-values.

FIGURE 13.5 *Overall Significance Test from SPSS Multiple Regression*

DEPENDENT VARIABLE.. PRICE

VARIABLE(S) ENTERED ON STEP NUMBER 1.. AGE
                                       SIZE

MULTIPLE R	0.97621		
R SQUARE	0.95298		
ADJUSTED R SQUARE	0.94745		
STANDARD ERROR	2.56194		

ANALYSIS OF VARIANCE

	DF	SUM OF SQUARES	MEAN SQUARE	F
REGRESSION	2.	2261.42030	1130.71015	172.27213
RESIDUAL	17.	111.57970	6.56351	

VARIABLES IN THE EQUATION

VARIABLE	B	BETA	STD ERROR B	F
AGE	-0.05652	-0.04847	0.06164	0.841
SIZE	12.83446	0.97019	0.69919	336.946
(CONSTANT)	10.04634			

ALL VARIABLES ARE IN THE EQUATION

DEPENDENT VARIABLE.. PRICE

SUMMARY TABLE

VARIABLE	MULTIPLE R	R SQUARE	RSQ CHANGE	SIMPLE R	B	BETA
AGE	0.14497	0.02102	0.02102	-0.14497	-0.05652	-0.04847
SIZE	0.97621	0.95298	0.93196	0.97501	12.83446	0.97019
(CONSTANT)					10.04634	

Recall also that BETA designates a beta coefficient or standardized regression coefficient.

We see in Figure 13.5 a summary table. The regression coefficients and beta weights are given again, as are the simple correlation coefficients. Under MULTIPLE R, R SQUARE, and RSQ CHANGE are what we might call sequential correlations. For example, the r^2-figure for age is simply r^2_{2y}, but that for size is $r^2_{xy} = r^2_{12,y}$, the multiple coefficient of determination. (The MULTIPLE R values are the positive square roots of these figures.) RSQ CHANGE is then the increase in r^2 obtained by the addition of each variable to the one(s) listed before it. The SPSS program will list the weakest variable first and other variables in ascending order of strength, as determined by the simple correlation coefficients.

RESIDUALS

Figure 13.6 shows the analysis of residuals that can be obtained from the SPSS package. The residuals are calculated using both predictors. The plot of the standardized residuals is the same as for the simple regression printout. The standardized residuals are

$$(Y_i - \hat{Y}_i)/s_{y\cdot x}.$$

FIGURE 13.6   *Residual Analysis from SPSS Multiple Regression (Part 1)*

SEQNUM	OBSERVED PRICE	PREDICTED PRICE	RESIDUAL	PLOT OF STANDARDIZED RESIDUAL
				−2.0 −1.0 0.0 1.0 2.0
1	32.00000	31.45265	.5473434	*
2	24.00000	21.01553	2.984468	| *
3	27.00000	30.45183	−3.451834	* |
4	47.00000	45.30452	1.695471	| *
5	35.00000	36.81253	−1.812538	* |
6	17.00000	18.90083	−1.900825	* |
7	52.00000	54.66771	−2.667726	* |
8	20.00000	22.52505	−2.525063	* |
9	38.00000	34.30217	3.697826	| *
10	45.00000	43.30286	1.697128	| *
11	44.00000	37.86987	6.130115	| *
12	19.00000	20.29732	−1.297317	*|
13	25.00000	24.76939	.2305935	*
14	50.00000	51.81821	−1.818222	* |
15	30.00000	31.80838	−1.808389	* |
16	43.00000	41.45419	1.545797	|*
17	27.00000	27.05367	−.5367964E-01	*
18	50.00000	51.49565	−1.495665	*|
19	37.00000	37.03862	−.3863144E-01	*
20	28.00000	27.65884	.3411473	*

DURBIN-WATSON TEST OF RESIDUAL DIFFERENCES COMPARED BY CASE ORDER (SEQNUM).

DURBIN-WATSON TEST 2.05248

FIGURE 13.6 *(Part 2)*

```
PLOT: STANDARDIZED RESIDUAL (DOWN) --
PREDICTED STANDARDIZED DEPENDENT VARIABLE (ACROSS)

       -2.0      -1.0       0.0       1.0       2.0
     .YX+---------+---------+---------+---------+XY.
     Y                      I                      Y
     X                      I                      X
 2.0 +                      I                      +
     I                      I                      I
     I                      I                      I
     I                      I                      I
     I                      I                      I
     I                      I                      I
     I                      I                      I
     I                      I                      I
     I                      I                      I
     I                      I                      I
 1.0 +                      I                      +
     I                      I                      I
     I                      I                      I
     I                      I                      I
     I                      I 27                   I
     I                      I  *                   I
     I         12           I                      I
     I          *           *      32              I
     I                      I    30 * *            I
     I            15 18 21  I 26  *  34            I
-0.0 +-------------*-**--*--I-*--------------------+
     I         *11  17   22 I            39* 42    I
     I      10*  *13   20 * I *            *  *    I
     I                  *   I                      I
     I                      I                      I
     I                      I                      I
     I                      I                      I
     I                      I                      I
     I                      I                      I
     I                      I                      I
-1.0 +                      I                      +
     I                      I                      I
     I                      I                      I
     I                      I                      I
     I                      I                      I
     I                      I                      I
     I                      I                      I
     I                      I                      I
     I                      I                      I
     I                      I                      I
-2.0 +                      I                      +
     X                      I                      X
     Y                      I                      Y
     .YX+---------+---------+---------+---------+XY.
       -2.0      -1.0       0.0       1.0       2.0
```

If any standardized residual is greater than $+3$ or less than -3, we might suspect that it is an outlier and would need to examine the residuals more closely, as was done in Chapter 9.

A value of the Durbin-Watson statistic for testing for serial correlation is also given. If the data are collected in time sequence and one is concerned about a time effect entering into the regression, the Durbin-Watson statistic as described in Chapter 9 can be used to test to see whether or not a time variable should be included in the model.

Finally one obtains a plot of the standardized residuals against the standardized $\hat{Y}$-values,

$$\frac{\hat{Y}_i - \bar{\hat{Y}}}{s_{\hat{y}}},$$

where $s_{\hat{y}}$ is the standard deviation of the predicted Y-values and $\bar{\hat{Y}}$ is the mean of the predicted Y-values. These standardized $\hat{Y}$-values have mean zero and standard deviation 1, so that most of the values will be between -3 and $+3$. This standardization provides a compact and easy-to-read scatter diagram. One looks for unusual patterns in the residuals (refer to Chapter 9) in order to determine if any of the assumptions of the model have been violated. Note that in a simple regression we plotted residuals against X, in most cases. Now that we have more than one X-variable, we plot the residuals against $\hat{Y}$ in order to look for patterns. This way, we avoid having to make several graphs.

When working Exercise 13.2 students sometimes see patterns in plots of residuals. However, one should not at this stage proceed to the examination of residuals, because the residuals here are based on all predictors in the model. One first determines which predictors to use (as discussed in Chapter 14). Then, once a reduced model is obtained, residuals from that model should be examined. For example, Figure 13.7 shows the plot of residuals from the AUTO data. We clearly see a pattern: first the residuals are positive, then negative, then positive again. However, these residuals are calculated from the equation

$$\hat{Y} = -0.84490 + 0.52528X_1 + 0.03994X_2,$$

where $Y = $ cost of upkeep, $X_1 = $ age of vehicle, and $X_2 = $ mileage. The partial F-value for mileage is 4.330 with 1 and 12 degrees of freedom, which is not significant at $\alpha = 0.05$. Thus, instead of the above regression equation, we might want to use an equation corresponding to the model

$$Y = \alpha + \beta_1 X_1 + \varepsilon.$$

The residuals from this regression equation could display an entirely different pattern. Thus we will hold off examination of residuals until we have decided on the appropriate model, in terms of which predictors to include.

FIGURE 13.7   *SPSS Plot of Residuals for Exercise 10.4*

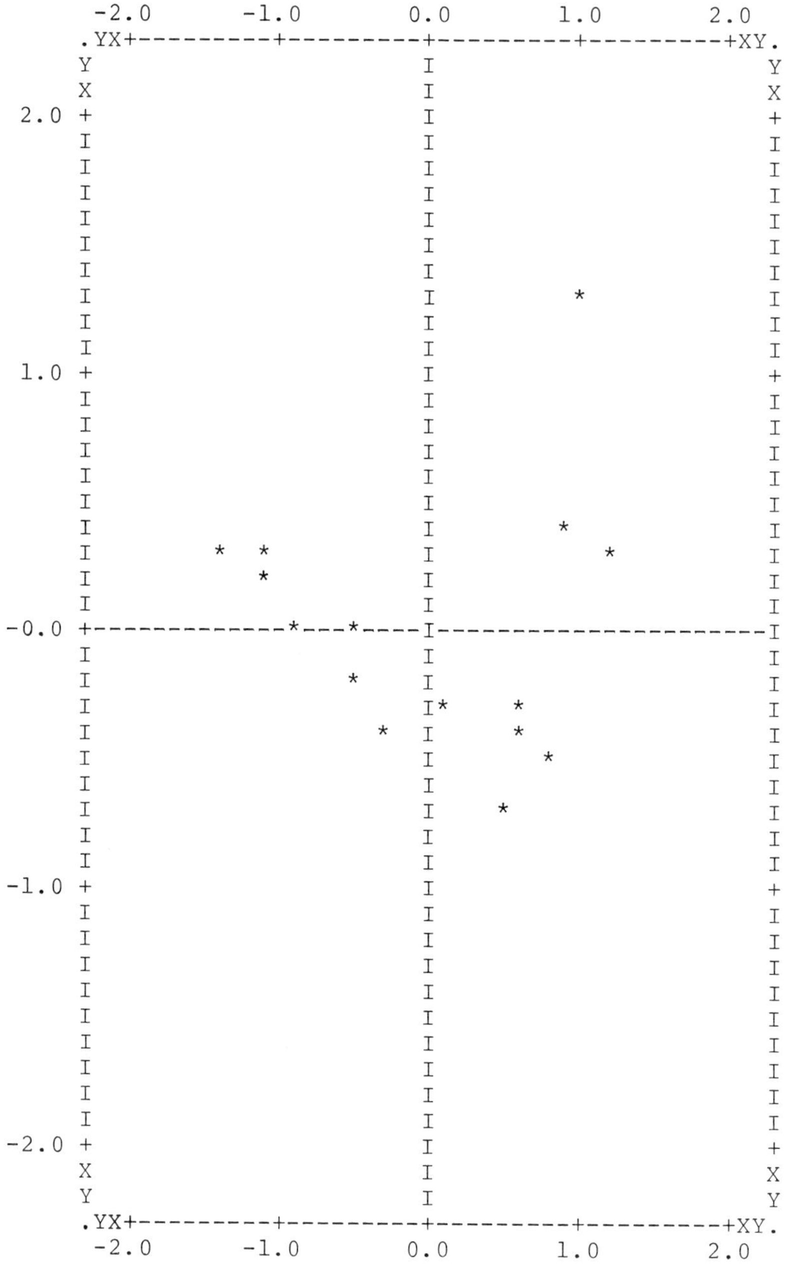

```
               -2.0        -1.0         0.0         1.0         2.0
              .YX+---------+---------+---------+---------+XY.
               Y                      I                      Y
               X                      I                      X
        2.0  +                        I                      +
               I                      I                      I
               I                      I                      I
               I                      I                      I
               I                      I                      I
               I                      I                      I
               I                      I                      I
               I                      I        *             I
               I                      I                      I
               I                      I                      I
        1.0  +                        I                      +
               I                      I                      I
               I                      I                      I
               I                      I                      I
               I                      I                      I
               I                      I                      I
               I                      I        *             I
               I         *     *      I             *        I
               I               *      I                      I
               I                      I                      I
       -0.0  +--------------*---*-----I----------------------I
               I                      I                      I
               I                  *   I                      I
               I                      I*       *             I
               I                    * I       *              I
               I                      I           *          I
               I                      I                      I
               I                      I    *                 I
               I                      I                      I
               I                      I                      I
       -1.0  +                        I                      +
               I                      I                      I
               I                      I                      I
               I                      I                      I
               I                      I                      I
               I                      I                      I
               I                      I                      I
               I                      I                      I
               I                      I                      I
       -2.0  +                        I                      +
               X                      I                      X
               Y                      I                      Y
              .YX+---------+---------+---------+---------+XY.
               -2.0        -1.0         0.0         1.0         2.0
```

EXERCISES — *Section 13.3*

13.2 Perform the general regression significance test using SPSS for the

p. 531 a. SALES II data,
p. 531 b. RESTAURANT data,
p. 532 c. DRUG data,
p. 532 d. AUTO data,
p. 533 e. ADS data,
p. 533 f. CHILDREN data,
p. 535 g. GLUE data,
p. 535 h. BOTTLES data,
p. 536 i. YIELD data.

Compare the results to the results of the previous analyses for the overall test and the individual tests.

13.4 THE SAS GLM PROCEDURE

RUNNING THE PROGRAM

Our final regression package is the SAS General Linear Model procedure. As Figure 13.8 shows, very few changes are required from the SAS program described in Chapter 4 (Figure 4.20) for performing a simple regression analysis. The few changes you see are largely matters of taste. For example, in the multiple regression program we omitted titles from the various sections of output. We also named the data set HØUSES, which was not done in the simple regression program. And instead of calling the variables Y, X_1, and X_2, we call them PRICE, SIZE, and AGE.

FIGURE 13.8 *Worksheet for SAS Multiple Regression*

```
         1-10                11-20                21-30
1 2 3 4 5 6 7 8 9 10 11 12 13 14 15 16 17 18 19 20 21 22 23 24 25 26 27 28 29 30 31 32 33 34
        {SYSTEM CARDS
DATA  HØUSES;
INPUT  PRICE  SIZE  AGE;
CARDS;
32  1.8  30
24  1.0  33
   :
28  1.5  29
PRØC  PRINT;
PRØC  CØRR;
PRØC  GLM;
MØDEL  PRICE=SIZE  AGE/P;
ØUTPUT  PREDICTED=YHAT  RESIDUAL=R;
PRØC  PLØT;
PLØT  R*YHAT;
```

The changes that do make a difference are the following. The PR$\emptyset$C C$\emptyset$RR; card will cause a table of means and standard deviations and a correlation matrix to be printed. The M$\emptyset$DEL statement now includes two independent variables, of course. If we want to plot the residuals versus the predicted Y-values, then we must add the residuals and predicted Y-values to the data set using the $\emptyset$UTPUT statement, and then ask the computer to plot YHAT on the horizontal axis and R on the vertical axis.

We might note here that in some very large regression problems, we cannot fit all the data for any observation into one card if we punch decimal points and leave a space between the values. In this case, the INPUT statement must tell the computer how to read the values. For example, suppose that we want to enter

$$X_1 = 237$$
$$X_2 = 198.4$$
$$X_3 = 0.505$$
$$X_4 = 1.972$$

and

$$X_5 = 12$$

on a card without leaving spaces or punching decimal points. The card would look like this

2371984505197212.

The input statement should then read

INPUT   X1   1–3   X2   4–7   1   X3   8–10   3   X4   11–14   3

X5   15–16;

This tells the computer that the value for X_1 is punched in columns 1–3. That for X_2 is in columns 4–7 and a decimal point is to be read to the left of the last digit in the field. Also, the value for X_3 is found in columns 8–10, and a decimal point is placed to the left of the third from the last digit in the field. X_4 is found in columns 11–14 and contains three digits to the right of the decimal, and X_5 is found in columns 15 and 16.

SIMPLE STATISTICS AND CORRELATIONS

To see what the printout looks like, refer to Figure 13.9. We omitted the printout of the data, so the first piece shown contains the results of the PR$\emptyset$C C$\emptyset$RR procedure. The number of observations mean, standard deviation, sum, and largest and smallest values for each variable are given. Then follows a correlation matrix, with two entries in each

FIGURE 13.9   *Correlation Matrix and Table of Simple Statistics from SAS*

VARIABLE	N	MEAN	STD DEV	SUM	MINIMUM	MAXIMUM
PRICE	20	34.50000000	11.17563138	690.00000000	17.00000000	52.00000000
SIZE	20	2.00000000	0.84479833	40.00000000	0.80000000	3.60000000
AGE	20	21.50000000	9.58342868	430.00000000	1.00000000	33.00000000

CORRELATION COEFFICIENTS / PROB > |R| UNDER HO:RHO=0 / N = 20

	PRICE	SIZE	AGE
PRICE	1.00000	0.97501	-0.14497
	0.0000	0.0001	0.5420
SIZE	0.97501	1.00000	-0.09946
	0.0001	0.0000	0.6765
AGE	-0.14497	-0.09946	1.00000
	0.5420	0.6765	0.0000

position. For example, in the row labeled PRICE and the column labeled SIZE we see

0.97501

0.0001 .

The top entry is the correlation coefficient $r_{1Y} = 0.97501$. The second is the smallest value of α at which this correlation is significant. That is, if we were to test

$$H_0 : \rho = 0$$
$$H_1 : \rho \neq 0,$$

then $r = 0.97501$ is significant at $\alpha = 0.0001$. From this we see that we can be almost certain that by itself, X_1 is an effective predictor of Y. However, since $r_{2Y} = -0.14497$ and its associated significance level is $\alpha = 0.5420$, there is almost a 50–50 chance that X_2 by itself carries no information about Y. We can also see that since $r_{12} = -0.09946$, there is multicollinearity present but since the associated α is 0.6765, we would not expect to see approximately the same value of r_{12} in another random sample drawn from the same population.

GENERAL REGRESSION SIGNIFICANCE TEST

Figure 13.10 shows the analysis from the GLM procedure. The overall test is at the top, along with other statistics that are now very familiar. Individual tests, both sequential (Type I) and partial (Type IV) are then given.

FIGURE 13.10   SAS GLM Multiple Regression

DEPENDENT VARIABLE: PRICE

SOURCE	DF	SUM OF SQUARES	MEAN SQUARE	F VALUE	PR > F	R-SQUARE	C.V.
MODEL	2	2261.42032584	1130.71016292	172.27	0.0001	0.952979	7.4259
ERROR	17	111.57967416	6.56351024			STD DEV	PRICE MEAN
CORRECTED TOTAL	19	2373.00000000				2.56193486	34.50000000

SOURCE	DF	TYPE I SS	F VALUE	PR > F	DF	TYPE IV SS	F VALUE	PR > F
SIZE	1	2255.90044248	343.70	0.0001	1	2211.54926567	336.95	0.0001
AGE	1	5.51988336	0.84	0.3719	1	5.51988336	0.84	0.3719

PARAMETER	ESTIMATE	T FOR H0: PARAMETER=0	PR > |T|	STD ERROR OF ESTIMATE
INTERCEPT	10.04633921	4.78	0.0002	2.09960300
SIZE	12.83445400	18.36	0.0001	0.69919316
AGE	-0.05652313	-0.92	0.3719	0.06163527

OBSERVATION	OBSERVED VALUE	PREDICTED VALUE	RESIDUAL
1	32.00000000	31.45266263	0.54733737
2	24.00000000	21.01553006	2.98446994
3	27.00000000	30.45183286	-3.45183286
4	47.00000000	45.30453289	1.69546711
5	35.00000000	36.81253673	-1.81253673
6	17.00000000	18.90082427	-1.90082427
7	52.00000000	54.66772608	-2.66772608
8	20.00000000	22.52506796	-2.52506796
9	38.00000000	34.30216906	3.69783094
10	45.00000000	43.30287334	1.69712666
11	44.00000000	37.86988963	6.13011037
12	19.00000000	20.29731592	-1.29731592
13	25.00000000	24.76940650	0.23059350
14	50.00000000	51.81821965	-1.81821965
15	30.00000000	31.80838787	-1.80838787
16	43.00000000	41.45419669	1.54580331
17	27.00000000	27.05368167	-0.05368167
18	50.00000000	51.49566738	-1.49566738
19	37.00000000	37.03862924	-0.03862924
20	28.00000000	27.65884956	0.34115044

SUM OF RESIDUALS	-0.00000000
SUM OF SQUARED RESIDUALS	111.57967416
SUM OF SQUARED RESIDUALS - ERROR SS	-0.00000000
FIRST ORDER AUTOCORRELATION	-0.02815929
DURBIN-WATSON D	2.05248566

SEQUENTIAL SUMS OF SQUARES

Recall from Chapter 4 that the Type I, or sequential, sums of squares take the predictors in the order given and show how much addition of a predictor adds to the regression sum of squares. That is, the Type I sum of squares for SIZE is simple

$$SSR(size) = 2255.9$$

and the Type I sum of squares for AGE is

$$SSR(age|size) = SSR(age,size) - SSR(size) = 5.5.$$

These two sequential sums of squares add up to the overall regression sum of squares:

$$SSR(size) \quad + \quad SSR(age|size) \quad = \quad SSR(age,size),$$

or

$$2255.9 \quad + \quad 5.5 \quad = \quad 2261.4.$$

Since these sequential sums of squares depend upon the order in which the variables are entered in the model statement, they are not too meaningful in a multiple regression. They are more meaningful in a polynomial regression, where one can see how much a cubic term improves the fit over a quadratic and linear term alone, for instance.

The researcher who has an order of preference for the variables, however, will find the sequential sums of squares quite useful. For example, suppose that someone has five predictor variables to investigate. Suppose further that X_3 is the cheapest and easiest to measure, followed in order by X_5, X_2, and X_1. One would really prefer not to use X_4 because it is expensive or difficult to measure. The researcher's model statement would read

MODEL Y = X3 X5 X2 X1 X4;

and the GLM procedure would show how much each variable in turn increases the regression sum of squares. If X_4 increases SSR very little over what was obtained using the other four variables, the researcher might feel justified in dropping X_4. Note that the researcher's order of preference is determined *outside* of the statistical analysis, because of a practical or personal preference for the order of the variables. In Chapter 14 we will examine techniques in which the order of the variables is determined by the analysis itself, based on the strengths of the variables alone or in combination with other variables.

As in Chapter 4, the F-values associated with the Type I sums of squares are obtained by dividing the Type I sums by the error mean square from the overall test. For example, the Type I F-value for the variable size is found by

$$\frac{2255.90044248}{111.57867416} = 343.70.$$

These sequential F-values have 1 and $n - (k + 1)$ degrees of freedom, or in this example, 1 and 17 degrees of freedom. The probabilities associated with these F-values indicate that if $\alpha = 0.05$, say, then size by itself explains a significant amount of variation in price because $0.0001 < 0.05$; but since $0.3719 \nless 0.05$, age does not add a significant amount to what size has already explained.

PARTIAL SUMS OF SQUARES

The Type IV sums of squares are our familiar partial sums of squares. A partial sum of squares is computed conditional on every other predictor in the model. That is, the Type IV sum of squares for the variable size is

$$\text{SSR(size|age)} = \text{SSR(size,age)} - \text{SSR(age)} = 2211.549$$

and for the variable age is

$$\text{SSR(age|size)} = \text{SSR(size,age)} - \text{SSR(size)} = 5.520.$$

Of course, these two partial sums of squares do not add up to the overall regression sum of squares. The F-values are obtained by dividing the sums of squares by the error mean square for the overall analysis: for example

$$F = 336.95 = \frac{2211.54926567}{111.57867416},$$

with 1 and $17 = n - (k + 1)$ degrees of freedom. From the probabilities given, we see that age does not add a significant amount to what size can tell alone, but size adds a significant amount to what age can explain about price. We are again brought to the conclusion that size should be used as a predictor and age should not. Things get more interesting when we have more than two predictors to consider.

REGRESSION COEFFICIENTS AND ASSOCIATED STATISTICS

The third block of printout in Figure 13.10 gives the regression constant and regression coefficients, their standard errors, and their associated partial t-values. Recall that to obtain a partial t-value, one divides the coefficient by its standard error. For example, for the variable size,

$$t = 18.36 = \frac{12.83445400}{0.69919316},$$

with $n - (k + 1) = 17$ degrees of freedom. Notice that this SAS GLM procedure provides for a test on whether or not the intercept is significantly different from zero.

One-sided alternative hypotheses can be tested using these t-values, although the probabilities given are those for testing *two-sided* alternatives. This can be seen by comparing the t-probabilities with the partial F-probabilities. Of course, if the t-values are squared, they will yield the partial (Type IV) F-values: for example, $(18.36)^2 = 336.95$.

RESIDUALS

Figure 13.10 also lists residuals calculated using both predictors, the first-order serial correlation coefficient, and the value of the Durbin-Watson statistic. In Figure 13.11, we see the requested plot of residuals versus the predicted Y-values. As mentioned before, one should not get too excited about the analysis of residuals at this stage. Instead, one should be reasonably certain which predictors to include in the model. The plot of residuals is included here primarily so that the reader can see how it is obtained.

FIGURE 13.11   *Plot of Residuals versus $\hat{Y}$ for SAS*

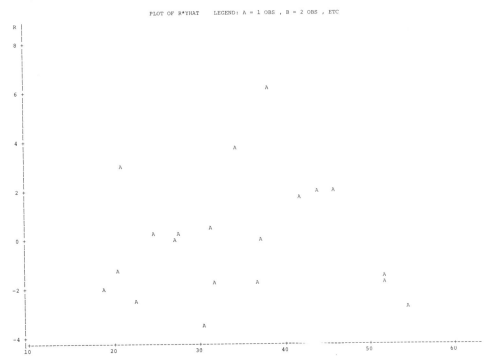

EXERCISES *Section 13.4*

13.3 Perform the overall regression significance test using SAS for the
p. 531   a. SALES II data,
p. 531   b. RESTAURANT data,
p. 532   c. DRUG data,
p. 532   d. AUTO data,
p. 533   e. ADS data,
p. 533   f. CHILDREN data,
p. 535   g. GLUE data,
p. 535   h. BOTTLES data,
p. 536   i. YIELD data.

Compare the results to the results of your previous analyses for the overall test and the individual tests.

13.5 ANALYSIS OF A LARGER PROBLEM

The house size-price-age problem has served us well, but it is actually too small to illustrate very well the myriad things that can happen in a regression problem. We now want to perform a multiple regression analysis of a larger example, we will then use this larger problem in Chapter 14 to illustrate various procedures for selecting variables.

CRAVENS DATA

For our larger example we will use data kindly provided by Professor David W. Cravens of the Department of Marketing and Transportation at the University of Tennessee, Knoxville. We title them CRAVENS data, for short.*

In a study to identify factors that influence territory sales and to assess the relative importances of the factors, eight predictor variables were chosen, as defined in Table 13.1. Data from a random sample of 25 sales territories of a firm's national sales organization were analyzed. These data are shown in Table 13.2. The original model for this problem is

$$Y = \alpha + \beta_1 X_1 + \beta_2 X_2 + \cdots + \beta_8 X_8 + \varepsilon.$$

The purpose is to see whether or not any predictors should be eliminated from the model as nonsignificant, and to find the appropriate regression equation, standard error, and coefficient of determination. We will run these data through the three multiple regression programs discussed earlier and make some preliminary observations. In the next chapter a more complete analysis is performed.

---

* These data were anlyzed in "An Analytical Approach for Evaluating Sales Territory Performance," by David W. Cravens, Robert B. Woodruff, and Joe C. Stamper, in the *Journal of Marketing*, **36** (January 1972), pp. 31–37.

TABLE 13.1   *Definitions of Variables in Sales Territory Performance Study*

Y = sales territory performance measured by aggregate sales, in units, credited to territory salesman*

Predictors	*How Measured*
X_1 = time with company	Length of time employed by company, in months.
X_2 = market potential	Industry sales, in units, of products sold in territory.*
X_3 = advertising (or company effort)	Dollar expenditures on advertising in territory.
X_4 = market share	The weighted average of past market share magnitudes for four previous years.
X_5 = market-share change	Change over four years previous to time period analyzed.
X_6 = accounts	Total number of accounts assigned to a salesman.*
X_7 = workload	Average workload per account using a weighted index based on annual purchases of accounts and concentration of accounts.
X_8 = rating	An aggregate rating on a 1–7 scale by applicable field sales manager on eight dimensions of performance.

* These data were coded by use of a linear transformation in order to preserve confidentiality of information for the respondents.

THREE COMPUTER ANALYSES OF CRAVENS DATA

Figure 13.12 shows the BMDP1R analysis of the CRAVENS data, while Figure 13.13 shows the SPSS analysis, and the SAS analysis of the CRAVENS data is shown in Figure 13.14. Except for rounding error, the results given by all three programs are essentially the same. No residuals or plots are given in Figures 13.12–13.14, but they could easily be obtained. The simple statistics and covariance and correlation matrices are given only for the BMDP1R printout.

PRELIMINARY ANALYSIS OF CRAVENS DATA

What can we tell from these preliminary analyses? We should first look at one of the correlation matrices obtained above. From the correlation matrix we can get an idea of (1) which predictors, by themselves, seem to be good ones, and (2) how highly correlated the predictors themselves are. For example, in the correlation matrix in Figure 13.12, we see that the number of accounts (ACCTS) is the strongest single predictor of sales, since $r_{6Y} = 0.7540$. Then TIME, PØTEN, and ADV are the next strongest predictors, with correlations around 0.6. The variables CHANGE, SHARE, and RATE comprise another group of predictors, all weaker than the preceding group, and finally WKLD is the weakest single predictor.

TABLE 13.2   Data for Sales Territory Performance Study

Y Sales	X_1 Time with Company	X_2 Market Potential	X_3 Advertising	X_4 Market Share	X_5 Market-Share Change	X_6 Accounts	X_7 Workload	X_8 Rating
3,669.88	43.10	74,065.11	4,582.88	2.51	0.34	74.86	15.05	4.9
3,473.95	108.13	58,117.30	5,539.78	5.51	0.15	107.32	19.97	5.1
2,295.10	13.82	21,118.49	2,950.38	10.91	-0.72	96.75	17.34	2.9
4,675.56	186.18	68,521.27	2,243.07	8.27	0.17	195.12	13.40	3.4
6,125.96	161.79	57,805.11	7,747.08	9.15	0.50	180.44	17.64	4.6
2,134.94	8.94	37,806.94	402.44	5.51	0.15	104.88	16.22	4.5
5,031.66	365.04	50,935.26	3,140.62	8.54	0.55	256.10	18.80	4.6
3,367.45	220.32	35,602.08	2,086.16	7.07	-0.49	126.83	19.86	2.3
6,519.45	127.64	46,176.77	8,846.25	12.54	1.24	203.25	17.42	4.9
4,876.37	105.69	42,053.24	5,673.11	8.85	0.31	119.51	21.41	2.8
2,468.27	57.72	36,829.71	2,761.76	5.38	0.37	116.26	16.32	3.1
2,533.31	23.58	33,612.67	1,991.85	5.43	-0.65	142.28	14.51	4.2
2,408.11	13.82	21,412.79	1,971.52	8.48	0.64	89.43	19.35	4.3
2,337.38	13.82	20,416.87	1,737.38	7.80	1.01	84.55	20.02	4.2
4,586.95	86.99	36,272.00	10,694.20	10.34	0.11	119.51	15.26	5.5
2,729.24	165.85	23,093.26	8,618.61	5.15	0.04	80.49	15.87	3.6
3,289.40	116.26	26,878.59	7,747.89	6.64	0.68	136.58	7.81	3.4
2,800.78	42.28	39,571.96	4,565.81	5.45	0.66	78.86	16.00	4.2
3,264.20	52.84	51,866.15	6,022.70	6.31	-0.10	136.58	17.44	3.6
3,453.62	165.04	58,749.82	3,721.10	6.35	-0.03	138.21	17.98	3.1
1,741.45	10.57	23,990.82	860.97	7.37	-1.63	75.61	20.99	1.6
2,035.75	13.82	25,694.86	3,571.51	8.39	-0.43	102.44	21.66	3.4
1,578.00	8.13	23,736.35	2,845.50	5.15	0.04	76.42	21.46	2.7
4,167.44	58.54	34,314.29	5,060.11	12.88	0.22	136.58	24.78	2.8
2,799.97	21.14	22,809.53	3,552.00	9.14	-0.74	88.62	24.96	3.9

How are the predictors themselves correlated? We see some large correlations,

$r_{16} = 0.7578$ between time with company and number of accounts
(that seems reasonable),

$r_{58} = 0.5493$ between market share change and rating,

and several other correlations around 0.3 and 0.4. We know that multicollinearity is present, and is quite serious in some instances. It would not be surprising to find that, for example, if ACCTS is used as a predictor, TIME will not be able to add very much more information.

Turning to the overall test for significant predictors, we find

$F = 23.65,$

which is significant at $\alpha = 0.0001$, so we can be quite confident that at least one predictor is significant. All predictors together explain about 92% of the variation in sales, and this is a very high coefficient of determination for real data. If all 8 predictors were used, the regression equation would be

$$\hat{Y} = -1507.81 + 2.01X_1 + 0.04X_2 + 0.15X_3 + 199.02X_4 + 290.86X_5$$
$$+ 5.55X_6 + 19.79X_7 + 8.19X_8,$$

and the standard error of estimate would be 449.02.

Before looking at the partial t- or F-values, let us examine the standard error of the regression coefficients. Recall that these measure how much we would expect $a, b_1, b_2, \ldots, b_8$ to vary from sample to sample of size 25 from this same population. If the standard errors are small, this means that we have pretty well zeroed in on the values of $\alpha, \beta_1, \beta_2, \ldots, \beta_8$. However, large standard errors mean that we have not estimated the parameters very well. Perhaps the sample size was too small, but more likely multicollinearity is throwing in a confusion factor. For example, for $X_3 = $ ADV, we have $b_3 = 0.15$ and its standard error is 0.05. Thus, in 68% of samples of size 25 drawn from this population, we'd expect to find a value of b_3 between 0.10 and 0.20.

FIGURE 13.12   *BMDP1R Analyses of CRAVENS Data (Part 1)*

VARIABLE	MEAN	STANDARD DEVIATION	ST.DEV/MEAN	MINIMUM	MAXIMUM
1 SALES	3374.56445	1313.06396	0.38911	1578.00000	6519.44922
2 TIME	87.64188	86.79675	0.99036	8.13000	365.03979
3 POTEN	38858.02344	15714.50391	0.40441	20416.86719	74065.06250
4 ADV	4357.37109	2689.09766	0.61714	402.43994	10694.19922
5 SHARE	7.56479	2.45979	0.32516	2.51000	12.88000
6 CHANGE	0.09560	0.62105	6.49633	-1.63000	1.24000
7 ACCTS	122.69910	45.57796	0.37146	74.85999	256.09985
8 WKLD	18.06070	3.67009	0.20321	7.81000	24.95999
9 RATING	3.74399	0.95920	0.25620	1.60000	5.50000

FIGURE 13.12   (Part 2)

COVARIANCE MATRIX

		SALES 1	TIME 2	POTEN 3	ADV 4	SHARE 5	CHANGE 6	ACCTS 7	WKLD 8	RATING 9
SALES	1	**********								
TIME	2	70993.8750	7533.6758							
POTEN	3	**********	**********	**********						
ADV	4	**********	58161.1250	**********	**********					
SHARE	5	1561.6716	22.6760	-8143.3633	1749.3037	6.0506				
CHANGE	6	398.9141	13.5561	2618.3333	628.8044	0.1306	0.3857			
ACCTS	7	45123.6641	2997.9321	**********	24517.3711	45.1824	9.2678	2077.3503		
WKLD	8	-564.9224	-57.1240	**********	-2686.7102	3.1538	-0.6557	-33.2626	13.4695	
RATING	9	506.1604	8.4192	5406.8047	1061.3218	-0.0556	0.3272	9.9945	-0.9748	0.9201

CORRELATION MATRIX

		SALES 1	TIME 2	POTEN 3	ADV 4	SHARE 5	CHANGE 6	ACCTS 7	WKLD 8	RATING 9
SALES	1	1.0000								
TIME	2	0.6229	1.0000							
POTEN	3	0.5978	0.4540	1.0000						
ADV	4	0.5962	0.2492	0.1741	1.0000					
SHARE	5	0.4835	0.1062	-0.2107	0.2645	1.0000				
CHANGE	6	0.4892	0.2515	0.2683	0.3765	0.0855	1.0000			
ACCTS	7	0.7540	0.7578	0.4786	0.2000	0.4030	0.3274	1.0000		
WKLD	8	-0.1172	-0.1793	-0.2588	-0.2722	0.3493	-0.2877	-0.1988	1.0000	
RATING	9	0.4019	0.1011	0.3587	0.4115	-0.0236	0.5493	0.2286	-0.2769	1.0000

FIGURE 13.12   (Part 3)

MULTIPLE R	0.9602	STD. ERROR OF EST.	449.0244
MULTIPLE R-SQUARE	0.9220		

ANALYSIS OF VARIANCE

	SUM OF SQUARES	DF	MEAN SQUARE	F RATIO	P(TAIL)
REGRESSION	38153296.000	8	4769162.000	23.654	0.00000
RESIDUAL	3225967.000	16	201622.938		

VARIABLE		COEFFICIENT	STD. ERROR	STD. REG COEFF	T	P(2 TAIL)
INTERCEPT		*******				
TIME	2	2.010	1.931	0.133	1.041	0.313
POTEN	3	0.037	0.008	0.445	4.536	0.000
ADV	4	0.151	0.047	0.309	3.205	0.006
SHARE	5	199.024	67.027	0.373	2.969	0.009
CHANGE	6	290.851	186.782	0.138	1.557	0.139
ACCTS	7	5.551	4.776	0.193	1.162	0.262
WKLD	8	19.793	33.676	0.055	0.588	0.565
RATING	9	8.192	128.505	0.006	0.064	0.950

This is pretty good. But on the other hand, for X_7 (WKLD), $b_7 = 19.79$ with a standard error of 33.68. Thus, approximately 68 % of all samples of size 25 would yield b_7 between -13.89 and 53.47. One would not consider that we have estimated β_7 very well.

The regression coefficients with relatively small standard errors are b_2(PØTEN) and b_3(ADV). Then b_1(TIME) and b_6(ACCTS) have reasonably small standard errors, while those for b_7(WKLD) and b_4(SHARE) are relatively large. The standard errors for b_8(RATE) and b_5(SHCHANGE) are quite large, as is the standard error for a.

Naturally, the sizes of the standard errors of the regression coefficients affect the sizes of the partial t- and F-values. If the standard error is too large relative to the size of the coefficient, then the partial t- and F-statistic will be too small (in absolute value) to show a significant predictor. For the partial t- or F-values, if $\alpha = 0.05$ we see that only X_2(PØTEN), X_3(ADV), and X_4(SHARE) turn out to be significant, given all the others. These results are in some ways not surprising and are in other ways quite surprising. These three variables all have reasonably high correlations with Y and reasonably small correlations with each other. However, our single strongest predictor, X_6(ACCTS), did not turn out to be significant—as a matter of fact, it had an extremely small partial t- or F-value. But X_6 was not extremely highly correlated with any of the three variables that ended up being significant. Recall, however, that the partial t- or F-tests answer the question, "Given *all seven* other variables, do I need this one?" Recall, for example, that X_6(ACCTS) is highly correlated with X_1(TIME): $r_{16} = 0.75782$. Thus, given time with company, number of accounts does not add much. But why are we interested in what X_6 can do, given X_1? How do we know that we want X_1 at all?

These partial t- and F-tests are just not that helpful to us if we want to determine which variables should be retained and which variables should be deleted from the model. Only if *all* partial t- and F-values were significant would these partial tests be very useful to us: in this case we would conclude that all predictors add information to

FIGURE 13.13   *SPSS Analysis of CRAVENS Data*

DEPENDENT VARIABLE.. SALES

VARIABLES ENTERED ON STEP NUMBER 1.. RATE
TIME
POTEN
ADV
SHARE
SHCHANGE
ACCTS
WKLD

MULTIPLE R	0.96023
R SQUARE	0.92204
ADJUSTED R SQUARE	0.88306
STANDARD ERROR	449.02534

ANALYSIS OF VARIANCE

	DF	SUM OF SQUARES	MEAN SQUARE	F
REGRESSION	8.	38153519.26158	4769189.90770	23.65391
RESIDUAL	16.	3225980.09106	201623.75569	

VARIABLES IN THE EQUATION

VARIABLE	B	BETA	STD ERROR B	F
RATE	8.18951	0.00598	128.50550	0.004
TIME	2.00957	0.13284	1.93065	1.083
POTEN	0.03720	0.44526	0.00820	20.574
ADV	0.15099	0.30922	0.04711	10.273
SHARE	199.02337	0.37283	67.02790	8.817
SHCHANGE	290.85496	0.13757	186.78192	2.425
ACCTS	5.55095	0.19268	4.77555	1.351
WKLD	19.79384	0.05532	33.67670	0.345
(CONSTANT)	-1507.81082			

VARIABLES NOT IN THE EQUATION

VARIABLE	BETA IN	PARTIAL	TOLERANCE	F

ALL VARIABLES ARE IN THE EQUATION

SUMMARY TABLE

DEPENDENT VARIABLE.. SALES

VARIABLE	MULTIPLE R	R SQUARE	RSQ CHANGE	SIMPLE R	B	BETA
RATE	0.40188	0.16151	0.16151	0.40188	8.18951	0.00598
TIME	0.70997	0.50406	0.34255	0.62292	2.00957	0.13284
POTEN	0.75250	0.56626	0.06220	0.59781	0.03720	0.44526
ADV	0.83846	0.70302	0.13676	0.59618	0.15099	0.30922
SHARE	0.94999	0.90249	0.19947	0.48351	199.02337	0.37283
SHCHANGE	0.95678	0.91543	0.01294	0.48918	290.85496	0.13757
ACCTS	0.95935	0.92036	0.00493	0.75399	5.55095	0.19268
WKLD	0.96023	0.92204	0.00168	-0.11722	19.79384	0.05532
(CONSTANT)					-1507.81082	

FIGURE 13.14   SAS Analysis of CRAVENS Data

GENERAL LINEAR MODELS PROCEDURE

DEPENDENT VARIABLE: SALES

SOURCE	DF	SUM OF SQUARES	MEAN SQUARE	F VALUE	PR > F	R-SQUARE	C.V.
MODEL	8	38153564.25210437	4769195.53151305	23.65	0.0001	0.922039	13.3062
ERROR	16	3225984.67475173	201624.04217198		STD DEV		SALES MEAN
CORRECTED TOTAL	24	41379548.92685610			449.02565870		3374.56760000

SOURCE	DF	TYPE I SS	F VALUE	PR > F	DF	TYPE IV SS	F VALUE	PR > F
TIME	1	16056475.09223126	79.64	0.0001	1	218442.90153750	1.08	0.3134
POTEN	1	5172529.51794887	25.65	0.0001	1	4148313.42001813	20.57	0.0003
ADV	1	7701444.95840667	38.20	0.0001	1	2071255.37925263	10.27	0.0055
SHARE	1	8144147.87563474	40.39	0.0001	1	1777623.07698216	8.82	0.0090
SHCHANGE	1	788061.45599777	3.91	0.0655	1	488906.44513135	2.42	.1390
ACCTS	1	219521.27264809	1.09	0.3123	1	272415.54793787	1.35	0.2621
WKLD	1	70565.26639185	0.35	0.5624	1	69653.96590204	0.35	0.5649
RATE	1	818.82284513	0.00	0.9500	1	818.82284513	0.00	0.9500

PARAMETER	ESTIMATE	T FOR HO: PARAMETER=0	PR > |T|	STD ERROR OF ESTIMATE
INTERCEPT	-1507.81372984	-1.94	0.0707	778.63493964
TIME	2.00956615	1.04	0.3134	1.93065421
POTEN	0.03720491	4.54	0.0003	0.00820230
ADV	0.15098890	3.21	0.0055	0.04710851
SHARE	199.02353635	2.97	0.0090	67.02792230
SHCHANGE	290.85513399	1.56	0.1390	186.78199574
ACCTS	5.55096065	1.16	0.2621	4.77554962
WKLD	19.79389189	0.59	0.5649	33.67669223
RATE	8.18928366	0.06	0.9500	128.50561301

what all others provide. This seldom happens, although it is not an impossibility. In most data sets, some variable selection must be performed. The stagewise procedures in Chapter 14 show various ways this can be done.

13.6 SUMMARY

BMD, SAS, and SPSS regression programs have been used to perform the general regression significance test, both for the house size-price-age example and for the larger CRAVENS example. All three programs are very similar in the information they give. An exception is that the SAS program gives sequential sums of squares and thus lets us perform sequential tests if desired.

The partial t- and F-statistics provided by all three programs have been shown to be disappointingly uninformative as to which of several candidate variables should be retained in the model. We thus turn to the study of variable selection procedures in order to help answer this question.

Chapter 14

SOME VARIABLE-SELECTION PROCEDURES

14.1 INTRODUCTION

In many multiple regression problems the researcher, having identified a large set of possible predictor variables, wants to narrow this set down. This was the case in the two multiple regression examples of Chapter 13. In the house size-price-age example, we started by using sizes and ages of houses to predict prices. Since that was such a small problem we were able to use the partial F-tests to conclude that size is a strong predictor of price, but age is not, especially when used in conjunction with size. Things were less clear-cut in the CRAVENS data. There we had eight potential predictors, all correlated with each other to some extent. By using partial F-tests, we detected three variables that are strong predictors, given all the others. But the question is, do we want to use all the others? For example, if X_8 (RATING) is not needed, given the other seven predictors, then perhaps we should throw it out. But if so, the partial F-values we computed are no longer relevant: we have no partial F for, say, X_3 given X_1, X_2, X_4, X_5, X_6, and X_7. The partial F that we have is for X_3 given X_1, X_2, X_4, X_5, X_6, X_7, *and* X_8.

Stagewise Procedure

What is needed is some selection procedure, in which predictor variables are entered one at a time. Such selection procedures are often referred to as stagewise procedures. We look at the increase in either the regression sum of squares or the multiple coefficient of determination to see if the variable entered adds anything worthwhile to what those variables previously entered were able to explain about the response. At some stage, a variable might be redundant with a variable previously entered and thus not really be needed. Also at some stage, a predictor might simply not be effective, either alone or in conjunction with other variables previously entered. Using a stagewise procedure, then, the analysis will terminate once ineffective predictors begin to be entered into the model.

"BEST" SUBSETS OF PREDICTORS

The order of entry of variables into the model can be determined by the experimenter's choice (based upon ease and cost of measuring the variable or theoretical considerations), or by the strength of the variable itself, or by a combination of these two factors. Thus, there is no best subset of predictors for any given set of data, even if the experimenter allows each variable to enter under its own strength and does not specify an order of preference. The *best* subset is determined by the criterion one uses to specify the order of entry. For example, one could add variables so that at each stage, the multiple coefficient of determination is increased the most. This criterion may or may not give the same subset of variables as that obtained by applying some other selection criterion.

SELECTION PROCEDURES

We will examine the selection procedures commonly available in the BMD, SPSS, and SAS packages. Not all procedures are available with all three packages, however. We will discuss how to run the programs and describe the procedures available in each. Then we will worry about what to do if these procedures result in different final reduced models. The CRAVENS data is used for illustration because the house size-price-age problem is too small for our purposes now. We discuss the SAS selection procedures first because its instructions are the most straightforward to the novice. We will first let the variables enter on their own merit and then move to an example in which the experimenter forces the order on some of the variables.

14.2 SAS SELECTION PROCEDURES

The five selection procedures available through the Statistical Analysis System are called forward selection, stepwise, backward elimination, maximum R^2 improvement, and minimum R^2 improvement. We will discuss them in order, comparing and contrasting what they do.

RUNNING THE PROGRAM

Figure 14.1 shows a worksheet for writing an SAS program to perform a forward selection procedure on the CRAVENS data. This program is very much like that used for the overall significance test in Figure 13.8. We have not asked for titles, a correlation matrix, a listing of data values, the GLM procedure, and so on, because these are already shown in Chapter 13. The PRØC STEPWISE; card asks the computer to perform a variable selection. (SAS uses the term *stepwise* in the sense that we use stagewise but we shall reserve the term *stepwise* to refer to a *particular* stagewise procedure.) The MØDEL card specifies the dependent variable (SALES) and the

FIGURE 14.1   *Program for SAS Forward Selection Procedure*

1-10	11-20	21-30	31-40	41-50	51-60	61-70
`{SYSTEM CARDS`						
`DATA CRAVENS;`						
`INPUT SALES TIME POTEN ADV SHARE SHCHANGE ACCTS WKLD RATE;`						
`CARDS;`						
`3669.88    43.10  74065.11  4582.88   2.51   .34   74.86  15.05  4.9`						
`3473.95   108.13  58117.30  5539.78   5.51   .15  107.32  19.97  5.1`						
`⋮`						
`1799.97    21.14  22809.53  3552.00   9.14  -.74   88.62  24.96  3.9`						
`PROC STEPWISE;`						
`MODEL SALES=TIME POTEN ADV SHARE SHCHANGE ACCTS WKLD RATE/FORWARD;`						

independent variables to use. The FØRWARD following the slash indicates that the forward selection procedure is to be used. In order to run a backward elimination procedure, replace FØRWARD with BACKWARD. In order to run the stepwise procedure, replace FØRWARD with STEPWISE, or simply place a semicolon after the last independent variable and drop the /FØRWARD. (This is called a default condition: unless you specify otherwise, the computer will perform a stepwise procedure.) To run the maximum R^2 improvement technique, replace FØRWARD with MAXR, and to use the minimum R^2 improvement, replace FØRWARD with MINR. Do not forget that each SAS statement must end with a semicolon.

It is often desirable to perform all five selection procedures on a set of data for the sake of comparison. The following model statement will allow all five procedures to be run at once:

MØDEL SALES=TIME...RATE/FØRWARD BACKWARD STEPWISE MAXR MINR;

Note that plots of residuals or predicted values can be made here just as they were before.

Now that we know how to run any or all of the five stagewise procedures using SAS, let us see how these procedures differ from one another.

FORWARD SELECTION

Forward
Selection
Procedure

Generally, a forward selection technique begins by running a simple regression analysis using the single strongest predictor, as measured by r or r^2. Then it finds the predictor with the largest partial F-value, given the strongest predictor, and denotes this as the second strongest predictor. A two-variable multiple regression analysis is performed using these two strongest predictors. The third strongest predictor, in terms of the variable yielding the largest partial F, given the two strongest, is then added and a three-predictor analysis is performed. The procedure continues either until all variables have been added, or until the largest remaining partial F-value is too small to bother with.

SIGNIFICANCE LEVEL FOR ENTRY

Significance
Level for
Entry

What determines when a partial F-value is "too small to bother with" is the significance level for entry. Unless otherwise specified, the process terminates when the largest partial F-value is no longer significant at $\alpha = 0.50$. At this point, no variable will be entered. Why such a high significance level? Why not make the significance level for entry small, say $\alpha = 0.05$ or $\alpha = 0.01$? The reason is that these partial F-values are true F-values *only* when we have found the correct model; that is, only if $s_{y \cdot x}^2 = MSE$ estimates true error, and not variation due to omitted predictors. If at any stage we do not have the true model, $s_{y \cdot x}^2$ will be inflated and thus the F-values will be deflated. Setting $\alpha = 0.05$ or $\alpha = 0.01$ would defeat our purposes and would wrongly prevent variables from having a chance to enter into the equation.

One rule of thumb is to set $\alpha = 0.25$. Though not as generous as the default condition of SAS, this rule still allows several variables to enter into the equation. Note that we would actually be using a significance level of 0.25 (if this were specified) *only* at that step in which *all* significant predictors and *only* significant predictors have been included. All the significance level for entry really does is provide a cutoff point: computer time is saved if extremely weak predictors are ignored. Nevertheless, the choice of the significance level for entry will usually have an emphatic effect on the final subset of variables.

To change the significance level for entry from the default condition of .50 is quite simple. If you want the significance level for entry to be $\alpha = 0.25$, then in the model statement in Figure 14.1, add

$$SLE = .25$$

before the semicolon. The model statement then reads

$$M\emptyset DEL\ SALES = TIME\ P\emptyset TEN \ldots RATE/F\emptyset RWARD\ SLE = .25;$$

FORWARD SELECTION PROCEDURE FOR CRAVENS EXAMPLE

Figure 14.2 shows the forward selection procedure with $\alpha = 0.50$ applied to the CRAVENS data. As we saw in Chapter 13, ACCTS is the single strongest predictor of SALES. Thus, a simple regression using ACCTS to predict SALES is run as the first step of the forward selection procedure. The F-value of 30.3 shows that ACCTS explains approximately 57% of the variation in SALES, and that using the regression equation

$$\hat{Y} = 709.32383372 + 21.72176971 X_6$$

our estimates will be off, on the average, by $[776,324.98847657]^{1/2} = 881.093$. The standard error of the regression coefficient, 3.946, is not unduly large, so the value $b_6 = 21.72176971$ can be trusted fairly well.

FIGURE 14.2   SAS Forward Selection Procedure for CRAVENS Data

FORWARD SELECTION PROCEDURE FOR DEPENDENT VARIABLE SALES

STEP 1   VARIABLE ACCTS ENTERED   R SQUARE = 0.56849518

	DF	SUM OF SQUARES	MEAN SQUARE	F	PROB>F
REGRESSION	1	23524074.19189501	23524074.19189501	30.30	0.0001
ERROR	23	17855474.73496109	776324.98847657		
TOTAL	24	41379548.92685610			

	B VALUE	STD ERROR	TYPE II SS	F	PROB>F
INTERCEPT	709.32383372				
ACCTS	21.72176971	3.94603304	23524074.19189501	30.30	0.0001

STEP 2   VARIABLE ADV ENTERED   R SQUARE = 0.77510077

	DF	SUM OF SQUARES	MEAN SQUARE	F	PROB>F
REGRESSION	2	32073320.05178234	16036660.02589117	37.91	0.0001
ERROR	22	9306228.87507376	423010.40341244		
TCTAL	24	41379548.92685610			

	B VALUE	STD ERROR	TYPE II SS	F	PROB>F
INTERCEPT	50.29906160				
ADV	0.22652657	0.05038842	8549245.85988733	20.21	0.0002
ACCTS	19.04824598	2.97291380	17365864.05826321	41.05	0.0001

STEP 3   VARIABLE POTEN ENTERED   R SQUARE = 0.82772280

	DF	SUM OF SQUARES	MEAN SQUARE	F	PROB>F
REGRESSION	3	34250796.02722474	11416932.00907491	33.63	0.0001
ERROR	21	7128752.89963136	339464.42379197		
TOTAL	24	41379548.92685610			

	B VALUE	STD ERROR	TYPE II SS	F	PROB>F
INTERCEPT	-327.23338939				
POTEN	0.02192192	0.00865564	2177475.97544240	6.41	0.0194
ADV	0.21607079	0.04532744	7713721.79142546	22.72	0.0001
ACCTS	15.55392188	2.99936692	9128825.32465885	26.89	0.0001

STEP 4   VARIABLE SHARE ENTERED   R SQUARE = 0.90044970

	DF	SUM OF SQUARES	MEAN SQUARE	F	PROB>F
REGRESSION	4	37260202.46282121	9315050.61570530	45.23	0.0001
ERROR	20	4119346.46403489	205967.32320174		
TOTAL	24	41379548.92685610			

	B VALUE	STD ERROR	TYPE II SS	F	PROB>F
INTERCEPT	-1441.93182868				
POTEN	0.03821753	0.00797694	4727717.33107410	22.95	0.0001
ADV	0.17499004	0.03690666	4630368.54356448	22.48	0.0001
SHARE	190.14429731	49.74415347	3009406.43559647	14.61	0.0011
ACCTS	9.21389567	2.86521038	2129962.08538966	10.34	0.0043

STEP 5   VARIABLE SHCHANGE ENTERED   R SQUARE = 0.91241574

	DF	SUM OF SQUARES	MEAN SQUARE	F	PROB>F
REGRESSION	5	37755351.59771667	7551070.31954333	39.59	0.0001
ERROR	19	3624197.32913943	190747.22784944		
TOTAL	24	41379548.92685610			

	B VALUE	STD ERROR	TYPE II SS	F	PROB>F
INTERCEPT	1285.94337067				
POTEN	0.03763121	0.00768517	4573489.56079583	23.98	0.0001
ADV	0.15443602	0.03773852	3194373.73424488	16.75	0.0006
SHARE	196.94952750	48.05692351	3203732.09272627	16.80	0.0006
SHCHANGE	262.50049338	162.92631286	495149.13489546	2.60	0.1236
ACCTS	8.23411280	2.82357962	1622152.88530014	8.50	0.0089

STEP 6   VARIABLE TIME ENTERED   R SQUARE = 0.92031405

	DF	SUM OF SQUARES	MEAN SQUARE	F	PROB>F
REGRESSION	6	38082180.17286742	6347030.02881124	34.65	0.0001
ERROR	18	3297368.75398868	183187.15299937		
TOTAL	24	41379548.92685610			

	B VALUE	STD ERROR	TYPE II SS	F	PROB>F
INTERCEPT	-1165.47855369				
TIME	2.26935112	1.69898362	326828.57515075	1.78	0.1983
POTEN	0.03827800	0.00754688	4712573.84321485	25.73	0.0001
ADV	0.14067029	0.03839221	2459312.09671656	13.43	0.0018
SHARE	221.60469221	50.58309112	3515945.91528346	19.19	0.0004
SHCHANGE	285.10928426	160.55965553	577623.53145916	3.15	0.0927
ACCTS	4.37770296	3.99903763	219521.27264808	1.20	0.2881

NO OTHER VARIABLES MET THE 0.5000 SIGNIFICANCE LEVEL FOR ENTRY INTO THE MODEL.

In Step 2 the variable X_3(ADV) is entered, because the partial F-value for ADV given ACCTS is larger than any of the other six partial F-values given ACCTS. (Imagine all the work the computer has done but not printed!) The overall F-value of 37.91 indicates that at least one of the two variables is a good predictor, and the multiple coefficient of determination of 0.775 shows that addition of ADV to ACCTS improves the percentage of explained variation by $77.5 - 56.8 = 20.7\%$. The regression constant and regression coefficients and their standard errors can also be found. Note that the variables are listed according to their order of appearance on the MØDEL statement; since ADV $= X_3$ and ACCTS $= X_6$, ADV is listed first.

What is new to us is the Type II sum of squares. By calling it a Type II sum of squares, SAS avoids having to name it. We might refer to it as a "mini-partial" sum of squares or a "sum-of-squares if last." The Type II sum of squares is a partial sum of squares, conditional on all other variables in the model *at that step*. As such, it measures the effect of the variables with which it is associated as if that variable were the last one entered in the model. Thus, the Type II sum of squares for ADV in Step 2 is

SSR(ADV|ACCTS)

and that for ACCTS is

SSR(ACCTS|ADV).

From the probabilities given, we see that if ADV has been included in the model we still need ACCTS, and if ACCTS were already in the model we would still need ADV.

Step 3 continues in much the same way. The variable PØTEN(X_2) is entered because

F(PØTEN|ADV,ACCTS)

is larger than any other partial F conditioned on ADV and ACCTS. The Type II sums of squares are

SSR(PØTEN|ADV,ACCTS) = 2,177,475.975,

SSR(ADV|PØTEN,ACCTS) = 7,713,721.791,

and

SSR(ACCTS|PØTEN,ADV) = 9,128,825.325.

The multiple has increased to about 83%, and all three partial F's are significant at $\alpha = 0.05$.

Step 4 adds SHARE(X_4) and shows all four variables significant at $\alpha = 0.05$. In Step 5, SCHANGE enters but is not significant at $\alpha = 0.05$, although all other variables are. Recall, however, that the computer is using $\alpha = 0.50$, so it goes on to the next step. Recall also that the probabilities printed out are valid at *only one step*: the step in which the correct model has been found.

In Step 6 a funny thing happens. When the variable TIME(X_1) is entered, it is significant at $\alpha = 0.05$, even though it is supposedly not as strong as the variable entered in Step 5 (SHCHANGE) which was found to be nonsignificant. Also, suddenly our single strongest predictor, ACCTS is no longer significant (and neither is SHCHANGE). The intercorrelations among predictors are, of course, to blame for these strange goings-on. Given that you decide to use TIME, PØTEN, ADV, SHARE, and SHCHANGE, you don't need to use ACCTS even though it is strong by itself. Refer back to the correlation matrices in Section 13.5. ACCTS is highly correlated with TIME ($r_{16} = 0.75782$) so if TIME is going to be used as a predictor, ACCTS is redundant.

Following Step 6 we see the message that no other variable met the 0.50 significance level for entry. That is, neither

$$F(\text{WKLD}|\text{TIME,PØTEN,ADV,SHARE,SHCHANGE,ACCTS})$$

nor

$$F(\text{RATE}|\text{TIME,PØTEN,ADV,SHARE,SHCHANGE,ACCTS})$$

is significant at $\alpha = 0.50$, so the procedure terminates without ever adding these variables to the model. Using the six variables TIME, PØTEN, ADV, SHARE, SHCHANGE, and ACCTS to predict sales, we can explain approximately 92% of the variation in SALES. It is interesting to compare the sizes of the r^2_{xy}-values in each step:

$$0.568, \ 0.775, \ 0.828, \ 0.900, \ 0.912, \ 0.920.$$

The first four steps give relatively large increases in r^2_{xy}, but the increases in the last two steps are rather small.

The Final Model

Which model do we want to adopt, then? Should we go with the results of Step 4 and use PØTEN, ADV, SHARE, and ACCTS to predict SALES? All four predictors there are significant at $\alpha = 0.05$. Or should we also use SHCHANGE and TIME? In Step 6, we noticed that once TIME was included in the model, ACCTS become superfluous. Would it be possible to drop ACCTS and replace it with either WKLD or RATE and have six predictors which have a multiple r^2 greater than 0.92? What we want to use as a model depends upon our criterion for entry of variables into the model. In this forward selection procedure, setting $\alpha = 0.50$ has allowed us to look at six predictors. We might now drop back to the step (Step 4) at which all predictors are significant at $\alpha = 0.50$ and say that our model is

$$Y = \alpha + \beta_2 X_2 + \beta_3 X_3 + \beta_4 X_4 + \beta_6 X_6 + \varepsilon.$$

We would then use the analysis associated with these four predictors. Our criterion is that all predictors are significant, given all the others. Not all data sets are as well-behaved as this one, however. It is possible, for instance, to have several predictors go nonsignificant at some step and then all become significant again at a later step.

It is always a good idea to look at the standard errors of the regression coefficients. If these are large, then we are not estimating the regression coefficients very well, and thus our predictions might not be too accurate. In both Steps 5 and 7 we see large standard errors for b_5 (SHCHANGE), and this is evidence that we might not want to go with a six-variable model. It can be further argued that b_4 (SHARE) also has a reasonably large standard error, so that we might want to drop back and use the three-variable model, in which all the standard errors are reasonably small. Again we stress that there is no single answer to the question, "What is the best model?" It all depends upon the criteria you use. Let us look at some other variable selection procedures which approach the question of which variables to include in different ways.

THE STEPWISE PROCEDURE

Stepwise Regression

One problem with the forward selection procedure is that once a variable is added to the model, you are stuck with it. You can never take it out again. For example, in Step 6 of Figure 14.2, once TIME was added, ACCTS became superfluous. We might want to see what a regression with the five variables

TIME, PØTEN, AD, SHARE, and SHCHANGE

looks like, but the forward selection would never allow us to do this, because ACCTS was entered on the very first step. Similarly, we might want to replace ACCTS with either WKLD or RATE and see if these six variables do a better job.

Stepwise Procedure

Significance Level for Staying In

The stepwise procedure lets us get rid of a previously entered variable that becomes redundant. The stepwise procedure begins just like the forward selection. In addition, at each step it looks at all partial F-values. Any variable having a partial F not meeting the significance level for staying in is dropped. Then the procedure continues (as in the forward selection) to add more variables, one at a time. Any variable deleted at any step goes back in the hopper and is eligible for reinclusion at a later step. The process terminates either when no new variables can be entered or when the one to be entered was the one dropped at the previous step. Unless otherwise specified, the significance level for staying in used by SAS is 0.10. To change this default condition to, say, 0.15, simply add

SLS = .15

before the semicolon on the model statement. Note that the significance level for staying in is considerably smaller than the significance level for entry—because we want to give any variable already entered a chance to stay in the model. Thus, while we give a variable a 50–50 chance to enter, we'll only give it a 0.10 or 0.15 probability of being dropped.

Figure 15.3 shows the CRAVENS data run through the stepwise procedure. The first five steps are identical to the first five steps of the forward selection procedure. In Step 6, however, we see a message that the variable SHCHANGE was removed from the list of variables in Step 5. Thus, Step 6 gives the analysis for a four-variable regression using X_2(PØTEN), X_3(ADV), X_4(SHARE), and X_6(ACCTS). We see in Step 5 that the F-value for SHCHANGE was 2.6, which was significant at $\alpha = 0.1236$. Since 0.1236 is greater than 0.10, SHCHANGE was dropped.

One of two things happened after Step 6. Either no partial F-value, given X_2, X_3, X_4, and X_6, was large enough to be significant at $\alpha = 0.50$, or the largest partial F-value was the one for SHCHANGE, which had just been dropped. Given the latter case, it would be useless to add SHCHANGE at Step 7, because it would have just been dropped again at Step 8. Therefore, the process terminates after Step 6 and gives us the message that all variables are significant at $\alpha = 0.10$. (Inspection shows that as a matter of fact, all variables are significant at $\alpha = 0.05$.) The "best" model arrived at by the stepwise procedure is seen to be different from the "best" forward selection model. We are still not through, however, for these are not the only possible criteria for "best" models.

BACKWARD ELIMINATION

Backward Elimination Procedure

The problem with both the forward selection and the stepwise procedures is that they can miss a "key set" of variables. For example, suppose that Y(SALES) were a linear combination of X_1(TIME), X_2(ADV), and X_8(RATE). That is, suppose that these three variables could together predict sales almost perfectly, although, for example, X_8 is almost worthless by itself. Neither the forward selection nor stepwise procedure would discover this magic trio. Both assume that if the variable to be entered at any step is nonsignificant, then any weaker variable cannot be significant.

Backward
Elimination
Procedure
The <u>backward elimination procedure</u> starts with all variables in the model. It removes variables one at a time according to which one gives the smallest partial F-value, given all the rest, provided that this partial F-value is not significant. The procedure terminates when all partial F-values are significant at the significance level for staying in. A large significance level is usually used in order to give each variable a chance to stay in. A rule of thumb is to use $\alpha = 0.20$ or 0.25, but unless otherwise specified, SAS will use $\alpha = 0.10$.

Backward Elimination Analysis for CRAVENS Example

Figure 14.4 shows the backward elimination procedure for the CRAVENS example. Step 0 shows all eight predictors and their partial F-values. The smallest F-value (associated with the largest probability) is that for X_8(RATE), so in Step 1 this variable is dropped. Note that this has hardly any effect on r_{xy}^2. In Step 2, the variable X_7(WKLD) has the smallest F-value, so it is dropped. This hardly affects r_{xy}^2, either. In

FIGURE 14.3   SAS Stepwise Analysis of CRAVENS Data

STEPWISE REGRESSION PROCEDURE FOR DEPENDENT VARIABLE SALES

STEP 1 VARIABLE ACCTS ENTERED R SQUARE = 0.56849518

	DF	SUM OF SQUARES	MEAN SQUARE	F	PROB>F
REGRESSION	1	23524074.19189501	23524074.19189501	30.30	0.0001
ERROR	23	17855474.73496109	776324.98847657		
TOTAL	24	41379548.92685610			

	B VALUE	STD ERROR	TYPE II SS	F	PROB>F
INTERCEPT	709.32583372				
ACCTS	21.72176971	3.94603304	23524074.19189501	30.30	0.0001

STEP 2 VARIABLE ADV ENTERED R SQUARE = 0.77510077

	DF	SUM OF SQUARES	MEAN SQUARE	F	PROB>F
REGRESSION	2	32073320.05178234	16036660.02589117	37.91	0.0001
ERROR	22	9306228.87507376	423010.40341244		
TOTAL	24	41379548.92685610			

	B VALUE	STD ERROR	TYPE II SS	F	PROB>F
INTERCEPT	50.29906160				
ADV	0.22652657	0.05038842	8549245.85988733	20.21	0.0002
ACCTS	19.04824598	2.97291380	17365864.05826321	41.05	0.0001

STEP 3 VARIABLE POTEN ENTERED R SQUARE = 0.82772280

	DF	SUM OF SQUARES	MEAN SQUARE	F	PROB>F
REGRESSION	3	34250796.02722474	11416932.00907491	33.63	0.0001
ERROR	21	7128752.89963136	339464.42379197		
TOTAL	24	41379548.92685610			

	B VALUE	STD ERROR	TYPE II SS	F	PROB>F
INTERCEPT	-327.23338939				
POTEN	0.02192192	0.00865564	2177475.97544240	6.41	0.0194
ADV	0.21607079	0.04532744	7713721.79142546	22.72	0.0001
ACCTS	15.55392188	2.99936692	9128825.32465885	26.89	0.0001

STEP 4   VARIABLE SHARE ENTERED   R SQUARE = 0.90044970

	DF	SUM OF SQUARES	MEAN SQUARE	F	PROB>F
REGRESSION	4	37260202.46282121	9315050.61570530	45.23	0.0001
ERROR	20	4119346.46403489	205967.32320174		
TOTAL	24	41379548.92685610			

	B VALUE	STD ERROR	TYPE II SS	F	PROB>F
INTERCEPT	-1441.93182868				
POTEN	0.03821753	0.00797694	4727717.33107410	22.95	0.0001
ADV	0.17499004	0.0369066	4630368.54356448	22.48	0.0001
SHARE	190.14429731	49.74415347	3009406.43559647	14.61	0.0011
ACCTS	9.21389567	2.86521038	2129962.08538966	10.34	0.0043

STEP 5   VARIABLE SHCHANGE ENTERED   R SQUARE = .91241574

	DF	SUM OF SQUARES	MEAN SQUARE	F	PROB>F
REGRESSION	5	37755351.59771667	7551070.31954333	39.59	0.0001
ERROR	19	3624197.32913943	190747.22784944		
TOTAL	24	41379548.92685610			

	B VALUE	STD ERROR	TYPE II SS	F	PROB>F
INTERCEPT	-1285.94337067				
POTEN	0.03763121	0.00768517	4573489.56079583	23.98	0.0001
ADV	0.15443602	0.0373852	3194373.73424488	16.75	0.0006
SHARE	196.94952750	48.05692351	3203732.09272627	16.80	0.0006
SHCHANGE	262.50049338	162.92631286	495149.13489546	2.60	0.1236
ACCTS	8.23411280	2.82357962	1622152.88530014	8.50	0.0089

STEP 6   VARIABLE SHCHANGE REMOVED   R SQUARE = 0.90044970

	DF	SUM OF SQUARES	MEAN SQUARE	F	PROB>F
REGRESSION	4	37260202.46282121	9315050.61570530	45.23	0.0001
ERROR	20	4119346.46403489	205967.32320174		
TOTAL	24	41379548.92685610			

	B VALUE	STD ERROR	TYPE II SS	F	PROB>F
INTERCEPT	-1441.93182868				
POTEN	0.03821753	0.00797694	4727717.33107410	22.95	0.0001
ADV	0.17499004	0.0369066	4630368.54356449	22.48	0.0001
SHARE	190.14429731	49.74415347	3009406.43559647	14.61	0.0011
ACCTS	9.21389567	2.86521038	2129962.08538966	10.34	0.0043

ALL VARIABLES IN THE MODEL ARE SIGNIFICANT AT THE 0.1000 LEVEL.

FIGURE 14.4 SAS *Backward Elimination Procedure for CRAVENS Data*

BACKWARD ELIMINATION PROCEDURE FOR DEPENDENT VARIABLE SALES

STEP 0 ALL VARIABLES ENTERED R SQUARE = 0.92203915

	DF	SUM OF SQUARES	MEAN SQUARE	F	PROB>F
REGRESSION	8	38153564.25210440	4769195.53151305	23.65	0.0001
ERROR	16	3225984.67475170	201624.04217198		
TOTAL	24	41379548.92685610			

	B VALUE	STD ERROR	TYPE II SS	F	PROB>F
INTERCEPT	-1507.81372984				
TIME	2.00956615	1.93065421	218442.90153750	1.08	0.3134
POTEN	0.03720491	0.00820230	4148313.42001816	20.57	0.0003
ADV	0.15098890	0.04710851	2071255.37925263	10.27	0.0055
SHARE	199.02353635	67.02792230	1777623.07692217	8.82	0.0090
SHCHANGE	290.85513399	186.78199574	488906.44513135	2.42	0.1390
ACCTS	5.55096065	4.77554962	272415.54793787	1.35	0.2621
WK_D	19.79389189	33.67669223	69653.96590204	0.35	0.5649
RATE	8.18928366	128.50561301	818.82284513	0.00	0.9500

STEP 1 VARIABLE RATE REMOVED R SQUARE = 0.92201937

	DF	SUM OF SQUARES	MEAN SQUARE	F	PROB>F
REGRESSION	7	38152745.42925927	5450392.20417990	28.71	0.0001
ERROR	17	3226803.49759683	189811.97044687		
TOTAL	24	41379548.92685610			

	B VALUE	STD ERROR	TYPE II SS	F	PROB>F
INTERCEPT	-1485.88075785				
TIME	1.97454330	1.79574963	229490.83764567	1.21	0.2868
POTEN	0.03729049	0.00785101	4282222.67815504	22.56	0.0002
ADV	0.15196094	0.04324545	2343725.58802753	12.35	0.0027
SHARE	198.30848767	64.11718234	1815755.40273993	9.57	0.0066
SH-CHANGE	295.86609393	164.38654930	614867.35490257	3.24	0.0897
ACCTS	5.61018822	4.54495654	289213.78714048	1.52	0.2339
WKLD	19.89903131	32.63610128	70565.25639185	0.37	0.5501

STEP 2 VARIABLE WKLD REMOVED   R SQUARE = 0.92031405

	DF	SUM OF SQUARES	MEAN SQUARE	F	PROB>F
REGRESSION	6	38082180.17286742	6347030.02881124	34.65	0.0001
ERROR	18	3297368.75398868	183187.15299937		
TOTAL	24	41379548.92685610			

	B VALUE	STD ERROR	TYPE II SS	F	PROB>F
INTERCEPT	-1165.47855369				
TIME	2.26935112	1.69898362	32828.57515075	1.78	0.1983
POTEN	0.03827800	0.00754688	4712573.84321485	25.73	0.0001
ADV	0.14067029	0.03839221	2459312.09671556	13.43	0.0018
SHARE	221.60469221	50.58309112	3515945.91528346	19.19	0.0004
SHCHANGE	285.10928426	160.55965553	577623.53145916	3.15	0.0927
ACCTS	4.37770296	3.99903763	219521.27264808	1.20	0.2881

STEP 3 VARIABLE ACCTS REMOVED   R SQUARE = 0.91500898

	DF	SUM OF SQUARES	MEAN SQUARE	F	PROB>F
REGRESSION	5	37862658.90021934	7572531.78004387	40.91	0.0001
ERROR	19	3516890.02663676	185099.47508615		
TOTAL	24	41379548.92685610			

	B VALUE	STD ERROR	TYPE II SS	F	PROB>F
INTERCEPT	-1113.78787943				
TIME	3.61210118	1.18169995	1729460.18780281	9.34	0.0065
POTEN	0.04208812	0.00673122	7236631.11940864	39.10	0.0001
ADV	0.12885675	0.03703609	2240625.11884974	12.10	0.0025
SHARE	256.95554016	39.13606967	7979335.79655743	43.11	0.0001
SHCHANGE	324.53344995	157.28308421	788061.45599777	4.26	0.0530

ALL VARIABLES IN THE MODEL ARE SIGNIFICANT AT THE 0.1000 LEVEL.

Step 3, X_6 (ACCTS) is dropped because its F-value is the smallest, having a probability of $0.2881 > 0.10$. Finally, in Step 3, the smallest F-value is 4.26, which has probability $0.0530 < 0.10$, so no variables are dropped and the process terminates. A message is given that all variables are significant at $\alpha = 0.10$.

Note that we get quite a different set of predictors with the backward elimination procedure than with either forward selection or stepwise. In the first two procedures, we had

$$X_6 = \text{ACCTS}, \quad X_3 = \text{ADV}, \quad X_2 = \text{PØTEN}, \quad X_4 = \text{SHARE}$$

and possibly also

$$X_5 = \text{SHCHANGE} \quad \text{and} \quad X_1 = \text{TIME}.$$

Now we have

$$X_1 = \text{TIME}, \quad X_2 = \text{PØTEN}, \quad X_3 = \text{ADV}, \quad X_4 = \text{SHARE}, \quad \text{and}$$

$$X_5 = \text{SHCHANGE}.$$

If we wanted our predictors to all be significant at $\alpha = 0.05$ instead of $\alpha = 0.10$, we would reduce the set obtained in the backward elimination procedure to just X_1, X_2, X_3, and X_4. This set of variables can also be defended on the basis that b_5 has an extremely large standard error. Notice that the six predictors obtained by forward selection give $r_{xy}^2 = 0.9203$, the four obtained by stepwise gave $r_{xy}^2 = 0.9004$, and the five obtained by backward elimination gave $r_{xy}^2 = 0.9150$. The five variables obtained using backward elimination thus explain almost as much variation as the six selected by forward selection; in addition, all five backward variables are significant at $\alpha = 0.10$. It seems that some relationships in the data detected by the backward elimination procedure escaped the forward selection and stepwise procedures. For this reason, many statisticians prefer the backward elimination procedure, especially if they have no access to the theoretical structure relating to the variables. Only the backward elimination procedure allows "key sets" of variables to all act together.

MAXIMUM R² IMPROVEMENT

Maximizing R² *at Each Step*

Maximum R²
Improvement
Procedure

The maximum R² improvement technique* does not give a final answer, as do the three previous techniques we have studied. Rather, it gives a best single-predictor regression, a best two-predictor regression, and so on until it gives the best k-predictor regression. The researcher uses the results to decide which regression to use. This technique is designed to look at a large number of regressions, although it falls quite a bit short

---

* The capital R in R² is often used to indicate a multiple coefficient of determination, while a lowercase r indicates a simple correlation.

(purposely) of looking at all possible regressions, since for a large number of predictors, looking at all possible regressions would be prohibitively expensive.

The procedure followed is this. In Step 1, the variable is entered which gives the largest r_{xy}^2 when acting alone. This is the best single-predictor regression. In Step 2, the variable which increases r_{xy}^2 the most, in the presence of the single strongest predictor, is added. This sounds like the forward selection procedure, but there is a difference. At each step, the program can perform switches. For example, suppose that the strongest predictor is X_1, and that given X_1, X_2 increases r_{xy}^2 the most. Variables X_3, X_4, and X_5 are still "in the hopper." The computer looks to see if replacing X_1 by X_3, X_4, or X_5 will yield a larger r_{xy}^2 than X_1 and X_2 will. If, for example, $r_{23,Y} > r_{12,Y}$, then the computer will replace X_1 by X_3. It will call this the BEST TWO-PREDICTOR MODEL FOUND. Note that this is not guaranteed to be the very best two-predictor model, but only the best one *found*. It may be that X_3 and X_5 give a higher r_{xy}^2 than do X_2 and X_3, but all the computer has found is that X_2 and X_3 together explain more than do X_1 and X_2.

Suppose that X_2 and X_3 comprise the best two-predictor model found. Now X_1, X_4, and X_5 are in the hopper. For Step 3, the computer looks at

$$X_2, X_3, X_1, \qquad X_2, X_3, X_4 \quad \text{and} \quad X_2, X_3, X_5$$

and selects the trio that gives the largest r_{xy}^2. Suppose this is

$$X_2, X_3, X_4.$$

The computer then holds the last variable to enter, X_4 in this example, and then says, "Can I switch X_2 or X_3 for X_1 or X_5 and get a larger r_{xy}^2?" In other words, the computer looks at

$$X_4, X_2, X_1, \qquad X_4, X_2, X_5$$

and

$$X_4, X_3, X_1, \qquad X_4, X_3, X_5$$

to see if any of these trios will beat X_2, X_3, and X_4. If so, the switch is made that increases the multiple coefficient of determination the most. If not, then the computer proceeds on to the next step. The process continues until all k-variables have been included.

Note that the computer does not make significance tests in this procedure. The significance levels for inclusion and deletion are not used.

Maximum R² Improvement Technique Applied to CRAVENS Data

The maximum R² improvement technique is applied to the CRAVENS data in Figure 14.5. Step 1 enters X_6 (ACCTS), since it has the highest simple correlation with Y (SALES). In Step 2, X_3 (ADV) is added, and in Step 3, X_2 (PØTEN) is added. The

FIGURE 14.5 SAS Maximum R² Improvement for CRAVENS Data

MAXIMUM R-SQUARE IMPROVEMENT FOR DEPENDENT VARIABLE SALES

STEP 1 VARIABLE ACCTS ENTERED R SQUARE = 0.56849518

	DF	SUM OF SQUARES	MEAN SQUARE	F	PROB>F
REGRESSION	1	23524074.19189501	23524074.19189501	30.30	0.0001
ERROR	23	17855474.73496109	776324.98847657		
TOTAL	24	41379548.92685610			

	B VALUE	STD ERROR	TYPE II SS	F	PROB>F
INTERCEPT	709.32383372				
ACCTS	21.72176971	3.94603304	23524074.19189501	30.30	0.0001

THE ABOVE MODEL IS THE BEST 1 VARIABLE MODEL FOUND.

STEP 2 VARIABLE ADV ENTERED R SQUARE = 0.77510077

	DF	SUM OF SQUARES	MEAN SQUARE	F	PROB>F
REGRESSION	2	32073320.05178234	16036660.02589117	37.91	0.0001
ERROR	22	9306228.87507376	423010.40341244		
TOTAL	24	41379548.92685610			

	B VALUE	STD ERROR	TYPE II SS	F	PROB>F
INTERCEPT	50.29906160				
ADV	0.22652657	0.05038842	8549245.85988733	20.21	0.0002
ACCTS	19.04824598	2.97291380	17365864.05826321	41.05	0.0001

THE ABOVE MODEL IS THE BEST 2 VARIABLE MODEL FOUND.

STEP 3 VARIABLE POTEN ENTERED R SQUARE = 0.82772280

	DF	SUM OF SQUARES	MEAN SQUARE	F	PROB>F
REGRESSION	3	34250796.02722474	11416932.00907491	33.63	0.0001
ERROR	21	7128752.89963136	339464.42379197		
TOTAL	24	41379548.92685610			

	B VALUE	STD ERROR	TYPE II SS	F	PROB>F
INTERCEPT	-327.23338939				
POTEN	0.02192192	0.00865564	2177475.97544240	6.41	0.0194
ADV	0.21607079	0.04532744	7713721.79142546	22.72	0.0001
ACCTS	15.55392158	2.99936692	9128825.32465885	26.89	0.0001

STEP 3 ACCTS REPLACED BY SHARE   R SQUARE = 0.84897591

	DF	SUM OF SQUARES	MEAN SQUARE	F	PROB>F
REGRESSION	3	35130240.37743155	11710080.12581052	39.35	0.0001
ERROR	21	6249308.54942455	297586.12140117		
TOTAL	24	41379548.92685610			

	B VALUE	STD ERROR	TYPE II SS	F	PROB>F
INTERCEPT	-1603.58091828				
POTEN	0.05428605	0.00747411	15698916.49197692	52.75	0.0001
ADV	0.16748034	0.04427318	4258521.38246777	14.31	0.0011
SHARE	282.74666591	48.75558026	10008269.67486566	33.63	0.0001

THE ABOVE MODEL IS THE BEST 3 VARIABLE MODEL FOUND.

STEP 4 VARIABLE ACCTS ENTERED   R SQUARE = 0.90044970

	DF	SUM OF SQUARES	MEAN SQUARE	F	PROB>F
REGRESSION	4	37260202.46282121	9315050.61570530	45.23	0.0001
ERROR	20	4119346.46403489	205967.32320174		
TOTAL	24	41379548.92685610			

	B VALUE	STD ERROR	TYPE II SS	F	PROB>F
INTERCEPT	-1441.93182868				
POTEN	0.03821753	0.00797694	4727717.33107410	22.95	0.0001
ADV	0.17499004	0.03690666	4630368.54356449	22.48	0.0001
SHARE	190.14429731	49.74415347	3009406.43559647	14.61	0.0011
ACCTS	9.21389567	2.86521038	2129962.08538966	10.34	0.0043

THE ABOVE MODEL IS THE BEST 4 VARIABLE MODEL FOUND.

STEP 5 VARIABLE SHCHANGE ENTERED   R SQUARE = 0.91241574

	DF	SUM OF SQUARES	MEAN SQUARE	F	PROB>F
REGRESSION	5	37755351.59771667	7551070.31954333	39.59	0.0001
ERROR	19	3624197.32913943	190747.22784944		
TOTAL	24	41379548.92685610			

	B VALUE	STD ERROR	TYPE II SS	F	PROB>F
INTERCEPT	-1285.94337067				
POTEN	0.03763121	0.00768517	4573489.56079583	23.98	0.0001
ADV	0.15443602	0.03773852	3194373.73424488	16.75	0.0006
SHARE	196.94952750	48.05692351	3203732.09272627	16.80	0.0006
SHCHANGE	262.50049338	162.92631286	495149.13489546	2.60	0.1236
ACCTS	8.23411280	2.82357962	1622152.88530014	8.50	0.0089

FIGURE 14.5   *(continued)*

MAXIMUM R-SQUARE IMPROVEMENT FOR DEPENDENT VARIABLE SALES

STEP 5   ACCTS REPLACED BY TIME   R SQUARE = 0.91500898

	DF	SUM OF SQUARES	MEAN SQUARE	F	PROB>F
REGRESSION	5	3762658.90021934	7572531.78004387	40.91	0.0001
ERROR	19	3516890.02663676	185099.47508615		
TOTAL	24	41379548.92685610			

	B VALUE	STD ERROR	TYPE II SS	F	PROB>F
INTERCEPT	-1113.78787943				
TIME	3.61210118	1.18169995	1729460.18780281	9.34	0.0065
POTEN	0.04208812	0.00673122	7236631.11940864	39.10	0.0001
ADV	0.12885675	0.03703609	2240625.11884974	12.10	0.0025
SHARE	256.9554016	39.13606967	7979335.79655744	43.11	0.0001
SHCHANGE	324.53344995	157.28308421	788061.45599777	4.26	0.0530

THE ABOVE MODEL IS THE BEST 5 VARIABLE MODEL FOUND.

STEP 6   VARIABLE ACCTS ENTERED   R SQUARE = 0.92031405

	DF	SUM OF SQUARES	MEAN SQUARE	F	PROB>F
REGRESSION	6	38082180.17288742	6347030.02881124	34.65	0.0001
ERROR	18	3297368.75398868	183187.15299937		
TOTAL	24	41379548.92685610			

	B VALUE	STD ERROR	TYPE II SS	F	PROB>F
INTERCEPT	-1165.47855369				
TIME	2.26935112	1.69898362	326828.57515075	1.78	0.1983
POTEN	0.03827800	0.00754688	4712573.84321485	25.73	0.0001
ADV	0.14067029	0.03839221	2459312.09671656	13.43	0.0018
SHARE	221.60469221	50.58309112	3515945.91528346	19.19	0.0004
SHCHANGE	285.10928426	160.55965553	577623.53145916	3.15	0.0927
ACCTS	4.3777296	3.99903763	219521.27264808	1.20	0.2881

THE ABOVE MODEL IS THE BEST 6 VARIABLE MODEL FOUND.

STEP 7 VARIABLE WKLD ENTERED R SQUARE = 0.92201937

	DF	SUM OF SQUARES	MEAN SQUARE	F	PROB>F
REGRESSION	7	38152745.42925927	5450392.20417990	28.71	0.0001
ERROR	17	3226803.49759683	189811.97044687		
TOTAL	24	41379548.92685610			

	B VALUE	STD ERROR	TYPE II SS	F	PROB>F
INTERCEPT	-1485.88075785				
TIME	1.97454330	1.79574963	229490.83764567	1.21	0.2868
POTEN	0.03729049	0.00785101	4282222.67815504	22.56	0.0002
ADV	0.15196094	0.04324545	2343725.58802753	12.35	0.0027
SHARE	198.30848767	64.11718234	1815755.40273993	9.57	0.0066
SHCHANGE	295.86609393	164.38654930	614867.35490257	3.24	0.0897
ACCTS	5.61018822	4.54495654	289213.78714048	1.52	0.2339
WKLD	19.89903131	32.63610128	70565.25639185	0.37	0.5501

THE ABOVE MODEL IS THE BEST 7 VARIABLE MODEL FOUND.

STEP 8 VARIABLE RATE ENTERED R SQUARE = 0.92203915

	DF	SUM OF SQUARES	MEAN SQUARE	F	PROB>F
REGRESSION	8	38153564.25210440	4769195.53151305	23.65	0.0001
ERROR	16	3225984.67475170	201624.04217198		
TOTAL	24	41379548.92685610			

	B VALUE	STD ERROR	TYPE II SS	F	PROB>F
INTERCEPT	-1507.81372984				
TIME	2.00956615	1.93065421	218442.90153751	1.08	0.3134
POTEN	0.03720491	0.00820230	4148313.42001816	20.57	0.0003
ADV	0.15098890	0.04710851	2071255.37925263	10.27	0.0055
SHARE	199.02353635	67.02792230	1777623.07698218	8.82	0.0090
SHCHANGE	290.85513399	186.78199574	488906.44513135	2.42	0.1390
ACCTS	5.55096065	4.77554962	272415.54793787	1.35	0.2621
WKLD	19.79389189	33.67669223	69653.96590204	0.35	0.5649
RATE	8.18928366	128.50561301	818.82284513	0.00	0.9500

THE ABOVE MODEL IS THE BEST 8 VARIABLE MODEL FOUND.

variables X_2, X_3, and X_6 together explain 82.77% of the variation in Y. But the computer finds that if X_6(ACCTS) is replaced by X_4(SHARE), then the trio X_2, X_3, and X_5 explain 84.90% of the variation in Y. So it calls the model containing X_2, X_3, and X_5 the "best three-variable model found." Note that X_6(ACCTS) enters right back in at Step 4, and is thrown out again in Step 5, only to enter again in Step 6. (Persistent little variable, isn't it?)

At the various steps we find the best, one-, two-, three-,…, eight-predictor models, in terms of which ones the computer could find that explained the most variation. Which model do we use? One should consider not only the size of r_{xy}^2 at each step, but also the significance levels given for the F-values. Practical and theoretical considerations also count in many actual problems.

We see in Figure 14.5 that all the variables in the best one-, two-, three-, and four-predictor models are significant at $\alpha = 0.05$. Of course, the multiple coefficient of determination is largest in the four-predictor model, using X_2(PØTEN), X_3(ADV), X_4(SHARE), and X_6(ACCTS). Note that these are the same variables found using the stepwise technique, and the same ones found in Step 4 of forward selection. In the stepwise procedure, the process terminated after four variables had been added, because the next strongest variable, X_5(SHCHANGE), was not significant at $\alpha = 0.10$. In the maximum R^2 improvement procedure, we see X_5 entering at Step 5 and note that it is not significant, given the others, at $\alpha = 0.10$. However, in Step 6 we see X_6(ACCTS) replaced by X_1(TIME), and then all five of these variables are conditionally significant at $\alpha = 0.10$. Thus, while the stepwise procedure found that for significance at $\alpha = 0.10$, the four variables X_2, X_3, X_4, and X_6 together explain $r_{xy}^2 = 90\%$ of the variation in Y(SALES), the maximum R^2 improvement technique found five variables, X_1, X_2, X_3, X_4, and X_5 that are all significant at $\alpha = 0.10$ and together explain more of the variation ($r_{xy}^2 = 91.5\%$). Since the maximum R^2 improvement technique can look at more regressions than stepwise, its chances of finding the strongest set of significant predictors are greater.

The disadvantage of the maximum R^2 improvement technique lies in the amount of computer time it takes. It can be quite expensive to run, especially if the original model contains a large number of predictors. There is no automatic cut-off for reducing the amount of output; the program will terminate only when the best k-predictor model, containing all predictors, is found. (However one can instruct the computer to terminate after a certain predetermined number of variables have been included. Refer to the SAS manual.) This procedure does not pay any attention to the significance level of any partial F-value that it computes.

Minimum R² IMPROVEMENT

The Technique Applied to CRAVENS Data

Minimum R^2
Improvement
Procedure

The minimum R^2 improvement technique may sound like the dumbest thing anyone would ever wish to do. It works like the maximum R^2 improvement technique with one exception: it adds variables and performs switches in a way that the multiple coefficient of determination is increased as *little* as possible at each step.

In Step 1, the single *weakest* predictor, X_7 (WKLD), is entered first. In subsequent pieces of Step 1, X_7 is replaced by the next weakest predictor, X_8(RATE); X_8 is replaced by the next weakest predictor, and so on. Finally, in the eighth part of Step 1 the strongest single predictor, X_6(ACCTS) is added. The single-predictor model using X_6 has the largest coefficient of determination and is thus called the best model found.

In Step 2, the computer looks for the variable that will improve r_{xy}^2 the *least*, when used in conjunction with X_6(ACCTS). This variable is X_7(WKLD). Switches with these variables are performed in such a way as to increase r_{xy}^2 by the *smallest amount possible* with each switch until two variables are found [in this case X_2(PØTEN) and X_4(SHARE)] that form the strongest pair.

This process of increasing r_{xy}^2 by the smallest amounts possible may seem quite inefficient, and indeed it is. It is designed to look at a very large number of regressions (but still not all possible regressions). This procedure is useful in that it allows the researcher to see a lot of information and increases the chances that *the* best pair, triple, or whatever, of variables will be found. It is so time consuming, however, that in a problem with many predictors the cost of performing this analysis could easily be prohibitive. In the interest of saving space, Figure 14.6 shows only the *final* 1-variable, 2-variable, ..., 8-variable models found by the minimum R^2 improvement technique. This is because there are eight parts to STEP 1, six parts to STEP 2, six parts to STEP 3, four parts to STEP 4, five parts to STEP 5, three parts to STEP 6, two parts to STEP 7, and one to STEP 8. All this takes up 12 pages of computer printout!

Notice that the maximum R^2 improvement and the minimum R^2 improvement techniques may or may not give the same best models with a designated number of predictors. For example, the best two-predictor model found by the MAXR procedure used the variables X_3(ADV) and X_6(ACCTS) and found a multiple R^2 of 77.5%. The MINR procedure, however, never considered these two variables together and arrived at a "best" two-predictor model using X_2(PØTEN) and X_4(SHARE), which together explain only 74.6% of the variation in sales. Both procedures do, however, find the same "best" models for $k = 3,4,5,6,7$, and of course, for $k = 8$. The MAXR technique finds these best models a lot more quickly than does the MINR procedure, however.

COMPARISON OF THE FIVE SAS PROCEDURES

We can see that the choice of variables to be included in our final model is very much a function of the criteria we apply in selecting variables. Only in extremely clear-cut problems would we expect all five techniques described here to produce the same subset of variables for the reduced model. Usually one wants a model in which all predictors are significant, in which the regression coefficients have small standard errors, and which gives a large multiple coefficient of determination. In addition, the researcher probably has some intuitive idea about which variables to include in the final model, because they make good sense to the problem. Some researcher favoritism might be based on convenience. If one is lucky, some one of these five analyses will yield a set of variables that satisfy all of these requirements. Then the researcher can be relatively happy with the final model.

FIGURE 14.6   SAS Minimum R^2 Improvement Technique for CRAVENS Data

MINIMUM R-SQUARE IMPROVEMENT FOR DEPENDENT VARIABLE SALES

STEP 1   TIME REPLACED BY ACCTS   R SQUARE = 0.56849518

	DF	SUM OF SQUARES	MEAN SQUARE	F	PROB>F
REGRESSION	1	23524074.19189501	23524074.19189501	30.30	0.0001
ERROR	23	17855474.73496109	76324.98847657		
TOTAL	24	41379548.92685610			

	B VALUE	STD ERROR	TYPE II SS	F	PROB>F
INTERCEPT	709.32383372				
ACCTS	21.72176971	3.94603304	23524074.19189501	30.30	0.0001

THE ABOVE MODEL IS THE BEST 1 VARIABLE MODEL FOUND.

STEP 2   ACCTS REPLACED BY SHARE   R SQUARE - 0.74606224

	DF	SUM OF SQUARES	MEAN SQUARE	F	PROB>F
REGRESSION	2	30871718.99496378	15435859.49748189	32.32	0.0001
ERROR	22	10507829.93189232	477628.63326783		
TOTAL	24	41379548.92685610			

	B VALUE	STD ERROR	TYPE II SS	F	PROB>F
INTERCEPT	-1578.08881206				
POTEN	0.06117826	0.00918324	21197905.35034394	44.38	0.0001
SHARE	340.44371373	58.66779312	16083534.29245808	33.67	0.0001

THE ABOVE MODEL IS THE BEST 2 VARIABLE MODEL FOUND.

STEP 3   TIME REPLACED BY ADV   R SQUARE = 0.84897591

	DF	SUM OF SQUARES	MEAN SQUARE	F	PROB>F
REGRESSION	3	35130240.37743154	11710080.12581051	39.35	0.0001
ERROR	21	6249308.54942456	297586.12140117		
TOTAL	24	41379548.92685610			

	B VALUE	STD ERROR	TYPE II SS	F	PROB>F
INTERCEPT	-1603.58091828				
POTEN	0.05428605	0.00747411	15698916.49197694	52.75	0.0001
ADV	0.16748034	0.04427318	4258521.38246776	14.31	0.0011
SHARE	282.74666591	48.75558026	10008269.67486566	33.63	0.0001

THE ABOVE MODEL IS THE BEST 3 VARIABLE MODEL FOUND.

STEP 4 WKLD REPLACED BY SHARE R SQUARE = 0.90044970

	DF	SUM OF SQUARES	MEAN SQUARE	F	PROB>F
REGRESSION	4	37260202.46282120	9315050.61570530	45.23	0.0001
ERROR	20	4119346.46403490	205967.32320174		
TOTAL	24	41379548.92685610			

	B VALUE	STD ERROR	TYPE II SS	F	PROB>F
INTERCEPT	-1441.93182868				
POTEN	0.03821753	0.00797694	4727717.33107411	22.95	0.0001
ADV	0.17499004	0.03690666	4630368.54356448	22.48	0.0001
SHARE	190.14429731	49.74415347	3009406.43559647	14.61	0.0011
ACCTS	9.21389567	2.86521038	2129962.08538966	10.34	0.0043

THE ABOVE MODEL IS THE BEST 4 VARIABLE MODEL FOUND.

STEP 5 ACCTS REPLACED BY TIME R SQUARE = 0.91500898

	DF	SUM OF SQUARES	MEAN SQUARE	F	PROB>F
REGRESSION	5	37862658.90021933	7572531.78004387	40.91	0.0001
ERROR	19	3516890.02663677	185099.47508615		
TOTAL	24	41379548.92685610			

	B VALUE	STD ERROR	TYPE II SS	F	PROB>F
INTERCEPT	-1113.78787943				
TIME	3.61210118	1.18169995	1729460.18780281	9.34	0.0065
POTEN	0.04208812	0.00673122	7236631.11940865	39.10	0.0001
ADV	0.12885675	0.03703609	2240625.11884974	12.10	0.0025
SHARE	256.95554016	39.13606967	7979335.79655745	43.11	0.0001
SHCHANGE	324.53344995	157.28308421	788061.45599777	4.26	0.0530

THE ABOVE MODEL IS THE BEST 5 VARIABLE MODEL FOUND.

STEP 6 RATE REPLACED BY ACCTS R SQUARE = 0.92031405

	DF	SUM OF SQUARES	MEAN SQUARE	F	PROB>F
REGRESSION	6	38082180.17286742	6347030.02881124	34.65	0.0001
ERROR	18	3297368.75398868	183187.15299937		
TOTAL	24	41379548.92685610			

	B VALUE	STD ERROR	TYPE II SS	F	PROB>F
INTERCEPT	-1165.47855369				
TIME	2.26935112	1.69898362	326828.57515075	1.78	0.1983
POTEN	0.03827800	0.00754688	4712573.84321486	25.73	0.0001
ADV	0.14067029	0.03839221	2459312.09671656	13.43	0.0018
SHARE	221.60469221	50.58309112	3515945.91528347	19.19	0.0004
SHCHANGE	285.10928426	160.55965553	577623.53145916	3.15	0.0927
ACCTS	4.37770296	3.99903763	219521.27264808	1.20	0.2881

THE ABOVE MODEL IS THE BEST 6 VARIABLE MODEL FOUND.

FIGURE 14.6 (continued)

STEP 7 RATE REPLACED BY WKLD R SQUARE = 0.92201937

	DF	SUM OF SQUARES	MEAN SQUARE	F	PROB>F
REGRESSION	7	38152745.42925926	5450392.20417989	28.71	0.0001
ERROR	17	3226803.49759684	189811.97044687		
TOTAL	24	41379548.92685610			

	B VALUE	STD ERROR	TYPE II SS	F	PROB>F
INTERCEPT	-1485.88075785				
TIME	1.97454330	1.79574963	229490.83764567	1.21	0.2868
POTEN	0.03729049	0.00785101	4282222.67815505	22.56	0.0002
ADV	0.15196094	0.0424545	2343725.58802754	12.35	0.0027
SHARE	198.30848767	64.11718234	1815755.40273993	9.57	0.0066
SHCHANGE	295.86609393	164.38654930	614867.35490257	3.24	0.0897
ACCTS	5.61018822	4.54495654	289213.78714048	1.52	0.2339
WKLD	19.89903131	32.63610128	70565.25639185	0.37	0.5501

THE ABOVE MODEL IS THE BEST 7 VARIABLE MODEL FOUND.

STEP 8 VARIABLE RATE ENTERED R SQUARE = 0.92203915

	DF	SUM OF SQUARES	MEAN SQUARE	F	PROB>F
REGRESSION	8	38153564.25210440	4769195.53151305	23.65	0.0001
ERROR	16	3225984.67475170	201624.04217198		
TOTAL	24	41379548.92685610			

	B VALUE	STD ERROR	TYPE II SS	F	PROB>F
INTERCEPT	-1507.81373984				
TIME	2.00956615	1.93065421	218442.90153751	1.08	0.3134
POTEN	0.03720491	0.00820230	4148313.42001817	20.57	0.0003
ADV	0.15098890	0.04710851	2071255.37925263	10.27	0.0055
SHARE	199.02353635	67.02792230	1777623.07698218	8.82	0.0090
SHCHANGE	290.85513399	186.78199574	488906.44513135	2.42	0.1390
ACCTS	5.55096065	4.77554962	272415.54793787	1.35	0.2621
WKLD	19.79389189	33.67669223	69653.96590204	0.35	0.5649
RATE	8.18928366	128.50561301	818.82284513	0.00	0.9500

THE ABOVE MODEL IS THE BEST 8 VARIABLE MODEL FOUND.

In most situations, however, there will be several models to choose from; all of them satisfying some criteria while falling short in others. The final choice of the model then is up to the researcher, rather than the statistician, because the researcher is the one familiar with the theory of the field. While the statistician is useful in explaining what the analysis means, the results of numerical analyses should never blindly take precedence over expertise in the field of application.

Now, a word of caution is in order. None of these five variable-selection techniques is guaranteed to result in a regression equation that will give good predictions when applied to new sets of data. All results of regression analysis are quite data-dependent: the regression line itself, r_{xy}^2, the standard error of estimate, and so on. Based on the conclusions of the significance tests, *provided all the assumptions of the model are met*, we can be reasonably certain that significant predictors are those which really should be used. But just because our predictors can explain, say, 92% of the variation in Y in one sample gives no assurance that these same predictors would do as well in another sample. If the objective is to make accurate predictions, then we need something beyond variable-selection techniques. This problem will be touched on lightly in Chapter 15.

We will next look at the variable-selection procedures available through BMD and SPSS. Then we will see how to force variables into the model using all three regression packages.

EXERCISES *Section 14.2*

p. 533 *14.1* Recall that in the analysis of the CHILDREN data, the overall regression significance test indicated that at least one of the predictors (AGE, HEIGHT, or WEIGHT) was significant. However, no partial F-test on one predictor, given the other two, indicated a significant predictor.

Run the data through the SAS forward selection or stepwise program and see if any model can be found in which all predictors are significant at $\alpha = 0.05$. (Use default conditions for the inclusion and deletion criteria.)

TENNESSEE Data

A simple random sample of 30 of the 95 counties in Tennessee was drawn, and values of eight variables were recorded for each of the sampled counties. Table 14.1 shows the information collected. Call these the TENNESSEE data,

14.2 Suppose it is desired to use the TENNESSEE data to estimate the percentage of families in a Tennessee county which have incomes falling below the poverty level. Use the five stagewise procedures available through SAS to analyze these data. Use a significance level for entry of 0.25 and a significance level for staying in of 0.15. You might also want to obtain a correlation matrix and other appropriate statistics. Compare the results of the five stagewise procedures and propose a final model. Justify your choice.

TABLE 14.1   Selected Data on a Random Sample of 30 of the 95 Tennessee Counties

County	% Population Change 1960–1970	No. Persons Employed in Agriculture	% Families below Poverty Level	% Residences with Telephones	Residential and Farm Property Tax Rate	% Rural Population	Median Age	Negro Population
Benton	13.7	400	19.0	82	1.09	74.8	33.5	360
Cannon	− 0.8	710	26.2	66	1.01	100.0	32.8	193
Carroll	9.6	1,610	18.1	80	0.40	69.7	33.4	3,080
Cheatheam	40.0	500	15.4	74	0.93	100.0	27.8	592
Cumberland	8.4	640	29.0	65	0.92	74.0	27.9	2
DeKalb	3.5	920	21.6	64	0.59	73.1	33.2	230
Dyer	3.0	1,890	21.9	82	0.63	52.3	30.8	3,978
Gibson	7.1	3,040	18.9	85	0.49	49.6	32.4	9,816
Greene	13.0	2,730	21.1	78	0.71	71.2	29.2	1,137
Hawkins	10.7	1,850	23.8	74	0.93	70.6	28.7	992
Haywood	−16.2	2,920	40.5	69	0.51	64.2	25.1	10,723
Henry	6.6	1,070	21.6	85	0.80	58.3	35.9	3,139
Houston	21.9	160	25.4	69	0.74	100.0	31.4	338
Humphreys	17.8	380	19.7	83	0.44	72.0	30.1	516
Jackson	−11.8	1,140	38.0	54	0.81	100.0	34.1	12
Johnson	7.5	690	30.1	65	1.05	100.0	30.5	104
Lawrence	3.7	1,170	24.8	76	0.73	69.5	30.0	430
McNairy	1.6	1,280	30.3	67	0.65	81.0	32.4	1,240
Madison	8.4	2,270	19.5	85	0.48	39.1	28.7	20,446
Marshall	2.7	960	15.6	84	0.72	58.4	33.4	1,863
Maury	5.6	1,710	17.2	84	0.62	42.4	29.9	8,035
Montgomery	12.7	1,410	18.4	86	0.84	36.4	23.3	10,620
Morgan	− 4.8	200	27.3	66	0.73	99.8	27.5	211
Sevier	16.5	960	19.2	74	0.45	90.6	29.5	133
Shelby	15.2	11,500	16.8	87	1.00	5.9	25.4	266,159
Sullivan	11.6	1,380	13.2	85	0.63	44.2	28.8	2,432
Trousdale	4.9	530	29.7	70	0.54	100.0	33.1	932
Unicoi	1.1	370	19.8	75	0.98	52.6	30.8	7
Wayne	3.8	440	27.7	48	0.46	100.0	28.4	208
Weakley	19.0	1,630	20.5	83	0.68	72.1	30.4	1,732

Source: Centre for Business and Economic Research, The University of Tennessee, Knoxville, *Tennessee Statistical Abstract,* 1974.

14.3 THE BMDP2R SELECTION PROCEDURES

With a bit of study, the BMDP2R Stepwise Regression package can be made to do almost everything that the SAS can do. If the program is run using default conditions on the method of selection, the BMD will perform a stepwise regression. (The previous section explained the stepwise procedure with reference to the SAS program.) The worksheet for using the BMDP2R shown in Figure 14.7 looks similar to those we used to run simple regressions and perform the general regression significance test using the BMDP1R. One might note that no spaces (except those resulting from right-justified values) were left between values in the data set, nor were any decimal points punched. The format card, of course, tells the computer how to read the data.

STEPWISE REGRESSION

Minimum Acceptable F to Enter

Maximum Acceptable F to Remove

Unless otherwise specified, the BMDP2R program will enter predictors based on which one has the largest partial F-values, given the other variables already in the equation, as long as this F-value is larger than the Minimum acceptable F to enter. Then, any variable that has a partial F-value less than the Maximum acceptable F-value to remove will be removed from the model. The output from the program described in the worksheet is shown in Figure 14.8. The simple statistics and correlation and covariance matrices for these data are the first pieces of information to be given. (They are not repeated here from Chapter 13.) The stepwise analysis of the data is shown in Figure 14.8.

FIGURE 14.7 *Worksheet for BMDP2R Using CRAVENS Data*

```
  {SYSTEM CARDS
PRØBLEM   TITLE='CRAVENS 2R'./
INPUT     VARIABLES=9.
          FØRMAT='(F6.2,F5.2,F7.2,F7.2,F4.2,F4.2,F5.2,F4.2,F2.1)'.
          CASES=25./
VARIABLE  NAMES=SALES,TIME,PØTEN,ADV,SHARE,CHANGE,ACCTS,WKLD,RATING./
PRINT     CØVAR.
          CØRR.
          DATA./
PLØT      RESID.
          VARIABLE=ACCTS./
REGRES    DEPENDENT=SALES./
END/
366988 43107406511 458288 251  34 7486150549
347395108135811730 553978 551  1510732199751
          :
279997 21142280953 355200 914 -74 8862249639
FINISH/
```

FIGURE 14.8   *BMDP2R Stepwise Regression for CRAVENS Data*

REGRESSION TITLE	CRAVENS 2R
STEPPING ALGORITHM	F
MAXIMUM NUMBER OF STEPS	18
DEPENDENT VARIABLE	1 SALES
MINIMUM ACCEPTABLE F TO ENTER	4.000, 4.000
MAXIMUM ACCEPTABLE F TO REMOVE	3.900, 3.900
MINIMUM ACCEPTABLE TOLERANCE	0.01000

STEP NO. 0

MULTIPLE R	0.0
MULTIPLE R-SQUARE	0.0
STD. ERROR OF EST.	1313.0637

ANALYSIS OF VARIANCE

	SUM OF SQUARES	DF	MEAN SQUARE	F RATIO
REGRESSION	0.0	0	0.0	0.0
RESIDUAL	41379264.	24	1724136.	

VARIABLES IN EQUATION

VARIABLE	COEFFICIENT	STD. ERROR OF COEFF	STD REG COEFF	F TO REMOVE	LEVEL
(Y-INTERCEPT)	3374.564)				

VARIABLES NOT IN EQUATION

VARIABLE		PARTIAL CORR.	TOLERANCE	F TO ENTER	LEVEL
TIME	2	0.62292	1.00000	14.583	1
POTEN	3	0.59781	1.00000	12.791	1
ADV	4	0.59618	1.00000	12.683	1
SHARE	5	0.48351	1.00000	7.018	1
CHANGE	6	0.48918	1.00000	7.235	1
ACCTS	7	0.75399	1.00000	30.302	1
WKLD	8	-0.11723	1.00000	0.320	1
RATING	9	0.40188	1.00000	4.430	1

STEP NO. 1
VARIABLE ENTERED   7 ACCTS

MULTIPLE R   0.7540
MULTIPLE R-SQUARE   0.5685
STD. ERROR OF EST.   881.0894

ANALYSIS OF VARIANCE

	SUM OF SQUARES	DF	MEAN SQUARE	F RATIO
REGRESSION	23523920.	1	0.2352392E 08	30.302
RESIDUAL	17855328.	23	776318.6	

VARIABLES IN EQUATION

VARIABLE	COEFFICIENT	STD. ERROR OF COEFF	STD REG COEFF	F TO REMOVE	LEVEL
(Y-INTERCEPT	709.328)				
ACCTS 7	21.722	3.946	0.754	30.302	1

VARIABLES NOT IN EQUATION

VARIABLE		PARTIAL CORR.	TOLERANCE	F TO ENTER	LEVEL
TIME	2	0.12024	0.42572	0.323	1
POTEN	3	0.41078	0.77090	4.466	1
ADV	4	0.69196	0.95998	20.210	1
SHARE	5	0.29882	0.83758	2.157	1
CHANGE	6	0.39040	0.89280	3.956	1
WKLD	8	0.05080	0.96046	0.057	1
RATING	9	0.35889	0.94774	3.253	1

STEP NO. 2
VARIABLE ENTERED   4 ADV

MULTIPLE R   0.8804
MULTIPLE R-SQUARE   0.7751
STD. ERROR OF EST.   650.3904

ANALYSIS OF VARIANCE

	SUM OF SQUARES	DF	MEAN SQUARE	F RATIO
REGRESSION	32073088.	2	0.16036544E 08	37.911
RESIDUAL	9306166.0	22	423007.5	

VARIABLES IN EQUATION

VARIABLE	COEFFICIENT	STD. ERROR OF COEFF	STD REG COEFF	F TO REMOVE	LEVEL
(Y-INTERCEPT	50.311)				
ADV 4	0.227	0.050	0.464	20.210	1
ACCTS 7	19.048	2.973	0.661	41.053	1

VARIABLES NOT IN EQUATION

VARIABLE		PARTIAL CORR.	TOLERANCE	F TO ENTER	LEVEL
TIME	2	0.02047	0.41579	0.009	1
POTEN	3	0.48372	0.76451	6.414	1
SHARE	5	0.22213	0.80238	1.090	1
CHANGE	6	0.23226	0.79203	1.197	1
WKLD	8	0.31166	0.90417	2.259	1
RATING	9	0.14034	0.80840	0.422	1

FIGURE 14.8   (continued)

STEP NO. 3
VARIABLE ENTERED   3 POTEN

MULTIPLE R   0.9098
MULTIPLE R-SQUARE   0.8277
STD. ERROR OF EST.   582.6335

ANALYSIS OF VARIANCE

	SUM OF SQUARES	DF	MEAN SQUARE	F RATIO
REGRESSION	34250544.	3	0.1141685E 08	33.632
RESIDUAL	7128700.0	21	339461.9	

VARIABLES IN EQUATION

VARIABLE	COEFFICIENT	STD. ERROR OF COEFF	STD REG COEFF	F TO REMOVE	LEVEL
(Y-INTERCEPT	-327.222)				
POTEN 3	0.022	0.009	0.262	6.414	1
ADV 4	0.216	0.045	0.443	22.723	1
ACCTS 7	15.554	2.999	0.540	26.892	1

VARIABLES NOT IN EQUATION

VARIABLE		PARTIAL CORR.	TOLERANCE	F TO ENTER	LEVEL
TIME	2	-0.05889	0.40672	0.070	1
SHARE	5	0.64973	0.57320	14.611	1
CHANGE	6	0.20542	0.78232	0.881	1
WKLD	8	0.45929	0.87679	5.347	1
RATING	9	0.00633	0.74542	0.001	1

STEP NO. 4
VARIABLE ENTERED   5 SHARE

MULTIPLE R   0.9489
MULTIPLE R-SQUARE   0.9005
STD. ERROR OF EST.   453.8337

ANALYSIS OF VARIANCE

	SUM OF SQUARES	DF	MEAN SQUARE	F RATIO
REGRESSION	37259952.	4	9314988.	45.226.
RESIDUAL	4119300.0	20	205965.0	

VARIABLES IN EQUATION

VARIABLE	COEFFICIENT	STD. ERROR OF COEFF	STD REG COEFF	F TO REMOVE	LEVEL
(Y-INTERCEPT	-1441.916)				
POTEN 3	0.038	0.008	0.457	22.954	1
ADV 4	0.175	0.037	0.358	22.481	1
SHARE 5	190.144	49.744	0.356	14.611	1
ACCTS 7	9.214	2.865	0.320	10.341	1

VARIABLES NOT IN EQUATION

VARIABLE		PARTIAL CORR.	TOLERANCE	F TO ENTER	LEVEL
TIME	2	0.24355	0.35494	1.198	1
CHANGE	6	0.34670	0.77627	2.596	1
WKLD	8	0.15438	0.60489	0.464	1
RATING	9	0.09459	0.73791	0.172	1

STEPWISE REGRESSION COEFFICIENTS

VARIABLES	0 Y-INTERCEPT	2 TIME	3 POTEN	4 ADV	5 SHARE	6 CHANGE	7 ACCTS	8 WKLD	9 RATING
STEP 0	3374.5645*	9.4235	0.0500	0.2911	258.1038	1034.2598	21.7217	-41.9408	550.1375
1	709.3284*	1.8314	0.0257	0.2265	114.4932	573.8323	21.7217*	12.1821	331.5022
2	50.3105*	0.2278	0.0219	0.2265*	62.7760	261.6736	19.0482*	55.6116	101.3282
3	-327.2222*	-0.5798	0.0219*	0.2161*	190.1441	203.8129	15.5539*	72.8388	4.1688
4	-1441.9160*	1.9512	0.0382*	0.1750*	190.1441*	262.4990	9.2139*	22.4066	47.5583

NOTE—
1) REGRESSION COEFFICIENTS FOR VARIABLES IN THE EQUATION ARE INDICATED BY AN ASTERISK
2) THE REMAINING COEFFICIENTS ARE THOSE WHICH WOULD BE OBTAINED IF THAT VARIABLE WERE TO ENTER IN THE NEXT STEP

SUMMARY TABLE

STEP NO.	VARIABLE ENTERED	REMOVED	MULTIPLE R	RSQ	INCREASE IN RSQ	F-TO-ENTER	F-TO-REMOVE	NUMBER OF INDEPENDENT VARIABLES INCLUDED
1	7 ACCTS		0.7540	0.5685	0.5685	30.3018		1
2	4 ADV		0.8804	0.7751	0.2066	20.2104		2
3	3 POTEN		0.9098	0.8277	0.0526	6.4145		3
4	5 SHARE		0.9489	0.9005	0.0727	14.6112		4

Default Conditions

Stepping
Algorithm

Prior to the actual analysis, we see some information concerning the conditions used in executing the program. The Stepping Algorithm, denoted by F, indicates that variables are entered into or removed from the model according to the sizes of their partial *F*-statistics. From

MINIMUM ACCEPTABLE F TO ENTER = 4.000,   4.000,

we see that unless a variable has partial *F* of 4, given the other variables already in the model, it will never enter into consideration. This criterion is used because most partial *F*'s must be greater than 3 to be significant. (See *F*-tables.) For

MAXIMUM ACCEPTABLE F TO REMOVE = 3.9,

we see that unless a variable has a partial *F* greater than 3.9, given the other variables in the model, it will not be permitted to remain in. We will see later how we can change these default conditions to obtain different selection procedures for entry and removal of the variables.

TOLERANCE

MINIMUM ACCEPTABLE TOLERANCE = 0.01 is included as a protection against multicollinearity. If the multicollinearity in a problem is too severe, $\mathbf{X'X}$ will be nearly singular and the estimates of the regression coefficients will be quite unreliable. Tolerance is the proportion of variation in the independent variable not explained by the variables already in the model. A tolerance of zero means that a predictor under consideration is a perfect linear combination of variables already in the model, and a tolerance of 1 means that this predictor is independent of the others already entered. Unless at least 1 % of the variation in the response variable remains unexplained by predictors already included, the predictor under consideration will not be allowed to enter into the model. This default condition on tolerance allows quite a bit of redundancy among predictors. However, one can change the default condition, by adding a sentence such as TØLERANCE = .05. to the REGRESSIØN paragraph. One might want to require a higher tolerance than that specified by the default condition, but it is not usually advisable to specify a tolerance less than 0.01. We saw earlier an instance of a variable not being allowed to enter because it failed the tolerance test. In Figure 4.9, the quadratic term in a third-degree polynomial regression, using BMDP1R, was summarily dropped from consideration.

Stepwise Analysis of CRAVENS Data

Let us now look at the steps in the stepwise regression analysis performed by the BMDP2R. At first, we see Step 0, in which no predictor variables are used. There is no multiple coefficient of determination, and no regression sum of squares. At this step, only the dependent variable is analyzed. The standard error of estimate is just the standard deviation of the *Y*-values, 1313.0637. The residual mean square is the variance of the *Y*-values: $(1313.0637)^2 = 1724,136$. Under VARIABLES IN EQUATION, the *Y*-intercept coefficient ($\bar{Y}$) is 3374.564.

Things get more interesting when we look under VARIABLES NOT IN EQUATION. Of course, all predictors are listed here. The variable names and numbers are listed. The partial correlations in this case are just the simple correlations of the various predictors with Y. (See the correlation matrix in Figure 14.8.) All tolerances are 1, since no predictors have been entered. In the last column, the levels are all 1, to indicate that all eight potential predictors are allowed to compete on equal footing (we are not forcing any variable into the equation). Under F TO ENTER we see what the F-values in the simple regression analysis would be for each variable, were it allowed to predict Y. Since the largest F to enter is 30.302, corresponding to ACCTS, and since the stepping algorithm is to enter the variable with the largest partial $F \geq 4$ having tolerance at least 0.01, a simple regression using ACCTS to predict SALES will be performed in Step 1. (Note that ACCTS also has the largest correlation with SALES.)

So, in Step 1, we see the simple regression analysis using ACCTS as a predictor. The overall regression significance test is given in the analysis of variance table, and under VARIABLES IN THE EQUATION we see the regression constant, regression coefficient, standard error of the regression coefficient, standardized regression (beta) coefficient, and partial F-statistic. (Only in Step 1, the partial F is just a plain old F, since there is only one predictor in the equation.) This partial F is labeled F TO REMOVE. If it were not larger than 3.9, then the predictor ACCTS would be removed from the equation.

Under VARIABLES NOT IN EQUATION are listed the names and numbers of the remaining candidates for entry into the model. Now the partial correlations computed are conditional on ACCTS being used as a predictor. For example, the partial correlation for ADV is 0.69196. This is $r_{4Y \cdot 7}$, where SALES $= Y = X_1$, ADV $= X_2$, and ACCTS $= X_7$. This first-order partial correlation can be calculated from quantities given in the correlation matrix in Figure 14.8:

$$0.69196 = r_{4Y \cdot 7} = \frac{r_{4Y} - r_{47}r_{7Y}}{\sqrt{(1 - r_{47}^2)(1 - r_{7Y}^2)}}$$

$$= \frac{0.596 - 0.2(0.754)}{\sqrt{[1 - (0.2)^2][1 - (0.754)^2]}} = \frac{0.4452}{0.6436} = 0.6917 \approx 0.69195.$$

The F-levels to enter in Step 1 are partial F's, conditional on ACCTS being used as a predictor. Since the largest partial F is that for ADV, ADV enters as a predictor in Step 2.

In Step 2 we again see the overall test and individual tests on the predictors. Since both F's-to-remove are greater than 3.9, both ADV and ACCTS will stay in the equation. Under VARIABLES NOT IN EQUATION, POTEN has the largest partial F, given ACCTS and ADV, and this partial F is greater than 4, the minimum F for entry. Thus, POTEN will enter in Step 3. The partial correlations given in Step 2 are second-order partials. For example, the partial correlation for POTEN (variable 3) is

$$r_{3Y \cdot 47} = 0.48372 = \frac{r_{3Y \cdot 7} - r_{4Y \cdot 7}r_{34 \cdot 7}}{\sqrt{(1 - r_{4Y \cdot 7}^2)(1 - r_{34 \cdot 7}^2)}}.$$

However, since we are not given a complete list of first-order partial correlations, we cannot compute these from values on the printout.

The procedure continues until we reach Step 4. We see there that four variables have entered, all of which had F-to-enter-values at least 4 and tolerances at least 0.01. At no point did any F-to-remove for any of these four variables fall below 3.9, so no variables were removed. Under VARIABLES NOT IN EQUATION, although all tolerances were greater than 0.01, none of the remaining four variables had partial F's-to-enter of 4 or more, so no more predictors were added to the model and the stepping process terminated.

Summary Tables

Finally, two summary tables are provided. The first shows what the regression coefficients for the predictors would be if that predictor had been added at the next step. This can be confusing, so consider Step 0. The table says that if, for example, ACCTS had entered in Step 1 (which it did), then the regression coefficient for ACCTS would be 21.7217. Looking back to Step 1, we see that, indeed the regression coefficient for ACCTS was 21.722. On the other hand, if you had wanted to use a simple regression predicting SALES from WKLD, then the regression coefficient on WKLD would have been −41.9408. Under Step 1, we see that the intercept and ACCTS were used (indicated by asterisks). If you wanted to add WKLD as a second predictor, then its partial (net) regression coefficient, given ACCTS, would be 12.1821. This does not say, however, that the best regression equation using ACCTS and WKLD would be $\hat{Y} = 709.3284 + 21.7217X_7 + 12.1821X_8$, because we know that adding X_8(WKLD) to the equation will change the intercept and will change the coefficient on X_7(ACCTS). We can, however, use this table to pick out the intercept and net regression coefficients for the equation appropriate at each step by looking for the values indicated by asterisks.

The second summary table shows which variables were entered and removed at each step, what the r_{xy}^2- and F-values were, and how many predictors were used.

We note that this stepwise analysis did not give the same steps as did the SAS stepwise analysis. Recall that in the SAS analysis (Figure 14.3), after SHARE had entered, then on the fifth step, SHCHANGE entered. It was removed, however, on the sixth step because its F-value was only 2.6, with an associated probability of greater than 0.10. In the BMD analysis, the default condition is that F must be at least 4 before a variable will enter. Since the partial F for SHCHANGE is less than 4 (see Step 4), this program never allowed SHCHANGE to enter, and the process terminated without ever having to remove any variable. It can be argued that the default condition for F-to-enter using the BMDP2R is too restrictive, in that it doesn't give some variables any chance to enter and prove their worth. It is easy to change the condition on F to enter: simply add ENTER = 2.0. to the REGRESSIØN paragraph, and the output from this program will look exactly like that from the SAS analysis. (It is equally easy to change the F-to-remove condition, as we'll soon see.)

Figure 14.9 shows a portion of the analysis of residuals for this problem. A printout of the data, predicted values, and residuals takes two pages, because we entered so many variables, so it is not reproduced here. Finally, we see plots of the

FIGURE 14.9   *Partial Analysis of Residuals from BMDP2R*

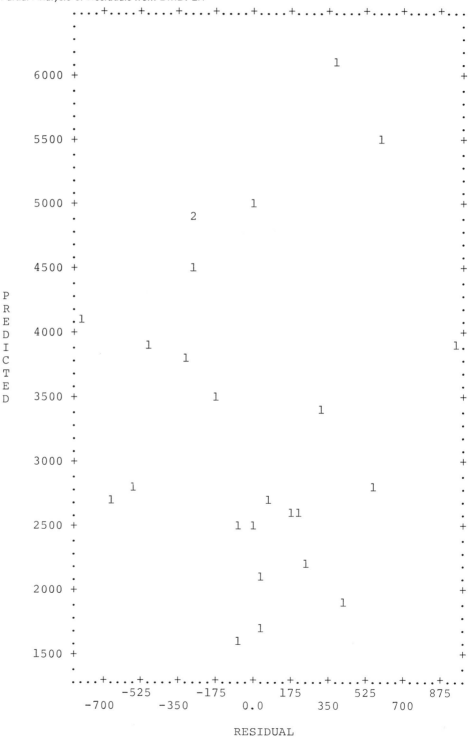

predicted values, $\hat{Y}$, from the variables in the final step of the analysis, versus (1) the residuals, and (2) the squares of the residuals (not shown). Note that these plots are sideways from the plots obtained using the BMDP1R to perform the general regression significance test: in that program, the residuals were plotted on the vertical axis and the predicted values on the horizontal axis. These plots are still useful in detecting outliers or patterns to the residuals and thus determining whether or not the assumptions of the model have been met.

The first graph in Figure 14.10 shows the Y and $\hat{Y}$ values plotted against ACCTS, the one independent variable selected. Note that all the points P do not lie on a straight line. Recall that the multiple regression equation obtained in the final step, using four predictors, is the equation of a four-dimensional hyperplane in five-dimensional space. The plot we are looking at is a projection of some of the points of this "plane" onto the plane formed by one of the predictor variables (ACCTS) and the dependent variable (SALES). (This might be easier to picture if you look at an object in the room in which you are sitting, pick some points on the object, and imagine those points splattering onto the wall behind the object. This is an example of the projection of a three-dimensional object onto a two-dimensional plane.)

The second graph in Figure 14.10 shows the residuals plotted against ACCTS. It is this plot that should be examined in order to determine whether or not the assumptions of the model hold with respect to this predictor. A more complete analysis would include plots of the residuals against the other three predictors in the equation at the final step.

FORWARD SELECTION

By changing the default condition on the maximum F-level to remove, we can make the BMDP2R perform a forward selection procedure, in which variables are entered one at a time, according to which one gives the largest partial F-value. The process will terminate when no variable which is not yet in the equation has a partial F greater than or equal to the F-level for entry, or when no variable not in the equation passes the tolerance test. There is no provision for removing a variable should its partial F become small when new predictors are added.

To prevent BMDP2R from removing any variable once it has entered into the equation, all we need to do is add the sentence REMØVE $= 0$. to the REGRESSIØN paragraph in Figure 14.7. Since F cannot be less than zero, no variable can be removed once it has been allowed to enter.

If the CRAVENS data were reanalyzed using BMDP2R with the default condition (4.0) on the F-to-enter and maximum F-to-remove set at zero, the analysis would look just like that in Figure 14.8, since in Figure 14.8 no variable with F-to-enter of 4.0 or more subsequently had a partial F less than 3.9. Had the F-to-enter value been lowered to 2 or less, more variables would have entered, but none would have been removed.

FIGURE 14.10   *Plots of Y, Ŷ, and Residuals against ACCTS (Part 1)*

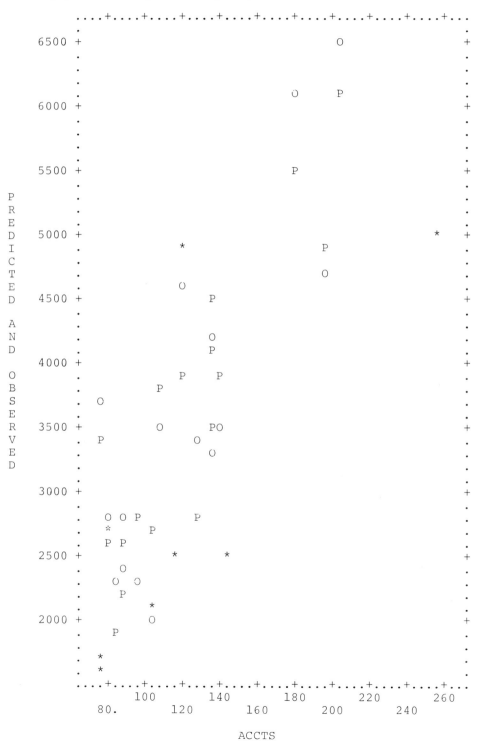

FIGURE 14.10   *(Part 2)*

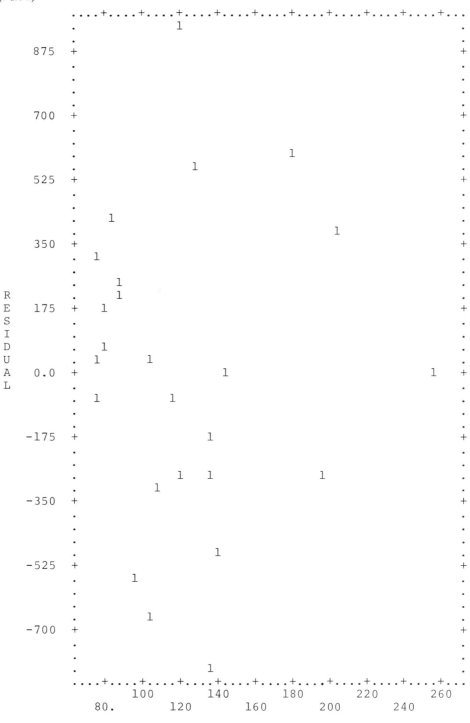

BACKWARD ELIMINATION

It is possible to perform a backward elimination procedure with the BMDP2R, but it is rather laborious, at least for the computer. We must first step in every variable. This is accomplished by specifying a very low F-to-enter and an F-to-remove of zero. Once all predictors have entered, the second set of F-to-enter and F-to-remove values come into play. If these are set at levels such as 4.0 and 4.0, then the procedure will backward step, removing variables one at a time until no remaining variable has an F-to-remove less than the second specified value.

Backward Stepwise for CRAVENS Data

Figure 14.11 shows the BMDP2R backward stepwise procedure beginning with Step 8, in which eight predictors have been entered, because the first F-to-enter value was set at 0.001 and the first F-to-remove value specified was 0. Then in Step 9, RATING was removed because its partial F was less than 4, which was the second F-to-remove specified. In subsequent steps, WKLD and ACCTS were removed, for the same reason. Finally, in Step 11, all partial F's are greater than 4, so the process terminates.

Note that whereas we obtained the same final model here as with the SAS backward elimination procedure, this BMD backward selection is actually a *stepwise* backward procedure. That is, if any variable not in the equation once the program began stepping backward had a partial F greater than 4 (the second F-to-enter value), then it could be reentered. The SAS backward elimination procedure did not allow any variable to enter again once it was removed.

When running a backward stepwise procedure using this program, you may want to change the default condition on the maximum number of steps allowed. Unless otherwise specified, the program will stop after $2k$ steps, where k is the number of variables. This might not be sufficient to fully analyze a problem using the backward procedure. Our example specified 25 as the maximum number of steps, instead of 18. This was accomplished by adding a sentence

STEPS = 25.

in the REGRESSI∅N paragraph.

OTHER SELECTION PROCEDURES AVAILABLE WITH BMDP2R

By adjusting the F-values to enter and remove, we have seen that the standard BMD stepwise selection program can be made to perform both forward selection and backward elimination. There are, however, other variable-selection criteria that can be employed with the BMDP2R.

F Method   The default condition on the METHOD of selection is called the F method in the BMD manual. It enters the variable with the largest F-to-enter greater than the specified minimum acceptable. Then, should any F-to-remove values fall below the

FIGURE 14.11   BMDP2R Backward Elimination Procedure

STEP NO. 8
VARIABLE ENTERED   9 RATING

MULTIPLE R	0.9602
MULTIPLE R-SQUARE	0.9220
STD. ERROR OF EST.	449.0247

ANALYSIS OF VARIANCE

	SUM OF SQUARES	DF	MEAN SQUARE	F RATIO
REGRESSION	38153280.	8	4769160.	23.654
RESIDUAL	3225972.0	16	201623.3	

VARIABLES IN EQUATION

VARIABLE	COEFFICIENT	STD. ERROR OF COEFF	STD REG COEFF	F TO REMOVE	LEVEL
(Y-INTERCEPT	-1507.787)				
TIME 2	2.010	1.931	0.133	1.083	1
POTEN 3	0.037	0.008	0.445	20.575	1
ADV 4	0.151	0.047	0.309	10.273	1
SHARE 5	199.024	67.027	0.373	8.817	1
CHANGE 6	290.851	186.782	0.138	2.425	1
ACCTS 7	5.551	4.776	0.193	1.351	1
WKLD 8	19.793	33.676	0.055	0.345	1
RATING 9	8.191	128.506	0.006	0.004	1

VARIABLES NOT IN EQUATION

VARIABLE	PARTIAL CORR.	TOLERANCE	F TO ENTER	LEVEL

STEP NO. 9
VARIABLE REMOVED   9 RATING

MULTIPLE R	0.9602
MULTIPLE R-SQUARE	0.9220
STD. ERROR OF EST.	435.6733

ANALYSIS OF VARIANCE

	SUM OF SQUARES	DF	MEAN SQUARE	F RATIO
REGRESSION	38152464.	7	5450352.	28.715
RESIDUAL	3226791.0	17	189811.2	

VARIABLES IN EQUATION

VARIABLE	COEFFICIENT	STD. ERROR OF COEFF	STD REG COEFF	F TO REMOVE	LEVEL
(Y-INTERCEPT	-1485.846)				
TIME 2	1.974	1.796	0.131	1.209	1
POTEN 3	0.037	0.008	0.446	22.560	1
ADV 4	0.152	0.043	0.311	12.348	1
SHARE 5	198.309	64.117	0.371	9.566	1
CHANGE 6	295.863	164.386	0.140	3.239	1
ACCTS 7	5.610	4.545	0.195	1.524	1
WKLD 8	19.898	32.636	0.056	0.372	1

VARIABLES NOT IN EQUATION

VARIABLE	PARTIAL CORR.	TOLERANCE	F TO ENTER	LEVEL
RATING 9	0.01593	0.55293	0.004	1

STEP NO. 10
VARIABLE REMOVED   8 WKLD
MULTIPLE R   0.9593
MULTIPLE R-SQUARE   0.9203
STD. ERROR OF EST.   428.0024

ANALYSIS OF VARIANCE

	SUM OF SQUARES	DF	MEAN SQUARE	F RATIO
REGRESSION	38081904.	6	6346984.	34.648
RESIDUAL	3297348.0	18	183186.0	

VARIABLES IN EQUATION

VARIABLE		COEFFICIENT	STD. ERROR OF COEFF	STD REG COEFF	F TO REMOVE	LEVEL
(Y-INTERCEPT		−1165.467)				
TIME	2	2.269	1.699	0.150	1.784	1
POTEN	3	0.038	0.008	0.458	25.725	1
ADV	4	0.141	0.038	0.288	13.425	1
SHARE	5	221.603	50.583	0.415	19.193	1
CHANGE	6	285.106	160.559	0.135	3.153	1
ACCTS	7	4.378	3.999	0.152	1.198	1

VARIABLES NOT IN EQUATION

VARIABLE		PARTIAL CORR.	TOLERANCE	F TO ENTER	LEVEL
WKLD	8	0.14628	0.55128	0.372	1
RATING	9	0.02291	0.55426	0.009	1

STEP NO. 11
VARIABLE REMOVED   7 ACCTS
MULTIPLE R   0.9566
MULTIPLE R-SQUARE   0.9150
STD. ERROR OF EST.   430.2314

ANALYSIS OF VARIANCE

	SUM OF SQUARES	DF	MEAN SQUARE	F RATIO
REGRESSION	37862368.	5	7572473.	40.910
RESIDUAL	3516882.0	19	185099.0	

VARIABLES IN EQUATION

VARIABLE		COEFFICIENT	STD. ERROR OF COEFF	STD REG COEFF	F TO REMOVE	LEVEL
(Y-INTERCEPT		−1113.771)				
TIME	2	3.612	1.182	0.239	9.343	1
POTEN	3	0.042	0.007	0.504	39.096	1
ADV	4	0.129	0.037	0.264	12.105	1
SHARE	5	256.955	39.136	0.481	43.108	1
CHANGE	6	324.531	157.283	0.153	4.257	1

VARIABLES NOT IN EQUATION

VARIABLE		PARTIAL CORR.	TOLERANCE	F TO ENTER	LEVEL
ACCTS	7	0.24985	0.22975	1.198	1
WKLD	8	0.01574	0.68721	0.004	1
RATING	9	0.07004	0.57575	0.089	1

specified maximum F-to-remove, the variable having the smallest F is removed. If, for example, five variables were entered at Step 5 and one of them had an F-to-remove less than the maximum specified, then it would be removed and there would be only four variables in the equation at Step 6.

FSWAP Method

FSWAP   A variation on this theme can be achieved by specifying METHØD = FSWAP. in the REGRESSIØN paragraph of the program. The selection procedure will basically be the same as the F method, except that at any step the program looks to see if any variable in the equation could be exchanged with any not in the equation to produce a larger multiple r_{xy}^2. If such an exchange can be made, it is done. Figure 14.12 shows the summary table from the CRAVENS data analyzed by the FSWAP method, with default conditions on the F-values to enter and remove. Compare it with the results shown in Figure 14.8. The first three steps in both analyses are identical. However, in the FSWAP procedure, instead of adding a fourth variable, SHARE, in Step 4, the program exchanges ACCTS for SHARE. Step 4 then uses three variables, PØTEN, ADV, and SHARE, for which $r_{xy}^2 = 0.8490$. Note that r_{xy}^2 for these three variables is larger than for the three-predictor model using PØTEN, ADV, and ACCTS; $r_{xy}^2 = 0.8277$. The procedure then finds a four-predictor equation in Step 5 and then terminates, because no more variables can be added or removed. A summary table is provided for the purpose of comparison with F procedure. While the same four-predictor model was obtained by both the F and FSWAP methods for the CRAVENS data, one can appreciate the fact that this will not always happen.

Note that the FSWAP method is not the same as the SAS maximum R^2 improvement method. In the SAS procedure, variables were added regardless of whether or not they had reasonably large F-values and they were not necessarily removed if their partial F's became small. Here, the procedure still wants to consider only those variables that have a chance of being significant.

R Method

R Method   A third procedure available through BMDP2R, called the R method, can be used by simply specifying METHØD = R. in the REGRESSIØN paragraph. Here, variables are entered in the usual way, according to which has the largest F-to-enter greater than the specified minimum. The removal criterion, however, is a bit different. To quote from the manual BMD (p. 505), "If two or more variables are in the regression equation, the one with the smallest F value is removed if its removal results in a larger multiple R than was obtained for the same number of variables previously." Let's see what happens when the CRAVENS data are analyzed by this procedure.

Figure 14.13 starts out looking like the regular stepwise procedure through Step 4. Then on Step 5, ACCTS is removed, leaving the three-predictor model using ADV, PØTEN, and SHARE. Note that for this model, $r_{xy}^2 = 0.8490$, but for the previously found three-predictor model (the one using ACCTS, ADV, and PØTEN) the multiple coefficient of determination was 0.8277.

FIGURE 14.12   Summary Table from BMDP2R FSWAP Method for CRAVENS Data

SUMMARY TABLE

STEP NO.	VARIABLE ENTERED	VARIABLE REMOVED	MULTIPLE R	RSQ	INCREASE IN RSQ	F-TO-ENTER	F-TO-REMOVE	NUMBER OF INDEPENDENT VARIABLES INCLUDED
1	7 ACCTS		0.7540	0.5685	0.5685	30.3018		1
2	4 ADV		0.8804	0.7751	0.2066	20.2104		2
3	3 POTEN		0.9098	0.8277	0.0526	6.4145		3
4	5 SHARE	7 ACCTS	0.9214	0.8490	0.0213	33.6315		3
5	7 ACCTS		0.9489	0.9005	0.0515	10.3413	10.3413	4

FIGURE 14.13   Summary Table from BMDP2R Analysis of CRAVENS Data Using R Method

SUMMARY TABLE

STEP NO.	VARIABLE ENTERED	VARIABLE REMOVED	MULTIPLE R	RSQ	INCREASE IN RSQ	F-TO-ENTER	F-TO-REMOVE	NUMBER OF INDEPENDENT VARIABLES INCLUDED
1	7 ACCTS		0.7540	0.5685	0.5685	30.3018		1
2	4 ADV		0.8804	0.7751	0.2066	20.2104		2
3	3 POTEN		0.9098	0.8277	0.0526	6.4145		3
4	5 SHARE		0.9489	0.9005	0.0727	14.6112		4
5		7 ACCTS	0.9214	0.8490	-0.0515		10.3413	3
6	7 ACCTS		0.9489	0.9005	0.0515	10.3413		4

This procedure is similar to the FSWAP, except that no variable will be removed only because its partial F becomes small. It will be removed if its partial F is the smallest of all in the equation *and* if its removal results in a multiple r_{xy}^2 larger than the one previously found for the same number of predictors.

RSWAP Method

RSWAP Finally, the BMDP2R provides the RSWAP method. Again, variables enter according to their F-to-enter values. As in the R method, they are removed if their removal results in a larger r_{xy}^2 than previously obtained for the resulting number of predictors. In addition, as in the FSWAP method, instead of removing a variable so that the number of predictors is reduced, any variable removed is exchanged for another. Figure 14.14 shows the RSWAP analysis for the CRAVENS data. In Step 5, SHARE is exchanged for ACCTS because the model using ADV, PØTEN, and SHARE has a larger r_{xy}^2 than does the model using ADV, PØTEN, and ACCTS.

Comparison of the Four Methods

Since all four procedures resulted in the same four-variable model at the final step (using minimum F-to-enter = 4.0 and maximum F-to-remove = 3.9), it may at first be hard to see how these four methods differ. It might help to think of them as combinations of just two principles:

1. Is a variable removed because its partial F becomes small (F method) or because its removal will increase the multiple coefficient of determination over that previously obtained for the resulting number of variables remaining (R method)?
2. When a variable is removed, does this result in fewer predictors being considered (regular method) or the same number (SWAP method)?

One can easily imagine that, depending upon the data being analyzed and the pattern of multicollinearity present, these four methods can give quite different final models.

SUMMARY OF BMDP2R STAGEWISE PROCEDURES

By now we can appreciate that there are just about as many criteria for determining what subset of variables to use as predictors as there are people trying to decide. The typically used selection procedures are the forward selection, backward elimination, and stepwise procedures, and both BMDP2R and SAS programs can perform all three. Variations on these themes can be provided with either program. The criteria for selection can rest either on the significance of the partial F-statistic or on the amount that the multiple coefficient of determination increases when a variable is to be added to the candidate list of predictors. Variables may be added and deleted one at a time or may be exchanged for other variables.

FIGURE 14.14   Summary Table from BMDP2R Analysis of CRAVENS Data Using RSWAP

SUMMARY TABLE

STEP NO.	VARIABLE ENTERED	VARIABLE REMOVED	MULTIPLE R	MULTIPLE RSQ	INCREASE IN RSQ	F-TO-ENTER	F-TO-REMOVE	NUMBER OF INDEPENDENT VARIABLES INCLUDED
1	7 ACCTS		0.7540	0.5685	0.5685	30.3018		1
2	4 ADV		0.8804	0.7751	0.2066	20.2104		2
3	3 POTEN		0.9098	0.8277	0.0526	6.4145		3
4	5 SHARE	7 ACCTS	0.9214	0.8490	0.0213	33.6315		3
5	7 ACCTS		0.9489	0.9005	0.0515	10.3413	10.3413	4

Since the final set of predictors is often determined by the selection criteria, there can be no single, universal best model for most sets of data, when variables are allowed to stand or fall on their own predictive ability when used in conjunction with other predictors. We will look at some practical matters in Section 14.5, after seeing what stagewise procedures SPSS has to offer.

EXERCISES   *Section 14.3*

p. 533   *14.3*   Recall that in the analysis of the CHILDREN data set, the overall regression significance test indicated that at least one of the predictors (AGE, HEIGHT, or WEIGHT) was significant. However, no partial F-test on one predictor, given the other two, indicated a significant predictor. Analyze the data using the BMDP2R stepwise procedure with default conditions on the method and the F-levels for inclusion and removal. See if any significant predictors can be found.

p. 536   *14.4*   Using the TENNESSEE data, run the BMDP2R stepwise regression analysis with F-to-enter of 1.00 and F-to-remove of 1.00. Use all four methods (F, FSWAP, R, and RSWAP) to predict percent below poverty level from the other eight variables. You might also want to obtain a correlation matrix and other pertinent information. Propose a final model to be used, and justify your choice.

14.4 THE SPSS SELECTION PROCEDURES

The only stagewise procedure available through the SPSS REGRESSI∅N routine is the forward selection procedure. Only one change is necessary to turn the general regression analysis, Figure 13.4, into a forward selection analysis. Change the 2 in parentheses in the regression model statement to any *odd* number between 1 and 99. Figure 14.15 shows the analysis resulting from this change.

THE SPSS FORWARD SELECTION PROCEDURE

In the various steps of the forward selection procedure, the variables are entered in the same order they were in the SAS forward selection procedure. However, since the default conditions in the SPSS program are more generous than those in the SAS program, its forward selection procedure goes one step further. The default conditions for SPSS are: the maximum number of independent variables which will be entered is 80, the tolerance is 0.001, and the F-value for inclusion is 0.01. Thus, as long as some variable not in the equation has tolerance greater than 0.001 and a partial F-value greater than 0.01, the program will continue. (Recall that SAS required F to be significant at $\alpha = 0.50$ before it could enter, which is much less lenient than SPSS.)

FIGURE 14.15   SPSS Forward Selection Procedure for CRAVENS Data

DEPENDENT VARIABLE.. SALES

VARIABLE(S) ENTERED ON STEP NUMBER 1.. ACCTS

MULTIPLE R	0.75399
R SQUARE	0.56850
ADJUSTED R SQUARE	0.54973
STANDARD ERROR	881.09256

ANALYSIS OF VARIANCE

	DF	SUM OF SQUARES	MEAN SQUARE	F
REGRESSION	1.	23524045.16953	23524045.16953	30.30184
RESIDUAL	23.	17855454.18311	776324.09492	

VARIABLES IN THE EQUATION

VARIABLE	B	BETA	STD ERROR B	F
ACCTS	21.72176	0.75399	3.94603	30.302
(CONSTANT)	709.32419			

VARIABLES NOT IN THE EQUATION

VARIABLE	BETA IN	PARTIAL	TOLERANCE	F
TIME	0.12106	0.12025	0.42572	0.323
POTEN	0.30733	0.41078	0.77090	4.466
ADV	0.46392	0.69196	0.95998	20.211
SHARE	0.21448	0.29882	0.83758	2.157
SHCHANGE	0.27141	0.39040	0.89280	3.956
WKLD	0.03405	0.05080	0.96046	0.057
RATE	0.24216	0.35889	0.94774	3.253

VARIABLE(S) ENTERED ON STEP NUMBER 2.. ADV

MULTIPLE R	0.88040
R SQUARE	0.77510
ADJUSTED R SQUARE	0.75466
STANDARD ERROR	650.39199

ANALYSIS OF VARIANCE

	DF	SUM OF SQUARES	MEAN SQUARE	F
REGRESSION	2.	32073284.91724	16036642.45862	37.91081
RESIDUAL	22.	9306214.43540	423009.74706	

VARIABLES IN THE EQUATION

VARIABLE	B	BETA	STD ERROR B	F
ACCTS	19.04824	0.66119	2.97291	41.053
ADV	0.22653	0.46392	0.05039	20.211
(CONSTANT)	50.29995			

VARIABLES NOT IN THE EQUATION

VARIABLE	BETA IN	PARTIAL	TOLERANCE	F
TIME	0.01506	0.02048	0.41579	0.009
POTEN	0.26236	0.48372	0.76451	6.414
SHARE	0.11760	0.22212	0.80237	1.090
SHCHANGE	0.12377	0.23226	0.79203	1.197
WKLD	0.15544	0.31167	0.90417	2.259
RATE	0.07402	0.14034	0.80840	0.422

VARIABLE(S) ENTERED ON STEP NUMBER 3.. POTEN

MULTIPLE R	0.90979
R SQUARE	0.82772
ADJUSTED R SQUARE	0.80311
STANDARD ERROR	582.63520

ANALYSIS OF VARIANCE

	DF	SUM OF SQUARES	MEAN SQUARE	F
REGRESSION	3.	34250759.99633	11416919.99878	33.63222
RESIDUAL	21.	7128739.35631	339463.77887	

VARIABLES IN THE EQUATION

VARIABLE	B	BETA	STD ERROR B	F
ACCTS	15.55391	0.53989	2.99937	26.892
ADV	0.21607	0.44250	0.04533	22.723
POTEN	0.02192	0.26236	0.00866	6.414
(CONSTANT)	-327.23235			

VARIABLES NOT IN THE EQUATION

VARIABLE	BETA IN	PARTIAL	TOLERANCE	F
TIME	-0.03833	-0.05889	0.40672	0.070
SHARE	0.35620	0.64973	0.57320	14.611
SHCHANGE	0.09640	0.20542	0.78232	0.881
WKLD	0.20359	0.45930	0.87679	5.347
RATE	0.00304	0.00633	0.74542	0.001

MULTIPLE R	0.94892	
R SQUARE	0.90045	
ADJUSTED R SQUARE	0.88054	
STANDARD ERROR	453.83589	

ANALYSIS OF VARIANCE

	DF	SUM OF SQUARES	MEAN SQUARE	F
REGRESSION	4.	37260159.03965	9315039.75991	45.22588
RESIDUAL	20.	4119340.31300	205967.01565	

VARIABLE(S) ENTERED ON STEP NUMBER 4.. SHARE

VARIABLES IN THE EQUATION

VARIABLE	B	BETA	STD ERROR B	F
ACCTS	9.21389	0.31982	2.86521	10.341
ADV	0.17499	0.35837	0.03691	22.481
POTEN	0.03822	0.45738	0.00798	22.954
SHARE	190.14409	0.35620	49.74412	14.611
(CONSTANT)	-1441.92922			

VARIABLES NOT IN THE EQUATION

VARIABLE	BETA IN	PARTIAL	TOLERANCE	F
TIME	0.12899	0.24355	0.35494	1.198
SHCHANGE	0.12416	0.34670	0.77627	2.596
WKLD	0.06263	0.15439	0.60489	0.464
RATE	0.03474	0.09458	0.73791	0.172

MULTIPLE R	0.95520
R SQUARE	0.91242
ADJUSTED R SQUARE	0.88937
STANDARD ERROR	436.74582

ANALYSIS OF VARIANCE

	DF	SUM OF SQUARES	MEAN SQUARE	F
REGRESSION	5.	37755308.01261	7551061.60252	39.58681
RESIDUAL	19.	3624191.34004	190746.91263	

VARIABLE(S) ENTERED ON STEP NUMBER 5.. SHCHANGE

VARIABLES IN THE EQUATION

VARIABLE	B	BETA	STD ERROR B	F
ACCTS	8.23410	0.28581	2.82358	8.504
ADV	0.15444	0.31628	0.03774	16.747
POTEN	0.03763	0.45036	0.00769	23.977
SHARE	196.94933	0.36895	48.05689	16.796
SHCHANGE	262.50053	0.12416	162.92623	2.596
(CONSTANT)	-1285.94081			

VARIABLES NOT IN THE EQUATION

VARIABLE	BETA IN	PARTIAL	TOLERANCE	F
TIME	0.15001	0.30030	0.35099	1.784
WKLD	0.08262	0.21524	0.59436	0.874
RATE	-0.02433	-0.06381	0.60225	0.074

MULTIPLE R	0.95933
R SQUARE	0.92031
ADJUSTED R SQUARE	0.89375
STANDARD ERROR	428.00336

ANALYSIS OF VARIANCE

	DF	SUM OF SQUARES	MEAN SQUARE	F
REGRESSION	6.	38082135.65120	6347022.60853	34.64780
RESIDUAL	18.	3297363.70144	183186.87230	

VARIABLE(S) ENTERED ON STEP NUMBER 6.. TIME

FIGURE 14.15 *(continued)*

VARIABLES IN THE EQUATION

VARIABLE	B	BETA	STD ERROR B	F
ACCTS	4.37770	0.15195	3.99904	1.198
ADV	0.14067	0.28809	0.03839	13.425
POTEN	0.03828	0.45810	0.00755	25.725
SHARE	221.60445	0.41513	50.58306	19.193
SHCHANGE	285.10928	0.13485	160.55958	3.153
TIME	2.26935	0.15001	1.69898	1.784
(CONSTANT)	-1165.47617			

VARIABLES NOT IN THE EQUATION

VARIABLE	BETA IN	PARTIAL	TOLERANCE	F
WKLD	0.05562	0.14629	0.55127	0.372
RATE	0.00869	0.02291	0.55426	0.009

VARIABLE(S) ENTERED ON STEP NUMBER 7.. WKLD

MULTIPLE R	0.96022
R SQUARE	0.92202
ADJUSTED R SQUARE	0.88991
STANDARD ERROR	435.67385

ANALYSIS OF VARIANCE	DF	SUM OF SQUARES	MEAN SQUARE	F
REGRESSION	7.	38152700.39302	5450385.77043	28.71470
RESIDUAL	17.	3226798.95962	189811.70351	

VARIABLES IN THE EQUATION

VARIABLE	B	BETA	STD ERROR B	F
ACCTS	5.61018	0.19474	4.54496	1.524
ADV	0.15196	0.31121	0.04325	12.348
POTEN	0.03729	0.44629	0.00785	22.560
SHARE	198.30830	0.37149	64.11717	9.566
SHCHANGE	295.86606	0.13994	164.38649	3.239
TIME	1.97454	0.13052	1.79575	1.209
WKLD	19.89898	0.05562	32.63611	0.372
(CONSTANT)	-1485.87722			

VARIABLES NOT IN THE EQUATION

VARIABLE	BETA IN	PARTIAL	TOLERANCE	F
RATE	0.00598	0.01593	0.55293	0.004

F-LEVEL OR TOLERANCE-LEVEL INSUFFICIENT FOR FURTHER COMPUTATION

SUMMARY TABLE

VARIABLE	MULTIPLE R	R SQUARE	RSQ CHANGE	SIMPLE R	B	BETA
ACCTS	0.75399	0.56850	0.56850	0.75399	5.61018	0.19474
ADV	0.88040	0.77510	0.20661	0.59618	0.15196	0.31121
POTEN	0.90979	0.82772	0.05262	0.59781	0.03729	0.44629
SHARE	0.94892	0.90045	0.07273	0.48351	198.30830	0.37149
SHCHANGE.	0.95520	0.91242	0.01197	0.48918	295.86606	0.13994
TIME	0.95933	0.92031	0.00790	0.62292	1.97454	0.13052
WKLD	0.96022	0.92202	0.00171	-0.11722	19.89898	0.05562
(CONSTANT)					-1485.87722	

To change the F-value for inclusion, the tolerance level, and the maximum number of variables in the model, refer to the regression design statement in Figure 13.4. Recall that this card, which begins in column 16, starts off with

REGRESSIØN = SALES WITH TIME ...

The changes in the default condition are inserted in parentheses before the word WITH. For example, to restrict the model to four predictors, the F-level for inclusion to 1.5, and the tolerance to 0.01, we would write, beginning in column 16,

REGRESSIØN = SALES(4,1.5,.01) WITH TIME. . .

The order of the numbers within the parentheses must always be first the maximum number of predictors, then the F-level for inclusion, and then the tolerance level. Another thing to worry about: You can specify only the maximum number of predictors, as (4); or both the maximum number of predictors and the F-level, as (4,1.5); or all three. But you cannot specify the F-level without also specifying the maximum number of predictors and you cannot specify tolerance without also specifying the other two. If all you are really interested in is, say, tolerance, then you can specify the maximum number of predictors large (say 80, the default condition) and the F-level small (say 0.01, the default condition) and then specify tolerance as you want it.

There is not much that is surprising about the SPSS forward selection to one who already understands the discussion in Section 14.3. At each step, variables are listed in the order they are entered. The only quantities not encountered on previous stagewise programs or explained in earlier chapters are the values labeled <u>BETA IN</u>. These are the beta weights, or standardized regression coefficients, for the candidate variables if they are added at the next step. For example, we see BETA IN = 0.05562 for the variable WKLD in Step 6. Then, when WKLD is added in Step 7, its beta coefficient is this same value.

A summary table follows the last step of the procedure. While this table gives beta coefficients for the variable entered at each step, it does not give partial F-values. Note also that the regression coefficients and beta weights given in the summary table are those found at the *last step* in the forward selection procedure. About the only thing this table really summarizes is the order in which the variables entered and how much each one increased $r^2_{\dot{x}y}$.

The residuals calculated are based on the predictors in the equation at the last step. They are then standardized and plotted, the plot being to the right of the list. Were all variables in the last step considered important, this plot could be examined, along with the value of the Durbin-Watson statistic, to see if there is any serial correlation when the data are given in order of their collection. Finally, the standardized residuals are plotted against the standardized values of $\hat{Y}$ (see Section 13.3).

BETA IN

EXERCISES   *Section 14.4*

14.5 Recall that in the analysis of the CHILDREN data set, the overall regression significance test indicated that at least one of the predictors (age, height, or weight) was significant.

However, no partial F-test on one predictor, given the other two, indicated a significant predictor. Analyze the data using the SPSS forward selection procedure with default condition on the inclusion criterion. See if any significant predictor(s) can be found.

14.6 Using the TENNESSEE data, run the SPSS forward selection analysis with F-level for inclusion of 1.00. You might also want to obtain a correlation matrix and other pertinent information. Propose a final model to be used, and justify your choice.

14.5 FORCING VARIABLES INTO THE MODEL

In all of our stagewise procedures, we have let the variables themselves determine the order in which they will enter the model, if at all, based on their relative strengths as predictors. There are instances, however, when the researcher specifies that one or more particular variables are to be included in the model. This decision can be based on theoretical considerations (total revenue from sales must be a function of number of items sold, for example), or on practical considerations (it may be very easy and inexpensive to collect information on a given variable or set of variables). Any of the three stagewise programs can provide for the forced entry of variables. They do differ in what they do with the variables once they have been forced in, however.

Suppose that for some reason, the researcher analyzing the CRAVENS data really wants to include time with company as a variable. Perhaps this is a very quick and easy piece of information to obtain from company records. Perhaps some company official needs to know the effect of time with the company on an employee's sales record.

We will look at the SAS, BMD, and SPSS programs and see what modifications are necessary to force the variable X_1 (TIME) into the model in the CRAVENS data. Then we will consider its impact on the various stagewise procedures.

SAS

Refer back to the MØDEL statement in Figure 14.1. In order to force X_1 into the model, ahead of the semicolon punch

    INCLUDE = 1 .

This will force X_1 to be included in every model considered, regardless of what stagewise procedure is being run. Thus, X_1 can never be dropped out in a stepwise or backward elimination procedure or switched off for another variable in a maximum or minimum R^2 improvement procedure. That is, SAS itself will perform no significance test on any variable forced into the model, although the appropriate statistics will be printed.

Note that if the forcing-in statement had read

    INCLUDE = 3,

then variables X_1, X_2, *and* X_3 would all be forced in. That is,

INCLUDE = V

for some integer v, will force in the first v independent variables listed in the model statement. Thus, one must be careful to list first the predictors one wants to force in. With the model statement given in Figure 14.1, we could not force in X_8 (RATE) and X_6 (ACCTS). We would have to rewrite the model statement so that X_6 and X_8 were listed first and second, and add INCLUDE = 2 before the semicolon.

Forward Selection

Let us see how all five SAS stagewise procedures are affected by forcing X_1 (TIME). Figure 14.16 shows the forward selection procedure with X_1 (TIME) forced into the model. Note that before Step 1, is a message saying that the first variable in each model was forced in. In each step, TIME is the first variable listed, and TIME is the variable entered in Step 0. This predictor, however, ceases to be significant at Step 2, in which ADV and ACCTS have both entered the model. Note also that the significance level associated with TIME at Step 2 is 0.9261, which is considerably greater than $\alpha = 0.50$, the default condition for inclusion. In addition, forcing X_1 into the model alters the order in which X_3 (ADV) and X_6 (ACCTS) enter.

    If one must have a regression model including the predictor TIME, then the question of which other variables to use with it becomes even more difficult than before. The pair TIME and ADV are both conditionally significant, while any use of any additional variables causes TIME not to be significant at $\alpha = 0.05$. The pair TIME and ADV together explain only 59.5 % of the variation in SALES. One might argue that since TIME must be used, it really doesn't matter whether or not it is conditionally significant; one would thus use the model indicated in Step 3, where all three other variables are conditionally significant and $r^2_y = 82.8$ %. Let us see what other stagewise procedures yield.

Stepwise

Figure 14.17 shows a summary of the analysis with X_1 (TIME) forced into the model using the stepwise procedure, which starts off very much like the forward selection procedure. Note in Step 2 that although TIME becomes nonsignificant, it is not dropped from the model, as would ordinarily happen in a stepwise regression. Also, in Step 5 the variable ACCTS is removed, yielding a four-predictor model in which all four predictors, including TIME, are significant at $\alpha = 0.05$. The five predictors included at Step 6 are also all significant at $\alpha = 0.05$, and these are the same predictors included in the last step, since ACCTS was entered and then removed again in Steps 7 and 8.

FIGURE 14.16 $X_1 = TIME$ Forced into SAS Using Forward Selection Procedure (Part 1)

FORWARD SELECTION PROCEDURE FOR DEPENDENT VARIABLE SALES

THE FIRST 1 VARIABLES IN EACH MODEL ARE INCLUDED VARIABLES.

STEP 0 INCLUDED VARIABLE ENTERED R SQUARE = 0.38802924

	DF	SUM OF SQUARES	MEAN SQUARE	F	PROB>F
REGRESSION	1	16056475.09223126	16056475.09223126	14.58	0.0009
ERROR	23	25323073.83462484	1101003.21020108		
TOTAL	24	41379548.92685610			

	B VALUE	STD ERROR	TYPE II SS	F	PROB>F
INTERCEPT	2548.66790165				
TIME	9.42356060	2.46765372	16056475.09223126	14.58	0.0009

STEP 1 VARIABLE ADV ENTERED R SQUARE = 0.59534345

	DF	SUM OF SQUARES	MEAN SQUARE	F	PROB>F
REGRESSION	2	24635043.52430164	12317521.76215082	16.18	0.0001
ERROR	22	16744505.40255446	761113.88193429		
TOTAL	24	41379548.92685610			

	B VALUE	STD ERROR	TYPE II SS	F	PROB>F
INTERCEPT	1703.67202604				
TIME	7.65124403	2.11853193	9927587.53078250	13.04	0.0016
ADV	0.22956997	0.06838053	8578568.43207038	11.27	0.0029

STEP 2 VARIABLE ACCTS ENTERED R SQUARE = 0.77519508

	DF	SUM OF SQUARES	MEAN SQUARE	F	PROB>F
REGRESSION	3	32077222.74134030	10692407.58044677	24.14	0.0001
ERROR	21	9302326.18551580	442967.91359599		
TOTAL	24	41379548.92685610			

	B VALUE	STD ERROR	TYPE II SS	F	PROB>F
INTERCEPT	72.85038428				
TIME	0.22784146	2.42737509	3902.68955796	0.01	0.9261
ADV	0.22517894	0.05217495	8294974.21232976	18.73	0.0003
ACCTS	18.72825955	4.56912976	7442179.21703867	16.80	0.0005

STEP 3   VARIABLE POTEN ENTERED   R SQUARE = 0.82832020

	DF	SUM OF SQUARES	MEAN SQUARE	F	PROB>F
REGRESSION	4	34275516.12135634	8568879.03033908	24.12	0.0001
ERROR	20	7104032.80549976	355201.64027499		
TOTAL	24	41379548.92685610			

	B VALUE	STD ERROR	TYPE II SS	F	PROB>F
INTERCEPT	-390.62720160				
TIME	-0.57978423	2.19775267	24720.09413160	0.07	0.7946
POTEN	0.02227078	0.00895220	2199293.38001603	6.19	0.0218
ADV	0.21780689	0.04683090	7683393.81248002	21.63	0.0002
ACCTS	16.31257897	4.20516993	5345066.55276952	15.05	0.0009

STEP 4   VARIABLE SHARE ENTERED   R SQUARE = 0.90635489

	DF	SUM OF SQUARES	MEAN SQUARE	F	PROB>F
REGRESSION	5	37504556.64140826	7500911.32828165	36.78	0.0001
ERROR	19	3874992.28544784	203946.96239199		
TOTAL	24	41379548.92685610			

	B VALUE	STD ERROR	TYPE II SS	F	PROB>F
INTERCEPT	-1349.90221560				
TIME	1.95130322	1.78268043	244354.17858705	1.20	0.2874
POTEN	0.03881710	0.00795659	4854104.95081998	23.80	0.0001
ADV	0.16467574	0.03791481	3847323.61999021	18.86	0.0004
SHARE	210.84007898	52.98769690	3229040.52005192	15.83	0.0008
ACCTS	5.97051932	4.11203788	429959.19718669	2.11	0.1628

STEP 5   VARIABLE SHCHANGE ENTERED   R SQUARE = 0.92031405

	DF	SUM OF SQUARES	MEAN SQUARE	F	PROB>F
REGRESSION	6	38082180.17286742	6347030.02881124	34.65	0.0001
ERROR	18	3297368.75398868	183187.15299937		
TOTAL	24	41379548.92685610			

	B VALUE	STD ERROR	TYPE II SS	F	PROB>F
INTERCEPT	-1165.47855369				
TIME	2.26935112	1.69898362	326828.57515075	1.78	0.1983
POTEN	0.03827800	0.00754688	4712573.84321485	25.73	0.0001
ADV	0.14067029	0.03839221	2459312.09671656	13.43	0.0018
SHARE	221.60469221	50.58309112	3515945.91528346	19.19	0.0004
SHCHANGE	285.10928426	160.55965553	577623.53145916	3.15	0.0927
ACCTS	4.37770296	3.99903763	219521.27264808	1.20	0.2881

NO OTHER VARIABLES MET THE 0.5000 SIGNIFICANCE LEVEL FOR ENTRY INTO THE MODEL.

FIGURE 14.17   *Summary of Steps Forcing TIME into Model Using SAS Stepwise Procedure*

DEPENDENT VARIABLE = SALES.

THE FIRST VARIABLE IN EACH MODEL IS AN INCLUDED VARIABLE.

STEP 0.   INCLUDED VARIABLE (TIME) ENTERED.   STEP 1.   ADV ENTERED.

$R^2 = .3880$ $R^2 = .5953$

	F	PROB > F		F	PROB > F
TIME	14.58	.0009	TIME	13.04	.0016
			ADV	11.27	.0029

STEP 2.   ACCTS ENTERED.   STEP 3.   POTEN ENTERED.

$R^2 = .7752$ $R^2 = .8283$

	F	PROB > F		F	PROB > F
TIME	.01	.9261	TIME	.07	.7946
ADV	18.73	.0003	POTEN	6.19	.0218
ACCTS	16.80	.0005	ADV	21.63	.0002
			ACCTS	15.05	.0009

STEP 4.   SHARE ENTERED.   STEP 5.   ACCTS REMOVED.

$R^2 = .9064$ $R^2 = .8960$

	F	PROB > F		F	PROB > F
TIME	1.20	.2874	TIME	9.03	.0070
POTEN	23.80	.0001	POTEN	38.47	.0001
ADV	18.86	.0004	ADV	16.12	.0007
SHARE	15.83	.0008	SHARE	37.84	.0001
ACCTS	2.11	.1628			

STEP 6.   SHCHANGE ENTERED.   STEP 7.   ACCTS ENTERED.

$R^2 = .9150$ $R^2 = .9203$

	F	PROB > F		F	PROB > F
TIME	9.34	.0065	TIME	1.78	.1983
POTEN	39.10	.0001	POTEN	25.73	.0001
ADV	12.10	.0025	ADV	13.43	.0018
SHARE	43.11	.0001	SHARE	19.19	.0004
SHCHANGE	4.26	.0530	SHCHANGE	3.15	.0927
			ACCTS	1.20	.2881

STEP 8.   ACCTS REMOVED.

$R^2 = .9150$

	F	PROB > F
TIME	9.34	.0065
POTEN	39.10	.0001
ADV	12.10	.0025
SHARE	43.11	.0001
SHCHANGE	4.26	.0530

We see that, at least for these data, the stepwise procedure does a much better job in finding conditionally significant predictors with a variable forced into the model than does the forward selection procedure.

Backward Elimination

Figure 14.18 shows how the backward elimination procedure handles a forced-in variable. There is no difference between this and the results of the backward elimination procedure with all variables competing, because the five variables TIME, PØTEN, ADV, SHARE, and SHCHANGE are all conditionally significant at $\alpha = 0.10$. However, forcing in a variable sometimes does make a difference. For example, if X_7(WKLD) had been forced in, our results would look quite different from those in Figures 14.18 or 14.4.

FIGURE 14.18   *Summary of Steps Forcing TIME into Model Using Backward Elimination*

DEPENDENT VARIABLE = SALES

THE FIRST VARIABLE IN EACH MODEL IS AN INCLUDED VARIABLE.

STEP 0.   ALL VARIABLES ENTERED   STEP 1.   RATE REMOVED

$R^2 = 0.9220$ $R^2 = 0.9220$

	F	PROB > F		F	PROB > F
TIME	1.08	0.3134	TIME	1.21	0.2868
POTEN	20.57	0.0003	POTEN	22.56	0.0002
ADV	10.27	0.0055	ADV	12.35	0.0027
SHARE	8.82	0.0090	SHARE	9.57	0.0066
SHCHANGE	2.42	0.1390	SHCHANGE	3.24	0.0897
ACCTS	1.35	0.2621	ACCTS	1.52	0.2339
WKLD	0.35	0.5649	WKLD	0.37	0.5501
RATE	0.00	0.9500			

STEP 2.   WKLD REMOVED   STEP 3.   ACCTS REMOVED

$R^2 = 0.9203$ $R^2 = 0.9150$

	F	PROB > F		F	PROB > F
TIME	1.78	.01983	TIME	9.34	0.0065
POTEN	25.73	0.0001	POTEN	39.10	0.0001
ADV	13.43	0.0018	ADV	12.10	0.0025
SHARE	19.19	0.0004	SHARE	43.11	0.0001
SHCHANGE	3.15	0.0927	SHCHANGE	4.26	0.0530
ACCTS	1.20	0.2881			

ALL VARIABLES IN THE MODEL ARE SIGNIFICANT AT THE 0.1000 LEVEL.

Maximum R² Improvement

Figure 14.19 summarizes the maximum R^2 improvement technique applied to the CRAVENS data with X_1 (TIME) forced into the model. Note that TIME is included at every step. We again see the five-predictor model including TIME, PØTEN, ADV, SHARE, and SHCHANGE in which all predictors are significant. Note that this is also the best five-predictor model found in Figure 14.5.

Minimum R² Improvement

Finally, the results of the minimum R^2 improvement technique are shown in Figure 14.20, where we see the same best five-predictor model as that found by the stepwise and maximum R^2 improvement methods. But it takes a lot more steps to get there. The variable TIME appears in all models considered: although many switches are made, TIME is never switched from another variable.

Let us now see how to force this same variable into the model if the problem must be analyzed using the BMDP2R program.

BMDP2R

We have actually already seen how to force variables into the model in a specified order using the BMDP2R. Recall that back in Chapter 4, we used the BMDP2R to perform polynomial regressions. In Figure 4.13, we forced the computer to first analyze a simple regression, then, a quadratic regression, and then to add in the cubic term by using the sentence LEVELS = 1,0,2,3. The first variable was X (SIZE), the second was Y(PRICE), the third was X^2, and the fourth was X^3. Since Y(PRICE) was the dependent variable, it was never allowed to enter as a predictor: hence the level zero. We wanted X to enter first (level = 1), then X^2 (level = 2), and then X^3 (level = 3).

Using LEVELS Assignments

The BMDP2R will first add all variables passing the tolerance and F-to-enter tests assigned level 1 before considering any variables assigned level 2. It will add all variables assigned level 2 that meet the entry criteria before considering any variables assigned level 3, and so on. Thus, if we had added the sentence

LEVELS = 0,1,2,2,2,2,2,2,2.

to the REGRESSIØN paragraph of Figure 14.7, then TIME would be the first variable used as a predictor. Then all six remaining predictors are allowed to compete for entry, according to the selection algorithm used. Figure 14.21 shows what happens when default conditions on method, F-to-enter, and F-to-remove are used, and the above LEVELS statement is included. Only the summary table is shown in Figure 14.21. But if we had the entire output, we would see in Step 0, under VARIABLES NOT IN EQUATION, that TIME has been assigned level 1 and all other predictors have been

FIGURE 14.19 *Summary of Steps Forcing TIME into Model Using Maximum R^2 Improvement Technique*

DEPENDENT VARIABLE = SALES

THE FIRST VARIABLE IN EACH MODEL IS AN INCLUDED VARIABLE.

STEP 1. INCLUDED VARIABLE ENTERED

R SQUARE = 0.38802924

	F	PROB > F
TIME	14.58	0.0009

STEP 2. VARIABLE ADV ENTERED

R SQUARE = 0.59534345

	F	PROB > F
TIME	13.04	0.0016
ADV	11.27	0.0029

THE ABOVE IS THE BEST 2 VARIABLE MODEL FOUND

STEP 3. VARIABLE ACCTS ENTERED

R SQUARE = 0.77519508

	F	PROB > F
TIME	0.01	0.9261
ADV	18.73	0.0003
ACCTS	16.80	0.0005

THE ABOVE IS THE BEST 3 VARIABLE MODEL FOUND.

STEP 4. VARIABLE POTEN ENTERED

R SQUARE = 0.82832020

	F	PROB > F
TIME	9.03	.0070
POTEN	38.47	.0001
ADV	16.12	.0007
SHARE	37.84	.0001

THE ABOVE IS THE BEST 4 VARIABLE MODEL FOUND

STEP 5. VARIABLE SHCHANGE ENTERED

R SQUARE = 0.91500898

	F	PROB > F
TIME	9.34	0.0065
POTEN	39.10	0.0001
ADV	12.10	0.0025
SHARE	43.11	0.0001
SHCHANGE	4.26	0.0530

THE ABOVE IS THE BEST 5 VARIABLE MODEL FOUND.

STEP 6. VARIABLE ACCTS ENTERED

R SQUARE = 0.92031405

	F	PROB > F
TIME	1.78	0.1983
POTEN	25.73	0.0001
ADV	13.43	0.0018
SHARE	19.19	0.0004
SHCHANGE	3.15	0.0927
ACCTS	1.20	0.2281

THE ABOVE IS THE BEST 6 VARIABLE MODEL FOUND

STEP 7. VARIABLE WKLD ENTERED

R SQUARE = 0.92201937

	F	PROB > F
TIME	1.21	0.2868
POTEN	22.56	0.0002
ADV	12.35	0.0027
SHARE	9.57	0.0066
SHCHANGE	3.24	0.0897
ACCTS	1.52	0.2339
WKLD	0.37	0.5501

THE ABOVE IS THE BEST 7 VARIABLE MODEL FOUND.

STEP 8. VARIABLE RATE ENTERED

R SQUARE = 0.92203915

	F	PROB > F
TIME	1.08	0.3134
POTEN	20.57	0.0003
ADV	10.27	0.0055
SHARE	8.82	0.0090
SHCHANGE	2.42	0.1390
ACCTS	1.35	0.2621
WKLD	0.35	0.5649
RATE	0.00	0.9500

THE ABOVE IS THE BEST 8 VARIABLE MODEL FOUND.

FIGURE 14.20   *Summary of Steps Forcing TIME into Model Using Minimum R² Improvement Techniq*

MINIMUM R-SQUARE IMPROVEMENT FOR DEPENDENT VARIABLE SALES

THE FIRST 1 VARIABLES IN EACH MODEL ARE INCLUDED VARIABLES.

STEP 0. INCLUDED VARIABLE ENTERED   STEP 2. VARIABLE WKLD ENTERED
         R SQUARE = 0.38802924   R SQUARE = 0.38806073

	F	PROB > F			F	PROB > F
TIME	14.58	0.0009		TIME	13.46	0.0014
				WKLD	0.00	0.9735

STEP 2. WKLD REPLACED BY RATE   STEP 2. RATE REPLACED BY SHCHANGE
         R SQUARE = 0.50405876   R SQUARE = 0.50606889

	F	PROB > F			F	PROB > F
TIME	15.20	0.0008		TIME	11.88	0.0023
RATE	5.15	0.0334		SHCHANGE	5.26	0.0318

STEP 2. SHCHANGE REPLACED BY POTEN   STEP 2. POTEN REPLACED BY SHARE
         R SQUARE = 0.51303132   R SQUARE = 0.56419737

	F	PROB > F			F	PROB > F
TIME	7.03	0.0146		TIME	16.68	0.0005
POTEN	5.65	0.0266		SHARE	8.89	0.0069

STEP 2. SHARE REPLACED BY ACCTS   STEP 2. ACCTS REPLACED BY ADV
         R SQUARE = 0.57473436   R SQUARE = 0.59534345

	F	PROB > F			F	PROB > F
TIME	0.32	0.5757		TIME	13.04	0.0016
ACCTS	9.66	0.0051		ADV	11.27	0.0029

THE ABOVE MODEL IS THE BEST 2 VARIABLE MODEL FOUND.

STEP 3. VARIABLE WKLD ENTERED   STEP 3. WKLD REPLACED BY RATE
         R SQUARE = 0.60662348   R SQUARE = 0.62512321

	F	PROB > F			F	PROB > F
TIME	13.29	0.0015		TIME	13.45	0.0015
ADV	11.67	0.0026		ADV	6.78	0.0166
WKLD	0.60	0.4464		RATE	1.67	0.2105

STEP 3. ADV REPLACED BY ACCTS   STEP 3. RATE REPLACED BY SHCHANGE
         R SQUARE = 0.63540384   R SQUARE = 0.64028201

	F	PROB > F			F	PROB > F
TIME	0.65	0.4282		TIME	0.35	0.5596
ACCTS	7.57	0.0120		SHCHANGE	3.83	0.0639
RATE	3.49	0.0756		ACCTS	7.84	0.0108

FIGURE 14.20   *(continued)*

STEP 3. SHCHANGE REPLACED BY POTEN			STEP 3. ACCTS REPLACED BY ADV		
R SQUARE = 0.64263925			R SQUARE = 0.69914850		
	F	PROB > F		F	PROB > F
TIME	0.08	0.7825	TIME	6.42	0.0193
POTEN	3.99	0.0589	POTEN	7.25	0.0137
ACCTS	7.62	0.0117	ADV	12.99	0.0017

STEP 3. POTEN REPLACED BY ACCTS			STEP 4. VARIABLE RATE ENTERED		
R SQUARE = 0.77519508			R SQUARE = 0.78004366		
	F	PROB > F		F	PROB > F
TIME	0.01	0.9261	TIME	0.05	0.8331
ADV	18.73	0.0003	ADV	13.15	0.0017
ACCTS	16.80	0.0005	ACCTS	14.09	0.0013
			RATE	0.44	0.5143

THE ABOVE MODEL IS THE BEST 3 VARIABLE
MODEL FOUND.

STEP 4. RATE REPLACED BY SHCHANGE			STEP 4. SHCHANGE REPLACED BY SHARE		
R SQUARE = 0.78746265			R SQUARE = 0.78904803		
	F	PROB > F		F	PROB > F
TIME	0.02	0.8846	TIME	0.27	0.6088
ADV	13.85	0.0013	ADV	15.03	0.0009
SHCHANGE	1.15	0.2954	SHARE	1.31	0.2653
ACCTS	14.21	0.0012	ACCTS	8.84	0.0075

STEP 4. SHARE REPLACED BY WKLD			STEP 4. WKLD REPLACED BY POTEN		
R SQUARE = 0.79706675			R SQUARE = 0.82832020		
	F	PROB > F		F	PROB > F
TIME	0.01	0.9148	TIME	0.07	0.7946
ADV	21.78	0.0001	POTEN	6.19	0.0218
ACCTS	18.77	0.0003	ADV	21.63	0.0002
WKLD	2.16	0.1576	ACCTS	15.05	0.0009

STEP 4. ACCTS REPLACED BY SHARE			STEP 5. VARIABLE WKLD ENTERED		
R SQUARE = 0.89596427			R SQUARE = 0.89648263		
	F	PROB > F		F	PROB > F
TIME	9.03	0.0070	TIME	8.30	0.0096
POTEN	38.47	0.0001	POTEN	36.44	0.0001
ADV	16.12	0.0007	ADV	12.59	0.0021
SHARE	37.84	0.0001	SHARE	30.85	0.0001
			WKLD	0.10	0.7611

THE ABOVE MODEL IS THE BEST 4 VARIABLE
MODEL FOUND.

STEP 5. WKLD REPLACED BY RATE			STEP 5. RATE REPLACED BY ACCTS		
R SQUARE = 0.90248609			R SQUARE = 0.90635489		
	F	PROB > F		F	PROB > F
TIME	10.00	0.0051	TIME	1.20	0.2874
POTEN	30.58	0.0001	POTEN	23.80	0.0001
ADV	10.61	0.0041	ADV	18.86	0.0004
SHARE	38.87	0.0001	SHARE	15.83	0.0008
RATE	1.27	0.2737	ACCTS	2.11	0.1628

FIGURE 14.20   *(continued)*

STEP 5.   ACCTS REPLACED BY SHCHANGE

R SQUARE = 0.91500898

	F	PROB > F
TIME	9.34	0.0065
POTEN	39.10	0.0001
ADV	12.10	0.0025
SHARE	43.11	0.0001
SHCHANGE	4.26	0.0530

THE ABOVE MODEL IS THE BEST 5 VARIABLE MODEL FOUND.

STEP 6.   VARIABLE WKLD ENTERED

R SQUARE = 0.91503007

	F	PROB > F
TIME	8.80	0.0083
POTEN	37.05	0.0001
ADV	10.58	0.0044
SHARE	32.48	0.0001
SHCHANGE	3.93	0.0629
WKLD	0.00	0.9474

STEP 6.   WKLD REPLACED BY RATE

R SQUARE = 0.91542586

	F	PROB > F
TIME	8.87	0.0081
POTEN	33.17	0.0001
ADV	9.98	0.0054
SHARE	41.12	0.0001
SHCHANGE	2.75	0.1143
RATE	0.09	0.7692

STEP 6.   RATE REPLACED BY ACCTS

R SQUARE = 0.92031405

	F	PROB > F
TIME	1.78	0.1983
POTEN	25.73	0.0001
ADV	13.43	0.0018
SHARE	19.19	0.0004
SHCHANGE	3.15	0.0927
ACCTS	1.20	0.2881

THE ABOVE MODEL IS THE BEST 6 VARIABLE MODEL FOUND.

STEP 7.   VARIABLE RATE ENTERED

R SQUARE = 0.92035586

	F	PROB > F
TIME	1.62	0.2204
POTEN	23.39	0.0002
ADV	11.05	0.0040
SHARE	17.74	0.0006
SHCHANGE	2.33	0.1449
ACCTS	1.05	0.3194
RATE	0.01	0.9258

STEP 7.   RATE REPLACED BY WKLD

R SQUARE = 0.92201937

	F	PROB > F
TIME	1.21	0.2868
POTEN	22.56	0.0002
ADV	12.35	0.0027
SHARE	9.57	0.0066
SHCHANGE	3.24	0.0897
ACCTS	1.52	0.2339
WKLD	0.37	0.5501

THE ABOVE MODEL IS THE BEST 7 VARIABLE MODEL FOUND.

STEP 8.   VARIABLE RATE ENTERED

R SQUARE = 0.92203915

	F	PROB > F
TIME	1.08	0.3134
POTEN	20.57	0.0003
ADV	10.27	0.0055
SHARE	8.82	0.0090
SHCHANGE	2.42	0.1390
ACCTS	1.35	0.2621
WKLD	0.35	0.5649
RATE	0.00	0.9500

THE ABOVE MODEL IS THE BEST 8 VARIABLE MODEL FOUND.

FIGURE 14.21 Summary Table from Forcing TIME into Model Using BMDP2R with LEVELS Assignment

SUMMARY TABLE

STEP NO.	VARIABLE ENTERED	VARIABLE REMOVED	MULTIPLE R	RSQ	INCREASE IN RSQ	F-TO-ENTER	F-TO-REMOVE	NUMBER OF INDEPENDENT VARIABLES INCLUDED
1	TIME		.6229	.3880	.3880	14.583		1
2	ADV		.7716	.5953	.2073	11.271		2
3	ACCTS		.8805	.7752	.1799	16.801		3
4		TIME	.8804	.7751	-.0001		.009	2
5	POTEN		.9098	.8277	.0526	6.414		3
6	SHARE		.9489	.9005	.0728	14.611		3

assigned level 2. In Step 1, then, TIME is used as the predictor. Then in Step 2, ADV enters because it has the largest partial *F*-to-enter. From now on, all predictors, including TIME, are on their own, remaining in the model only if the associated *F*-to-remove stays above 3.9. In Step 3, we see that the partial *F* for TIME, given ADV and ACCTS, is only 0.009, so TIME is removed in Step 4 and never enters again.

Forcing TIME into the Model

How can we keep TIME from disappearing from the equation? One way is to specify REMØVE = 0 in the REGRESSIØN paragraph. However, this would not allow any of the other variables to be removed, either, should their partial *F*'s become small. To keep TIME in, in addition to the LEVELS statement, add

FØRCE = 1.

to the REGRESSIØN paragraph. This means that all variables assigned level 1 will be forced into the equation, and others will be allowed to compete on the basis of their performances with respect to the selection algorithm. The summary table in Figure 14.22 shows how this works for the CRAVENS data, using the F method, *F*-to-enter = 4.0 and *F*-to-remove = 3.9.

Of course, the LEVELS assignments and FØRCE statement can be used with the three other selection methods available through BMDP2R, and with different values for *F*-to-enter and *F*-to-remove. The final models obtained will be determined, to some extent, by the conditions that you place on the selection procedure.

Let us now see what SPSS can do to force TIME into the model.

SPSS

Refer back to Figure 13.4, p. 371, and the regression design statement beginning in column 16. Recall that including an even number in parentheses after the last predictor variable requests the computer to perform the overall regression significance test. Inclusion of an odd number requests the forward selection stagewise procedure. The number in parentheses is called an *inclusion number*, or *inclusion level*. Not only can it be used to tell the computer whether or not to run a stagewise regression, but it also designates the order in which the predictors are to enter. The higher the inclusion level assigned to any variable, the sooner it will be entered.

To force X_1 (TIME) into the model first and then allow all other predictors to compete, the regression design statement might read, beginning in column 16,

REGRESSIØN = SALES WITH TIME(5),PØTEN,... RATE(1)/

Note that any odd number greater than 1 could have been included in the parentheses

FIGURE 14.22   Summary Table from Using BMDP2R with FØRCE Statement

SUMMARY TABLE

STEP NO.	VARIABLE ENTERED	REMOVED	MULTIPLE R	RSQ	INCREASE IN RSQ	F-TO-ENTER	F-TO-REMOVE	NUMBER OF INDEPENDENT VARIABLES INCLUDED
1	2 TIME		0.6229	0.3880	0.3880	14.5834		1
2	4 ADV		0.7716	0.5953	0.2073	11.2710		2
3	7 ACCTS		0.8805	0.7752	0.1799	16.8009		3
4	3 POTEN		0.9101	0.8283	0.0531	6.1889		4
5	5 SHARE		0.9520	0.9064	0.0780	15.8328		5
6		7 ACCTS	0.9466	0.8960	-0.0104		2.1083	4
7	6 CHANGE		0.9566	0.9150	0.0190	4.2575		5

following TIME; also the 1 in parentheses following RATE applies to all variables preceding it for which no inclusion level is designated.

Since the SPSS performs only a forward selection, a variable that has been forced into the analysis can never be deleted. The analysis produced by the SPSS forward selection procedure shown in Figure 14.23 does not differ substantially from what we saw from the SAS forward selection procedure or the BMDP2R forward selection procedure when TIME was forced in.

SUMMARY

If it is important from either a theoretical or practical standpoint to include some predictors in the model, any one of the three programs we have studied can do the job. Only the BMDP2R, however, will allow a forced predictor to enter the model and then leave again should it become weak in conjunction with other predictors. This is a nice option if one wants to give a particular predictor first chance to be included but also to get rid of it if it turns out to be of little use.

Forcing variables into the model can have a very great effect on the final model found by the selection procedure being used. Since there is no universally "best" model, the best model for a given set of data is a function of the criteria used in the variable selection procedures.

EXERCISES   *Section 14.5*

14.7 Refer to the TENNESSEE data. Suppose that the percentage of families below the poverty level is to be predicted, and the client whose problem it is requires that the Negro population be used as a predictor. Use an available forward selection or stepwise procedure to come up with an acceptable model. Use inclusion criterion that F must be 1.0 or significant at $\alpha = 0.25$ and deletion criterion that F must be 2.0 or significant at $\alpha = 0.15$.

14.8 Repeat Exercise 14.7 using the SAS backward elimination procedure.

14.6 SUMMARY

In this chapter we considered several different variable-selection techniques. Our objective has been to eliminate from a set of potential predictors any variables that are not effective either alone or in conjunction with other predictors. Ideally, we seek a set of predictors that is as small as possible while still explaining a large percentage of the variation in the response.

The choice of predictors for the final reduced model is seldom clear-cut. We typically want all of them to be statistically significant. Only then can we be reasonably

FIGURE 14.23 X_1 = TIME forced in by SPSS Forward Selection Procedure

DEPENDENT VARIABLE.. SALES

VARIABLE(S) ENTERED ON STEP NUMBER 1.. TIME

VARIABLES IN THE EQUATION

VARIABLE	B	BETA	STD ERROR B	F
TIME	9.42356	0.62292	2.46765	14.583
(CONSTANT)	2548.66766			

VARIABLES NOT IN THE EQUATION

VARIABLE	BETA IN	PARTIAL	TOLERANCE	F
POTEN	0.39680	0.45195	0.79391	5.647
ADV	0.47015	0.58204	0.93791	11.271
SHARE	0.42211	0.53654	0.98872	8.893
SHCHANGE	0.35498	0.43919	0.93676	5.258
ACCTS	0.66224	0.55235	0.42572	9.659
WKLD	-0.00570	-0.00717	0.96784	0.001
RATE	0.34239	0.43543	0.98977	5.147

VARIABLE(S) ENTERED ON STEP NUMBER 2.. ADV

VARIABLES IN THE EQUATION

VARIABLE	B	BETA	STD ERROR B	F
TIME	7.65124	0.50577	2.11853	13.043
ADV	0.22957	0.47015	0.06838	11.271
(CONSTANT)	1703.67202			

VARIABLES NOT IN THE EQUATION

VARIABLE	BETA IN	PARTIAL	TOLERANCE	F
POTEN	0.36250	0.50648	0.78995	7.246
SHARE	0.32904	0.49838	0.92833	6.940
SHCHANGE	0.22239	0.31884	0.83173	2.376
ACCTS	0.65008	0.66667	0.42558	16.801
WKLD	0.11117	0.16696	0.91264	0.602
RATE	0.18934	0.27128	0.83069	1.668

VARIABLE(S) ENTERED ON STEP NUMBER 3.. ACCTS

VARIABLES IN THE EQUATION

VARIABLE	B	BETA	STD ERROR B	F
TIME	0.22784	0.01506	2.42737	0.009
ADV	0.22578	0.46238	0.05217	18.726
ACCTS	18.72825	0.65008	4.56913	16.801
(CONSTANT)	72.85149			

VARIABLES NOT IN THE EQUATION

VARIABLE	BETA IN	PARTIAL	TOLERANCE	F
POTEN	0.26653	0.48612	0.74783	6.189
SHARE	0.14188	0.24824	0.68820	1.313
SHCHANGE	0.12460	0.23360	0.79011	1.154
WKLD	0.15554	0.31192	0.90411	2.156
RATE	0.07886	0.14686	0.77967	0.441

VARIABLE(S) ENTERED ON STEP NUMBER 4.. POTEN

VARIABLES IN THE EQUATION

VARIABLE	B	BETA	STD ERROR B	F
TIME	-0.57978	-0.03833	2.19775	0.070
ADV	0.21781	0.44606	0.04683	21.631
ACCTS	16.31256	0.56623	4.20517	15.048
POTEN	0.02227	0.26653	0.00895	6.189
(CONSTANT)	-390.62595			

VARIABLES NOT IN THE EQUATION

VARIABLE	BETA IN	PARTIAL	TOLERANCE	F
SHARE	0.39497	0.67419	0.50022	15.833
SHCHANGE	0.09498	0.20229	0.77883	0.811
WKLD	0.20413	0.46123	0.87650	5.134
RATE	-0.00404	-0.00817	0.70179	0.001

VARIABLE(S) ENTERED ON STEP NUMBER 5.. SHARE

VARIABLES IN THE EQUATION

VARIABLE	B	BETA	STD ERROR B	F
TIME	1.95130	0.12899	1.78268	1.198
ADV	0.16468	0.33725	0.03791	18.864
ACCTS	5.97051	0.20724	4.11204	2.108
POTEN	0.03882	0.46456	0.00796	23.801
SHARE	210.83983	0.39497	52.98766	15.833
(CONSTANT)	-1349.89975			

VARIABLES NOT IN THE EQUATION

VARIABLE	BETA IN	PARTIAL	TOLERANCE	F
SHCHANGE	0.13485	0.38609	0.76765	3.153
WKLD	0.03800	0.09273	0.55770	0.156
RATE	0.06750	0.18082	0.67203	0.608

VARIABLE(S) ENTERED ON STEP NUMBER 6.. SHCHANGE

VARIABLES IN THE EQUATION

VARIABLE	B	BETA	STD ERROR B	F
TIME	2.26935	0.15001	1.69898	1.784
ADV	0.14067	0.28809	0.03839	13.425
ACCTS	4.37770	0.15195	3.99904	1.198
POTEN	0.03828	0.45810	0.00755	25.725
SHARE	221.60445	0.41513	50.58306	19.193
SHCHANGE	285.10928	0.13485	160.55958	3.153
(CONSTANT)	-1165.47617			

VARIABLES NOT IN THE EQUATION

VARIABLE	BETA IN	PARTIAL	TOLERANCE	F
WKLD	0.05562	0.14629	0.55127	0.372
RATE	0.00869	0.02291	0.5426	0.009

VARIABLE(S) ENTERED ON STEP NUMBER 7.. WKLD

VARIABLES IN THE EQUATION

VARIABLE	B	BETA	STD ERROR B	F
TIME	1.97454	0.13052	1.79575	1.209
ADV	0.15196	0.31121	0.04325	12.348
ACCTS	5.61018	0.19474	4.54496	1.524
POTEN	0.03729	0.44629	0.00785	22.560
SHARE	198.30830	0.37149	64.11717	9.566
SHCHANGE	295.86606	0.13994	164.38649	3.239
WKLD	19.89898	0.05562	32.63611	0.372
(CONSTANT)	-1485.87722			

VARIABLES NOT IN THE EQUATION

VARIABLE	BETA IN	PARTIAL	TOLERANCE	F
RATE	0.00598	0.01593	0.55293	0.004

F-LEVEL OR TOLERANCE-LEVEL INSUFFICIENT FOR FURTHER COMPUTATION certain that a different independent sample drawn from the same population would yield the same predictors.

We would also like the multiple coefficient of determination for the final model to be near 100 %, although in many practical applications this is not possible. We realize that just because our final set of k predictors can explain, for example, 92 % of the variation in the response in this sample, these same predictors might not explain 92 % of the variation in response when applied to a new sample. Any predictor included in the regression equation should have an estimated regression coefficient with a small standard error. Then, we would not expect the value of the regression coefficient to change very much if another analysis were run on another sample. In consequence, $\hat{Y}$ should also be a pretty good estimate of the true mean response value. On the other hand, if $\hat{Y}$ is based on a regression equation for which the standard errors of the regression coefficients are large, then $\hat{Y}$ also becomes untrustworthy. A final consideration is that theoretical or practical matters may dictate that some predictors should be included in the model.

Very seldom in an actual regression problem are the four criteria—significant predictors, large r_{xy}^2, small standard errors, and theoretically or practically appealing variables—all met at some step of a stagewise regression procedure. Ideally, one should look at the results of several stagewise techniques and select several alternative possible final models. Then the final model should not be selected until an analysis of residuals shows that the model meets as closely as possible the assumptions made in multiple regression analysis: the errors should be independent, normally distributed random variables with mean zero and common variance in order for the results of the significance tests to be valid.

The final chapter looks briefly at residual analysis for multiple regression, considering also what can and cannot be done when the data are only categorical (0 = nonsmoker, 1 = smoker, for example). We will also consider a method for choosing the model that will make the most accurate predictions based on new observations.

Chapter 15 OTHER CONSIDERATIONS IN MULTIPLE REGRESSION

15.1 INTRODUCTION

In Chapter 14, we looked at some methods for determining which variables to include in a regression model. The various stagewise procedures proved to be helpful in indicating effective predictors when the researcher has no theoretical grounds for including one predictor rather than another. Although these selection procedures produce candidates for a final model, they do not necessarily come up with a final, appropriate model. Before deciding on a final model, one should look at the residuals to make sure that the relationship is truly linear, that no important predictors have been omitted, and that none of the assumptions of the model have been too seriously violated. If the researcher intends to use the final regression equation to make estimates based on data not included in the sample, then it is important to examine how well various models might be expected to perform.

We begin with the analysis of residuals in multiple regression, applying the same kinds of checks on the assumptions as were made in Chapter 9 for simple linear regression. Some adjustments must be made when using these checks for multiple regression problems.

Examination of residuals leads to a study of *dummy variables*, which are used in assigning numerical values to observations that can only be categorized. These dummy variables are helpful if we find that the regression model should take into account the classification of an observation.

Finally, we consider a way of determining how well various candidate models will actually estimate values of Y for future values of $X_1, X_2, \ldots, X_k$.

15.2 EXAMINATION OF RESIDUALS IN MULTIPLE REGRESSION

As in simple linear regression, we have several assumptions to check out before using our regression equation. Just because the computer will perform an analysis on a set of data and come up with a model containing all significant predictors does not mean that we have necessarily done the right thing. For example, $s_{y \cdot x}^2$ can always be calculated, but it is not meaningful unless we can make the assumption that regardless of the values of $X_1, X_2, \ldots, X_k$, the scatter of the data points around the regression line is uniform. We also need to be reasonably certain that the relationship of Y to $X_1, X_2, \ldots, X_k$ is really linear, and not curvilinear. The presence of outliers or an unusual pattern in the residuals also might indicate that we have omitted a predictor which should be taken into account. Finally, the assumption that the errors are normally distributed must be satisfied if we are going to apply significance tests.

Most of the assumptions can be checked out in multiple regression in pretty much the same way that they were in simple regression. We will briefly consider these tests of assumptions, noting the changes to be made for a multiple regression model.

UNIFORM SCATTER

Recall that in simple regression, to see if the scatter were uniform, we either plotted Y against X and looked at the pattern of the points around the regression line, or we plotted the residuals against X. However, in multiple regression, we have many X-variables; and since we cannot make a two-dimensional plot with more than one predictor, we have a bit of a dilemma. We could plot residuals versus each of the X's, and you will recall that the BMDP2R program can be instructed to do this, as can the SAS program and the SPSS. Perhaps the easiest thing to do, however, is to plot the residuals versus $\hat{Y}$.

Suppose we consider the model using the predictors PØTEN, ADV, SHARE, and ACCTS, which was found by the forward selection and stepwise procedures. There, all predictors were significant at $\alpha = 0.05$. The regression equation was

$$Y = -1441.916 + 0.038X_2 + 0.175X_3 + 190.144X_4 + 9.214X_6,$$

where $X_2 = $ PØTEN, $X_3 = $ ADV, $X_4 = $ SHARE, and $X_6 = $ ACCTS. Table 15.1 shows the residuals from this model, and Figure 15.1 shows a plot of the residuals versus $\hat{Y}$.

The plot of residuals versus $\hat{Y}$ shows that there appears to be slightly larger scatter toward the center than at either end. If the data are split into two groups, one containing the six largest and six smallest Y-values and the other containing the middle 13 Y-values, and s^2 is calculated for the residuals in each group, then

$$F = \frac{s_2^2}{s_1^2} = 2.91.$$

This F-value is significant at $\alpha = 0.05$ but not at $\alpha = 0.01$. If one must be 99 % sure that the scatter is not homogeneous in order to apply remedial measures, then the evidence falls short.

On the other hand, if one uses $\alpha = 0.05$, then the evidence is strong enough to indicate nonhomogeneous scatter and some remedial measure must be taken. One

TABLE 15.1   *Calculation of $\hat{Y}$ and Residuals Using*
$\hat{Y} = -1441.916 + 0.038X_2 + 0.175X_3 + 190.144X_4 + 9.214X_6$

Y	$\hat{Y}$	$Y - \hat{Y}$
3669.88	3357.661156	312.218844
3473.95	3785.131639	−311.181639
2295.10	2847.380336	−552.280336
4675.56	4939.637611	−264.077611
6125.96	5525.301106	600.658894
2134.94	2087.445268	47.494732
5031.66	5037.800094	− 6.140094
3367.45	2796.684212	570.765788
6519.45	6127.991101	391.458899
4876.37	3941.926155	934.443845
2468.27	2543.090277	− 74.820277
2533.31	2534.667681	− 1.357681
2408.11	2157.840822	250.269178
2337.38	1904.544659	432.835341
4586.95	4882.934267	−295.984267
2729.24	2669.686565	59.553435
3289.40	3462.109443	−172.799443
2800.78	2632.294688	168.485312
3264.20	4052.445395	−788.245395
3453.62	3935.392933	−481.772933
1741.45	1723.636616	17.813384
2035.75	2704.235023	−668.485023
1578.00	1646.527102	− 68.527102
4167.44	4462.452816	−295.012816
2799.97	2605.821608	194.148392

could perhaps transform the data so that the scatter in the transformed data is uniform. In Chapter 9, in cases in which $\sigma^2_{y\cdot x}$ was proportional to X, we used the transformation

$$Y' = \alpha' + \beta'X' + \varepsilon',$$

where

$$Y' = \frac{Y}{X}, \quad X' = \frac{1}{X}, \quad \alpha' = \beta, \quad \beta' = \frac{\alpha}{X}, \quad \text{and} \quad \varepsilon' = \frac{\varepsilon}{X}.$$

In the multiple regression case, the fact that we have more than one X, however, makes this transformation inappropriate. The technique of *weighted least squares*, of which the above transformation is a special case, must be used in order to make the error variances equal. This involves finding a matrix of weights $\mathbf{W}$, related to the matrix of variances and covariances of the errors, so that when the normal equations

$$\mathbf{X'W^{-1}X\hat{\beta} = X'W^{-1}y}$$

are solved, the residuals resulting from the regression equation found will have homogeneous variance. This is not an easy task.

FIGURE 15.1   *Residuals Versus $\hat{Y}$ for a Multiple Regression Model (CRAVENS Data)*

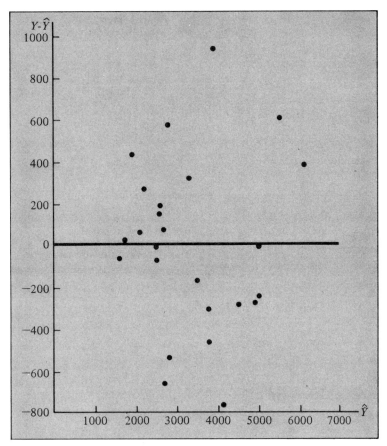

OUTLIERS

Further examination of the residual plot shows the possibility of an outlier in our data. All but one of the residuals are between -800 and $+800$. We could test to see if this residual indicates an outlier in the same way that we did in Chapter 9: hold out this observation, rerun the analysis using the remaining observations, form a confidence interval using the remaining observations. If the withheld observation does not lie in the confidence interval, we have some justification for calling it an outlier. Then we can either throw it out as an erroneous measurement or we can search for an omitted predictor that should have been taken into account. (Note that throwing out this extreme residual as an error in measurement will probably help resolve the question of whether or not the scatter is uniform.)

Considerably more work is involved in holding out a suspected observation and using the remaining $n - 1$ observations to calculate a confidence interval than was involved in simple regression, however. One might not have easy or cheap access to a computer to reanalyze the problem, for one thing. Several other outlier-detection techniques are much easier to use (in simple regression as well as in multiple regression). One very quick test is the *Dixon criterion*, which can be used as long as $n \leq 25$.

Dixon
Criteria

The <u>Dixon criteria</u> is a small-sample test for outliers that are either extremely large or extremely small. The procedure is as follows. First, choose α as the largest risk you are willing to take of throwing out an observation which really does belong in the set. That is,

H_0: observation is not an outlier,

H_1: observation is an outlier.

Arrange the set of values V_i in order from smallest to largest. Call the smallest $V_{(1)}$ and the largest $V_{(n)}$, where n is the number of values in the set.

Next, compute a comparison of ranges, z_{ij}, depending upon the size of n: if

$$3 \leq n \leq 7, \quad \text{compute } z_{10},$$

$$8 \leq n \leq 10, \quad \text{compute } z_{11},$$

$$11 \leq n \leq 13, \quad \text{compute } z_{21},$$

$$14 \leq n \leq 25, \quad \text{compute } z_{22},$$

where z_{ij} values are computed as follows:

z_{ij}	*If Smallest Value Is Suspect*	*If Largest Value Is Suspect*
z_{10}	$\dfrac{V_{(2)} - V_{(1)}}{V_{(n)} - V_{(1)}}$	$\dfrac{V_{(n)} - V_{(n-1)}}{V_{(n)} - V_{(1)}}$
z_{11}	$\dfrac{V_{(2)} - V_{(1)}}{V_{(n-1)} - V_{(1)}}$	$\dfrac{V_{(n)} - V_{(n-1)}}{V_{(n)} - V_{(2)}}$
z_{21}	$\dfrac{V_{(3)} - V_{(1)}}{V_{(n-1)} - V_{(1)}}$	$\dfrac{V_{(n)} - V_{(n-2)}}{V_{(n)} - V_{(2)}}$
z_{22}	$\dfrac{V_{(3)} - V_{(1)}}{V_{(n-2)} - V_{(1)}}$	$\dfrac{V_{(n)} - V_{(n-2)}}{V_{(n)} - V_{(3)}}$

Critical values for selected values of α are shown in Table 15.2. If the computed value for z_{ij} exceeds the tabulated value, you can be $100(1 - \alpha)\%$ sure that the observation is an outlier.

TABLE 15.2   *Criteria for Rejection of Outlying Observations*

Statistic	n	$\alpha = 0.10$	$\alpha = 0.04$	$\alpha = 0.02$	$\alpha = 0.01$
z_{10}	3	0.941	0.976	0.988	0.994
	4	0.765	0.846	0.889	0.926
	5	0.642	0.729	0.780	0.821
	6	0.560	0.644	0.698	0.740
	7	0.507	0.586	0.637	0.680
z_{11}	8	0.554	0.631	0.683	0.725
	9	0.512	0.587	0.635	0.677
	10	0.477	0.551	0.597	0.639
z_{21}	11	0.576	0.638	0.679	0.713
	12	0.546	0.605	0.642	0.675
	13	0.521	0.578	0.615	0.649
z_{22}	14	0.546	0.602	0.641	0.674
	15	0.525	0.579	0.616	0.647
	16	0.507	0.559	0.595	0.624
	17	0.490	0.542	0.577	0.605
	18	0.475	0.527	0.561	0.589
	19	0.462	0.514	0.547	0.575
	20	0.450	0.502	0.535	0.562
	21	0.440	0.491	0.524	0.551
	22	0.430	0.481	0.514	0.541
	23	0.421	0.472	0.505	0.532
	24	0.413	0.464	0.497	0.524
	25	0.406	0.457	0.489	0.516

Adapted by permission from *Introduction to Statistical Analysis* (3rd ed.) by J. W. Dixon and F. J. Massey, Jr., 1969, McGraw Hill Book Company, Inc.

Let us test to see if the one residual in Figure 15.1 is an outlier. Table 15.3 shows the residuals ordered from smallest to largest.

TABLE 15.3   *Residuals from Table 15.1 in Order from Smallest to Largest*

$Y - \hat{Y}$
$-788.25 = V_{(1)}$
$-668.49 = V_{(2)}$
$-552.28 = V_{(3)}$
$\vdots$
$570.77 = V_{(n-2)}$
$600.66 = V_{(n-1)}$
$934.44 = V_{(n)}$

Since $n = 25$ and since we are testing to the largest value, we calculate

$$t_{22} = \frac{V_{(n)} - V_{(n-2)}}{V_{(n)} - V_{(3)}} = \frac{934.44 - 570.77}{934.44 - (-552.28)} = \frac{363.67}{1486.72} = 0.245.$$

If α is chosen to be 0.01, then we conclude that this residual is an outlier if $t_{22} > 0.516$. Since $0.245 \not> 0.516$, we cannot be 99% sure that we have an outlier.

LACK OF FIT

Sometimes a plot of residuals indicates an unusual pattern. As in simple regression, if we have repeated measurements at any set of $(X_1, X_2, \ldots, X_k)$-values, we can test lack of fit of the model. This can often be done in designed experiments, but seldom if $X_1, X_2, \ldots, X_k$ as well as Y are random variables. Very seldom does chance let us observe the same values on $X_1, X_2, \ldots, X_k$ so that we can see "pure" error by comparing their corresponding Y-values. If such a test can be made, the only changes from the procedure outlined in Section 9.3 is that the degrees of freedom for error will be $n - p$ instead of $n - 2$. Thus the degrees of freedom for lack of fit will be the number of different $(X_1, X_2, \ldots, X_k)$-values minus p.

Even without repeated measurements, however, plots of residuals can often provide a clue to what is wrong with the fitted model, especially if the true relationship is not linear with respect to the variables in the model. For example, suppose that the true model is

$$Y = 4 + X_1 + 2X_1^2 + X_2,$$

and that this is a deterministic relationship (no error is involved). Suppose that 10 observations give the following results:

X_1	2	2	3	6	4	6	5	1	1	6
X_2	9	0	3	7	4	6	9	7	4	2
Y	23	14	28	89	44	88	68	14	11	84

The reader can verify that each Y-value is found by plugging the given values of X_1 and X_2 into the above equation.

Now, if we are ignorant of the presence of the quadratic term in X_1 and fit a linear model

$$Y = \alpha + \beta_1 X_1 + \beta_2 X_2 + \varepsilon,$$

then we will find the regression equation

$$\hat{Y} = 15.2026 + 15.4122 X_1 + 1.1802 X_2.$$

Predicted values and residuals from fitting this model are shown below:

X_1	X_2	Y	$\hat{Y}$	$Y - \hat{Y}$
2	9	9	26.2436	−3.2436
2	0	0	15.6218	−1.6218
3	3	3	34.5746	−6.5746
6	7	7	85.5320	3.4680
4	4	4	51.1670	−7.1670
6	6	6	84.3518	3.6482
5	9	9	72.4802	−4.4802
1	7	7	8.4710	5.5290
1	4	4	4.9304	6.0696
6	2	2	79.6310	4.3690

Figure 15.2 shows the residuals plotted against the predicted values.

FIGURE 15.2   *Plot of Residuals Versus $\hat{Y}$ for Incorrectly Specified Model*

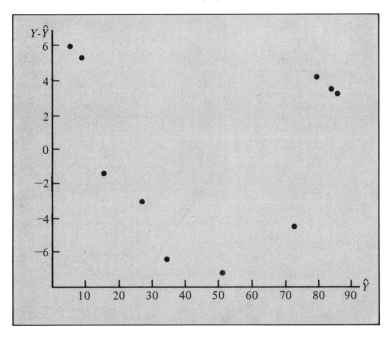

There is obviously a pattern to these residuals—one indicating that a quadratic term has been omitted from the model. But how would we know whether the quadratic term involves X_1 or X_2? Part a of Figure 15.3 shows the residuals plotted against X_1, and part b shows the residuals plotted against X_2. Since the pattern emerges again in the plot of the residuals against X_1, we know that an X_1^2 term has been omitted from the model. This example is admittedly over-simplified, but it illustrates how one can detect quadratic terms omitted from the model.

FIGURE 15.3   *(a) Plot of Residuals Versus X₁*

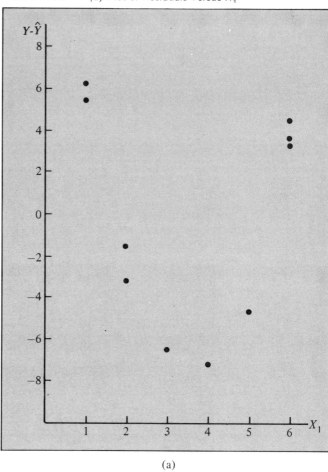

(a)

TIME DEPENDENCIES

If the data were collected over time, making it possible that a time effect is causing the errors not to be independent, then a test for serial correlation can be performed. The procedure is exactly the same as that for simple regression, in Section 9.4. Most computer programs give serial correlation coefficients, along with Durbin-Watson statistics. One need only compare these to the appropriate critical values from the tables in order to discover any serial correlation present.

NORMALITY

Finally, a quick check on the assumption that the errors are normally distributed can be made by computing the standardized residuals, to see if 68 % of them lie between ± 1, and so on. The SPSS program is especially handy for this check, since it computes and plots standardized residuals.

(b) Plot of Residuals Versus X_2

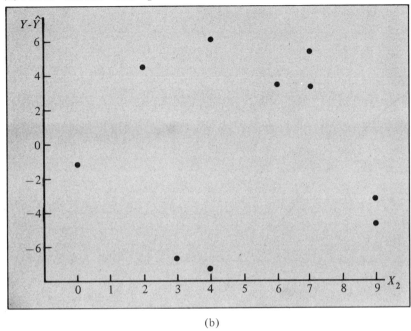

(b)

EXERCISES   *Section 15.2*

15.1   A diving coach at a large university wants to determine some variables that affect his athletes' performances. An Olympic hopeful is measured, or rated. on 6 variables over 192 consecutive practice sessions and his performance at each session is rated by the coach. The variables are

Y = performance rating
X_1 = weight (pounds)
X_2 = number of hours sleep the night before
X_3 = number of dives
X_4 = mental rating
X_5 = physical rating
X_6 = coach's attitude

The rating scale is: 1 = excellent, 2 = above average, 3 = average, 4 = below average, 5 = poor.

Because of the likelihood of a practice effect, a Durbin-Watson statistic is calculated from the model containing all six predictors. Its value is found to be 1.4863. What does this tell you about any practice effect?

p. 536   15.2   Refer to the TENNESSEE data. Suppose we consider the model found by the backward elimination procedure which uses population change, agricultural employment, percent rural population, and median age to predict the percentage of families below the poverty level. (Refer to the solution of Exercise 14.2.) Plot the residuals against $\hat{Y}$ and analyze the plot.

15.3   Using the same model as in Exercise 15.2, find the standardized residuals and check to see whether the normality assumption can be justified.

15.3 DUMMY VARIABLES

TWO CATEGORIES

Recall from Chapter 9 our example of a regression relating age and earnings for a group of people. Figure 9.11, showing the relationship, is reproduced as Figure 15.4. The dots on the scatter diagram represent males and the circles represent females. Here, it became apparent that an important predictor, namely sex, had been omitted from the model. In order to make accurate predictions of earnings from age, we should take sex into account.

FIGURE 15.4   *Scatter Diagram and Regression Line Relating Earnings to Age*

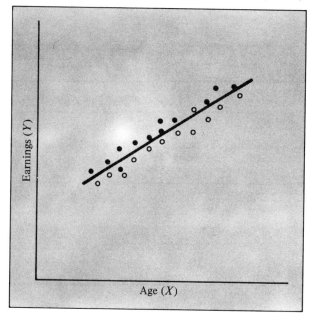

If the omitted predictor were, say, years on the job, our task would be simple, because this variable is easily measured. But how does one assign a meaningful numerical measurement to the variable sex?

Perhaps we could divide the group of people according to sex and perform a separate simple regression analysis for each sex. This would solve our problem but might not produce as good a regression equation as the combined group. If you consider just the dots (males) in Figure 15.4, you can draw a regression line through the dots. Another regression line can be drawn in for the circles (females). Both lines have approximately the same slope; but the line for the males is higher. Since the two groups together would give us a more accurate estimate of the common slope, we would like to avoid running separate analyses if at all possible.

This brings us to the question of how to assign a meaningful numerical measurement to the values male and female of the variable "sex." For this purpose, we employ a dummy, or *indicator*, variable. For example, we could call male = 1 and female = 0. Note that there is no inherent ordering in "male-female," as there is in "high-low," for example. Further, we could use male = 3, female = 5 if we want to; but it is conventional—because it is easier—to stick with 0 and 1.

Dummy
Variable

What does this do for us? We have a model

$$Y = \alpha + \beta_1 X_1 + \beta_2 X_2 + \varepsilon,$$

where Y = earnings; X_1 = age; and X_2 = sex, where

$$X_2 = \begin{cases} 0 & \text{if female} \\ 1 & \text{if male.} \end{cases}$$

Then, if we are predicting earnings for a female ($X_2 = 0$), our regression equation will be

$$\hat{Y} = a + b_1 X_1 + b_2(0)$$
$$= a + b_1 X_1,$$

which is essentially a simple regression equation for the females; and to predict earnings for a male ($X_2 = 1$), the regression equation is

$$\hat{Y} = a + b_1 X_1 + b_2(1)$$
$$= (a + b_2) + b_1 X_1.$$

The regression coefficient relating income to age is the same for both males and females. The intercept for the males, however, which is b_2 units greater than that for the females, says that the regression line for males is b_2 units higher than that for females. We have incorporated all features of two separate simple regressions into one multiple regression, and an F-test on the regression coefficient β_2 tells us whether there is any significant difference between the two groups male and female. How to make this test would be less obvious had we used two separate simple regressions.

MORE THAN TWO CATEGORIES

Suppose we have a variable that can be divided into more than two categories. For example, suppose that the strength of a manufactured part is to be related to the percentage of hardening agent used (X_1) as well as to which one of four machines turned out the part. In this case, we need *three* dummy variables. Let the model be

$$Y = \alpha + \beta_1 X_1 + \beta_2 X_2 + \beta_3 X_3 + \beta_4 X_4,$$

where

$$Y = \text{strength},$$

$$X_1 = \text{percent of hardening agent},$$

$$X_2 = \begin{cases} 1 & \text{if made on machine 1} \\ 0 & \text{if not made on machine 1,} \end{cases}$$

$$X_3 = \begin{cases} 1 & \text{if made on machine 2} \\ 0 & \text{if not made on machine 2,} \end{cases}$$

$$X_4 = \begin{cases} 1 & \text{if made on machine 3} \\ 0 & \text{if not made on machine 3.} \end{cases}$$

What about machine 4? It has been taken care of automatically. If the part were made on machine 1, then

$$(X_2, X_3, X_4) = (1, 0, 0)$$

indicates this. If the part were made on machine 2, we have

$$(X_2, X_3, X_4) = (0, 1, 0).$$

If the part were made on machine 3, then

$$(X_2, X_3, X_4) = (0, 0, 1).$$

If the part were made on machine 4, then

$$(X_1, X_2, X_3) = (0, 0, 0)$$

says it was not made on machine 1 nor on machine 2 nor on machine 3; thus, it must have been made on machine 4.

Why could we not just add on an

$$X_5 = \begin{cases} 1 & \text{if made on machine 4} \\ 0 & \text{if not made on machine 4?} \end{cases}$$

Suppose that two parts are to be made from each machine, so that $n = 8$. Consider the **X** matrix for this situation.

$$\mathbf{X} = \begin{array}{c} \\ \\ \\ \\ \\ \\ \\ \\ \end{array} \begin{array}{cccccc} (a) & (b_1) & (b_2) & (b_3) & (b_4) & (b_5) \\ \left[\begin{array}{cccccc} 1 & X_{11} & 1 & 0 & 0 & 0 \\ 1 & X_{21} & 1 & 0 & 0 & 0 \\ 1 & X_{31} & 0 & 1 & 0 & 0 \\ 1 & X_{41} & 0 & 1 & 0 & 0 \\ 1 & X_{51} & 0 & 0 & 1 & 0 \\ 1 & X_{61} & 0 & 0 & 1 & 0 \\ 1 & X_{71} & 0 & 0 & 0 & 1 \\ 1 & X_{81} & 0 & 0 & 0 & 1 \end{array} \right] \end{array}$$

Note that the last four columns of **X** sum to the first column. We have thus introduced a linear dependency in our data, and **X′X** will be singular. Remember: *if there are c categories to be made, c − 1 dummy variables are required.*

For our model with four categories and three indicator variables, if the part is made by machine 1, then the regression equation is

$$\hat{Y} = a + b_1 X_1 + b_2(1) + b_3(0) + b_4(0)$$
$$= (a + b_2) + b_1 X_1.$$

For a part made by machine 2, the regression equation is

$$\hat{Y} = a + b_1 X_1 + b_2(0) + b_3(1) + b_4(0)$$
$$= (a + b_3) + b_1 X_1.$$

If it is made by machine 3,

$$\hat{Y} = (a + b_4) + b_1 X_1;$$

and if it is made by machine 4,

$$\hat{Y} = a + b_1 X_1 + b_2(0) + b_3(0) + b_4(0)$$
$$= a + b_1 X_1.$$

Here, our basis of reference is machine 4. The regression coefficient b_2 shows how much stronger or weaker the parts made by machine 1 were than those made by machine 4, and so on.

SEVERAL CATEGORICAL VARIABLES

If more than one predictor is a categorical variable, all can be handled in the same way. For example, consider

$$Y = \text{degree of political conservatism,}$$

$$X_1 = \text{income,}$$

$$X_2 = \text{age,}$$

$$X_3 = \text{race} \begin{cases} 1 = \text{white} \\ 0 = \text{nonwhite,} \end{cases}$$

$$X_4 = \text{party affiliation} \begin{cases} 1 = \text{belongs to a political party} \\ 0 = \text{does not belong to a political party.} \end{cases}$$

(Note that only two classifications of both race and party affiliation are being considered.) Then the model is

$$Y = \alpha + \beta_1 X_1 + \beta_2 X_2 + \beta_3 X_3 + \beta_4 X_4;$$

and if the person considered is nonwhite and does not belong to a political party, the regression equation becomes

$$\hat{Y} = a + b_1 X_1 + b_2 X_2.$$

A nonwhite who belongs to a political party has regression equation

$$\hat{Y} = a + b_1 X_1 + b_2 X_2 + b_4(1)$$
$$= (a + b_4) + b_1 X_1 + b_2 X_2,$$

so that b_4 indicates the effect of political party membership. A white belonging to a political party has regression equation

$$\hat{Y} = a + b_1 X_1 + b_2 X_2 + b_3(1) + b_4(1)$$
$$= (a + b_3 + b_4) + b_1 X_1 + b_2 X_2,$$

so that $b_3 + b_4$ indicates the effect of being white and belonging to a political party, where b_3 indicates the effect of being white.

SUMMARY

In this section and the one before it, we have seen how to examine the multiple regression equation for its conformation to the assumptions of the model, and how to fix up the model if the data do not support it. Assuming that we have obtained a regression equation which adequately describes the data, how well can this equation be expected to predict Y based on new observed or assumed values of $X_1, X_2, \ldots, X_k$? We shall address this question in the next section.

EXERCISES   *Section 15.3*

15.4   In a survey to be taken of undergraduate students at a university, it is desired to indicate which class an interviewed student belongs to (freshman, sophomore, junior, senior). It will be too much trouble, however, to record the number of hours completed in order to indicate class, so dummy variables are to be used. Show how you would define dummy variables in order to take class into account.

15.5   Suppose that in a survey, values of a variable "education" were defined in the following manner:

1 = no formal schooling,
2 = some formal schooling, but no high school diploma,
3 = graduated high school only,
4 = attended college, but no degree,
5 = undergraduate degree only,
6 = schooling past undergraduate degree.

Show an alternative method of indicating education, using dummy variables.

15.6   Suppose that in a study relating the amount spent on medical expenses to the age, sex, and income of a patient, the following regression equation was found:

$$\hat{Y} = 400 + 2X_1 - 20X_2 - 4X_3,$$

where

Y = medical expenses, in dollars, for a given year,

X_1 = age of patient in years,

$X_2 = \text{sex} \begin{cases} 1 & \text{if female} \\ 0 & \text{if male,} \end{cases}$

X_3 = annual family income, in thousands of dollars.

What is the meaning of the regression coefficient $b_2 = -20$? Illustrate by comparing the medical expenses of a male and a female patient, both 60 years of age and both having $30 thousand annual family income.

15.7 Consider the following data:

Weekly Food Expenditure ($)	Annual Income ($000)	Children?	Area
40	14	no	city
72	38	yes	suburb
48	25	no	rural
76	24	yes	suburb
70	19	yes	suburb
59	10	no	city
77	23	yes	suburb
59	12	yes	rural
91	26	yes	city
69	24	yes	suburb

Suppose it is desired to predict food expenditures from the other three variables. Show the vector **y** and the matrix **X** necessary to set up the problem.

15.4 THE MODEL THAT PREDICTS BEST

THE DIFFICULTY IN CHOOSING AMONG CANDIDATE MODELS

The various stagewise regression procedures discussed in Chapter 14 can all give best models according to some criteria. One can find several different models in which all predictors are significant, depending upon which variable selection procedure is used. Thus we sometimes have to choose among various candidate models.

If the objective is to determine which predictors are effective in describing a real-world relationship, then the researcher can probably choose from among various candidate models the one that makes the best sense in terms of the theoretical framework of the problem. On the other hand, if the researcher wants to use a regression equation to make estimates based on future observed values of $X_1, X_2, \ldots, X_k$, then the question of which alternative model to use still remains.

One might think that the model (containing all significant predictors) having the largest r_{xy}^2 is best, but this is not necessarily the case. The problem is that $b_1, b_2, \ldots, b_k$, and thus r_{xy}^2, are very data-dependent because the least squares criterion found the line that best fits *these* data. A different sample will give different values for the regression coefficients and the multiple coefficient of determination. If any of the regression coefficients have large standard errors, then we could expect their values to change quite a bit from sample to sample. We are still $100(1 - \alpha)\%$ sure that these predictors are important, but we do not completely trust our estimates of their effects. The regression equation obtained from one set of data may make rather poor predictions when applied to a new set of data.

For example, let us take the CRAVENS data and split it in half, randomly choosing 13 of the 25 observations for one-half of the split. We will use as possible predictors X_2, X_3, X_4, and X_6 and run the 13 observations and four predictors through the SAS GLM regression program. The regression equation thus found will then be made to predict values of Y for the remaining 12 of the 25 observations.

Table 15.4 shows the 13 randomly selected observations to which a model is to be fitted. Figure 15.5 shows the results of the multiple regression analysis.

TABLE 15.4   *Thirteen Randomly Selected Observations from CRAVENS Data*

Y SALES	X_2 POTEN	X_3 ADV	X_4 SHARE	X_6 ACCTS
3473.95	58,117.30	5539.78	5.51	107.32
6125.96	57,805.11	7747.08	9.15	180.44
5031.66	50,935.26	3140.62	8.54	256.10
6519.45	46,176.77	8846.25	12.54	203.25
4876.37	42,053.24	5673.11	8.85	119.51
2468.27	36,829.71	2761.76	5.38	116.26
2533.31	33,612.67	1991.85	5.43	142.28
2337.38	20,416.87	1737.38	7.80	84.55
2729.24	23,093.26	8618.61	5.15	80.49
2800.78	39,571.96	4565.81	5.45	78.86
3453.62	58,749.82	3721.10	6.35	138.21
1741.45	23,990.82	860.97	7.37	75.61
2035.75	25,694.86	3571.51	8.39	102.44

We note that for this analysis, all predictors are significant at $\alpha = 0.0625$, and that $r_{xy}^2 = 0.94$. The regression equation is

$$\hat{Y} = -1935.675893 + 0.035458X_2 + 0.214951X_3$$
$$+ 272.004922X_4 + 8.446730X_6,$$

with $s_{y\cdot x}^2 = 469.635665$. Compare the regression coefficients in the above equation to those found using the same predictors for all 25 observations:

$$\hat{Y} = -1{,}441.931829 + 0.038218X_2 + 0.174990X_3$$
$$+ 190.144297X_4 + 9.213896X_6.$$

We see some considerable differences in the regression coefficients, especially b_4, illustrating that these coefficients are very data-dependent.

Using the equation found in Figure 15.5, let us calculate $\hat{Y}$-values for the remaining 12 observations and see how well this regression equation predicts for a set of data not used in constructing the equation itself. Table 15.5 shows values of $\hat{Y}$ and $Y - \hat{Y}$ for the 12 other observations. Taking the square root of $\Sigma(Y_i - \hat{Y}_i)^2 = 327{,}684.818949$ found there, we have 572.437611. Comparing this to $s_{y\cdot x} = 469.635665$ from Figure 15.5, we see that the regression equation found for 13 of the observations does not predict so accurately for the remaining 12 observations.

FIGURE 15.5   *SAS Multiple Regression Analysis Using 13 Randomly Selected Observations from CRAVENS Data*

DEPENDENT VARIABLE: SALES

SOURCE	DF	SUM OF SQUARES	MEAN SQUARE	F VALUE	PR > F	R-SQUARE	C.V.
MODEL	4	28131377.74179306	7032844.4354826	31.89	0.0001	0.940980	13.2357
ERROR	8	1764461.26333002	220557.65791625			STD DEV	SALES MEAN
CORRECTED TOTAL	12	29895839.00512308				469.63566508	3548.24538462

SOURCE	DF	TYPE I SS	F VALUE	PR > F	DF	TYPE IV SS	F VALUE	PR > F
POTEN	1	13608399.51206864	61.70	0.0001	1	1539529.8365616	6.98	0.0296
ADV	1	6341582.49673924	28.75	0.0007	1	2865518.01855294	12.99	0.0069
SHARE	1	7149617.26921522	32.42	0.0005	1	2241123.7896977	10.16	0.0128
ACCTS	1	1031778.46376996	4.68	0.0625	1	1031778.46376996	4.68	0.0625

PARAMETER	ESTIMATE	T FOR H0: PARAMETER=0	PR > |T|	STD ERROR OF ESTIMATE
INTERCEPT	-1935.67589269	-3.24	0.0119	598.08515819
POTEN	0.03545828	2.64	0.0296	0.01342099
ADV	0.21495099	3.60	0.0069	0.05963469
SHARE	272.00492231	3.19	0.0128	85.33062203
ACCTS	8.44672961	2.16	0.0625	3.90531827

TABLE 15.5   *Calculation of Residuals for 12 Observations, Using*
$\hat{Y} = -1935.675893 + 0.035458X_2 + 0.214951X_3 + 272.004922X_4 + 8.446730X_6$

Y_i	$\hat{Y}_i$	$Y_i - \hat{Y}_i$
3669.88	2990.673978	679.206022
2295.10	3232.125483	−937.025483
4675.56	4873.708102	−198.148102
2134.94	1876.027628	258.912372
3367.45	2769.498403	597.951597
2408.11	2309.351814	98.758186
4586.95	5471.185262	−884.235262
3289.40	3642.568919	−353.168919
3264.20	4067.984883	−803.784883
1578.00	1563.935130	14.064870
4167.44	5025.793685	−858.353685
2799.97	2871.284574	−71.314574
$\Sigma(\cdot)$		−2457.137861
$\Sigma(\cdot)^2$		4,107,660.214107

$$\sum_{i=1}^{12} (Y_i - \hat{Y}_i)^2 = \frac{12(4,107,660.214107) - (-2457.13861)^2}{12(11)}$$

$$= 327,684.818949$$

The point is: just because a regression equation having a large r_{xy}^2 is found for a given set of data, we cannot be sure that it will predict as well for a new set of data. We have found several models which could be used. If our interest is in making accurate predictions of Y for new observed values of $X_1, X_2, \ldots, X_k$, then we need not only look for significant predictors, but also need to find out which model predicts best.

THE PRESS STATISTIC

In order to see how well a model predicts, consider the following procedure. Suppose we have some number, say ℓ, candidate models. Pick one. For this model, hold out one of the observations. Fit the model to the remaining $n - 1$ observations and then see how well the regression equation predicts the withheld observation by looking at its residual $Y_i - \hat{Y}_i$. Do this for each observation in turn. Thus, n regressions, each based on $n - 1$ observations will be run. By summing the n residuals thus obtained, we then see how this model does on the whole set of data. Let $\hat{y}_i^{(\ell)}$ denote the prediction of the ith observation using the ℓth model. Then the PRESS statistic for the ℓth model,

PRESS
Statistic

$$\text{PRESS}_\ell = \sum_{i=1}^{n} [\hat{y}_i^{(\ell)} - y_i]^2,$$

is a measure of how well this ℓth model predicts the set of data on which it is based. PRESS stands for prediction error sum of squares. If we did this procedure for all ℓ models under consideration, then we could find the model giving the smallest value of the PRESS statistic. This model would then be considered best, in that it gives the most accurate prediction.

While this is a reasonable enough notion, it is obviously too much work, even using a computer; each of the ℓ models would need to have n different regressions run. It would be prohibitive to do this except that it can be shown that nowhere near this much work is actually needed. The computational form of the PRESS statistic for the ℓth model is

$$\text{PRESS}_\ell = \sum_{i=1}^{n} \left[\frac{\delta_i^{(\ell)}}{Q_i} \right]^2,$$

where

$$\delta_i^{(\ell)} = Y_i - \hat{Y}_i,$$

the ordinary residual, using all n observations, associated with the ith observation and the ℓth model and

$$Q_i = 1 - \mathbf{x}_i'(\mathbf{X}'\mathbf{X})^{-1}\mathbf{x}_i,$$

where $\mathbf{x}_i$ is the ith set of $(X_1, X_2,\ldots, X_k)$-values.

It turns out that this statistic is reasonably easy to calculate. For one thing, the $\delta_i^{(\ell)}$-values are given from the usual regression programs and thus need not be calculated by hand. Also, note that Q_i is related to the width of the confidence interval on the mean Y-value at the point $\mathbf{x}_i = (X_{1i}, X_{2i},\ldots, X_{ki})$. Recall that the width of the confidence interval at x_i is

$$2t_{\alpha/2, n-p} s_{y\cdot x} \sqrt{\mathbf{x}_i'(\mathbf{X}'\mathbf{X})^{-1}\mathbf{x}_i} = w_i.$$

Then

$$\sqrt{\mathbf{x}_i'(\mathbf{X}'\mathbf{X})^{-1}\mathbf{x}_i} = \frac{w_i}{2t_{\alpha/2, n-p} s_{y\cdot x}},$$

or

$$\mathbf{x}_i'(\mathbf{X}'\mathbf{X})^{-1}\mathbf{x}_i = \left[\frac{w_i/2}{t_{\alpha/2, n-p} s_{y\cdot x}} \right]^2.$$

Thus,

$$Q_i = 1 - \left[\frac{(\text{width of } CI \text{ at } \mathbf{x}_i)/2}{t_{\alpha/2, n-p} s_{y\cdot x}} \right]^2.$$

Further, the SAS computer program can be instructed to give confidence intervals for the mean response at the point x_i. So while a PRESS statistic does call for a considerable amount of calculation, it is simpler than it appears at first glance. Unfortunately, the PRESS statistic is too new for the computer packages to have a program to calculate it.

Looking back at the steps of the various stagewise procedures applied to the CRAVENS data, several models present themselves for consideration. Let us look at nine models, all with significant or almost significant predictors at $\alpha = 0.05$, and compare them by using the PRESS statistic. The models we will consider are

k	Predictors
1	ACCTS
2	ACCTS, ADV
	POTEN, SHARE
3	POTEN, ADV, ACCTS
	POTEN, ADV, SHARE
4	POTEN, ADV, SHARE, ACCTS
5	POTEN, ADV, SHARE, CHANGE, ACCTS
	TIME, POTEN, ADV, SHARE, CHANGE
6	TIME, POTEN, ADV, SHARE, CHANGE, ACCTS

The residuals and confidence intervals for each of these models were obtained by the SAS GLM analysis. After listing the variables in the model statement, a slash followed by a P and CLM; direct the computer to print (P) the residuals and calculate confidence limits on the mean response (CLM). Of course, we also obtain regression coefficients, $s_{y \cdot x}^2$, r_{xy}^2, and so on, for each model, as well as the partial F-values.

Figure 15.6 shows the analysis from the model using ACCTS to predict SALES. For each of the 25 observations we have residual, $\delta_i^{(1)}, i = 1, 2, \ldots, 25$, resulting from this model. We take the difference between the upper 95% confidence limit for the mean response and the lower 95% confidence limit for the mean response, to obtain the length of the confidence interval. Also, $s_{y \cdot x} = 881.093065$ and $t_{0.025,23} = 2.069$. With this information we can find the value of $PRESS_1$ statistic, since

$$Q_i = 1 - \left[\frac{\text{width}}{2(2.069)(881.093065)} \right]^2$$

$$= 1 - \left[\frac{\text{width}}{3645.9633103} \right]^2 .$$

FIGURE 15.6 *SAS Analysis of One-Predictor Model Giving Confidence Intervals on the Mean Response*

DEPENDENT VARIABLE: SALES

SOURCE	DF	SUM OF SQUARES	MEAN SQUARE	F VALUE	PR > F	R-SQUARE	C.V.
MODEL	1	23524074.19189502	23524074.19189502	30.30	0.0001	0.568495	26.1098
ERROR	23	17855474.73496108	776324.98847657			STD DEV	SALES MEAN
CORRECTED TOTAL	24	41379548.92685610				881.09306459	3374.56760000

SOURCE	DF	TYPE I SS	F VALUE	PR > F
ACCTS	1	23524074.19189502	30.30	0.0001

SOURCE	DF	TYPE IV SS	F VALUE	PR > F
ACCTS	1	23524074.19189502	30.30	0.0001

PARAMETER	ESTIMATE	T FOR HO: PARAMETER=0	PR > |T|	STD ERROR OF ESTIMATE
INTERCEPT	709.32383372	1.38	0.1819	515.24608170
ACCTS	21.72176971	5.50	0.0001	3.94603304

OBSERVATION	OBSERVED VALUE	PREDICTED VALUE	RESIDUAL	LOWER 95% CL FOR MEAN	UPPER 95% CL FOR MEAN
1	3669.88000000	2335.41551438	1334.46448562	1801.20540422	2869.62562453
2	3473.95000000	3040.50415924	433.44584076	2654.95998680	3426.04833169
3	2295.10000000	2810.90505338	-515.80505338	2389.29803748	3232.51206928
4	4675.56000000	4947.67553998	-272.11553998	4253.15364349	5642.19743647
5	6125.96000000	4628.79996060	1497.16003940	4032.94795327	5224.65196794
6	2134.94000000	2987.50304114	-852.56304114	2595.02134911	3379.98473318
7	5031.66000000	6272.26905704	-1240.60905704	5123.93321540	7420.60489868
8	3367.45000000	3464.29588633	-96.84588633	3098.20680701	3830.38496564
9	6519.45000000	5124.27352774	1395.17647226	4372.45588585	5876.09116964
10	4876.37000000	3305.29253203	1571.07746797	2939.83125745	3670.75380662
11	2468.27000000	3234.69678047	-766.42678047	2866.39384507	3602.99971587
12	2533.31000000	3799.89722838	-1266.58722838	3401.86213469	4197.93232207
13	2408.11000000	2651.90169909	-243.79169909	2197.32872545	3106.47467272
14	2337.38000000	2545.89946289	-208.51946289	2066.46256078	3025.33636500
15	4586.95000000	3305.29253203	1281.65746797	2939.83125745	3670.75380662
16	2729.24000000	2457.70907786	271.53092214	1956.11243204	2959.30572368
17	3289.40000000	3676.08314102	-386.68314102	3294.34646061	4057.81982144
18	2800.78000000	2422.30259323	378.47740677	1911.47472890	2933.13045755
19	3264.20000000	3676.08314102	-411.88314102	3294.34646061	4057.81982144
20	3453.62000000	3711.48962565	-257.86962565	3325.59432702	4097.38492429
21	1741.45000000	2351.70684166	-610.25684166	1821.95558129	2881.45810203
22	2035.75000000	2934.50192304	-898.75192304	2534.21094862	3334.79289747
23	1578.00000000	2369.30147513	-791.30147513	1844.32811557	2894.27483469
24	4167.44000000	3676.08314102	491.35685898	3294.34646061	4057.81982144
25	2799.97000000	2634.30706562	165.66293438	2175.75327230	3092.86085894

SUM OF RESIDUALS	0.00000000
SUM OF SQUARED RESIDUALS	17855474.73496115
SUM OF SQUARED RESIDUALS - ERROR SS	0.00000007
FIRST ORDER AUTOCORRELATION	0.18085477
DURBIN-WATSON D	1.55579444

Calculations are shown in Table 15.6.

TABLE 15.6   *Calculation of PRESS*₁

Obs	$\delta_i^{(1)}$	Width of CI	Q_i	$[\delta_i^{(1)}/Q_i]^2$
1	1334.46	1068.42	0.91	2,150,446.27
2	433.45	771.09	0.96	230,861.28
3	−515.81	843.21	0.95	294,805.56
4	−272.12	1389.04	0.85	102,489.62
5	1497.16	1191.70	0.89	2,829,796.84
6	−852.56	784.96	0.95	805,380.60
7	−1240.61	2296.67	0.60	4,275,300.58
8	−96.85	732.17	0.96	10,178.79
9	1395.18	1503.63	0.83	2,825,559.28
10	1571.08	730.92	0.96	2,678,263.17
11	−766.43	736.61	0.96	637,378.69
12	−1266.59	796.07	0.95	1,777,555.56
13	−243.79	909.14	0.94	67,262.42
14	−208.52	958.88	0.93	50,274.61
15	1281.66	730.92	0.96	1,782,385.20
16	271.53	1003.20	0.92	87,107.62
17	−386.68	763.47	0.96	162,239.78
18	378.48	1021.66	0.92	169,241.73
19	−411.88	763.47	0.96	184,075.32
20	257.87	771.79	0.96	72,151.33
21	−610.26	1059.50	0.92	440,006.69
22	−898.74	800.58	0.95	895,010.60
23	−791.30	1049.94	0.92	739,789.21
24	491.36	763.47	0.96	261.969.95
25	165.66	917.11	0.94	31,057.01
				$PRESS_1 = 23,533,587.71$

PRESS statistics can be calculated for each of the other eight models, as well. Of course, the value for $s_{y \cdot x}$ will be different for each different candidate model, and $t_{0.025, n-p}$ will depend upon the number of predictors in the model under consideration. For purposes of comparison, Table 15.7 shows calculation of $PRESS_2$ for the model using ACCTS and ADV to predict SALES. In this case, we will be using the results of the printout shown in Figure 15.7, from which we find $s_{y \cdot x} = 650.3925$ and $t_{0.025, 22} = 2.074$. In this case,

$$Q_i = 1 - \left[\frac{\text{width}}{2(2.074)(650.3925)} \right]^2$$

$$= 1 - \left[\frac{\text{width}}{2697.83} \right]^2 .$$

FIGURE 15.7 SAS Analysis of Model Using ACCTS and ADV and Showing Confidence Limits

DEPENDENT VARIABLE: SALES

SOURCE	DF	SUM OF SQUARES	MEAN SQUARE	F VALUE	PR > F
MODEL	2	32073320.05178237	16036660.02589118	37.91	0.0001
ERROR	22	9306228.87507373	423010.40341244		
CORRECTED TOTAL	24	41379548.92685610			

R-SQUARE	C.V.
0.775101	19.2734
STD DEV	SALES MEAN
650.39249951	3374.56760000

SOURCE	DF	TYPE I SS	F VALUE	PR > F	DF	TYPE IV SS	F VALUE	PR > F
ACCTS	1	23524074.19189502	55.61	0.0001	1	17365864.05826323	41.05	0.0001
ADV	1	8549245.85988735	20.21	0.0002	1	8549245.85988735	20.21	0.0001

PARAMETER	ESTIMATE	T FOR H0: PARAMETER=0	PR > |T|	STD ERROR OF ESTIMATE
INTERCEPT	50.29906160	0.12	0.9029	407.60969367
ACCTS	19.04824598	6.41	0.0001	2.97291380
ADV	0.22652657	4.50	0.0002	0.05038842

FIGURE 15.7 *(continued)*

OBSERVATION	OBSERVED VALUE	PREDICTED VALUE	RESIDUAL	LOWER 95% CL FOR MEAN	UPPER 95% CL FOR MEAN
1	3669.88000000	2514.39486262	1155.48513738	2110.53541499	2918.25431026
2	3473.95000000	3349.46420605	124.48579395	3030.53320533	3668.39520677
3	2295.10000000	2561.55633436	-266.45633436	2229.02767142	2894.08499729
4	4675.56000000	4275.10777971	400.45222029	3674.75693853	4875.45862090
5	6125.96000000	5242.28405948	883.67594052	4718.33248094	5766.23563803
6	2134.94000000	2139.24245419	-4.30245419	1651.92137058	2626.56353781
7	5031.66000000	5639.98874619	-608.32874619	4741.52728507	6538.45020731
8	3367.45000000	2938.75877720	428.69122280	2575.20749308	3302.31006133
9	6519.45000000	5925.76576497	593.68423503	5257.74978760	6593.78174235
10	4876.37000000	3611.88511266	1264.50488734	3306.66918111	3917.06104421
11	2468.27000000	2890.46017089	-422.19017089	2575.01964990	3205.90069189
12	2533.31000000	3211.69045631	-678.38045631	2611.20186886	3612.17904376
13	2408.11000000	2200.38537120	207.72462880	1804.72631170	2596.04443071
14	2337.38000000	2054.39099871	282.98900129	1633.33326639	2475.44873103
15	4586.95000000	4749.27543006	-162.32543006	4030.34633841	5468.20452171
16	2729.24000000	3535.83657953	-806.59657953	2915.23967047	4156.43348859
17	3289.40000000	4407.01147749	-1117.61147749	3967.13014965	4846.89280532
18	2800.78000000	2586.72103790	214.05896210	2201.16087131	2972.28120450
19	3264.20000000	4016.21009662	-752.01009662	3693.06595210	4339.35424114
20	3453.62000000	3525.88517395	-72.26517395	3227.75313548	3824.01721241
21	1741.45000000	1685.56952462	55.88047538	1187.45602022	2183.68302903
22	2035.75000000	2810.64330511	-774.89330511	2508.95729956	3112.32931067
23	1578.00000000	2150.54738654	-572.54738654	1749.16081534	2551.93395773
24	4167.44000000	3798.15788138	369.28211862	3510.10432022	4086.21144254
25	2799.97000000	2542.97701225	256.99298775	2201.02959654	2884.92442796

SUM OF RESIDUALS	0.00000000
SUM OF SQUARED RESIDUALS	9306228.87507376
SUM OF SQUARED RESIDUALS - ERROR SS	0.00000003
FIRST ORDER AUTOCORRELATION	0.19713929
DURBIN-WATSON D	1.48583131

TABLE 15.7   *Calculation of PRESS$_2$*

Obs	$\delta_i^{(2)}$	Width of CI	Q_i	$[\delta_i^{(2)}/Q_i]^2$
1	1155.49	807.71	0.91	1,612,315.85
2	124.49	637.87	0.94	17,540.35
3	−266.46	665.05	0.94	80,355.24
4	400.45	1200.70	0.80	250,560.31
5	883.68	1047.91	0.85	1,080,809.74
6	−4.30	974.64	0.87	24.40
7	−608.33	1796.92	0.55	1,223,346.60
8	428.69	727.10	0.93	212,484.12
9	593.68	1336.03	0.75	626,583.06
10	1264.50	610.39	0.95	1,771,694.10
11	−422.19	630.88	0.95	197,500.25
12	−678.38	800.98	0.91	555,725.52
13	207.72	791.31	0.92	50,976.61
14	282.99	842.12	0.90	98,866.22
15	−162.33	1437.85	0.72	50,832.21
16	−806.60	1241.19	0.79	1,042,461.42
17	−1117.61	879.76	0.89	1,576,822.95
18	214.06	771.12	0.92	54,135.33
19	−752.01	646.28	0.94	640,016.00
20	−72.27	596.27	0.95	5,786.64
21	55.88	996.22	0.86	4,222.40
22	−774.89	603.37	0.95	665,317.55
23	−572.55	802.77	0.91	395,867.47
24	369.28	576.11	0.96	147,971.01
25	256.99	683.89	0.94	74,742.09
				PRESS$_2$ = 12,437,016.44

Comparing PRESS$_1$ = 23,533,587.71 to PRESS$_2$ = 12,437,016.44, we see that since PRESS$_2$ < PRESS$_1$, the model using ACCTS and ADV as predictors is a better predicting model than is the one using just ACCTS. Table 15.8 shows PRESS statistics for all of the nine models under consideration.

TABLE 15.8   *PRESS Statistics for Nine Models under Consideration*

Model No.	Predictors	PRESS
1	ACCTS	23,533,587.71
2	ACCTS, ADV	12,437,016.44
3	POTEN, SHARE	14,199,601.98
4	POTEN, ADV, ACCTS	9,990,021.32
5	POTEN, ADV, SHARE	8,601,908.91
6	POTEN, ADV, SHARE, ACCTS	5,794,837.54
7	POTEN, ADV, SHARE, CHANGE, ACCTS	5,338,063.76
8	TIME, POTEN, ADV, SHARE, CHANGE	5,440,757.72
9	TIME, POTEN, ADV, SHARE, CHANGE, ACCTS	6,089,047.96

From Table 15.8 we see that model number 8 has the smallest PRESS statistic. The analysis for this five-predictor model using independent variables PØTEN, ADV, SHARE, CHANGE, and ACCTS is shown in Figure 15.8. There we see that all predictors are significant at $\alpha = 0.1236$, and that this model has $r_{xy}^2 = 0.912416$. The

FIGURE 15.8   SAS Analysis for Model with the Smallest PRESS Statistic (Part 1)

DEPENDENT VARIABLE: SALES

SOURCE	DF	SUM OF SQUARES	MEAN SQUARE	F VALUE	PR > F	R-SQUARE	C.V.
MODEL	5	37755351.59771664	7551070.31954333	39.59	0.0001	0.912416	12.9423
ERROR	19	3624197.32913946	190747.22784945		STD DEV		SALES MEAN
CORRECTED TOTAL	24	41379548.92685610			436.74618241		

SOURCE	DF	TYPE I SS	F VALUE	PR > F	DF	TYPE IV SS	F VALUE	PR > F
POTEN	1	14788184.70250567	77.53	0.0001	1	4573489.56079580	23.98	0.0001
ADV	1	10333786.00006020	54.18	0.0001	1	3194373.73424489	16.75	0.0006
SHARE	1	10008269.67486565	52.47	0.0001	1	3203732.09272625	16.80	0.0006
CHANGE	1	1002958.33498497	5.26	0.0334	1	495149.13489546	2.60	0.1236
ACCTS	1	1622152.88530016	8.50	0.0089	1	1622152.88530016	8.50	0.0089

PARAMETER	ESTIMATE	T FOR HO: PARAMETER = 0	PR > |T|	STD ERROR OF ESTIMATE
INTERCEPT	-1285.94337067	-3.07	0.0063	418.97088874
POTEN	0.03763121	4.90	0.0001	0.00768517
ADV	0.15443602	4.09	0.0006	0.03773852
SHARE	196.94952750	4.10	0.0006	48.05692351
CHANGE	262.50049338	1.61	0.1236	162.9263286
ACCTS	8.23411280	2.92	0.0089	2.82357962

FIGURE 15.8   (Part 2)

OBSERVATION	OBSERVED VALUE	PREDICTED VALUE	RESIDUAL	LOWER 95% CL FOR MEAN	UPPER 95% CL FOR MEAN
1	3669.88000000	3408.97747225	260.90252775	2762.11930334	4055.83564116
2	3473.95000000	3764.87466208	-290.92466208	3404.18867674	4125.56064741
3	2295.10000000	2720.78537826	-425.68537826	2275.06530719	3166.50544934
4	4675.56000000	4919.04369728	-243.48369728	4418.62544340	5419.46195115
5	6125.96000000	5504.86298926	621.09701074	5128.00451980	5881.72145871
6	2134.94000000	2187.08958704	-52.14958704	1822.13161228	2552.04756181
7	5031.66000000	5050.91762456	-19.25762456	4393.50201388	5708.33323525
8	3367.45000000	2684.12477590	683.32522410	2392.00152873	2976.24802308
9	6519.45000000	6286.75527283	232.69472717	5742.34713119	6831.16341447
10	4876.37000000	3981.14089308	895.22910692	3723.70687225	4238.57491391
11	2468.27000000	2640.53010968	-172.26010968	2333.54964589	2947.51057347
12	2533.31000000	2356.91574225	176.39425775	1897.29865283	2816.53283167
13	2408.11000000	2398.82861166	9.28138834	1948.04319751	2849.61402582
14	2337.38000000	2248.20831708	89.17168292	1695.07697926	2801.33965490
15	4586.95000000	4779.97769129	-193.02769129	4253.86177317	5306.09360941
16	2729.24000000	2601.66169247	127.57830753	2040.19960956	3163.12377538
17	3289.40000000	3532.94421540	-243.54421540	3040.15356249	4025.73486831
18	2800.78000000	2804.29040245	-3.51040245	2439.67860965	3168.90219525
19	3264.20000000	3937.08119004	-672.88119004	3623.77664757	4250.38573250
20	3453.62000000	3880.34670818	-426.72670818	3557.88811188	4202.80530447
21	1741.45000000	1396.04854678	345.40145322	840.71854906	1951.37854450
22	2035.75000000	2615.58901438	-579.83901438	2341.97453284	2889.20349591
23	1578.00000000	1700.77295813	-122.77295813	1351.36155187	2050.18436440
24	4167.44000000	4505.88339348	-338.44339348	4018.90102847	4992.86575850
25	2799.97000000	2456.53905419	343.43094581	2091.14964707	2821.92846130

SUM OF RESIDUALS	0.00000000
SUM OF SQUARED RESIDUALS	3624197.32913956
SUM OF SQUARED RESIDUALS - ERROR SS	0.00000010
FIRST ORDER AUTOCORRELATION	0.00206333
DURBIN-WATSON D	1.94465351

error sum of squares in Figure 15.6 is 3,624,197.33. We can compare this to the value of PRESS for this model: PRESS = 5,338,063.76. This comparison says that by fitting this model to these 25 observations, we get *SSE*-3,624,197.33; but if we took another sample of 25 observations and fit this same model to them, we would expect to find an *SSE* of approximately 5,338,000.

Note that when calculating a PRESS statistic we are weighting each residual before squaring and summing them. The weights have the effect of increasing the absolute values of the residuals, so that PRESS will be larger than the residual sum of squares. How much larger any residual becomes is determined by its weight, which is in turn determined by the width of the associated confidence interval. You will recall that the farther $(X_{1i}, X_{2i}, \ldots, X_{ki})$ is from $(\overline{X}_1, \overline{X}_2, \ldots, \overline{X}_k)$, the wider the confidence interval. The wider the confidence interval, the smaller the weight (Q_i) and thus the more the residual is increased. Thus PRESS takes into account observations that cannot be predicted very well by the model under consideration and assesses a "penalty" for each such observation.

Note that the other five-predictor model we considered—the one using TIME, PØTEN, ADV, SHARE, and CHANGE arrived at through the backward elimination procedure—has all predictors significant at $\alpha = 0.0530$ and also has $r_{xy}^2 = 0.915009$, which is greater than the r_{xy}^2 for the model selected by PRESS. Usually PRESS will pick the same model as arrived at by the backward elimination procedure. That it did not in this case is probably due to the fact that $n = 25$ is relatively small, and this affects the results of the backward elimination procedure. It could be argued that since PRESS_7 and PRESS_8 are so close, one would prefer to use the model in which all predictors are significant at $\alpha = 0.0530$, that is, the one arrived at by the backward elimination procedure. In any case, if your interest is in choosing a model which will make the most accurate possible predictions for future observations, the PRESS statistic provides a method for choosing among various models suggested by the different stagewise analyses.

THE C_p STATISTIC

C_p-Statistic

Another widely used model selection procedure uses the C_p-statistic. This statistic is considerably easier to calculate than PRESS, but it does not end up with an estimate of the error sum of squares for new data, as the PRESS statistic did. It is possible, however, to test for significance using C_p.

The C_p-statistic is defined as

$$C_p = p + \frac{s_p^2 - \sigma^2}{\sigma^2}(n - p).$$

Here, p is the number of parameters in a candidate model ($p = k + 1$), and s_p^2 is the error mean square associated with the candidate model. The variance σ^2 is the error mean square associated with the true best model, however many predictors it might contain. Now, if the candidate model under consideration is indeed the true best model, then s_p^2 for this model should be very close to σ^2, so that $s_p^2 - \sigma^2$ is very close to zero and C_p is approximately equal to p. Of course, even if we find the true best model, s_p^2 may not exactly equal σ^2, because s_p^2 is based on a sample and is therefore subject to random variability. It may be that s_p^2 is even less than σ^2, so that $C_p < p$ by a little bit. On the other hand, if our candidate model is not the correct one, then s_p^2 will be quite a bit larger than σ^2, since s_p^2 will contain error due to an incorrectly specified model, and C_p will be quite a bit larger than p.

In order to choose from among several candidate models, then, we calculate the C_p-statistic for each and pick the model for which C_p is closest to p.

All this supposes that σ^2, the error mean square for the true best model, is known. But in order to know σ^2, we must already know what the true best model is, and we don't; this is precisely what we are trying to find out. Thus, in order to use the C_p-statistic, we must come up with some method by which σ^2 can be estimated independently. There are generally three methods available for obtaining an independent estimate of σ^2.

The first method can be used if the researcher is quite familiar with the process being studied and can come up with an estimate of the random variability inherent in the process based on past experience. Recall that if the model is correct, $\sigma^2 = \sigma_{y \cdot x}^2$ does not contain any error due to omitted predictors or to an incorrectly specified model: it contains only pure random error. More often than one might expect, especially in engineering problems, the researcher can give a reasonably accurate estimate of the "noise" in his system. This estimate can then be substituted for σ^2 and the C_p-statistic can be calculated.

The second method for estimating σ^2 is to replicate the experiment or some parts of it, in order to break $s_{y \cdot x}^2$ into pure error and lack of fit, as we did in Chapter 9. Then the pure error component, which estimates only random variability, is used as an estimate of σ^2. The researcher who anticipates a problem in model selection should take care to have replicated observations, if possible, in order to obtain such an estimate of σ^2.

The least preferred but most common method of estimating σ^2 uses $s_p^2 = s_{y \cdot x}^2$ from the model containing all candidate predictors. Of course, this means that C_p will always equal p for the candidate model using all predictors. If many of the candidate predictors are not significant, $s_{y \cdot x}^2$ for the full model is likely to be larger than σ^2. This will lead to C_p being less than p for the "good" candidate models.

Let us illustrate the calculation of C_p for the CRAVENS data. Recall from the maximum R^2 improvement technique from SAS, one model found in which all predictors were significant at $\alpha = 0.01$ used ADV and ACCTS to predict SALES. For this model, the error mean square was found to be 423.010. From Step 0 of the backward elimination procedure, in which all eight predictors were used to predict SALES, the error mean square was 201,624. Thus we have

$$n = 25,$$

$$\sigma^2 = 201,624,$$

$$p = k + 1 = 3,$$

$$s_p^2 = 423,010.$$

Then

$$C_p = 3 + \frac{423{,}010 - 201{,}624}{201{,}624}(25 - 3)$$

$$= 3 + 1.0980(22)$$

$$= 27.1560.$$

Since C_p is considerably greater than p, we have evidence that the model using just ADV and ACCTS is not the best.

Table 15.9 shows C_p-statistics for the nine models we previously considered. In all nine calculations, σ^2 is estimated by the error mean square for the full model: $\sigma^2 = 201.624$. Comparing C_p to p for each of the nine, it seems that we can pretty quickly eliminate Models 1–5 from further consideration, since for each C_p is quite a bit larger than p. In Model 6, C_p is only slightly larger than p, and in Models 7–9, C_p is less than p.

HYPOTHESIS TESTS BASED ON C_p

There is also available an hypothesis test to eliminate models from consideration. Let

$k =$ total number of predictors under consideration $= 8$ in our example,

$p =$ number of terms in candidate model (may or may not include intercept term),

$q =$ number of predictors omitted in candidate model (may or may not include intercept),

so that

$$p + q = k + 1.$$

Let $\beta_p =$ vector of parameters for candidate model, with β_i's (and perhaps α) replaced by zeros for omitted terms.

For example, the parameter vector for the full model is

$$\boldsymbol{\beta}_k = \begin{bmatrix} \alpha \\ \beta_1 \\ \beta_2 \\ \beta_3 \\ \beta_4 \\ \beta_5 \\ \beta_6 \\ \beta_7 \\ \beta_8 \end{bmatrix}.$$

TABLE 15.9 C_p-Statistics for Nine Models under Consideration

Model No.	Selection Technique(s)	Variable(s)	Prob F	R^2	$s^2_{y \cdot x}$	No. Predictors k	$p = k + 1$	C_p
1	F. S. MAXR. MINR	ACCTS	0.0001	0.57	776,325	1	2	67.6
2	MAXR	ADV ACCTS	0.0002 0.0001	0.76	423,010	2	3	27.2
3	MINR	POTEN SHARE	0.0001 0.0001	0.75	477,629	2	3	33.1
4	MAXR	POTEN ADV ACCTS	0.0194 0.0001 0.0001	0.83	339,464	3	4	18.4
5	MAXR, MINR	POTEN ADV SHARE	0.0001 0.0011 0.0001	0.85	297,586	3	4	14.0
6	F. MAXR, MINR	POTEN ADV SHARE ACCTS	0.0001 0.0001 0.0011 0.0043	0.90	205,967	4	5	5.4
7	B, MAXR, MINR	TIME POTEN ADV SHARE CHANGE	0.0065 0.0001 0.0025 0.0001 0.0530	0.92	185,099	5	6	4.4
8	MAXR, MINR	POTEN ADV SHARE CHANGE ACCTS	0.0001 0.0006 0.0006 0.1236 0.0089	0.91	190,747	5	6	5.0
9	MAXR, MINR	TIME POTEN ADV SHARE CHANGE ACCTS	0.1983 0.0001 0.0018 0.0004 0.0927 0.2881	0.92	183,187	6	7	5.3

In Model 2, using just ADV $= X_3$ and ACCTS $= X_6$, and including an intercept,

$$\boldsymbol{\beta}_p = \begin{bmatrix} \alpha \\ 0 \\ 0 \\ \beta_3 \\ 0 \\ 0 \\ \beta_6 \\ 0 \\ 0 \end{bmatrix},$$

and $p = 3$, $q = 6$ and $3 + 6 = 8 + 1$.
Then to test

$$H_0 : \boldsymbol{\beta} = \boldsymbol{\beta}_p$$

$$H_1 : \boldsymbol{\beta} \neq \boldsymbol{\beta}_p$$

where $\boldsymbol{\beta}$ is the parameter vector for the correct model, calculate the test statistic

$$F_p = C_p - p,$$

and reject H_0 if $F_p > q[F_{\alpha.q.v} - 1]$ where v = number of degrees of freedom for error in the candidate model: $v = n - p = 25 - p$ in our example, and $F_{\gamma.q.v}$ is read from F-tables.

Table 15.10 shows the results of these significance tests for each of the nine candidate models.

TABLE 15.10   *Tests of Significance for Nine C_p-Statistics*

Model No.	p	$v = 25 - p$	$q = 9 - p$	$F_{0.05.q.v}$	$q[F_{0.05.q.v} - 1]$	$F_p = C_p - p$
1	2	23	7	2.44	10.08	65.6
2	3	22	6	2.55	9.30	24.2
3	3	22	6	2.55	9.30	30.1
4	4	21	5	2.68	8.40	14.4
5	4	21	5	2.68	8.40	10.0
6	5	20	4	2.87	7.48	0.4
7	6	19	3	3.13	6.39	−1.6
8	6	19	3	3.13	6.39	−1.0
9	7	18	2	3.55	5.10	−1.7

Table 15.10 shows that Models 1–5 can be eliminated from further consideration since in each case $C_p - p$ is greater than the critical value, so that we can be more than 95% sure that these models are not appropriate. However, we cannot be 95% sure about any

of the remaining four models. While we reached the same conclusion without aid of the hypothesis test in this example, it will not always be so easy to eliminate models just by looking.

Final choice of a model is narrowed down to a choice among Models 6, 7, 8, and 9. Selection should be based on which model makes the best sense to the researcher, whether all variables in the model are significant at the desired significance level, the multiple coefficient of determination for the model, which variables are cheaper or easier to measure, and possibly which one has smallest predictor error.

EXERCISES   *Section 15.4*

15.8   Some candidate models which came out of the various stagewise analyses of the
p. 536   TENNESSEE data were:

Model	Predictors
1	percent of households with telephones
2	percent of households with telephones
	population change
3	population change
	agricultural employment
	percent rural population
4	population change
	agricultural employment
	percent rural population
	median age

In all four models, all predictors are significant at $\alpha = 0.05$. Use the PRESS statistic to determine which model will make the best predictions when applied to different data.

15.9   Calculate the C_p-statistic for each of the four candidate models in Exercise 15.8. Use the C_p-statistic to select an appropriate model.

15.5 SUMMARY

In this chapter we considered alternative final models arrived at by various variable selection procedures in terms of how well they will predict the response based on future values of the predictors. Even with these considerations, it is possible that no uniformly best model will be found; rather it is likely that two or three candidate models will be judged about equally appropriate from a statistical point of view. Choice of a final model is considerably facilitated if the researcher has a theoretical or practical framework within which to operate and can choose the model that makes the best sense to him or her.

FURTHER STUDY

We have presented only the classical least squares approach to regression analysis in this text. Of interest to the student might also be nonlinear regression, in which a model such as

$$y = \frac{1}{1 + e^{x\beta}}$$

is to be fitted. (This *logistic model* is sometimes used when y is a probability to be estimated, $0 \leq y \leq 1$.) While this model can be linearized by taking logarithms, violations of assumptions can be brought about by such linearization. Iterative computer procedures have been developed to fit such nonlinear models.

The concept of *weighted least squares* was mentioned in Section 15.2, in connection with transforming the data so as to achieve uniform scatter in a multiple regression. This topic is also worthy of further study.

It can happen that unusual multicollinearity structures cause the standard errors of some of the regression coefficients to be quite large, resulting in very unstable estimates of the corresponding parameters. In such cases, the researcher may note that regression coefficients are far out of line with what was expected and may even have the opposite sign from what they should have. It is possible in such instances to employ *biased estimation* procedures, such as *ridge regression* or *eigenvalue regression*, which will give estimates with smaller variance than those obtained using least squares. When the multicollinearity structure is such that $\mathbf{X'X}$ is nearly singular and you run into difficulty in trying to find its inverse, then *regression on principal components* is sometimes helpful.

It is hoped that with the basis in least squares regression provided by this book, the reader will be able to proceed to these more advanced topics.

CONTENTS

SELECTED REFERENCES AND SUGGESTIONS FOR FURTHER READING

Here is a short list of some commonly available texts that might be useful to the student who wishes to pursue topics in regression analysis. While this list is far from a complete set of references for regression analysis, it includes:

1. Elementary-level statistics survey texts and books of readings that might help supplement and illustrate concepts, especially simple regression;
2. Texts aimed toward a specialized audience, such as sociologists or economists, that show how regression techniques are applied in these fields;
3. Regression texts that explain more advanced topics, such as weighted regression or ridge regression, that are alluded to in this text;
4. Manuals for running regression programs;
5. Theoretical statistics texts for the rigorous mathematical and statistical basis for regression.

A short description of the reference is provided as an aid to directing the student to potentially useful references for his or her particular interests.

Barr, Anthony J., et al. 1976. *A User's Guide to SAS76*. SAS Institute, Inc., Raleigh, N.C.

————. 1977. *SAS Supplemental Library User's Guide*. SAS Institute, Inc., Raleigh, N.C.
Instructions for running SAS76 programs.

Blalock, Hubert M. 1960. *Social Statistics*. McGraw-Hill, New York.
A statistical survey text designed for use by sociologists. Uses many examples taken from sociology and includes detailed discussions. Of interest to the reader might be rank correlations as well as interpretations and use of correlations and beta coefficients in sociological studies.

Chatterjee, Samprit, and Betram Price. 1977. *Regression Analysis by Example*. John Wiley & Sons, New York.
Many actual data sets are analyzed to introduce concepts and also to illustrate failure of the assumptions of the linear model. More advanced topics, such as ridge regression and regression on principal components, are also motivated and illustrated. Relatively mathematically sophisticated.

Daniel, Cuthbert, and Fred S. Wood. 1971. *Fitting Equations to Data: Computer Analysis of Multifactor Data for Scientists and Engineers*. Wiley-Interscience, New York.
Use of computer to analyze data.

Dixon, W. J., ed. 1973. *BMD Biomedical Computer Programs.* University of California Press, Berkeley.

————. 1975. *BMDP Biomedical Computer Programs.* University of California Press, Berkeley. Instructions for running BMD and BMDP programs, and explanation of algorithms used in executing the various programs.

Draper, N. R., and H. Smith. 1966. *Applied Regression Analysis.* John Wiley & Sons, New York. A useful text to those with good training in mathematics and statistics. Covers several advanced concepts and contains many solved examples. Provides a more rigorous and thorough treatment of some concepts presented here and should be the next step in the student's study.

Dutta, M. 1975. *Econometric Methods.* South-Western Publishing Co., Cincinnati. An introductory regression text geared toward the student of economics. Considers some statistical properties and covers several more advanced topics, mostly dealing with time-dependent data.

Edwards, Allen L. 1976. *An Introduction to Linear Regression and Correlation.* W. H. Freeman and Co., San Francisco. Written for students of the behavorial sciences, this text includes such topics as nonparametric measures of association, regression with standardized variables, and other specialized topics.

Freedman, David, Robert Pisani, and Roger Purves. 1978. *Statistics.* W. W. Norton, New York. A statistics survey text written for nonmathematically oriented students. Chapters on correlation and regression are well illustrated with historical notes and actual studies.

Graybill, Franklin A. 1961. *An Introduction to Linear Statistical Models, Volume I.* McGraw-Hill, New York. The bible for statisticians, this very mathematical text provides all the basic statistical theory of regression. Not recommended for the nonstatistician. See also: ————. 1976. *Theory and Application of the Linear Model.* Duxbury Press, North Scituate, Mass.

Hamburg, Morris. 1977. *Statistical Analysis for Decision Making.* Harcourt Brace Jovanovich, New York. A statistical survey text intended for use by business students. The chapter on regression and correlation is especially good for a business statistics text and might be helpful in supplementing various topics.

Kerlinger, Fred N., and Elazar J. Pedhazur. 1973. *Multiple Regression in Behavioral Research.* Holt, Rinehart & Winston, New York. Elements of simple and multiple regression are covered in the first two chapters, with the rest of the book considering more advanced topics. Examples are taken from the behavioral sciences. Quite useful for advanced work in social sciences.

Kleinbaum, David G., and Lawrence L. Kupper. 1978. *Applied Regression Analysis and Other Multivariate Methods.* Duxbury Press, North Scituate, Mass.

A relatively complete treatment of simple and multiple regression and variable selection techniques, including extensions to experimental design models as well as factor and discriminant analysis. Extensive references to output from computer programs.

Merrill, William C., and Karl A. Fox. 1970. *Introduction to Economic Statistics.* John Wiley & Sons, New York.

A statistical survey text intended for economics and business students. Chapters on regression are useful to relate regression techniques to economics applications and to investigate problems of time-dependent data.

Mosteller, Fredrick, and John W. Tukey. 1977. *Data Analysis and Regression.* Addison-Wesley Publishing Co., Reading, Mass.

Emphasis on appropriate techniques for data sets displaying particular properties. More advanced topics, such as weighted least squares, are well illustrated.

Neter, John, and William Wasserman. 1974. *Applied Linear Statistical Models.* Richard D. Irwin, Homewood, Ill.

A very complete text in the use of regression techniques. Especially useful in relating regression to various experimental design models.

Nile, Norman H. 1975. *SPSS: Statistical Package for Social Sciences.* McGraw-Hill, New York.

Instructions for running SPSS programs. Also a brief introduction to regression.

Ryan, Thomas A., Brian L. Joiner, and Barbara F. Ryan. 1976 *MINITAB Student Handbook.* Duxbury Press, North Scituate, Mass.

A good introduction to regression using the MINITAB computing system.

Spurr, William A., and Charles P. Bonini. 1973. *Statistical Analysis for Business Decisions.* Richard D. Irwin, Homewood, Ill.

A very complete statistical survey text intended for business students. Provides very good introductory regression chapters with well-illustrated examples and information on the use of the BMDP2R.

Tanur, G. M., ed. 1972. *Statistics: A Guide to the Unknown.* Holden-Day, San Francisco.

A delightful book of nontechnical essays on the use of statistics in actual applications from astronomy to zoology.

Wonnacott, Thomas H., and Ronald J. Wonnacott. 1972. *Introductory Statistics for Business and Economics.* John Wiley & Sons, New York.

A statistical survey text intended for business majors. Chapters on regression could be useful as a supplement to this text by providing explanations in terms of business-type problems.

APPENDICES

Appendix A MINIMIZATION OF $\sum_{i=1}^{n} (Y_i - \hat{Y}_i)^2$

In order to find a minimum value for a function, one takes a first derivative, equates this to zero, and solves for the unknown quantity. Since this gives either the maximum or minimum, a second derivative check is necessary in order to determine which has been found. Our house size-price example, however, has *two* unknown quantities, a and b, so that we can determine the values in the equation $\hat{Y} = a + bX$. Note that X and Y are *known variables* where respective values are the observed sizes and prices of houses given in Table 1.1. The quantities a and b are *unknown constants*. Our example has *one* unknown a-value and *one* unknown b-value; there are 20 X-values which we do know, and 20 Y-values which are also known.

Since we have two unknowns, we must take partial derivatives. A partial derivative with respect to a will simply ignore the fact that b is also unknown, treating it as a constant; and similarly for a partial derivative with respect to b. Thus to minimize

$$Q = \sum_{i=1}^{n} (Y_i - \hat{Y}_i)^2 = \sum_{i=1}^{n} (Y_i - a - bX_i)^2$$

we first take the first partial derivative of Q with respect to a:

$$\frac{\partial Q}{\partial a} = 2 \sum_{i=1}^{n} (Y_i - a - bX_i)(-1) = -2 \sum_{i=1}^{n} (Y_i - a - bX_i).$$

Equating to zero, we have

$$-2 \sum_{i=1}^{n} (Y_i - a - bX_i) = 0,$$

$$\sum_{i=1}^{n} Y_i - \sum_{i=1}^{n} a - \sum_{i=1}^{n} bX_i = 0,$$

$$\sum_{i=1}^{n} Y_i - na - b\sum_{i=1}^{n} X_i = 0,$$

$$\sum_{i=1}^{n} Y_i = na + b\sum_{i=1}^{n} X_i.$$

Note that this is one equation in two unknowns, a and b. In order to solve for a and b, we need another equation involving these two variables, so we take a first partial derivative of Q with respect to b:

$$\frac{\partial Q}{\partial b} = 2\sum_{i=1}^{n} (Y_i - a - bX_i)(-X_i)$$

$$= -2\sum_{i=1}^{n} (X_iY_i - aX_i - bX_i^2).$$

Equating to zero,

$$-2\sum_{i=1}^{n} (X_iY_i - aX_i - bX_i^2) = 0,$$

$$\sum_{i=1}^{n} X_iY_i - a\sum_{i=1}^{n} X_i - b\sum_{i=1}^{n} X_i^2 = 0,$$

$$\sum_{i=1}^{n} X_iY_i = a\sum_{i=1}^{n} X_i + b\sum_{i=1}^{n} X_i^2,$$

and this is the second equation in two unknowns. Thus, solving these two equations,

$$\sum_{i=1}^{n} Y_i = na + b\sum_{i=1}^{n} X_i$$

and

$$\sum_{i=1}^{n} X_iY_i = a\sum_{i=1}^{n} X_i + b\sum_{i=1}^{n} X_i^2,$$

simultaneously for a and b will give either a maximum or a minimum. Thus, the line $\hat{Y} = a + bX$ will be the line which is, on the average, either the farthest from or the closest to all the (X, Y) points, in terms of squared deviations. The second derivative check will tell us which. Since there are two unknowns, we must see if the determinant of the matrix of coefficients on a and b is positive in order for a minimum to have been found:

$$\begin{vmatrix} n & \sum_{i=1}^{n} X_i \\ \sum_{i=1}^{n} X_i & \sum_{i=1}^{n} X_i^2 \end{vmatrix} = n\sum_{i=1}^{n} X_i^2 - \left(\sum_{i=1}^{n} X_i\right)^2.$$

(Those unfamiliar with matrices and determinants may want to look back to this after completion of Chapter 3.) You will recall that this determinant is the numerator of the calculating formula for the sum of squares,

$$\sum_{i=1}^{n} (X_i - \overline{X})^2 = \frac{n \sum\limits_{i=1}^{n} X_i^2 - \left(\sum\limits_{i=1}^{n} X_i\right)^2}{n}.$$

Thus, since a sum of squares can never be negative, being the sum of squared values, the quantity

$$n \sum_{i=1}^{n} X_i^2 - \left(\sum_{i=1}^{n} X_i\right)^2$$

must always be positive or zero, regardless of what the X-values are. The only way this quantity can be zero is for all X_i-values to be equal to $\overline{X}$, and thus all equal to each other. Thus we have shown that the solution to the normal equations will yield values a and b so that the line $\hat{Y} = a + bX$ is the one which is, on the average, closest to all data points, in terms of squared deviations.

Appendix B THE BINARY NUMBER SYSTEM

Because human beings have five fingers on each of two hands, most cultures use the decimal (base 10) system. There is really nothing sacred about it, and indeed there exist and have existed cultures which use different systems. Cultures which counted on their toes as well as their fingers developed base 20 number systems, and time is measured in a base 60 system. Whatever the system, there are three main features:

1. The number of symbols required is the same as the base figure.
2. The fewer the symbols, the more of them required to represent any given value.
3. The value is calculated, starting at the decimal point and moving to the left, by multiplying each digit by the base number raised to the zeroth, first, second, etc., power.

In the decimal system, then, we have the ten symbols 0, 1, 2, 3, 4, 5, 6, 7, 8, 9. The number 123 is

$$1(10)^2 + 2(10) + 3(10)^0 = 100 + 2(10) + 3(1)$$
$$= 100 + 20 + 3,$$

or one-hundred twenty-three. If we were working in base 5 (suppose people counted on only one hand), then the number 123 would be

$$1(5)^2 + 2(5)^1 + 3(5)^0 = 25 + 10 + 3,$$

or thirty-eight. To represent the value one-hundred twenty-three to base 5, we would need to write

$$4(5)^2 + 4(5)^1 + 3(5)^0, \quad \text{or} \quad 443.$$

Computers depend not on ten fingers, but on such things as switches that are open or closed, positions on cards that are or are not punched. Thus, the computer recognizes only two symbols—off/on or no/yes—which we usually denote by 0 and 1, respectively. In the base two system then, when a number is moved one place to the left, the number it represents is multiplied by two; moving a digit one place to the right signifies division by two. Thus,

$$1 = 1(2)^0 = 1,$$
$$10 = 1(2)^1 + 0(2)^0 = 2,$$
$$100 = 1(2)^2 + 0(2)^1 + 0(2)^0 = 4,$$

and similarly,

$$0.1 = 1(\tfrac{1}{2})^1 = \tfrac{1}{2},$$
$$0.01 = 0(\tfrac{1}{2})^1 + 1(\tfrac{1}{2})^2 = \tfrac{1}{4},$$
$$0.001 = 0(\tfrac{1}{2})^1 + 0(\tfrac{1}{2})^2 + 1(\tfrac{1}{2})^3 = \tfrac{1}{8},$$

and so on.

Thus, the binary number 101 represents the decimal number 5, since

$$1(2)^2 + 0(2)^1 + 1(2)^0 = 4 + 0 + 1 = 5.$$

To convert a decimal number to a binary number, repeatedly divide the decimal number by 2 until the quotient is less than 1. Reading the remainders, in reverse order, gives the binary representation. For example, to convert 553 to the binary system:

$$553 \div 2 = 276, \quad \text{remainder} = 1,$$
$$276 \div 2 = 138, \quad \text{remainder} = 0,$$
$$138 \div 2 = 69, \quad \text{remainder} = 0,$$
$$69 \div 2 = 34, \quad \text{remainder} = 1,$$
$$34 \div 2 = 17, \quad \text{remainder} = 0,$$
$$17 \div 2 = 8, \quad \text{remainder} = 1,$$
$$8 \div 2 = 4, \quad \text{remainder} = 0,$$
$$4 \div 2 = 2, \quad \text{remainder} = 0,$$
$$2 \div 2 = 1, \quad \text{remainder} = 0,$$
$$1 \div 2 = 0, \quad \text{remainder} = 1.$$

Thus, the binary representation of the decimal number 553 is

1000101001.

To convert fractions to the binary system, instead of repeatedly dividing by two, we need to multiply by two. To convert the decimal number 0.625 to a binary number,

$$0.625 \times 2 = 1.250,$$

$$0.250 \times 2 = 0.500,$$

$$0.500 \times 2 = 1.000.$$

Note that at each step, we took only the fractional part of the number and multiplied by 2. We stopped when the fractional part became zero. The binary equivalent of the decimal number 0.625 is then

$$0.101,$$

reading the integral parts of the products in order, or down the above column. The decimal part of the products will not always turn out to be zero. In this case, the decimal part will sooner or later repeat, as in the case below:

$$0.55 \times 2 = 1.10,$$

$$0.10 \times 2 = 0.20,$$

$$0.20 \times 2 = 0.40,$$

$$0.40 \times 2 = 0.80,$$

$$0.80 \times 2 = 1.60,$$

$$0.60 \times 2 = 1.20.$$

Notice that the decimal portion is beginning to repeat. Thus, the binary representation of the decimal number 0.55 is

$$0.10001100110011\ldots = 0.1\overline{00011}.$$

Binary arithmetic simplifies the circuitry necessary to perform calculations in a computer. The tables below give the rules for multiplication and addition:

Addition	0	1
0	0	1
1	1	10

Multiplication	0	1
0	0	0
1	0	1

Note that if any value converted to a binary number results in a repeating decimal, we have a source of rounding error when the value is reconverted to the decimal system.

Appendix C   ORTHOGONAL POLYNOMIALS

Suppose one desires to fit a polynomial equation of the form

$$Y = a + b_1 X + b_2 X^2 + \cdots + b_k X^k$$

to a set of data. Even if $k = 2$, solution of the normal equations

$$\mathbf{X'X}\hat{\boldsymbol{\beta}} = \mathbf{X'y}$$

can be quite difficult if $\mathbf{X'X}$ must be inverted by hand. If, however, X is a controlled variable with values at evenly spaced intervals, one can use orthogonal polynomials and rewrite the equation so that $\mathbf{X'X}$ will be a diagonal matrix and therefore easy to invert, regardless of its dimension. Suppose, for example, that an experiment has been run using the following X-values:

X
60
80
100
120
140
160
180
200
210

For these values, we can show that if our model is $Y = \alpha + \beta X + \varepsilon$, then

$$\mathbf{X'X} = \begin{bmatrix} 9 & 1,250 \\ 1,250 & 196,100 \end{bmatrix}.$$

While this matrix is not difficult to invert, we saw in Section 2.4 that we can reduce labor if we scale the X-values by

$$Z = \frac{X - \bar{X}}{d},$$

where d is the spacing between levels of X. Since $\bar{X} = 140$ and $d = 20$, we code the X-values to

$$
\begin{array}{c}
Z \\
\hline
-4 \\
-3 \\
-2 \\
-1 \\
0 \\
1 \\
2 \\
3 \\
4
\end{array}
$$

and then

$$
\mathbf{Z'Z} = \begin{bmatrix} 9 & 0 \\ 0 & 60 \end{bmatrix},
$$

which is diagonal and quite simple to invert.

Let us call this Z-value the *first orthogonal polynomial* and give it a subscript to denote that it is the first one:

$$
Z_1 = \frac{X - \bar{X}}{d}.
$$

We have seen that the simple linear regression equation

$$
\hat{Y} = a + bX
$$

can be equivalently written as

$$
\hat{Y} = a' + b'Z_1.
$$

What if we have a quadratic regression equation

$$
\hat{Y} = a + b_1 X + b_2 X^2 ?
$$

Can we find a second orthogonal polynomial so that the regression equation can be written

$$
\hat{Y} = a' + b_1' Z_1 + b_2' Z_2
$$

and such that $\mathbf{Z'Z}$ is a diagonal? The reader has probably guessed that this appendix would not be here unless the answer to the above question were yes.

If we wanted to fit a quadratic equation using the original data, then

$$\mathbf{X} = \begin{bmatrix} 1 & 60 & 3{,}600 \\ 1 & 80 & 6{,}400 \\ 1 & 100 & 10{,}000 \\ 1 & 120 & 14{,}400 \\ 1 & 140 & 19{,}600 \\ 1 & 160 & 25{,}600 \\ 1 & 180 & 32{,}400 \\ 1 & 200 & 40{,}000 \\ 1 & 210 & 44{,}100 \end{bmatrix}$$

and $\mathbf{X}'\mathbf{X}$ will not be diagonal. Coding the X-values using the first orthogonal polynomial transforms the second column of $\mathbf{X}$ into $-4, -3, \ldots, 3, 4$. Define the *second orthogonal polynomial* as

$$Z_2 = \left(\frac{X - \bar{X}}{d}\right)^2 - \frac{n^2 - 1}{12}.$$

In our example, since $n = 9$, $(n^2 - 1)/12 = 80/12 = 6.67$, then

X	Z_1	Z_2	
60	-4	$(-4)^2 - 6.67 =$	9.33
80	-3	$(-3)^2 - 6.67 =$	2.33
100	-2	$(-2)^2 - 6.67 =$	-2.67
120	-1	$(-1)^2 - 6.67 =$	-5.67
140	0	$0^2 - 6.67 =$	-6.67
160	1	$(1)^2 - 6.67 =$	-5.67
180	2	$(2)^2 - 6.67 =$	-2.67
200	3	$(3)^2 - 6.67 =$	2.33
210	4	$(4)^2 - 6.67 =$	9.33
	0	$\approx$	0

and

$$\mathbf{Z}'\mathbf{Z} = \begin{bmatrix} 9 & 0 & 0 \\ 0 & 60 & 0 \\ 0 & 0 & 308 \end{bmatrix},$$

since it is easy to see that $\sum Z_1 Z_2 = 0$; the reader can verify that $\sum Z_2^2 = 308$. Thus it is an easy matter to find a', b'_1, and b'_2 for the regression equation

$$Y = a' + b'_1 Z_1 + b'_2 Z_2.$$

One can then transform back to find a, b_1, and b_2 for the original equation

$$Y = a + b_1 X + b_2 X.$$

The third, fourth, ... orthogonal polynomials, used in fitting cubic or higher-degree polynomial equations can be found using a recursive formula:

$$Z_{i+1} = Z_i Z_1 - \frac{i^2(n^2 - i^2)}{4(4i^2 - 1)} Z_{i-1},$$

where $Z_0 = 1$. For the third orthogonal polynomial, let $i + 1 = 3$, so $i = 2$. Then

$$Z_3 = Z_2 Z_1 - \frac{2^2(n^2 - 2^2)}{4[4(2^2) - 1]} Z_1$$

$$= \left[\left(\frac{X - \bar{X}}{d} \right)^2 - \frac{n^2 - 1}{12} \right] \left[\frac{X - \bar{X}}{d} \right] - \frac{n^2 - 4}{15} \left[\frac{X - \bar{X}}{d} \right]$$

$$= \left(\frac{X - \bar{X}}{d} \right)^3 - \left(\frac{n^2 - 1}{12} + \frac{n^2 - 4}{15} \right) \left(\frac{X - \bar{X}}{d} \right)$$

$$= \left(\frac{X - \bar{X}}{d} \right)^3 - \frac{3n^2 - 7}{20} \left(\frac{X - \bar{X}}{d} \right).$$

While it can be time-consuming to calculate orthogonal polynomials, a great saving in labor can be achieved with them if one has no access to a computer to solve polynomial equations. They are of most use, however, in designed experiments in which X-values are at evenly-spaced intervals; otherwise they are usually more trouble than they are worth.

Appendix D   FITTING THE REGRESSION LINE THROUGH THE ORIGIN

We want to find a value for b^* such that

$$Q_1 = \sum_{i=1}^{n} (Y_i - \hat{Y}_i)^2 = \sum_{i=1}^{n} (Y_i - b^* X_i)^2$$

is minimized. Taking the first derivative with respect to b^*, we obtain

$$\frac{dQ_1}{db^*} = 2\sum (Y_i - b^* X_i)(-X_i)$$

$$= -2\sum (X_i Y_i - b^* X_i^2).$$

Equating to zero,

$$-2\sum (X_i Y_i - b^* X_i^2) = 0,$$

$$\sum X_i Y_i - b^* \sum X_i^2 = 0,$$

$$\frac{\sum X_i Y_i}{\sum X_i^2} = b^*.$$

For a second-derivative check for a minimum, note that the first derivative can be written as

$$-2\sum X_i Y_i + 2b^* \sum X_i^2.$$

Then

$$\frac{dQ_2}{db^{*2}} = 2\sum X_i^2 > 0,$$

and a minimum has been found.

Table I   CRITICAL VALUES OF THE STUDENT'S t-DISTRIBUTION

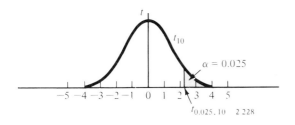

Table gives values of t which cut off $100\alpha\%$ of area under upper tail.

Example: $t_{.025,10} = 2.228$.

For a t with 10 degrees of freedom, the area under the curve over the interval from $t = 2.228$ to infinity is $2\frac{1}{2}\%$ of the total area.

υ	α				
	0.10	0.05	0.025	0.01	0.005
1	3.078	6.314	12.706	31.821	63.657
2	1.886	2.920	4.303	6.965	9.925
3	1.638	2.353	3.182	4.541	5.841
4	1.533	2.132	2.776	3.747	4.604
5	1.476	2.015	2.571	3.365	4.032
6	1.440	1.943	2.447	3.143	3.707
7	1.415	1.895	2.365	2.998	3.499
8	1.397	1.860	2.306	2.896	3.355
9	1.383	1.833	2.262	2.821	3.250
10	1.372	1.812	2.228	2.764	3.169
11	1.363	1.796	2.201	2.718	3.106
12	1.356	1.782	2.179	2.681	3.055
13	1.350	1.771	2.160	2.650	3.012
14	1.345	1.761	2.145	2.624	2.977
15	1.341	1.753	2.131	2.602	2.947
16	1.337	1.746	2.120	2.583	2.921
17	1.333	1.740	2.110	2.567	2.898
18	1.330	1.734	2.101	2.552	2.878
19	1.328	1.729	2.093	2.539	2.861
20	1.325	1.725	2.086	2.528	2.845
21	1.323	1.721	2.080	2.518	2.831
22	1.321	1.717	2.074	2.508	2.819
23	1.319	1.714	2.069	2.500	2.807
24	1.318	1.711	2.064	2.492	2.797
25	1.316	1.708	2.060	2.485	2.787
26	1.315	1.706	2.056	2.479	2.779
27	1.314	1.703	2.052	2.473	2.771
28	1.313	1.701	2.048	2.467	2.763
29	1.311	1.699	2.045	2.462	2.756
30	1.310	1.697	2.042	2.457	2.750
40	1.303	1.684	2.021	2.423	2.704
60	1.296	1.671	2.000	2.390	2.660
120	1.289	1.658	1.980	2.358	2.617
∞	1.282	1.645	1.960	2.326	2.576

Table I is taken from Table III of Fisher and Yates: *Statistical Tables for Biological, Agricultural, and Medical Research*, published by Longman Group, Ltd., London (previously published by Oliver and Boyd, Edinburgh), and by permission of the authors and publishers; and from Table 12 of *Biometrika Tables for Statisticians*, Vol. 1, 1966, by permission of the Biometrika Trustees.

Table II   CRITICAL VALUES OF THE F-DISTRIBUTION

$$F_{.05,\nu_1,\nu_2}$$

$$F_{.05,\nu_1,\nu_2}$$

ν_2	1	2	3	4	5	6	7	8	9	10	12	15	20	24	30	40	60	120	∞
1	161.4	199.5	215.7	224.6	230.2	234.0	236.8	238.9	240.5	6056	6106	6157	6209	6235	6261	6287	6313	6339	6366
2	18.51	19.00	19.16	19.25	19.30	19.33	19.35	19.37	19.38	99.40	99.42	99.43	99.45	99.46	99.47	99.47	99.48	99.49	99.50
3	10.13	9.55	9.28	9.12	9.01	8.94	8.89	8.85	8.81	27.23	27.05	26.87	26.69	26.60	26.50	26.41	26.32	26.22	26.13
4	7.71	6.94	6.59	6.39	6.26	6.16	6.09	6.04	6.00	14.55	14.37	14.20	14.02	13.93	13.84	13.75	13.65	13.56	13.46
5	6.61	5.79	5.41	5.19	5.05	4.95	4.88	4.82	4.77	10.05	9.89	9.72	9.55	9.47	9.38	9.29	9.20	9.11	9.02
6	5.99	5.14	4.76	4.53	4.39	4.28	4.21	4.15	4.10	7.87	7.72	7.56	7.40	7.31	7.23	7.14	7.06	6.97	6.88
7	5.59	4.74	4.35	4.12	3.97	3.87	3.79	3.73	3.68	6.62	6.47	6.31	6.16	6.07	5.99	5.91	5.82	5.74	5.65
8	5.32	4.46	4.07	3.84	3.69	3.58	3.50	3.44	3.39	5.81	5.67	5.52	5.36	5.28	5.20	5.12	5.03	4.95	4.86
9	5.12	4.26	3.86	3.63	3.48	3.37	3.29	3.23	3.18	5.26	5.11	4.96	4.81	4.73	4.65	4.57	4.48	4.40	4.31
10	4.96	4.10	3.71	3.48	3.33	3.22	3.14	3.07	3.02	4.85	4.71	4.56	4.41	4.33	4.25	4.17	4.08	4.00	3.91
11	4.84	3.98	3.59	3.36	3.20	3.09	3.01	2.95	2.90	4.54	4.40	4.25	4.10	4.02	3.94	3.86	3.78	3.69	3.60
12	4.75	3.89	3.49	3.26	3.11	3.00	2.91	2.85	2.80	4.30	4.16	4.01	3.86	3.78	3.70	3.62	3.54	3.45	3.36
13	4.67	3.81	3.41	3.18	3.03	2.92	2.83	2.77	2.71	4.10	3.96	3.82	3.66	3.59	3.51	3.43	3.34	3.25	3.17
14	4.60	3.74	3.34	3.11	2.96	2.85	2.76	2.70	2.65	3.94	3.80	3.66	3.51	3.43	3.35	3.27	3.18	3.09	3.00
15	4.54	3.68	3.29	3.06	2.90	2.79	2.71	2.64	2.59	3.80	3.67	3.52	3.37	3.29	3.21	3.13	3.05	2.96	2.87
16	4.49	3.63	3.24	3.01	2.85	2.74	2.66	2.59	2.54	3.69	3.55	3.41	3.26	3.18	3.10	3.02	2.93	2.84	2.75
17	4.45	3.59	3.20	2.96	2.81	2.70	2.61	2.55	2.49	3.59	3.46	3.31	3.16	3.08	3.00	2.92	2.83	2.75	2.65
18	4.41	3.55	3.16	2.93	2.77	2.66	2.58	2.51	2.46	3.51	3.37	3.23	3.08	3.00	2.92	2.84	2.75	2.66	2.57
19	4.38	3.52	3.13	2.90	2.74	2.63	2.54	2.48	2.42	3.43	3.30	3.15	3.00	2.92	2.84	2.76	2.67	2.58	2.49
20	4.35	3.49	3.10	2.87	2.71	2.60	2.51	2.45	2.39	3.37	3.23	3.09	2.94	2.86	2.78	2.69	2.61	2.52	2.42
21	4.32	3.47	3.07	2.84	2.68	2.57	2.49	2.42	2.37	3.31	3.17	3.03	2.88	2.80	2.72	2.64	2.55	2.46	2.36
22	4.30	3.44	3.05	2.82	2.66	2.55	2.46	2.40	2.34	3.26	3.12	2.98	2.83	2.75	2.67	2.58	2.50	2.40	2.31
23	4.28	3.42	3.03	2.80	2.64	2.53	2.44	2.37	2.32	3.21	3.07	2.93	2.78	2.70	2.62	2.54	2.45	2.35	2.26
24	4.26	3.40	3.01	2.78	2.62	2.51	2.42	2.36	2.30	3.17	3.03	2.89	2.74	2.66	2.58	2.49	2.40	2.31	2.21
25	4.24	3.39	2.99	2.76	2.60	2.49	2.40	2.34	2.28	3.13	2.99	2.85	2.70	2.62	2.54	2.45	2.36	2.27	2.17
26	4.23	3.37	2.98	2.74	2.59	2.47	2.39	2.32	2.27	3.09	2.96	2.81	2.66	2.58	2.50	2.42	2.33	2.23	2.13
27	4.21	3.35	2.96	2.73	2.57	2.46	2.37	2.31	2.25	3.06	2.93	2.78	2.63	2.55	2.47	2.38	2.29	2.20	2.10
28	4.20	3.34	2.95	2.71	2.56	2.45	2.36	2.29	2.24	3.03	2.90	2.75	2.60	2.52	2.44	2.35	2.26	2.17	2.06
29	4.18	3.33	2.93	2.70	2.55	2.43	2.35	2.28	2.22	3.00	2.87	2.73	2.57	2.49	2.41	2.33	2.23	2.14	2.03
30	4.17	3.32	2.92	2.69	2.53	2.42	2.33	2.27	2.21	2.98	2.84	2.70	2.55	2.47	2.39	2.30	2.21	2.11	2.01
40	4.08	3.23	2.84	2.61	2.45	2.34	2.25	2.18	2.12	2.80	2.66	2.52	2.37	2.29	2.20	2.11	2.02	1.92	1.80
60	4.00	3.15	2.76	2.53	2.37	2.25	2.17	2.10	2.04	2.63	2.50	2.35	2.20	2.12	2.03	1.94	1.84	1.73	1.60
120	3.92	3.07	2.68	2.45	2.29	2.17	2.09	2.02	1.96	2.47	2.34	2.19	2.03	1.95	1.86	1.76	1.66	1.53	1.38
∞	3.84	3.00	2.60	2.37	2.21	2.10	2.01	1.94	1.88	2.32	2.18	2.04	1.88	1.79	1.70	1.59	1.47	1.32	1.00

Table II (continued)

$$F_{.01, v_1, v_2} \qquad\qquad F_{.01, v_1, v_2}$$

v_2	v_1=1	2	3	4	5	6	7	8	9	v_1=10	12	15	20	24	30	40	60	120	∞
1	4052	4999.5	5403	5625	5764	5859	5928	5981	6022	241.9	243.9	245.9	248.0	249.1	250.1	251.1	252.2	253.3	254.3
2	98.50	99.00	99.17	99.25	99.30	99.33	99.36	99.37	99.39	19.40	19.41	19.43	19.45	19.45	19.46	19.47	19.48	19.49	19.50
3	34.12	30.82	29.46	28.71	28.24	27.91	27.67	27.49	27.35	8.79	8.74	8.70	8.66	8.64	8.62	8.59	8.57	8.55	8.53
4	21.20	18.00	16.69	15.98	15.52	15.21	14.98	14.80	14.66	5.96	5.91	5.86	5.80	5.77	5.75	5.72	5.69	5.66	5.63
5	16.26	13.27	12.06	11.39	10.97	10.67	10.46	10.29	10.16	4.74	4.68	4.62	4.56	4.53	4.50	4.46	4.43	4.40	4.36
6	13.75	10.92	9.78	9.15	8.75	8.47	8.26	8.10	7.98	4.06	4.00	3.94	3.87	3.84	3.81	3.77	3.74	3.70	3.67
7	12.25	9.55	8.45	7.85	7.46	7.19	6.99	6.84	6.72	3.64	3.57	3.51	3.44	3.41	3.38	3.34	3.30	3.27	3.23
8	11.26	8.65	7.59	7.01	6.63	6.37	6.18	6.03	5.91	3.35	3.28	3.22	3.15	3.12	3.08	3.04	3.01	2.97	2.93
9	10.56	8.02	6.99	6.42	6.06	5.80	5.61	5.47	5.35	3.14	3.07	3.01	2.94	2.90	2.86	2.83	2.79	2.75	2.71
10	10.04	7.56	6.55	5.99	5.64	5.39	5.20	5.06	4.94	2.98	2.91	2.85	2.77	2.74	2.70	2.66	2.62	2.58	2.54
11	9.65	7.21	6.22	5.67	5.32	5.07	4.89	4.74	4.63	2.85	2.79	2.72	2.65	2.61	2.57	2.53	2.49	2.45	2.40
12	9.33	6.93	5.95	5.41	5.06	4.82	4.64	4.50	4.39	2.75	2.69	2.62	2.54	2.51	2.47	2.43	2.38	2.34	2.30
13	9.07	6.70	5.74	5.21	4.86	4.62	4.44	4.30	4.19	2.67	2.60	2.53	2.46	2.42	2.38	2.34	2.30	2.25	2.21
14	8.86	6.51	5.56	5.04	4.69	4.46	4.28	4.14	4.03	2.60	2.53	2.46	2.39	2.35	2.31	2.27	2.22	2.18	2.13
15	8.68	6.36	5.42	4.89	4.56	4.32	4.14	4.00	3.89	2.54	2.48	2.40	2.33	2.29	2.25	2.20	2.16	2.11	2.07
16	8.53	6.23	5.29	4.77	4.44	4.20	4.03	3.89	3.78	2.49	2.42	2.35	2.28	2.24	2.19	2.15	2.11	2.06	2.01
17	8.40	6.11	5.18	4.67	4.34	4.10	3.93	3.79	3.68	2.45	2.38	2.31	2.23	2.19	2.15	2.10	2.06	2.01	1.96
18	8.29	6.01	5.09	4.58	4.25	4.01	3.84	3.71	3.60	2.41	2.34	2.27	2.19	2.15	2.11	2.06	2.02	1.97	1.92
19	8.18	5.93	5.01	4.50	4.17	3.94	3.77	3.63	3.52	2.38	2.31	2.23	2.16	2.11	2.07	2.03	1.98	1.93	1.88
20	8.10	5.85	4.94	4.43	4.10	3.87	3.70	3.56	3.46	2.35	2.28	2.20	2.12	2.08	2.04	1.99	1.95	1.90	1.84
21	8.02	5.78	4.87	4.37	4.04	3.81	3.64	3.51	3.40	2.32	2.25	2.18	2.10	2.05	2.01	1.96	1.92	1.87	1.81
22	7.95	5.72	4.82	4.31	3.99	3.76	3.59	3.45	3.35	2.30	2.23	2.15	2.07	2.03	1.98	1.94	1.89	1.84	1.78
23	7.88	5.66	4.76	4.26	3.94	3.71	3.54	3.41	3.30	2.27	2.20	2.13	2.05	2.01	1.96	1.91	1.86	1.81	1.76
24	7.82	5.61	4.72	4.22	3.90	3.67	3.50	3.36	3.26	2.25	2.18	2.11	2.03	1.98	1.94	1.89	1.84	1.79	1.73
25	7.77	5.57	4.68	4.18	3.85	3.63	3.46	3.32	3.22	2.24	2.16	2.09	2.01	1.96	1.92	1.87	1.82	1.77	1.71
26	7.72	5.53	4.64	4.14	3.82	3.59	3.42	3.29	3.18	2.22	2.15	2.07	1.99	1.95	1.90	1.85	1.80	1.75	1.69
27	7.68	5.49	4.60	4.11	3.78	3.56	3.39	3.26	3.15	2.20	2.13	2.06	1.97	1.93	1.88	1.84	1.79	1.73	1.67
28	7.64	5.45	4.57	4.07	3.75	3.53	3.36	3.23	3.12	2.19	2.12	2.04	1.96	1.91	1.87	1.82	1.77	1.71	1.65
29	7.60	5.42	4.54	4.04	3.73	3.50	3.33	3.20	3.09	2.18	2.10	2.03	1.94	1.90	1.85	1.81	1.75	1.70	1.64
30	7.56	5.39	4.51	4.02	3.70	3.47	3.30	3.17	3.07	2.16	2.09	2.01	1.93	1.89	1.84	1.79	1.74	1.68	1.62
40	7.31	5.18	4.31	3.83	3.51	3.29	3.12	2.99	2.89	2.08	2.00	1.92	1.84	1.79	1.74	1.69	1.64	1.58	1.51
60	7.08	4.98	4.13	3.65	3.34	3.12	2.95	2.82	2.72	1.99	1.92	1.84	1.75	1.70	1.65	1.59	1.53	1.47	1.39
120	6.85	4.79	3.95	3.48	3.17	2.96	2.79	2.66	2.56	1.91	1.83	1.75	1.66	1.61	1.55	1.50	1.43	1.35	1.25
∞	6.63	4.61	3.78	3.32	3.02	2.80	2.64	2.51	2.41	1.83	1.75	1.67	1.57	1.52	1.46	1.39	1.32	1.22	1.00

Table II is reproduced from Table 18 of *Biometrika Tables for Statisticians*, Vol. I, 1966, by permission of E. S. Pearson and the Biometrika Trustees.

Table III DURBIN-WATSON TEST BOUNDS

Level of Significance α = 0.05

n	p=2 d_L	p=2 d_U	p=3 d_L	p=3 d_U	p=4 d_L	p=4 d_U	p=5 d_L	p=5 d_U	p=6 d_L	p=6 d_U
15	1.08	1.36	0.95	1.54	0.82	1.75	0.69	1.97	0.56	2.21
16	1.10	1.37	0.98	1.54	0.86	1.73	0.74	1.93	0.62	2.15
17	1.13	1.38	1.02	1.54	0.90	1.71	0.78	1.90	0.67	2.10
18	1.16	1.39	1.05	1.53	0.93	1.69	0.82	1.87	0.71	2.06
19	1.18	1.40	1.08	1.53	0.97	1.68	0.86	1.85	0.75	2.02
20	1.20	1.41	1.10	1.54	1.00	1.68	0.90	1.83	0.79	1.99
21	1.22	1.42	1.13	1.54	1.03	1.67	0.93	1.81	0.83	1.96
22	1.24	1.43	1.15	1.54	1.05	1.66	0.96	1.80	0.86	1.94
23	1.26	1.44	1.17	1.54	1.08	1.66	0.99	1.79	0.90	1.92
24	1.27	1.45	1.19	1.55	1.10	1.66	1.01	1.78	0.93	1.90
25	1.29	1.45	1.21	1.55	1.12	1.66	1.04	1.77	0.95	1.89
26	1.30	1.46	1.22	1.55	1.14	1.65	1.06	1.76	0.98	1.88
27	1.32	1.47	1.24	1.56	1.16	1.65	1.08	1.76	1.01	1.86
28	1.33	1.48	1.26	1.56	1.18	1.65	1.10	1.75	1.03	1.85
29	1.34	1.48	1.27	1.56	1.20	1.65	1.12	1.74	1.05	1.84
30	1.35	1.49	1.28	1.57	1.21	1.65	1.14	1.74	1.07	1.83
31	1.36	1.50	1.30	1.57	1.23	1.65	1.16	1.74	1.09	1.83
32	1.37	1.50	1.31	1.57	1.24	1.65	1.18	1.73	1.11	1.82
33	1.38	1.51	1.32	1.58	1.26	1.65	1.19	1.73	1.13	1.81
34	1.39	1.51	1.33	1.58	1.27	1.65	1.21	1.73	1.15	1.81
35	1.40	1.52	1.34	1.58	1.28	1.65	1.22	1.73	1.16	1.80
36	1.41	1.52	1.35	1.59	1.29	1.65	1.24	1.73	1.18	1.80
37	1.42	1.53	1.36	1.59	1.31	1.66	1.25	1.72	1.19	1.80
38	1.43	1.54	1.37	1.59	1.32	1.66	1.26	1.72	1.21	1.79
39	1.43	1.54	1.38	1.60	1.33	1.66	1.27	1.72	1.22	1.79
40	1.44	1.54	1.39	1.60	1.34	1.66	1.29	1.72	1.23	1.79
45	1.48	1.57	1.43	1.62	1.38	1.67	1.34	1.72	1.29	1.78
50	1.50	1.59	1.46	1.63	1.42	1.67	1.38	1.72	1.34	1.77
55	1.53	1.60	1.49	1.64	1.45	1.68	1.41	1.72	1.38	1.77
60	1.55	1.62	1.51	1.65	1.48	1.69	1.44	1.73	1.41	1.77
65	1.57	1.63	1.54	1.66	1.50	1.70	1.47	1.73	1.44	1.77
70	1.58	1.64	1.55	1.67	1.52	1.70	1.49	1.74	1.46	1.77
75	1.60	1.65	1.57	1.68	1.54	1.71	1.51	1.74	1.49	1.77
80	1.61	1.66	1.59	1.69	1.56	1.72	1.53	1.74	1.51	1.77
85	1.62	1.67	1.60	1.70	1.57	1.72	1.55	1.75	1.52	1.77
90	1.63	1.68	1.61	1.70	1.59	1.73	1.57	1.75	1.54	1.78
95	1.64	1.69	1.62	1.71	1.60	1.73	1.58	1.75	1.56	1.78
100	1.65	1.69	1.63	1.72	1.61	1.74	1.59	1.76	1.57	1.78

Level of Significance α = 0.01

n	p=2 d_L	p=2 d_U	p=3 d_L	p=3 d_U	p=4 d_L	p=4 d_U	p=5 d_L	p=5 d_U	p=6 d_L	p=6 d_U
15	0.81	1.07	0.70	1.25	0.59	1.46	0.49	1.70	0.39	1.96
16	0.84	1.09	0.74	1.25	0.63	1.44	0.53	1.66	0.44	1.90
17	0.87	1.10	0.77	1.25	0.67	1.43	0.57	1.63	0.48	1.85
18	0.90	1.12	0.80	1.26	0.71	1.42	0.61	1.60	0.52	1.80
19	0.93	1.13	0.83	1.26	0.74	1.41	0.65	1.58	0.56	1.77
20	0.95	1.15	0.86	1.27	0.77	1.41	0.68	1.57	0.60	1.74
21	0.97	1.16	0.89	1.27	0.80	1.41	0.72	1.55	0.63	1.71
22	1.00	1.17	0.91	1.28	0.83	1.40	0.75	1.54	0.66	1.69
23	1.02	1.19	0.94	1.29	0.86	1.40	0.77	1.53	0.70	1.67
24	1.04	1.20	0.96	1.30	0.88	1.41	0.80	1.53	0.72	1.66
25	1.05	1.21	0.98	1.30	0.90	1.41	0.83	1.52	0.75	1.65
26	1.07	1.22	1.00	1.31	0.93	1.41	0.85	1.52	0.78	1.64
27	1.09	1.23	1.02	1.32	0.95	1.41	0.88	1.51	0.81	1.63
28	1.10	1.24	1.04	1.32	0.97	1.41	0.90	1.51	0.83	1.62
29	1.12	1.25	1.05	1.33	0.99	1.42	0.92	1.51	0.85	1.61
30	1.13	1.26	1.07	1.34	1.01	1.42	0.94	1.51	0.88	1.61
31	1.15	1.27	1.08	1.34	1.02	1.42	0.96	1.51	0.90	1.60
32	1.16	1.28	1.10	1.35	1.04	1.43	0.98	1.51	0.92	1.60
33	1.17	1.29	1.11	1.36	1.05	1.43	1.00	1.51	0.94	1.59
34	1.18	1.30	1.13	1.36	1.07	1.43	1.01	1.51	0.95	1.59
35	1.19	1.31	1.14	1.37	1.08	1.44	1.03	1.51	0.97	1.59
36	1.21	1.32	1.15	1.38	1.10	1.44	1.04	1.51	0.99	1.59
37	1.22	1.32	1.16	1.38	1.11	1.45	1.06	1.51	1.00	1.59
38	1.23	1.33	1.18	1.39	1.12	1.45	1.07	1.52	1.02	1.58
39	1.24	1.34	1.19	1.39	1.14	1.45	1.09	1.52	1.03	1.58
40	1.25	1.34	1.20	1.40	1.15	1.46	1.10	1.52	1.05	1.58
45	1.29	1.38	1.24	1.42	1.20	1.48	1.16	1.53	1.11	1.58
50	1.32	1.40	1.28	1.45	1.24	1.49	1.20	1.54	1.16	1.59
55	1.36	1.43	1.32	1.47	1.28	1.51	1.25	1.55	1.21	1.59
60	1.38	1.45	1.35	1.48	1.32	1.52	1.28	1.56	1.25	1.60
65	1.41	1.47	1.38	1.50	1.35	1.53	1.31	1.57	1.28	1.61
70	1.43	1.49	1.40	1.52	1.37	1.55	1.34	1.58	1.31	1.61
75	1.45	1.50	1.42	1.53	1.39	1.56	1.37	1.59	1.34	1.62
80	1.47	1.52	1.44	1.54	1.42	1.57	1.39	1.60	1.36	1.62
85	1.48	1.53	1.46	1.55	1.43	1.58	1.41	1.60	1.39	1.63
90	1.50	1.54	1.47	1.56	1.45	1.59	1.43	1.61	1.41	1.64
95	1.51	1.55	1.49	1.57	1.46	1.59	1.43	1.62	1.42	1.64
100	1.52	1.56	1.50	1.58	1.48	1.60	1.46	1.63	1.44	1.65

Reprinted from J. Durbin and G. S. Watson, "Testing for Serial Correlations in Least Squares Regression. II," Biometrika, Vol. 38 (1951), pp. 159–78, with permission of the authors and the Biometrika Trustees.

ANALYSES OF DATA SETS

TEACHERS Data Set
(Exercise 1.4)

State	Salary ($000) X	No. Teachers (000) Y
AL	4.0	27
AR	3.3	14
FL	5.1	35
GA	3.9	32
KY	3.3	23
LA	5.0	24
MS	3.3	17
NC	4.2	36
SC	3.4	20
TN	3.9	26
VA	4.3	30
WV	4.0	15
$\sum(\cdot)$	47.7	299
Mean	3.975	24.9167
$\sum(\cdot)^2$	193.79	8065
$\sum XY$	1219.1	
$S_{(\cdot)(\cdot)}$	4.1825	614.9167
S_{XY}	30.575	
r_{xy}^2	0.3635	
r_{xy}	0.6029	

Normal equations
$$299 = 12a + 47.7b$$
$$1219.1 = 47.7a + 193.79b$$

Regression equation
$$\hat{Y} = -4.1414 + 7.3102X$$

$$(\mathbf{X'X})^{-1} = \frac{1}{50.19}\begin{bmatrix} 193.79 & -47.7 \\ -47.7 & 12 \end{bmatrix}$$

$s_{y\cdot x}^2 = 39.1407$, $s_{y\cdot x} = 6.2562$, $CV = 0.2511$

For testing H_0:
$\beta = 0$, $t = 2.3896$ with 10 df
$F = 5.7104$ with 1 and 10 df

For testing H_0:
$\alpha = 0$, $t = -0.3369$ with 10 df

FIGURE A1.1 *Scatter Diagram for Salaries and Numbers of Teachers with Regression Line*

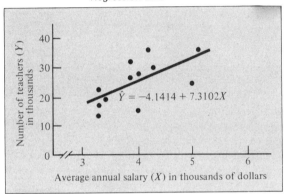

FIGURE A1.2 *Scatter Diagram for Salaries and Numbers of Teachers with Regression Line Fitted Through the Origin*

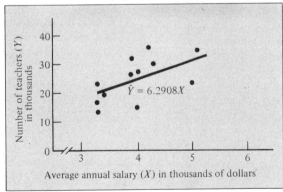

For testing H_0:
$\beta = 0$ if regression line is fitted through the origin,
$t = 6.2908$ with 10 df
95% CI on β: $0.4945 < \beta < 14.1259$
95% CI on α: $-31.5308 < \alpha < 23.2480$
For testing H_0:
$\rho = 0$, $t = 2.3897$ with 10 df

FIGURE A1.3   *Residuals for TEACHERS Data*

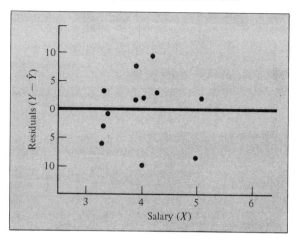

For testing H_0:
$\sigma_1^2 = \sigma_2^2$,
$F = 0.2224$ with 4 and 4 df or
$F = 4.4954$ with 4 and 4 df

PRECIPITATION Data Set
(Exercise 1.4)

City	Precipitation (Inches) X	Temperature (Degrees F) Y
A	69.2	57.3
B	51.2	55.8
Ch	64.5	57.7
Cl	63.8	57.7
Ga	67.4	55.1
Gr	53.6	57.6
J	62.7	56.6
Ki	54.0	57.6
Kn	58.0	58.1
L	65.9	57.0
Ne	55.0	56.3
No	63.5	58.9
O	64.9	57.3
R	65.5	54.6
S	56.0	55.1
$\sum(\cdot)$	915.2	852.7
Mean	61.0133	56.8467
$\sum(\cdot)^2$	52,306.10	48,494.57
$\sum XY$	52,028.77	
$S_{(\cdot)(\cdot)}$	466.6973	21.4173
S_{XY}		2.7007
r_{xy}^2		0.0007
r_{xy}		0.0270

Normal Equations:
$$852.7 = 15a + 915.2b$$
$$52,028.77 = 915a + 56,306.10b$$
Regression equation:
$$\hat{Y} = 56.4928 + 0.0058X$$

FIGURE A2.1   *Scatter Diagram for Temperature and Precipitation with Regression Line*

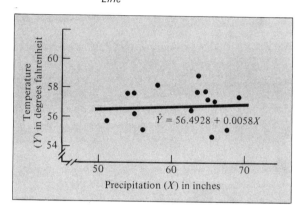

$$(\mathbf{X'X})^{-1} = \frac{1}{7000.46}\begin{bmatrix} 56,306.1 & -915.2 \\ -915.2 & 15 \end{bmatrix}$$

FIGURE A2.2   *Scatter Diagram for Temperature and Precipitation, with Regression Line Fitted Through the Origin*

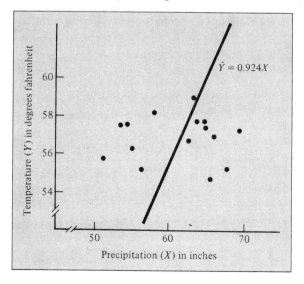

$s_{y \cdot x}^2 = 1.6463, \quad s_{y \cdot x} = 1.2831, \quad CV = 0.0226$

For testing $H_0: \beta = 0$,

$\quad t = 0.0976$ with 13 df

$\quad F = 0.0095$ with 1 and 13 df

For testing $H_0: \alpha = 0$,

$\quad t = 15.5247$ with 13 df

95% CI on β: $-0.1225 < \beta < 0.1341$

95% CI on α: $45.5323 < \alpha < 67.4533$

For testing $H_0: \rho = 0$,

$\quad t = 0.0974$ with 13 df

POLICE Data Set
(Exercise 1.4)

City	No. of Police Officers X	No. of Robberies Y
1	64	625
2	53	750
3	67	560
4	52	690
5	82	515
6	59	680
7	67	630
8	90	510
9	50	800
10	77	550
11	88	550
12	71	525
13	58	625
$\sum(\cdot)$	878	8,010
$\sum(\cdot)^2$	61,470	
$\sum(\cdot)^3$	4,454,930	
$\sum(\cdot)^4$	333,321,846	
$\sum XY$		527,885
$\sum X^2Y$		36,104,535

Normal equations for fitting a straight line:

$\quad 8,010 = 13a + 878b$

$\quad 527,885 = 878a + 61,470b$

Straight line regression equation:

$\quad \hat{Y} = 1,023.5864 - 6.0326X$

Using this equation,

$$\frac{1}{n}\sum |Y_i - \hat{Y}_i| = \frac{467.5}{13} = 35.9615$$

Normal equation for fitting a curve:

$\quad 8010 = 13a + 878b + 61,470c$

$\quad 527,885 = 878a + 61,470b + 4,454,930c$

$\quad 36,104,535 = 61,470a + 4,454,930b + 333,321,846c$

Second-degree regression equation:

$\quad \hat{Y} = 2,010.2296 - 35.3156X + 0.2096X^2$

FIGURE A3.1   Scatter Diagram for Number of Policemen and Number of Robberies with Straight Line and Curve Superimposed

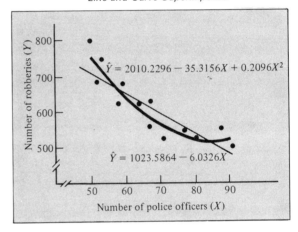

Using this equation,

$$\frac{1}{n}\sum |Y_i - \hat{Y}_i| = \frac{360.6436}{13}$$

$$= 27.7418$$

$$(\mathbf{X}'\mathbf{X})^{-1} = \frac{1}{8,011,646,096}$$

$$\begin{bmatrix} 642,892,568,720 & & \text{Sym.} \\ -18,812,033,688 & 554,623,098 & \\ 132,867,640 & -3,943,430 & 28,226 \end{bmatrix}$$

To test $H_0: \beta_1 = \beta_2 = 0$,

$\quad F = 38.6787$ with 2 and 10 df

To test $H_{01}: \beta_1 = 0$,

$\quad F = 66.8153$ with 1 and 10 df

To test $H_{02}: \beta_2 | \beta_1 = 0$,

$\quad F = 10.5420$ with 1 and 10 df

SALES Data Set
(Exercise 1.4)

	No. of Employees X	Avg. Weekly Retail Sales ($000) Y
	17	7
	39	17
	32	10
	17	5
	25	7
	43	15
	25	11
	32	13
	48	19
	10	3
	48	17
	42	15
	36	14
	30	12
	19	8
$\sum(\cdot)$	463	173
Mean	30.8667	11.5333
$\sum(\cdot)^2$	16,275	2315
$\sum XY$	6102	
$S_{(\cdot)(\cdot)}$	1983.7338	319.7333
S_{XY}	762.0609	
r_{XY}^2	0.9156	
r_{XY}	0.9569	

Normal equations:

$$173 = 15a + 463b$$
$$6,102 = 463a + 16,275b$$

Regression equation: $\hat{Y} = -0.3256 + 0.3842X$

FIGURE A4.1   Scatter Diagram for Number of Employees and Average Weekly Retail Sales with Regression Line

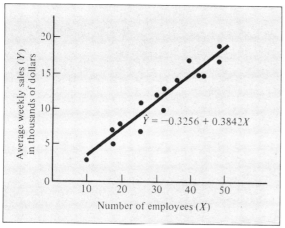

$$(X'X)^{-1} = \frac{1}{29,756}\begin{bmatrix} 16,275 & -463 \\ -463 & 15 \end{bmatrix}$$

FIGURE A4.2   Scatter Diagram for Number of Employees and Sales Revenue, with Regression Line Fitted Through the Origin

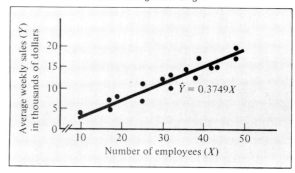

$s_{y\cdot x}^2 = 2.0729$, $s_{y\cdot x} = 1.4397$, $CV = 0.1248$

For testing $H_0: \beta = 0$,

$t = 11.8853$ with 13 df

$F = 141.2446$ with 1 and 13 df

For testing $H_0: \alpha = 0$,

$t = -0.3058$ with 13 df

For testing $H_0: \beta = 0$,

if regression equation is fitted through the origin,

$t = 10.3198$ with 13 df

99% CI on β: $0.2868 < \beta < 0.4816$

90% CI on α: $-2.2113 < \alpha < 1.4601$

For testing $H_0: \rho = 0$, $t = 11.8759$ with 13 df

VALUE ADDED Data Set
(Exercise 1.4)

Industry	Number of Employees (000) X	Value Added (mil $) Y
Food	2.8	29
Textiles	5.5	44
Apparel	1.5	10
Lumber	3.1	42
Furniture	5.5	26
Paper	1.8	18
Printing	5.0	35
Chemicals	5.8	70
Rubber	1.1	17
Plastics	4.3	48
Stone	5.4	54

Table continues

VALUE ADDED Data Set (cont.)

Industry	Number of Employees (000) X	Value Added (mil $) Y
Ceramics	4.1	24
Machinery	4.9	59
Electrical	2.4	18
Transportation	3.3	31
$\sum(\cdot)$	56.50	525
Mean	3.7667	35
$\sum(\cdot)^2$	248.41	22,617
$\sum XY$	2274.5	
$S_{(\cdot)(\cdot)}$	35.5933	4242
S_{XY}	297	

Regression equation:

$$\hat{Y} = 3.5695 + 8.3443X$$

FIGURE A5.1   *Scatter Diagram for Number of Employees and Value Added*

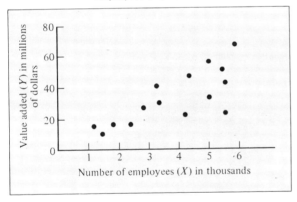

FIGURE A5.2   *Plot of Residuals for VALUE ADDED Data*

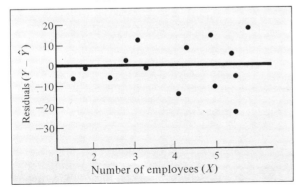

For testing $H_0: \sigma_1^2 = \sigma_2^2$,

$F = 1.745$ with 5 and 6 df

or

$F = 5.7292$ with 6 and 5 df

If data transformed by $X' = 1/X$ and $Y' = Y/X$,

$\hat{Y}' = 7.9831 + 4.6808X'$;   and to test $H_0: \sigma_1^2 = \sigma_2^2$,

$F = 1.1269$ with 5 and 6 df   or

$F = 0.8874$ with 6 and 5 df

FIGURE A5.3   *Plot of Transformed Data and Regression Line for VALUE ADDED Data*

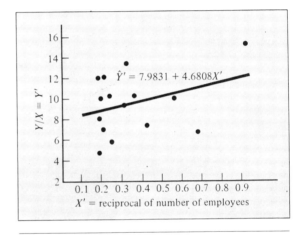

HEATING Data Set
(Exercise 1.4)

Day	Average Daily Temperature (°F) X	Daily Oil Bill ($) Y
1	50	0.30
2	4	0.70
3	5	0.65
4	19	0.65
5	33	0.50
6	48	0.30
7	6	0.65
8	11	0.60
9	27	0.50
10	−2	0.80
11	6	0.70
12	26	0.55
13	49	0.40
14	0	0.70
15	2	0.75
$\sum(\cdot)$	284	8.75
Mean	18.9333	0.5833
$\sum(\cdot)^2$	10,302	5.4425
$\sum XY$		126.30
$S_{(\cdot)(\cdot)}$	4924.9326	0.3383
S_{XY}		−39.3661
r^2_{XY}		0.9302
r_{XY}		−0.9644

Normal equations:

$$8.75 = 15a + 284b$$
$$126.3 = 284a + 10,302b$$

Regression equation:

$$\hat{Y} = 0.7348 - 0.008X$$

FIGURE A6.1   *Scatter Diagram for Temperature and Oil Bill with Regression Line*

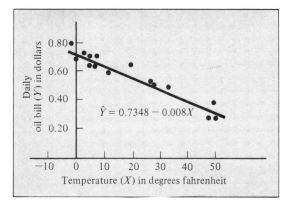

FIGURE A6.2   *Scatter Diagram for Temperature and Oil Bill with Regression Line Fitted Through the Origin*

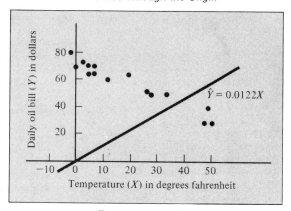

$$(\mathbf{X'X})^{-1} = \frac{1}{73,874}\begin{bmatrix} 10,302 & -284 \\ -284 & 15 \end{bmatrix}$$

$s^2_{y \cdot x} = 0.0018, \quad s_{y \cdot x} = 0.0424, \quad CV = 0.0727$

For testing $H_0: \beta = 0$,

$t = -13.2329$ with 13 df

$F = 174.9444$ with 1 and 13 df

95% CI on β: $-0.0092 < \beta < -0.0068$

98% CI on α: $0.6928 < \alpha < 0.7768$

For testing $H_0: \rho = 0 \quad t = -13.1614$ with 13 df

For testing $H_0: \rho_{t, t-1} = 0$,

$r_{t, t-1} = 0.0175$ and $d = 1.8613$.

FERTILIZER Data Set
(Exercise 2.15)

Pounds of Fertilizer X	Coded Pounds of Fertilizer Z	Bushels Yield Y
15	−1	45
15	−1	43
15	−1	46
20	0	58
20	0	62
20	0	64
25	1	70
25	1	78
25	1	76
$\sum(\cdot)$	180	0	542
$\sum(\cdot)^2$	3750	6	—
$\sum ZY$			90
$S_{(\cdot)(\cdot)}$		6	1413.5555
S_{ZY}			90

Normal equations:

$$542 = 9a' + 0b'$$
$$90 = 0a' + 6b'$$

Regression equation:

$$\hat{Y} = 60.2222 + 15Z, \text{ where } Z = \frac{X - 20}{5},$$

$$= 0.2222 + 3X$$

FIGURE A7.1 *Scatter Diagram and Regression Equation for Fertilizer and Yield*

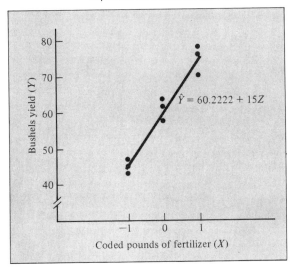

$$(\mathbf{X'X})^{-1} = \frac{1}{1350}\begin{bmatrix} 3750 & -180 \\ -180 & 9 \end{bmatrix}$$

$$(\mathbf{Z'Z})^{-1} = \text{diag}(\tfrac{1}{9}, \tfrac{1}{6})$$

$$s_{y \cdot x}^2 = 9.0794, \quad s_{y \cdot x} = 3.0132, \quad CV = 0.0500$$

To test $H_0 : \beta = 0$,

$$t = 12.1938 \text{ with 7 df}$$

Normal equations: $\quad -53 \quad = 20b'$
$$508 = 10a' + 20c'$$
$$1249 = 20a' + 68c'$$

Regression equation:

$$\hat{Y} = 34.1572 - 2.65Z + 8.3214Z^2$$

DEPRIVATION Data Set
(Exercise 2.16)

	Coded Number Hours Deprived Z	Time to Learn Task (min) Y
	−2	73
	−2	68
	−1	49
	−1	54
	0	32
	0	25
	1	42
	1	40
	2	60
	2	65
$\sum(\cdot)$	0	508
$\sum(\cdot)^2$	20	—
$\sum(\cdot)^3$		
$\sum(\cdot)^4$	68	
$\sum ZY$	−53	
$\sum Z^2 Y$	1249	

FIGURE A8.1 *Scatter Diagram and Regression Equation for Hours Deprivation and Time to Learn Task*

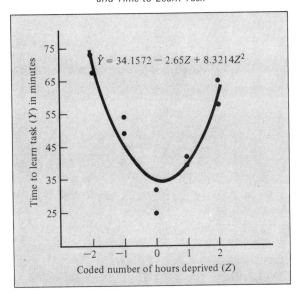

$$(\mathbf{Z}'\mathbf{Z})^{-1} = \frac{1}{5600} \begin{bmatrix} 1360 & 0 & -400 \\ 0 & 280 & 0 \\ -400 & 0 & 200 \end{bmatrix}$$

Simple linear regression equation:

$\hat{Y} = 50.8 - 2.65Z$

FIGURE A8.2

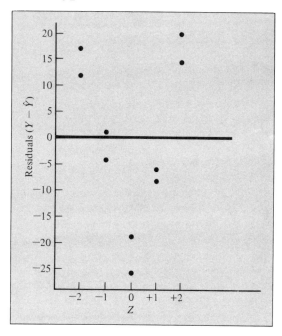

For testing $H_0: \beta = 0$ in simple linear regression,

$F = 0.5199$ with 1 and 8 df

For testing H_0: simple linear model is adequate,

$F = 54.6133$ with 3 and 5 df

For testing $H_0: \beta_1 = \beta_2 = 0$ for second-degree polynomial model,

$F = 32.5779$ with 2 and 7 df

For testing $H_{01}: \beta_1 = 0$ in polynomial model,

$F = 4.4032$ with 1 and 7 df

For testing $H_{02}: \beta_2|\beta_1 = 0$ in polynomial model,

$I = 60.7527$ with 1 and 7 df

For testing H_0: second-degree polynomial model is adequate,

$F = 6.2220$ with 2 and 5 df

SECRETARIES Data Set
(Exercise 9.1)

Employee	Rating X	Salary ($) Y	Residual
Adams	3.9	790	20.9210
Bright	3.6	740	-18.0603
Collins	3.1	750	10.3042
Dale	4.9	820	14.1920
Evans	3.1	720	-19.6958
Fisher	2.4	700	-13.9855
Faster	4.9	790	-15.8080
Greene	3.6	770	11.9397
Hill	4.4	800	12.5565
Hustings	2.2	720	13.3603
Irving	4.4	770	-17.4435
Jacobs	4.8	780	-22.1351
Knight	2.3	700	-10.3126
Lawley	2.5	730	12.3416
Moore	3.6	850	91.9397
Owen	2.5	690	-27.6584
Paul	3.0	720	-16.0229
Priestly	2.3	720	9.6874
Williams	3.2	730	-13.3687
Wilson	4.0	750	-22.7519
$\sum(\cdot)$	68.7	15,040	
Mean	3.4350	752	
$\sum(\cdot)^2$	252.01	11,345,400	
$\sum XY$		52,251	
$S_{(\cdot)(\cdot)}$	16.0255		
S_{XY}		588.6	

$\hat{Y} = 625.8359 + 36.7290X$

FIGURE A9.1   Plot of Residuals for Secretaries'
Ratings and Salaries

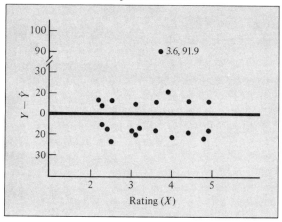

If Ms. Moore omitted, a 99% CI on her salary is
$703.1074 < \hat{Y}_{2.3} < 802.9898$.

TREES Data Set
(Exercise 9.5)

Tree	Circumference (ft) X	Volume (cu ft) Y	Residual
1	4.7	114	−6.2961
2	2.7	96	−2.1979
3	3.5	100	−7.0372
4	2.5	110	14.0119
5	4.0	112	−0.5617
6	3.6	108	−0.1421
7	2.7	100	1.8021
8	5.5	132	2.8646
9	5.0	118	−5.6108
10	2.6	100	2.9070
11	2.5	94	−1.9881
12	6.0	144	9.3401
13	3.8	110	−0.3519
14	5.8	134	1.5499
15	3.9	106	−5.4568
16	2.8	102	2.6972
17	5.8	138	5.5499
18	3.5	102	−5.0372
19	3.6	102	−6.1421
20	2.5	100	4.0119
21	3.5	106	−1.0372
22	4.7	118	−2.2961
23	3.6	106	−2.1421
24	2.5	98	2.0119
25	5.8	132	−0.4501

$\sum(\cdot)$	97.1	2782	
Mean	3.8840	111.2800	
$\sum(\cdot)^2$	411.81	314,412	
$\sum X^3$	1888.5350		
$\sum X^4$	9226.5789		
$\sum XY$		11,188.4	
$\sum X^2 Y$		49,120.48	
$S_{(\cdot)(\cdot)}$	34.6736	4831.04	
S_{XY}		338.112	

Simple linear regression equation:
$$\hat{Y} = 68.356 + 11.0491X$$

FIGURE A10.1   *Plot of Residuals for TREES Data*

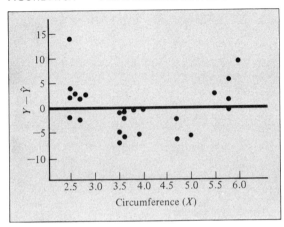

To test H_0: no significant lack of fit to straight line,
   $F = 1.6812$ with 12 and 11 df
Second-degree polynomial regression equation:
$$\hat{Y} = 116.3789 - 14.0954X + 3.0200X^2$$
To test $H_0: \beta_1 = \beta_2 = 0$,
   $F = 7{,}554.0495$ with 2 and 22 df
To test $H_{01}: \beta_1 = 0$,
   $F = 13{,}257.2590$ with 1 and 22 df
To test $H_{02}: \beta_2 | \beta_1 = 0$,
   $F = 1{,}850.84$ with 1 and 22 df

TYPIST Data Set
(Exercise 9.7)

Page	Number of Words X	Number of Errors Y
1	228	10
2	230	11
3	221	9
4	208	5
5	222	8
6	197	4
7	180	1
8	229	10
9	212	7
10	190	3
11	209	6
12	207	6
13	285	4
14	209	9
15	193	6
16	209	8
17	207	8
18	208	9
19	229	13
20	206	8
21	208	9
22	207	7
23	185	5
$\sum(\cdot)$	4,779	166
Mean	207.7826	7.2134
$\sum(\cdot)^2$	997,661	1,368
$\sum XY$	35,273	
$S_{(\cdot)(\cdot)}$	4667.9130	
S_{XY}	781.0870	

$\hat{Y} = -27.5486 + 0.1673X$

FIGURE A11.1   *Data and Regression Line for TYPIST Data*

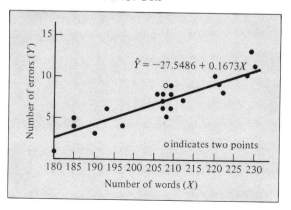

For given data:

$r_{t,t-1} = 0.7739$

    For testing $H_0: \rho_{t,t-1} = 0$, $d = 0.4788$

For first differences, $r_{t,t-1} = -0.2366$

For testing $H_0: \rho_{t,t-1} = 0$, $d^- = 1.7985$

FIGURE A11.2   *Plot of Residual, in Order, for TYPIST Data*

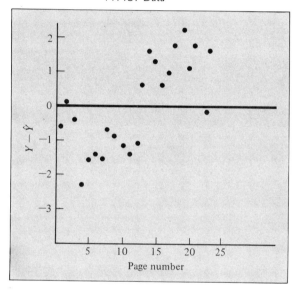

IQ Data Set
(Exercise 9.8)

Student	IQ (X)	Score (Y)	Teacher
1	114	90	A
2	118	90	B
3	110	84	B
4	109	88	A
5	94	78	A
6	113	90	B
7	114	88	B
8	110	88	B
9	107	80	B
10	106	82	B
11	106	84	B
12	109	84	B
13	109	90	A
14	109	86	B
15	96	80	A
16	130	98	A
17	123	96	A
18	108	82	B
19	106	86	B

Table continues

IQ Data Set (cont.)

Student	IQ (X)	Score (Y)	Teacher
20	113	92	A
21	125	94	A
22	124	96	A
23	111	86	B
24	120	94	A
25	104	82	A
26	117	92	A
$\sum(\cdot)$	2905	2280	
Mean	111.7308	87.6923	
$\sum(\cdot)^2$	326,307		
$\sum XY$		225,790	
$S_{(\cdot)(\cdot)}$		1729.1154	
S_{XY}		1043.8462	

$$\hat{Y} = 20.2404 + 0.6037X$$

FIGURE A12.1 *Scatter Diagram and Regression Line for IQ Data*

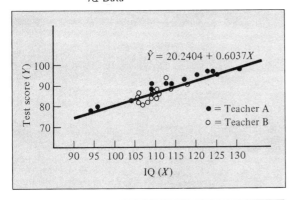

SALES II Data Set
(Exercise 10.1)

	Y= Average Weekly Retail Sales ($000)	X₁ = Number of Employees	X₂ = Size (000 sq ft)
	7	17	7
	17	39	9
	10	32	8
	5	17	4
	7	25	5
	15	43	9
	11	25	8
	13	32	10
	19	48	12

	Y = Average Weekly Retail Sales ($000)	X₁ = Number of Employees	X₂ = Si (000 sq ft)
	3	10	5
	17	48	12
	15	42	10
	14	36	10
	12	30	10
	8	19	8
$\sum(\cdot)$	173	463	127
$\sum(\cdot)^2$	2315	16,275	1157
$\sum(\cdot)Y$		6102	1611
$\sum X_1 X_2$			4260
$S_{(\cdot)(\cdot)}$	319.7333	1983.7333	81.7333
$S_{(\cdot)Y}$		762.0667	146.2667
S_{12}			339.9333
$r_{(\cdot)Y}$		0.9569	0.9048
r_{12}			0.8442

Simple linear regression equation relating Y to X_2:

$$\hat{Y} = -3.6183 + 1.7896X_2$$

$$s_{y \cdot x}^2 = 4.4596, \qquad s_{y \cdot x} = 2.1180$$

Multiple linear regression equation:

$$\hat{Y} = -1.447907 + 0.269825X_1 + 0.667393X_2$$

$$s_{y \cdot x}^2 = 1.3743$$

$$s_{y \cdot x} = 1.1723$$

$$r_{xy}^2 = 0.9484$$

$$r_{1Y \cdot 2} = 0.8457$$

$$r_{2Y \cdot 1} = 0.6232$$

To test $H_0 : \beta_1 = \beta_2 = 0$,

$F = 110.1992$ with 2 and 12 df

To test $H_{01} : \beta_1 | \beta_2 = 0$,

$F = 30.1577$ with 1 and 12 df

To test $H_{02} : \beta_2 | \beta_1 = 0$,

$F = 7.6022$ with 7.6022

95% *CI* on regression coefficients:

$0.162 \ < \beta_1 < 0.3769$

$0.1310 < \beta_2 < 1.1948$

RESTAURANT Data Set
(Exercise 10.2)

Restaurant	Y	X_1	X_2
1	4.5	3.3	4.0
2	2.0	1.7	1.5
3	1.7	1.5	2.0
4	2.3	1.7	3.0
5	4.0	3.0	3.0
6	4.1	3.2	3.7
7	2.5	1.6	2.0
8	3.7	3.5	4.0
9	3.6	3.0	3.5
10	3.0	3.0	3.5
$\sum (\cdot)$	31.14	25.5	30.2
$\sum (\cdot)^2$	167.14	70.97	98.44
$\sum (\cdot) Y$		86.58	101.37
$\sum X_1 X_2$		82.89	
$S_{(\cdot)(\cdot)}$	8.544	5.945	7.236
$S_{(\cdot)Y}$		6.51	6.542
S_{12}		5.88	
$r_{(\cdot)Y}$		0.9134	0.8320
r_{12}		0.8965	

$\hat{Y} = 0.311472 + 1.023199 X_1 + 0.072631 X_2$

$s_{y \cdot x}^2 = 0.2011$

$s_{y \cdot x} = 0.4484$

$r_{xy}^2 = 0.8350$

$r_{1Y \cdot 2} = 0.6819$

$r_{2Y \cdot 1} = 0.0721$

To test $H_0 : \beta_1 = \beta_2 = 0$,

　　$F = 17.7213$ with 2 and 7 df

To test $H_{01} : \beta_1 | \beta_2 = 0$,

　　$F_1 = 6.0686$ with 1 and 7 df

To test $H_{02} : \beta_2 | \beta_1 = 0$,

　　$F_2 = 0.0373$ with 1 and 7 df

DRUG Data Set
(Exercise 10.3)

Patient	Time (Y)	Dosage (X_1)	Age (X_2)
1	11	2	59
2	3	2	57
3	20	2	22
4	25	2	12
5	27	2	18
6	15	5	40
7	10	5	64
8	34	5	27
9	14	5	54
10	34	5	22
11	35	7	33
12	28	7	49
13	23	7	29
14	21	7	32
15	33	7	20
16	27	10	43
17	8	10	61
18	3	10	69
19	12	10	62
20	14	10	61
$\sum (\cdot)$	397	120	834
$\sum (\cdot)^2$	9927	890	41,158
$\sum (\cdot) Y$		2327	13,688
$\sum X_1 X_2$		5472	
$S_{(\cdot)(\cdot)}$	2046.55	170	6380.2
$S_{(\cdot)Y}$		−55	−2866.9
S_{12}		468	
$r_{(\cdot)Y}$		−0.0933	−0.7668
r_{12}		0.4494	

$\hat{Y} = 35.221155 + 1.143994 X_1 - 0.533289 X_2$

$s_{x \cdot y}^2 = 34.1520$

$s_{y \cdot x} = 5.8440$

$r_{xy}^2 = 0.7163$

$r_{1Y \cdot 2} = 0.4385$

$r_{2Y \cdot 1} = -0.8150$

To test $H_0 : \beta_1 = \beta_2 = 0$,

　　$F = 21.4640$ with 2 and 17 df

To test $H_{01} : \beta_1 | \beta_2 = 0$,

　　$F_1 = 5.1992$ with 1 and 17 df

To test $H_{02} : \beta_2 | \beta_1 = 0$,

　　$F_2 = 42.403$ with 1 and 17 df

95% *CI*s on regression coefficients:

　　$0.854 < \beta_1 < 2.2026$

　　$-0.7061 < \beta_2 < -0.3605$

AUTO Data Set
(Exercise 10.4)

Vehicle No.	Age (X_1)	Mileage During the Year (000) (X_2)	Cost ($000) (Y)
1	3	12	1.2
2	3	36	1.5
3	2	10	0.9
4	4	14	1.5
5	3	16	1.0
6	5	22	4.0
7	2	14	0.8
8	2	10	0.8
9	2	24	1.0
10	5	14	2.0
11	5	18	2.0
12	5	21	3.0
13	5	28	3.2
14	0*	27	0.5
15	5	14	1.9
$\sum(\cdot)$	51	280	25.3
$\sum(\cdot)^2$	209	6018	57.13
$\sum(\cdot)Y$	104.6	502.3	
$\sum X_1 X_2$	949		
$S_{(\cdot)(\cdot)}$	35.6	791.3333	14.4573
$S_{(\cdot)Y}$	18.58	30.0333	
S_{12}	−3		
$r_{(\cdot)Y}$	0.8190	0.2807	
r_{12}	−0.0179		

* Age 0 denotes a car less than 1 year old.

$\hat{Y} = -0.84486 + 0.525260X_1 + 0.39944X_2$

$s_{y\cdot x}^2 = 0.2915$

$s_{y\cdot x} = 0.5399$

$r_{xy}^2 = 0.7580$

$r_{1Y\cdot 2} = 0.8587$

$r_{2Y\cdot 1} = 0.5148$

To test $H_0: \beta_1 = \beta_2 = 0$,

   $F = 18.7901$ with 2 and 12 df

To test $H_{01}: \beta_1 | \beta_2 = 0$,

   $F_1 = 33.6725$ with 1 and 12 df

To test $H_{02}: \beta_2 | \beta_1 = 0$,

   $F_2 = 4.3405$ with 1 and 12 df

ADS Data Set
(Exercise 10.5)

Store	Revenues (Y)	TV (X_1)	Radio (X_2)	Paper (X_3)
1	84	13	5	2
2	84	13	7	1
3	80	8	6	3
4	50	9	5	3
5	20	9	3	1
6	68	13	5	1
7	34	12	7	2
8	30	10	3	2
9	54	8	5	2
10	40	10	5	3
11	57	5	6	2
12	46	5	7	2
$(\cdot)$	647	115	64	24
$(\cdot)^2$	39,973	1191	362	54
$(\cdot)Y$		6393	3600	1292
$(\cdot)X_3$		222	129	
$X_1 X_2$	610			
$S_{(\cdot)(\cdot)}$	5088.9167	88.9167	20.6667	6
$S_{(\cdot)Y}$		192.5833	149.3333	−2
$S_{(\cdot)3}$		−8	1	
S_{12}		−3.3333		
$r_{(\cdot)Y}$		0.2864	0.4604	−0.01
$r_{(\cdot)3}$		−0.3464	0.0898	
r_{12}		−0.0778		

$\hat{Y} = -15.293983 + 2.620384X_1 + 7.556194X_2$
$\qquad + 1.900656X_3$

$\qquad s_{y\cdot x}^2 = 384.4094$

$\qquad s_{y\cdot x} = 19.6064$

$\qquad r_{xy}^2 = 0.3202$

$\qquad r_{1Y\cdot 2} = 0.3640 \qquad r_{2Y\cdot 1} = 0.5054$

$\qquad r_{13\cdot 2} = -0.3419 \qquad r_{23\cdot 1} = 0.0673$

$\qquad r_{3Y\cdot 2} = -0.0580 \qquad r_{3Y\cdot 1} = 0.0993$

$\qquad r_{1Y\cdot 23} = 0.3670 \qquad r_{2Y\cdot 13} = 0.5024$

$\qquad r_{3Y\cdot 12} = 0.0759$

To test $H_0: \beta_1 = \beta_2 = \beta_3 = 0$,

   $F = 1.2558$ with 3 and 8 df

To test $H_{01}: \beta_1 | \beta_2, \beta_3 = 0$,

   $F_1 = 1.2393$ with 1 and 8 df

To test $H_{02}: \beta_2 | \beta_1, \beta_3 = 0$,

   $F_2 = 2.6998$ with 1 and 8 df

To test $H_{03}: \beta_3 | \beta_1, \beta_2 = 0$,

   $F_3 = 0.0439$ with 1 and 8 df

CHILDREN Data Set
(Exercise 10.6)

Child	Score (Y)	Age (X_1)	Height (X_2)	Weight (X_3)
1	58	7	47.5	53
2	54	7	45.0	50
3	55	9	52.5	85
4	74	7	48.0	52
5	86	9	55.0	76
6	98	8	51.0	64
7	96	9	53.0	75
8	70	7	46.0	75
9	40	7	48.0	68
10	67	9	50.5	74
11	41	6	45.0	40
12	41	7	48.5	66
13	47	8	50.5	65
14	45	8	49.0	70
15	92	9	51.5	70
16	50	7	46.5	60
17	98	9	53.5	77
18	42	8	45.0	65
19	64	8	52.5	65
20	70	8	51.5	67
$\sum (\cdot)$	1288	157	990.0	1317
$\sum (\cdot)^2$	90,910	1249	49,185.5	88,909
$\sum (\cdot) Y$		10,329	64,537.0	86,350
$\sum (\cdot) X_3$		10,489	65,612.0	
$\sum X_1 X_2$		7816.5		
$S_{(\cdot)(\cdot)}$	7862.8	16.55	180.5	2184.55
$S_{(\cdot)Y}$		218.2	781.0	1535.2
$S_{(\cdot)3}$		150.55	420.5	
S_{12}		45		
$r_{(\cdot)Y}$		0.6011	0.6514	0.3681
$r_{(\cdot)3}$		0.7918	0.6696	
r_{12}		0.8233		

$$\hat{Y} = -136.52606 + 9.527225X_1 + 3.348974X_2 - 0.596901X_3$$

$$s_{Y \cdot X}^2 = 246.1631$$
$$s_{Y \cdot X} = 15.6896$$
$$r_{xy}^2 = 0.4745$$
$$r_{1Y \cdot 2} = 0.1503 \qquad r_{2Y \cdot 1} = 0.3452$$
$$r_{1Y \cdot 23} = 0.2690$$
$$r_{13 \cdot 2} = 0.5706 \qquad r_{23 \cdot 1} = 0.0511$$
$$r_{2Y \cdot 12} = 0.3661$$
$$r_{3Y \cdot 2} = -0.1209 \qquad r_{3Y \cdot 1} = -0.2208$$
$$r_{3Y \cdot 12} = -0.2544$$

To test $H_0 : \beta_1 = \beta_2 = \beta_3 - 0$,
  $F = 4.8150$ with 3 and 16 df
To test $H_{01} : \beta_1 | \beta_2, \beta_3 = 0$,
  $F_1 = 1.2481$ with 1 and 16 df
To test $H_{02} : \beta_2 | \beta_1, \beta_3 = 0$,
  $F_2 = 2.4867$ with 1 and 16 df
To test $H_{03} : \beta_3 | \beta_1, \beta_2 = 0$,
  $F_3 = 1.1075$ with 1 and 16 df

FIGURE A18.1   *SAS Forward Selection Procedure for Exercise 10.6.*

FORWARD SELECTION PROCEDURE FOR DEPENDENT VARIABLE SCORE

STEP 1 VARIABLE HT ENTERED R SQUARE = 0.42438405

	DF	SUM OF SQUARES	MEAN SQUARE	F	PROB>F
REGRESSION	1	3379.28531856	3379.28531856	13.27	0.0019
ERROR	18	4583.51468144	254.63970452		
TOTAL	19	7962.80000000			

	B VALUE	STD ERROR	TYPE II SS	F	PROB>F
INTERCEPT	-149.78005540				
HT	4.32686981	1.18774841	3379.28531856	13.27	0.0019

NO OTHER VARIABLES MET THE 0.5000 SIGNIFICANCE LEVEL FOR ENTRY INTO THE MODEL.

GLUE Data Set
(Exercise 10.7)

Strength (Y)	Tempera-ture	Coded Temp. (X_1)	Humidity	Coded Humidity (X_2)
190	80	−1	40	−1
189	80	−1	40	−1
192	90	1	40	−1
190	90	1	40	−1
196	80	−1	60	0
193	80	−1	60	0
195	90	1	60	0
196	90	1	60	0
201	80	−1	80	1
200	80	−1	80	1
203	90	1	80	1
205	90	1	80	1
$\sum(\cdot)$	2350	0		0
$\sum(\cdot)^2$	460,526	12		8
$\sum(\cdot)Y$		12		48
$\sum X_1 X_2$			0	
$S_{(\cdot)(\cdot)}$	317,6667	12		8
$S_{(\cdot)}$		12		48
S_{12}			0	
r_{12}			0	
$r_{(\cdot)}$		0.1944		0.9522

$\hat{Y} = 195.833333 + X_1 + 6X_2$

$s_{y.x}^2 = 1.9630$

$s_{y.x} = 1.4011$

$r_{xy}^2 = 0.9444$

To test $H_0: \beta_1 = \beta_2 = 0$,

$F = 76.4137$ with 2 and 9 df

To test $H_{01}: \beta_1|\beta_2 = 0$,

$F = 6.1131$ with 1 and 9 df

To test $H_{02}: \beta_2|\beta_1 = 0$,

$F = 146.7142$ with 1 and 9 df

95% CIs on regression coefficients:

$0.0849 < \beta_1 < 1.9151$

$4.8793 < \beta_2 < 7.1207$

BOTTLES Data Set
(Exercise 10.8)

Time (X_1)	Temperature (X_2)	Impurities (Y)
−2	−1	4
−2	−1	5
−2	0	6
−2	0	4
−2	1	5
−2	1	5
−1	−1	3
−1	−1	5
−1	0	4
−1	0	4
−1	1	2
−1	1	3
1	−1	4
1	−1	3
1	0	3
1	0	3
1	1	2
1	1	1
2	−1	2
2	−1	1
2	0	1
2	0	0
2	1	0
2	1	0
$\sum(\cdot)$	0	0	70
$\sum(\cdot)^2$	60	16	276
$\sum(\cdot)Y$	−55	−9	
$\sum X_1 X_2$	0		
$S_{(\cdot)(\cdot)}$	60	16	71.8333
$S_{(\cdot)Y}$	−55	−9	
S_{12}	0		
$r_{(\cdot)Y}$	−0.8378	−0.2655	

$\hat{Y} = 2.916667 - 0.916667X_1 - 0.562500X_2$

$s_{y.x}^2 = 0.7788$

$s_{y.x} = 0.8825$

$r_{xy}^2 = 0.7724$

To test $H_0: \beta_1 = \beta_2 = 0$,

$F = 35.6241$ with 2 and 21 df

To test $H_{01}: \beta_1|\beta_2 = 0$,

$F = 64.747$ with 1 and 21 df

To test $H_{02}: \beta_2|\beta_1 = 0$,

$F = 6.5012$ with 1 and 21 df

95% CIs on regression coefficients:

$-1.1536 < \beta_1 < -0.6797$

$-1.0214 < \beta_2 < -0.1036$

YIELD Data Set
(Exercise 10.9)

Yield (Y)	Nitro-gen (X_1)	Potas-sium (X_2)	Phos-phorus (X_3)
539	−1	−1	−1
319	−1	−1	0
164	−1	−1	1
228	−1	0	−1
207	−1	0	0
178	−1	0	1
180	−1	1	−1
250	−1	1	0
340	−1	1	1
491	0	−1	−1
305	0	−1	0
514	0	−1	1
380	0	0	−1
364	0	0	0
366	0	0	1
192	0	1	−1
171	0	1	0
187	0	1	1
354	1	−1	−1
415	1	−1	0
326	1	−1	1
239	1	0	−1
292	1	0	0
446	1	0	1
250	1	1	−1
527	1	1	0
477	1	1	1

	Yield (Y)	Nitro-gen (X_1)	Potas-sium (X_2)	Phos-phorus (X_3)
$\sum(\cdot)$	8701	0	0	0
$\sum(\cdot)^2$	3,179,299	18	18	18
$\sum(\cdot)Y$		921	−853	145
$\sum(\cdot)X_3$		0	0	
$\sum X_1 X_2$		0		
$S_{(\cdot)(\cdot)}$	375,321,1852	18	18	18
$S_{(\cdot)Y}$		921	−853	145
$S_{(\cdot)3}$		0	0	
S_{12}		0		
$r_{(\cdot)Y}$		0.3544	−0.3282	0.0557

$\hat{Y} = 322.259259 + 51.166667X_1 - 47.388889X_2$
$+ 8.055556X_3$

$s_{y\cdot x}^2 = 11,941.9128$

$s_{y\cdot x} = 109.2790$

$r_{xy}^2 = 0.2364$

To test $H_0: \beta_1 = \beta_2 = \beta_3 = 0$,

$F = 2.3731$ with 3 and 23 df

To test $H_{01}: \beta_1 | \beta_2, \beta_3 = 0$,

$F_1 = 3.7817$ with 1 and 23 df

To test $H_{02}: \beta_2 | \beta_1, \beta_3 = 0$,

$F_2 = 3.2439$ with 1 and 23 df

To test $H_{03}: \beta_3 | \beta_1, \beta_2 = 0$,

$F_3 = 1,168.0557$ with 1 and 23 df

TENNESSEE Data Set
(Exercise 14.2)

County	% Population Change 1960–70	No. Persons Employed in Agri-culture	% Families below Poverty Level	Residen-tial and Farm Property Tax Rate	% of Residences with Telephones	% Rural Popula-tion	Median Age	Negro Popula-tion
Benton	13.7	400	19.0	1.09	82	74.8	33.5	360
Cannon	−0.8	710	26.2	1.01	66	100.0	32.8	193
Carroll	9.6	1,610	18.1	0.40	80	69.7	33.4	3,080
Cheatheam	40.0	500	15.4	0.93	74	100.0	27.8	592
Cumberland	8.4	640	29.0	0.92	65	74.0	27.9	2
DeKalb	3.5	920	21.6	0.59	64	73.1	33.2	230
Dyer	3.0	1,890	21.9	0.63	82	52.3	30.8	3,978
Gibson	7.1	3,040	18.9	0.49	85	49.6	32.4	9,816
Greene	13.0	2,730	21.1	0.71	78	71.2	29.2	1,137

Table continues

County	% Population Change 1960–70	No. Persons Employed in Agriculture	% Families below Poverty Level	Residential and Farm Property Tax Rate	% of Residences with Telephones	% Rural Population	Median Age	Negro Population
Hawkins	10.7	1,850	23.8	0.93	74	70.6	28.7	992
Haywood	−16.2	2,920	40.5	0.51	69	64.2	25.1	10,723
Henry	6.6	1,070	21.6	0.80	85	58.3	35.9	3,139
Houston	21.9	160	25.4	0.74	69	100.0	31.4	338
Humphreys	17.8	380	19.7	0.44	83	72.0	30.1	516
Jackson	−11.8	1,140	38.0	0.81	54	100.0	34.1	12
Johnson	7.5	690	30.1	1.05	65	100.0	30.5	104
Lawrence	3.7	1,170	24.8	0.73	76	69.5	30.0	430
McNairy	1.6	1,280	30.3	0.65	67	81.0	32.4	1,240
Madison	8.4	2,270	19.5	0.48	85	39.1	28.7	20,446
Marshall	2.7	960	15.6	0.72	84	58.4	33.4	1,863
Maury	5.6	1,710	17.2	0.62	84	42.4	29.9	8,035
Montgomery	12.7	1,410	18.4	0.84	86	36.4	23.3	10,620
Morgan	−4.8	200	27.3	0.73	66	99.8	27.5	211
Sevier	16.5	960	19.2	0.45	74	90.6	29.5	133
Shelby	15.2	11,500	16.8	1.00	87	5.9	25.4	266,159
Sullivan	11.6	1,380	13.2	0.63	85	44.2	28.8	2,432
Trousdale	4.9	530	29.7	0.54	70	100.0	33.1	932
Unicoi	1.1	370	19.8	0.98	75	52.6	30.8	7
Wayne	3.8	440	27.7	0.46	48	100.0	28.4	208
Weakley	19.0	1,630	20.5	0.68	83	72.1	30.4	1,732

Some Models Considered Are Forward selection, stepwise, maximum R^2 improvement, minimum R^2 improvement:

	B	Std. Error	F	Prob > F
INTERCEPT	58.2588			
PHONE	−0.4710	0.0825	32.60	0.0001

$r^2_{xy} = 0.5380 \qquad s^2_{y \cdot x} = 19.7636$

PRESS = 669.66 $\qquad C_p = 22.64$

Forward selection, stepwise, maximum R^2 improvement, minimum R^2 improvement:

	B	Std. Error	F	Prob > F
INTERCEPT	52.4945			
PHONE	−0.2698	0.0709	14.47	0.0007
POPCH	−0.3656	0.0732	24.93	0.0001

$r^2_{xy} = 0.6992 \qquad s^2_{y \cdot x} = 13.3457$

PRESS = 493.54 $\qquad C_p = 7.67$

Maximum R^2 improvement, minimum R^2 improvement.

	B	Std. Error	F	Prob > F
INTERCEPT	10.9868			
POPCHG	−0.4037	0.0630	41.07	0.0001
AGRIEMP	0.0010	0.0004	6.05	0.0208
RURAL	0.1921	0.0359	28.56	0.0001

$r^2_{xy} = 0.7358 \qquad s^2_{y \cdot x} = 12.6588$

PRESS = 514.03 $\qquad C_p = 6.02$

Backward elimination, maximum R^2 improvement, minimum R^2 improvement:

	B	Std. Error	F	Prob > F
INTERCEPT	25.3280			
POPCHG	−0.4215	0.0600	49.32	0.0001
AGRIEMP	0.0009	0.0004	4.42	0.0459
RURAL	0.1998	0.0341	34.34	0.0001
AGE	−0.4777	0.2319	4.24	0.0500

$r^2_{xy} = 0.7725 \qquad s^2_{y \cdot x} = 10.9011$

PRESS = 482.57 $\qquad C_p = 3.95$

FIGURE A22.1 *Preliminary Statistics for TENNESSEE Data.*

STATISTICAL ANALYSIS SYSTEM

VARIABLE	N	MEAN	STD DEV	MINIMUM	MAXIMUM	SUM
POPCHG	30	7.8666667	10.33225059	-16.20000000	40.00000000	236.00000000
AGRIEMP	30	1548.6666667	2038.38633497	160.00000000	11500.00000000	46460.00000000
POVERTY	30	23.01000000	6.42658004	13.20000000	40.50000000	690.30000000
TAXRT	30	0.71866667	0.20270130	0.40000000	1.09000000	21.56000000
PHONE	30	74.83333333	10.00718133	48.00000000	87.00000000	2245.00000000
RURAL	30	70.72666667	24.02156981	5.90000000	100.00000000	2121.80000000
AGE	30	30.28000000	2.88484745	23.30000000	35.90000000	908.40000000
NEGRO	30	11655.33333333	48289.86298017	2.00000000	266159.00000000	349660.33333333

CORRELATION COEFFICIENTS / PROB > |R| UNDER H0:RHO=0 / N = 30

	POPCHG	AGRIEMP	POVERTY	TAXRT	PHONE	RURAL	AGE	NEGRO
POPCHG	1.00000 0.0000	0.04031 0.8325	-0.64914 0.0001	0.13095 0.4903	0.37837 0.0392	-0.01877 0.9216	-0.14688 0.4386	0.12115 0.5237
AGRIEMP	0.04031 0.8325	1.00000 0.0000	-0.16769 0.3758	0.10355 0.5861	0.35549 0.0539	-0.65774 0.0001	-0.36371 0.0482	0.94176 0.0001
POVERTY	-0.64914 0.0001	-0.16769 0.3758	1.00000 0.0000	0.00898 0.9624	-0.73347 0.0001	0.51260 0.0038	0.02068 0.9136	-0.19208 0.3092
TAXRT	0.13095 0.4903	0.10355 0.5861	0.00898 0.9624	1.00000 0.0000	-0.03802 0.8419	0.02330 0.9027	-0.04669 0.8065	0.22795 0.2257
PHONE	0.37837 0.0392	0.35549 0.0539	-0.73347 0.0001	-0.03802 0.8419	1.00000 0.0000	-0.74858 0.0001	-0.07836 0.6807	0.27032 0.1485
RURAL	-0.01877 0.9216	-0.65774 0.0001	0.51260 0.0038	0.02330 0.9027	-0.74858 0.0001	1.00000 0.0000	0.31442 0.0906	-0.56116 0.0013
AGE	-0.14688 0.4386	-0.36371 0.0482	0.02068 0.9136	-0.04669 0.8065	-0.07836 0.6807	0.31442 0.0906	1.00000 0.0000	-0.34873 0.0589
NEGRO	0.12115 0.5237	0.94176 0.0001	-0.19208 0.3092	0.22795 0.2257	0.27032 0.1485	-0.56116 0.0013	-0.34873 0.0589	1.00000 0.0000

Forward selection, forcing in NEGRO:

	B	Std. Error	F	Prob > F
INTERCEPT	52.6675			
NEGRO	0.0000	0.0000	0.02	0.8906
POPCHG	−0.2700	0.0723	13.96	0.0009
PHONE	−0.3683	0.0769	22.91	0.0001

$$r^2_{xy} = 0.6994 \qquad s^2_{y \cdot x} = 13.8487$$

Stepwise, forcing in NEGRO:

	B	Std. Error	F	Prob > F
INTERCEPT	13.8034			
NEGRO	0.0000	0.0000	3.73	0.0646
POPCHG	−0.4149	0.0660	39.58	0.0001
RURAL	0.1709	0.0340	25.24	0.0001

$$r^2_{xy} = 0.7130 \qquad s^2_{y \cdot x} = 13.2205$$

or

	B	Std. Error	F	Prob > F
INTERCEPT	28.5599			
NEGRO	0.0000	0.0000	2.50	0.1263
POPCHG	−0.4312	0.0624	47.74	0.0001
RURAL	0.1819	0.0324	31.59	0.0001
AGE	−0.5062	0.2391	4.48	0.0444

$$r^2_{xy} = 0.7566 \qquad s^2_{y \cdot x} = 11.6598$$

Backward elimination, forcing in NEGRO:

	B	Std. Error	F	Prob > F
INTERCEPT	24.0985			
NEGRO	0.0000	0.0000	0.55	0.4651
POPCHG	−0.4100	0.0625	43.00	0.0001
AGRIEMP	0.0016	0.0010	2.26	0.1458
RURAL	0.2059	0.0354	33.86	0.0001
AGE	−0.4794	0.2340	4.20	0.0516

$$r^2_{xy} = 0.7776 \qquad s^2_{y \cdot x} = 11.1005$$

FIGURE A22.2   *Plot of Residuals for TENNESSEE Data Using Four Predictors*

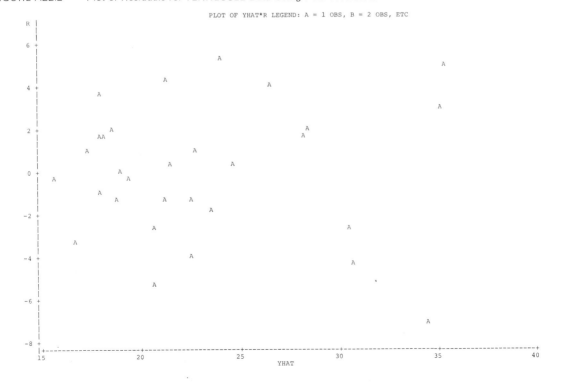

SELECTED ANSWERS TO EXERCISES

Chapter 1

Section 1.2

1.2.

	Y	X
a.	insurance	income
b.	sales	commercials
c.	time	number open
d.	customers	reservations
e.	femininity	size of family
f.	sessions	intensity
g.	mutation	radiation
h.	house	dosage
i.		arbitrary
j.	height	number of trees

1.3. random: a, c, d, e, i, j.
controlled: b, f, g, h.

Section 1.3

1.4. TEACHERS Data: b. direct; c. possibly curvilinear; d. not very strong; e. reasonably uniform.
PRECIPITATION Data: b. possibly a weak direct; c. linear if it exists; d. not very strong; e. not very uniform.
POLICE Data: b. inverse; c. curvilinear; d. reasonably strong; e. reasonably uniform.
SALES Data: b. direct; c. linear; d. reasonably strong; e. uniform.
VALUE ADDED Data: b. direct; c. linear; d. strong only for industries employing small numbers because e is not uniform.
HEATING Data: b. inverse; c. linear; d. strong; e. uniform.

1.6. a. The more fertilizer applied, the greater the yield tends to be.
b. The longer the time spent playing intramural sports, the lower students' grade averages tend to be.
c. Regardless of how many jars are sold, the price tends to remain constant.
d. The longer one cooks the vegetable, typically the lower the vitamin content.

e. It has been found that families that consume a large amount of electricity also tend to consume a relatively large amount of beer.

Section 1.4

1.8. In deterministic models, once X is given, Y is determined without error by applying the rules of arithmetic. In probabilistic models, Y can only be estimated from a given value of X, since probabilistic models take individual differences into account.

1.9. a.–c. There are other variables that need to be taken into account.
d. No
e. He relates the average time elapsed to the degree of incline.

Chapter 2

Sections 2.2 and 2.3

2.1. (X_i, Y_i) is a point on the scatter diagram, representing observed values of X and Y. $(X_i, \hat{Y}_i)$ is a point on the regression line, representing an estimated value of Y for an observed value of X.

2.2. a. $\sum X_i^2 - \sum X_i + 4n$.
b. $3 \sum X_i + \sum Y_i$.
c. $\sum X_i^2 - 8 \sum X_i + 16n$.
d. $9 \sum X_i^2 - 24 \sum X_i Y_i + 16 \sum Y_i^2$.
e. $12 \sum X_i^2 - 11 \sum X_i Y_i - 5 \sum Y_i^2$.
f. $\sum Y_i - na - b \sum X_i$.
g. $\sum X_i Y_i - a \sum X_i + b \sum X_i^2$.
h. $\sum X_i Y_i - a \sum X_i - b \sum X_i^2$.
i. $\sum X_i^2 - 2\bar{X} \sum X_i + n\bar{X}^2$.
j. $\sum X_i Y_i - \bar{X} \sum Y_i - \bar{Y} \sum X_i + n\bar{X}\bar{Y}$.

2.3. a. $\sum(X_i - \bar{X}) = \sum X_i - nX$. Substitute $\dfrac{\sum X_i}{n}$ for $\bar{X}$.
b. $\sum(X_i - \bar{X})^2 = \sum X_i^2 - 2\bar{X} \sum X_i + n\bar{X}^2$. Substitute $\dfrac{\sum X_i}{n}$ for $\bar{X}$.

542SELECTED ANSWERS TO EXERCISES

c. Divide result in part b by $n - 1$.

d. Refer to part h of Problem 2.2. Substitute
$$\frac{\sum X_i}{n} = \bar{X} \text{ and } \frac{\sum Y_i}{n} = \bar{Y}.$$

e. Same as part b but with X replaced by Y.

2.4. a. $\sum X_i = 15$, $\bar{X} = 3$, $\sum(X_i - \bar{X}) = 0$.

b. $MD = 6/5$.

c. $s_x^2 = 5/2$.

2.5. a. $(\sum X_i)^2 = 225 \neq \sum X_i^2 = 55$.

b. $s_x^2 = 5/2$.

c. False.

2.6. The least squares line is, on the average, closest to all the points, in terms of squared deviations.

2.7. a. $\sum Q_i = na + b\sum P_i$,
$\sum P_i Q_i = a\sum P_i + b\sum P_i^2$.

b. $\sum E_i = na + b\sum T_i + c\sum T_i^2$,
$\sum T_i E_i = a\sum T_i + b\sum T_i^2 + c\sum T_i^3$,
$\sum T_i^2 E_i = a\sum T_i^2 + b\sum T_i^3 + c\sum T_i^4$.

c. $\sum Y_i = na + b\sum X_i + c\sum X_i^2 + d\sum X_i^3$,
$\sum X_i Y_i = a\sum X_i + b\sum X_i^2 + c\sum X_i^3 + d\sum X_i^4$,
$\sum X_i^2 Y_i = a\sum X_i^2 + b\sum X_i^3 + c\sum X_i^4 + d\sum X_i^5$,
$\sum X_i^3 Y_i = a\sum X_i^3 + b\sum X_i^4 + c\sum X_i^5 + d\sum X_i^6$.

d. $\sum Y_i = na + b\sum X_i + c\sum Z_i$,
$\sum X_i Y_i = a\sum X_i + b\sum X_i^2 + c\sum X_i Z_i$,
$\sum Z_i Y_i = a\sum Z_i + b\sum X_i Z_i + c\sum Z_i^2$.

e. $\sum Y_i = nb_0 + b_1\sum X_{1i} + b_2\sum X_{2i}$,
$\sum X_{1i} Y_i = b_0\sum X_{1i} + b_1\sum X_{1i}^2 + b_2\sum X_{1i}X_{2i}$,
$\sum X_{2i} Y_i = b_0\sum X_{2i} + b_1\sum X_{1i}X_{2i} + b_2\sum X_{2i}^2$.

2.8. TEACHERS Data: $\hat{Y} = -4.1414 + 7.3102X$.
PRECIPITATION Data: $\hat{Y} = 56.4928 + 0.0058X$.
SALES Data: $\hat{Y} = -0.3256 + 0.3842X$.
HEATING Data: $\hat{Y} = 0.7348 - 0.008X$.

2.10. TEACHERS Data: $\bar{X} = 3.975$, $\bar{Y} = 24.9167$.
PRECIPITATION Data: $\bar{X} = 61.0133$, $\bar{Y} = 56.8467$.
SALES Data: $\bar{X} = 30.8667$, $\bar{Y} = 11.5333$.
HEATING Data: $\bar{X} = 18.9333$, $\bar{Y} = 0.5833$.

2.11. a. $\hat{Y} = 1023.5864 - 6.0326X$.

b. $\hat{Y} = 2010.2296 - 35.3156X + 0.2096X^2$.

c. $\sum|Y_i - \hat{Y}_i| = 467.5000$.

d. $\sum|Y_i - \hat{Y}_i| = 360.6436$.

e. $MD = 35.9615$.

f. $MD = 27.7418$.

g. The curve gives the better description.

Section 2.4

2.12. a. $S_{TT} = \dfrac{n\sum T_i^2 - (\sum T_i)^2}{n}$.

b. $S_{RR} = \dfrac{n\sum R_i^2 - (\sum R_i)^2}{n}$.

c.–d. $S_{TR} = S_{RT} = \dfrac{n\sum R_i T_i - (\sum R_i)(\sum T_i)}{n}$.

2.13. TEACHERS Data: $S_{XY} = 30.575$, $S_{XX} = 4.1825$.
PRECIPITATION Data: $S_{XY} = 2.7007$, $S_{XX} = 466.6973$.
SALES Data: $S_{XY} = 762.0667$, $S_{XX} = 1983.7333$.
HEATING Data: $S_{XY} = -39.3667$, $S_{XX} = 4924.9333$.

Section 2.5

TEACHERS Data: a. $\sum X_i = 47.7$, $\sum U_i = 0$, $\sum U_i^2 = 4.1825$; $\sum Y_i = 299$, $\sum U_i Y_i = 30.575$;

b. $a' = 24.9167 = \bar{Y}$, $b = 7.3102$.

c. $S_{UY} = 30.575$, $S_{UU} = 4.1825$.

d. and e. Intercept is shifted to $\bar{Y}$; slope is the same.

PRECIPITATION Data: a. $\sum X_i = 915.2$, $\sum U_i \approx 0$, $\sum U_i^2 = 466.6975$, $\sum Y_i = 852.7$, $\sum U_i Y_i = 2.7291$.

b. $a' = 56.8467$, $b = 0.0058$.

c. $S_{UY} = 2.7291$, $S_{UU} = 466.6975$.

d. and e. See answer to TEACHERS.

SALES Data: a. $\sum X_i = 463$, $\sum U_i \approx 0$, $\sum U_i^2 = 1983.7338$, $\sum Y_i = 173$, $\sum U_i Y_i = 762.0609$.

b. $a' = 11.53333$, $b = 0.3842$.

c. $S_{UY} = 762.0609$, $S_{UU} = 1983.7338$.

d. and e. See answer to TEACHERS.

HEATING Data: a. $\sum X_i = 284$, $\sum U_i \approx 0$, $S_{UU} = 4924.9326$, $\sum Y_i = 8.75$, $\sum U_i Y_i = -39.3661$.

b. $a' = 0.5833$, $\quad b = -0.008$.

c. $S_{UY} = -39.3661$, $\quad S_{UU} = 4924.9326$.

d. and e. See answer to **TEACHERS**.

2.15. a. $\mu = 20$, $\quad d = 5$, $\quad X_1 = -1$, $\quad X_2 = 0$,
$X_3 = 1$.

c. $\hat{Y} = 60.2222 + 15Z$.

d. 45.2222.

e. $\hat{Y} = 0.2222 + 3X$.

f. 45.2222.

g. Inference must be restricted to the three levels at which the experiment was run unless one is sure that the relationship is linear throughout.

2.16. a. $\mu = 72$, $\quad d = 24$, $\quad X_1 = 24$, $\quad X_2 = 48$,
$X_3 = 72$, $\quad X_4 = 96$, $\quad X_5 = 120$.

b. Time to learn task is not a linear function of hours deprivation.

c. $\sum Z_i = 0$, $\sum Y_i = 508$, $\sum Z_i Y_i = -53$,
$\sum Z_i^2 = 20$, $\sum Z_i^2 Y_i = 1249$, $\sum Z_i^3 = 0$,
$\sum Z_i^4 = 68$,
$\hat{Y} = 34.1572 - 2.65Z + 8.3214Z^2$.

d. Lowest point occurs at $Z = 0.1592$, so $\hat{Y} = 33.9458$.

e. (1) Yes, (2) no.

Chapter 3

Section 3.2

3.1. a.–d. Yes.

3.2. a. 3×4.   e. 12×12.

b. 4×3.   f. 1×12.

c. 3×3.   g. 4×1.

d. 6×9.   h. 1×1.

3.3. a. Fourth row, seventh column.

b. Seventh row, fourth column.

c. Seventh row, seventh column.

d. ith row, ith column.

3.4. a. Yes.

b.–e. No.

3.5. a. $\begin{bmatrix} 1 \\ 2 \\ 3 \\ 4 \\ 5 \end{bmatrix}$.   b. $[5 \ 4 \ 3 \ 2 \ 1]$.   c. $\begin{bmatrix} 6 & 2 \\ 9 & 4 \\ 0 & 6 \\ 1 & 0 \end{bmatrix}$.

d. $\begin{bmatrix} a & c & e \\ b & d & f \end{bmatrix}$.

e. $\begin{bmatrix} a_{11} & a_{21} & a_{31} \\ a_{12} & a_{22} & a_{32} \\ a_{13} & a_{23} & a_{33} \end{bmatrix}$.

f. $\begin{bmatrix} 1 & 2 \\ 2 & 3 \end{bmatrix}$.   g. $\begin{bmatrix} 9 & 8 & 7 \\ 8 & 6 & 5 \\ 7 & 5 & 4 \end{bmatrix}$.

3.6. a. $a_{11}, a_{22}, a_{33}, a_{44}$.

b. 9, 6, 4.

c. 1, 1.

d. 1, 1, 2, 4.

e. Matrix not square.

3.7. a. $\begin{bmatrix} 1 & 0 & 0 & 0 & 0 \\ 0 & 1 & 0 & 0 & 0 \\ 0 & 0 & 2 & 0 & 0 \\ 0 & 0 & 0 & 3 & 0 \\ 0 & 0 & 0 & 0 & 5 \end{bmatrix}$.

b. $\begin{bmatrix} 7 & 0 & 0 & 0 & 0 & 0 \\ 0 & 0 & 0 & 0 & 0 & 0 \\ 0 & 0 & 3 & 0 & 0 & 0 \\ 0 & 0 & 0 & 4 & 0 & 0 \\ 0 & 0 & 0 & 0 & 1 & 0 \\ 0 & 0 & 0 & 0 & 0 & 0 \end{bmatrix}$

c. $\begin{bmatrix} 1 & 0 & 0 & 0 \\ 0 & 1 & 0 & 0 \\ 0 & 0 & 1 & 0 \\ 0 & 0 & 0 & 1 \end{bmatrix}$.   d. $\begin{bmatrix} 5 & 0 & 0 \\ 0 & 5 & 0 \\ 0 & 0 & 5 \end{bmatrix}$.

e. $\begin{bmatrix} a_1 & 0 & \cdots & 0 \\ 0 & a_2 & \cdots & 0 \\ \vdots & \vdots & & \vdots \\ 0 & 0 & \cdots & a_n \end{bmatrix}$.

3.8. a. $\begin{bmatrix} 5 & 2 & 7 \\ 2 & 1 & 1 \\ 7 & 1 & 2 \end{bmatrix}$.   b. $\begin{bmatrix} 9 & 9 & 2 & 9 \\ 9 & 2 & 1 & 1 \\ 2 & 1 & 2 & 4 \\ 9 & 1 & 4 & 0 \end{bmatrix}$.

c. $\begin{bmatrix} a & b & d \\ b & c & e \\ d & e & f \end{bmatrix}$.

d. $\begin{bmatrix} a_{11} & a_{12} & a_{13} & a_{14} \\ a_{12} & a_{22} & a_{23} & a_{24} \\ a_{13} & a_{23} & a_{33} & a_{34} \\ a_{14} & a_{24} & a_{34} & a_{44} \end{bmatrix}$

e. $\begin{bmatrix} 0 & 0 & 9 & 3 \\ 0 & 6 & 2 & 7 \\ 9 & 2 & 4 & 9 \\ 3 & 7 & 9 & 1 \end{bmatrix}$.

f. Not a square matrix.

Section 3.3

3.9. a. $\begin{bmatrix} 7 & 11 \\ 3 & 7 \end{bmatrix}$.

b. diag (9, 10, 2, 11).

c. $\begin{bmatrix} 2 & -8 & 4 & 9 \\ 8 & 5 & 2 & -2 \\ 2 & -3 & 9 & 6 \\ 0 & 14 & 9 & 9 \end{bmatrix}$.

d. $\begin{bmatrix} 0 & 2 & 0 & 5 \\ -6 & -5 & -10 & 4 \\ 0 & 13 & -5 & -10 \\ -10 & -2 & 3 & -1 \end{bmatrix}$.

e. Sum cannot be formed.

f. Sum cannot be formed.

g. $\begin{bmatrix} 7 \\ 7 \\ 1 \\ 12 \\ 10 \end{bmatrix}$.

h. Sum cannot be formed.

i. [0 6 -6 0 -2].

j. Sum cannot be formed.

3.10. a. $\begin{bmatrix} 16 & -28 & 16 \\ 20 & 32 & 4 \\ -20 & 32 & 8 \end{bmatrix}$.

b. $\begin{bmatrix} 1 & -7/4 & 1 \\ 5/4 & 2 & 1/4 \\ -5/4 & 2 & 1/2 \end{bmatrix}$.

c. $\begin{bmatrix} -18 & -15 \\ -21 & -12 \\ -15 & -6 \end{bmatrix}$.

d. [56 28 63 14 21].

e. $\begin{bmatrix} -35 \\ -10 \\ -5 \\ 10 \\ 0 \\ -35 \end{bmatrix}$.

3.11. a. $(2 \times 3)(3 \times 2) = (2 \times 2)$.
b. $(3 \times 2)(2 \times 3) = (3 \times 3)$.
c. $(3 \times 3)(3 \times 3) = (3 \times 3)$.
d. $(3 \times 3)(3 \times 3) = (3 \times 3)$.
e. No.
f. $(1 \times 4)(4 \times 1) = (1 \times 1)$.
g. $(4 \times 1)(1 \times 4) = (4 \times 4)$.
h. No.
i. Scalar times a vector (3×1).

3.12. a. $\begin{bmatrix} 131 & 140 \\ 90 & 95 \end{bmatrix}$.

b. $\begin{bmatrix} 62 & 45 & 117 \\ 46 & 29 & 72 \\ 72 & 52 & 135 \end{bmatrix}$.

c. $\begin{bmatrix} 42 & 84 & 76 \\ 17 & 35 & 23 \\ 23 & 28 & 27 \end{bmatrix}$.

d. $\begin{bmatrix} 32 & 14 & 24 \\ 46 & 16 & 23 \\ 90 & 23 & 56 \end{bmatrix}$.

e. Cannot be formed. f. 17.

g. $\begin{bmatrix} 0 & 6 & 2 & 3 \\ 0 & 12 & 4 & 6 \\ 0 & 6 & 2 & 3 \\ 0 & 6 & 2 & 3 \end{bmatrix}$.

h. Cannot be formed.

i. $\begin{bmatrix} 4 \\ 8 \\ 4 \end{bmatrix}$.

3.13. a. The products are not equal—the product matrices are not the same dimensions.
b. Even though the products are the same dimension, they are not equal.

3.14. a. $\mathbf{SS'} = \begin{bmatrix} 4 & 18 & 14 \\ 18 & 110 & 81 \\ 14 & 81 & 70 \end{bmatrix}$.

$\mathbf{S'S} = \begin{bmatrix} 3 & 9 & 16 & 8 \\ 9 & 41 & 63 & 27 \\ 16 & 63 & 114 & 53 \\ 8 & 27 & 53 & 26 \end{bmatrix}$.

b. Both are symmetric.

c. $\mathbf{SS}' \neq \mathbf{S}'\mathbf{S}$.

3.15. For any of the given matrices $\mathbf{M}$, $\mathbf{IM} = \mathbf{MI} = \mathbf{M}$.

Section 3.4

3.16. a. -25.   f. 0.

  b. 7.   g. 42.

  c. $a_{11}a_{22} - a_{12}a_{21}$.   h. Matrix not square.

  d. $n \sum X_i^2 - (\sum X_i)^2$.   i. 8.

3.17. a. Matrix in 3.16 f is singular.

  b. Matrices in 3.16 a, b, e, g, i, are nonsingular. If $a_{11}a_{22} - a_{12}a_{21} \neq 0$, then the matrix in c is nonsingular. Matrix in d is nonsingular as long as $X_i \neq 0$ for all i.

3.18. a. 320.   f. Matrix not square.

  b. 12.   g. Matrix not square.

  c. -347.   h. 280.

  d. 0.   i. 105.

  e. 0.   j. $(d_1)(d_2)(d_3)$.

3.19. a. Singular: d and e.

  b. Nonsingular: a, b, c, h, and i and j as long as no d_i, $i = 1, 2, 3$, is zero.

3.20. a. -28.   e. 1179.   i. Matrix not square.

  b. 20.   f. 504.   j. 0.

  c. 0.   g. 1960.   k. 14,222.

  d. 0.   h. $(a_1)(a_2)\dots(a_n)$.

3.21. a. Singular: c, d, j.

  b. Nonsingular: a, b, e, f, g, k, and h as long as $a_i \neq 0$, $i = 1, 2, \dots, n$.

Section 3.5

3.22. a. (a) $\begin{bmatrix} -1/25 & 3/25 \\ 9/25 & -2/25 \end{bmatrix}$.   (b) $\begin{bmatrix} 1 & 9/7 \\ 0 & 1/7 \end{bmatrix}$.

  (c) $\dfrac{1}{a_{11}a_{22} - a_{12}a_{21}} \begin{bmatrix} \sum x_i^2 & -\sum x_i \\ -\sum x_i & n \end{bmatrix}$.

  (d) $\dfrac{1}{n \sum x_i^2 - (\sum x_i)^2} \begin{bmatrix} \sum x_i^2 & -\sum x_i \\ -\sum x_i & n \end{bmatrix}$.

  (e) $\begin{bmatrix} -2 & 1 \\ 3/2 & -1/2 \end{bmatrix}$.

  (g) $\begin{bmatrix} 1/6 & 0 \\ 0 & 1/7 \end{bmatrix}$.   (h) 1/8.

  b.

  (a) $\dfrac{1}{320} \begin{bmatrix} -16 & 42 & -14 \\ -56 & 7 & 51 \\ 48 & -6 & 2 \end{bmatrix}$.

(b) $\dfrac{1}{2} \begin{bmatrix} 12 & 0 & 0 \\ 32 & 4 & -4 \\ -43 & -2 & 5 \end{bmatrix}$.

(c) $\dfrac{1}{-347} \begin{bmatrix} -40 & -21 & 49 \\ -21 & 15 & -35 \\ 49 & -35 & -34 \end{bmatrix}$.

(h) Diag $(1/7, 1/5, 1/8)$.   (i) Diag $(1/7, 1/5, 1/3)$.

(j) Diag $(1/d_1, 1/d_2, 1/d_3)$.

c. (a) $\dfrac{1}{-28} \begin{bmatrix} -28 & 0 & 0 & 0 \\ 0 & 8 & 4 & -8 \\ 0 & -10 & -12 & 10 \\ 84 & 16 & 36 & -30 \end{bmatrix}$.

  (b) $\dfrac{1}{20} \begin{bmatrix} 4 & -18 & 10 & 20 \\ 0 & 15 & -15 & -10 \\ 0 & 40 & -20 & -60 \\ 0 & -20 & 20 & 20 \end{bmatrix}$.

  (e) $\dfrac{1}{1179} \begin{bmatrix} 99 & -171 & -342 & 306 \\ -132 & 228 & 63 & -15 \\ 172 & -47 & -225 & 91 \\ 32 & -91 & 342 & -175 \end{bmatrix}$.

  (f) Diag $(1/3, 1/6, 1/7, 1/4)$.

  (g) Diag $(1/7, 1/7, 1/8, 1/5)$.

  (h) Diag $(1/a_1, 1/a_2, \dots, 1/a_n)$.

3.23. a. $\begin{bmatrix} 1 & 0 \\ 0 & 1/6 \end{bmatrix}$.

  b. If $\mathbf{A} = \begin{bmatrix} 1 & 0 & 0 \\ \hline 0 & \mathbf{M} \end{bmatrix}$

  then $\mathbf{A}^{-1} = \begin{bmatrix} 1 & 0 & 0 \\ \hline 0 & & \\ 0 & \mathbf{M}^{-1} & \end{bmatrix}$

  c. $\begin{bmatrix} 1 & 0 & 0 & 0 \\ 0 & -\frac{16}{320} & \frac{42}{320} & -\frac{14}{320} \\ 0 & -\frac{56}{320} & \frac{7}{320} & \frac{51}{320} \\ 0 & \frac{48}{320} & -\frac{6}{320} & \frac{2}{320} \end{bmatrix}$.

d. $\begin{bmatrix} 1 & \vdots & 0 & \cdots & 0 \\ \hline 0 & \vdots & & & \\ \vdots & \vdots & & \mathbf{M}^{-1} & \\ 0 & \vdots & & & \end{bmatrix}.$

e. $\begin{bmatrix} 1/d & \vdots & 0 & \cdots & 0 \\ \hline 0 & \vdots & & & \\ \vdots & \vdots & & \mathbf{M}^{-1} & \\ 0 & \vdots & & & \end{bmatrix}.$

Section 3.6

3.24. a. $\mathbf{x} = \begin{bmatrix} 14/25 \\ -1/25 \end{bmatrix}.$ b. $\mathbf{x} = \begin{bmatrix} 1 \\ 11/7 \end{bmatrix}.$

c. $\mathbf{x} = \begin{bmatrix} -6 \\ 4 \end{bmatrix}.$

d. No solution.   e. $\mathbf{x} = \begin{bmatrix} 7/6 \\ 3/7 \end{bmatrix}.$

f. No solution.   g. $x = 5/8.$

3.25 a. $x = \dfrac{1}{320} \begin{bmatrix} -300 \\ 270 \\ 180 \end{bmatrix}.$

b. $\mathbf{x} = \dfrac{1}{12} \begin{bmatrix} 36 \\ 100 \\ -113 \end{bmatrix}.$ c. $\mathbf{x} = \dfrac{1}{347} \begin{bmatrix} -324 \\ 281 \\ -193 \end{bmatrix}.$

d. No solution.   e. No solution.

f. $\mathbf{x} = \begin{bmatrix} 5/7 \\ 1 \\ 5/8 \end{bmatrix}.$

3.26. a. $\mathbf{x} = \dfrac{1}{28} \begin{bmatrix} 196 \\ 44 \\ -34 \\ -486 \end{bmatrix}.$ b. $\mathbf{x} = \dfrac{1}{20} \begin{bmatrix} -56 \\ 40 \\ 140 \\ -40 \end{bmatrix}.$

c. No solution.

d. $\mathbf{x} = \dfrac{1}{1179} \begin{bmatrix} -5409 \\ 2889 \\ -3300 \\ 2265 \end{bmatrix}.$

e. No solution.

Section 3.7

3.27.

TEACHERS Data:

a. $\mathbf{y} = \begin{bmatrix} 27 \\ 14 \\ \vdots \\ 15 \end{bmatrix}$; $\hat{\boldsymbol{\beta}} = \begin{bmatrix} a \\ b \end{bmatrix}$; $\mathbf{x} = \begin{bmatrix} 1 & 4.0 \\ 1 & 3.3 \\ \vdots & \vdots \\ 1 & 4.0 \end{bmatrix}.$

b. $\mathbf{X'y} = \begin{bmatrix} 299 \\ 1219.1 \end{bmatrix}$; $\mathbf{X'X} = \begin{bmatrix} 12 & 47.7 \\ 47.7 & 193.79 \end{bmatrix}.$

c. $(\mathbf{X'X})^{-1} = \begin{bmatrix} 193.79 & -47.7 \\ -47.7 & 12 \end{bmatrix}.$

d. $\hat{\boldsymbol{\beta}} = \begin{bmatrix} -4.1415 \\ 7.3102 \end{bmatrix}.$

e. (a) $\hat{\mathbf{y}}$ as before; $\hat{\boldsymbol{\beta}} = \begin{bmatrix} a' \\ b \end{bmatrix}$; $\mathbf{X} = \begin{bmatrix} 1 & 0.025 \\ 1 & -0.675 \\ \vdots & \vdots \\ 1 & 0.025 \end{bmatrix}.$

(b) $\mathbf{X'y} = \begin{bmatrix} 299 \\ 30.575 \end{bmatrix}$; $\mathbf{X'X} = \begin{bmatrix} 12 & 0 \\ 0 & 4.1825 \end{bmatrix}.$

(c) $(\mathbf{X'X})^{-1} = \begin{bmatrix} 1/12 & 0 \\ 0 & 1/4.1825 \end{bmatrix}.$

(d) $\hat{\boldsymbol{\beta}} = \begin{bmatrix} 24.9167 \\ 7.3102 \end{bmatrix}.$

PRECIPITATION Data:

a. $\mathbf{y} = \begin{bmatrix} 57.3 \\ 55.8 \\ \vdots \\ 55.1 \end{bmatrix}$; $\hat{\boldsymbol{\beta}} = \begin{bmatrix} a \\ b \end{bmatrix}$; $\mathbf{X} = \begin{bmatrix} 1 & 69.2 \\ 1 & 51.2 \\ \vdots & \vdots \\ 1 & 56.0 \end{bmatrix}.$

b. $\mathbf{X'X} = \begin{bmatrix} 15 & 915.2 \\ 915.2 & 56,306.1 \end{bmatrix}$; $\mathbf{X'y} = \begin{bmatrix} 852.7 \\ 52,028.77 \end{bmatrix}.$

c. $(\mathbf{X'X})^{-1} = \dfrac{1}{700.46} \begin{bmatrix} 56,306.1 & -915.2 \\ -915.2 & 15 \end{bmatrix}.$

d. $\hat{\boldsymbol{\beta}} = \begin{bmatrix} 56.4936 \\ 0.0058 \end{bmatrix}$.

e. (a) y as above; $\hat{\boldsymbol{\beta}} = \begin{bmatrix} a \\ b \end{bmatrix}$; $\mathbf{X} = \begin{bmatrix} 1 & 8.1867 \\ 1 & -9.8133 \\ \vdots & \vdots \\ 1 & -5.0133 \end{bmatrix}$.

(b) $\mathbf{X'X} = \begin{bmatrix} 15 & 0 \\ 0 & 466.6973 \end{bmatrix}$.

(c) $(\mathbf{X'X})^{-1} = \begin{bmatrix} 1/15 & 0 \\ 0 & 1/466.6973 \end{bmatrix}$.

(d) $\hat{\boldsymbol{\beta}} = \begin{bmatrix} 56.8467 \\ 0.0058 \end{bmatrix}$.

SALES Data:

a. $\mathbf{y} = \begin{bmatrix} 7 \\ 17 \\ \vdots \\ 8 \end{bmatrix}$; $\hat{\boldsymbol{\beta}} = \begin{bmatrix} a \\ b \end{bmatrix}$; $\mathbf{X} = \begin{bmatrix} 1 & 17 \\ 1 & 39 \\ \vdots & \vdots \\ 1 & 19 \end{bmatrix}$.

b. $\mathbf{X'X} = \begin{bmatrix} 15 & 463 \\ 463 & 16,275 \end{bmatrix}$; $\mathbf{X'y} = \begin{bmatrix} 173 \\ 6102 \end{bmatrix}$.

c. $(\mathbf{X'X})^{-1} = \dfrac{1}{29,756} \begin{bmatrix} 16,275 & -463 \\ -463 & 15 \end{bmatrix}$.

d. $\hat{\boldsymbol{\beta}} = \begin{bmatrix} -0.3243 \\ 0.3842 \end{bmatrix}$.

e. (a) $\mathbf{y}$ as above, $\hat{\boldsymbol{\beta}} = \begin{bmatrix} a' \\ b \end{bmatrix}$; $\mathbf{X} = \begin{bmatrix} 1 & -13.8667 \\ 1 & 8.1333 \\ \vdots & \vdots \\ 1 & -11.8667 \end{bmatrix}$.

(b) $\mathbf{X'X} = \begin{bmatrix} 15 & 0 \\ 0 & 1983.7333 \end{bmatrix}$; $\mathbf{X'y} = \begin{bmatrix} 173 \\ 762.0667 \end{bmatrix}$.

(c) $(\mathbf{X'X})^{-1} = \begin{bmatrix} 1/15 & 0 \\ 0 & 1/1983.7333 \end{bmatrix}$.

(d) $\hat{\boldsymbol{\beta}} = \begin{bmatrix} 11.5333 \\ 0.3842 \end{bmatrix}$.

HEATING Data:

a. $\mathbf{y} = \begin{bmatrix} 0.30 \\ 0.70 \\ \vdots \\ 0.75 \end{bmatrix}$; $\hat{\boldsymbol{\beta}} = \begin{bmatrix} a \\ b \end{bmatrix}$; $\mathbf{X} = \begin{bmatrix} 1 & 50 \\ 1 & 4 \\ \vdots & \vdots \\ 1 & 2 \end{bmatrix}$.

b. $\mathbf{X'X} = \begin{bmatrix} 15 & 284 \\ 284 & 10,302 \end{bmatrix}$; $\mathbf{X'y} = \begin{bmatrix} 8.75 \\ 126.3 \end{bmatrix}$.

c. $(\mathbf{X'X})^{-1} = \dfrac{1}{73,874} \begin{bmatrix} 10,302 & -284 \\ -284 & 15 \end{bmatrix}$.

d. $\hat{\boldsymbol{\beta}} = \begin{bmatrix} 0.7347 \\ -0.008 \end{bmatrix}$.

e. (a) $\mathbf{y}$ as above; $\hat{\boldsymbol{\beta}} = \begin{bmatrix} a' \\ b \end{bmatrix}$; $\mathbf{X} = \begin{bmatrix} 1 & 31.0667 \\ 1 & -14.9333 \\ \vdots & \vdots \\ 1 & 76.9333 \end{bmatrix}$.

(b) $\mathbf{X'X} = \begin{bmatrix} 15 & 0 \\ 0 & 4924.9333 \end{bmatrix}$; $\mathbf{X'y} = \begin{bmatrix} 8.75 \\ -39.3667 \end{bmatrix}$.

(c) $(\mathbf{X'X})^{-1} = \begin{bmatrix} 1/15 & 0 \\ 0 & 1/4924.9333 \end{bmatrix}$.

(d) $\hat{\boldsymbol{\beta}} = \begin{bmatrix} 0.5833 \\ -0.008 \end{bmatrix}$.

3.28. a. $\mathbf{y} = \begin{bmatrix} 625 \\ 750 \\ \vdots \\ 625 \end{bmatrix}$; $\hat{\boldsymbol{\beta}} = \begin{bmatrix} a \\ b \\ c \end{bmatrix}$; $\mathbf{X} = \begin{bmatrix} 1 & 64 & 4096 \\ 1 & 53 & 2809 \\ \vdots & \vdots & \vdots \\ 1 & 58 & 3364 \end{bmatrix}$.

b. $\mathbf{X'X} = \begin{bmatrix} 13 & 878 & 61,470 \\ 878 & 61,470 & 4,454,930 \\ 61,470 & 4,454,930 & 333,321,846 \end{bmatrix}$.

$\mathbf{X'y} = \begin{bmatrix} 8010 \\ 527.885 \\ 36,104,535 \end{bmatrix}$

c. $(\mathbf{X'X})^{-1} = \frac{1}{8,011,646,096}$.

$\begin{bmatrix} 642,892,568,720 & & \text{sym.} \\ -18,812,033.688 & 554,623,098 & \\ 132.867,640 & -3,943,430 & 28,226 \end{bmatrix}$.

d. $\hat{\boldsymbol{\beta}} = \begin{bmatrix} 2010.0028 \\ -35.3089 \\ 0.2096 \end{bmatrix}$.

3.29. a. $\mathbf{y} = \begin{bmatrix} 45 \\ 43 \\ \vdots \\ 76 \end{bmatrix}$; $\hat{\boldsymbol{\beta}} = \begin{bmatrix} a \\ b \end{bmatrix}$; $\mathbf{X} = \begin{bmatrix} 1 & 15 \\ 1 & 15 \\ \vdots & \vdots \\ 1 & 25 \end{bmatrix}$.

b. $\mathbf{X'X} = \begin{bmatrix} 9 & 180 \\ 180 & 3750 \end{bmatrix}$; $\mathbf{X'y} = \begin{bmatrix} 542 \\ 11{,}290 \end{bmatrix}$.

c. $(\mathbf{X'X})^{-1} = \dfrac{1}{1350} \begin{bmatrix} 3750 & -180 \\ -180 & 9 \end{bmatrix}$.

d. $\hat{\boldsymbol{\beta}} = \begin{bmatrix} 0.2222 \\ 3 \end{bmatrix}$.

3.30. a. $\mathbf{y}$ as in 3.28, $\hat{\boldsymbol{\beta}} = \begin{bmatrix} a' \\ b' \end{bmatrix}$, $\mathbf{Z} = \begin{bmatrix} 1 & -1 \\ 1 & -1 \\ \vdots & \vdots \\ 1 & 1 \end{bmatrix}$.

b. $\mathbf{Z'Z} = \text{diag}\,(9, 6)$; $\mathbf{Z'y} = \begin{bmatrix} 542 \\ 90 \end{bmatrix}$.

c. $(\mathbf{Z'Z})^{-1} = \text{diag}\,(1/9,\ 1/6)$.

d. $\hat{\boldsymbol{\beta}} = \begin{bmatrix} 60.2222 \\ 15 \end{bmatrix}$.

3.31. a. $\mathbf{Z} = \begin{bmatrix} 1 & -2 & 4 \\ 1 & -2 & 4 \\ 1 & -1 & 1 \\ \vdots & \vdots & \vdots \\ 1 & 2 & 4 \end{bmatrix}$; $\hat{\boldsymbol{\beta}} = \begin{bmatrix} a' \\ b' \\ c' \end{bmatrix}$; $\mathbf{y} = \begin{bmatrix} 73 \\ 68 \\ \vdots \\ 65 \end{bmatrix}$.

b. $\mathbf{Z'Z} = \begin{bmatrix} 10 & 0 & 20 \\ 0 & 20 & 0 \\ 20 & 0 & 68 \end{bmatrix}$; $\mathbf{Z'y} = \begin{bmatrix} 508 \\ -53 \\ 1249 \end{bmatrix}$.

c. $(\mathbf{Z'Z})^{-1} = \dfrac{1}{5600} \begin{bmatrix} 1360 & 0 & -400 \\ 0 & 280 & 0 \\ -400 & 0 & 200 \end{bmatrix}$.

d. $\hat{\boldsymbol{\beta}} = \begin{bmatrix} 34.1571 \\ -2.65 \\ 8.3214 \end{bmatrix}$.

Chapter 4

Section 4.2

4.1. a. TEACHERS Data: $\hat{Y} = 4.141 + 7.310X$.
PRECIPITATION Data:
$\hat{Y} = 56.494 + 0.006X$.
SALES Data: $\hat{Y} = 0.324 + 0.384X$.
HEATING Data: $\hat{Y} = 0.735 - 0.008X$.

b. In all cases, the plots are the same as Problem 2.9.

4.2. a.–d. $a' = \overline{Y}$, b as before. Residuals are the same as in Problem 2.14. Plot is the same as in Problem 2.14 except that X-axis is labeled differently.

Section 4.3

4.3. a. 1.9659.
b. 1.6255.
c. Note that $1.6255 < 1.9659$.

4.4. a. $s_{y \cdot x}^2 = 6.5055$.
b. See mean square for residual.

4.5. a. $a = 1023.582$. b. $b = -6.033$.
c. 35.9615.

4.6. a. $a = 1023.582$.
b. $b = b_1 = -6.033$; note message of redundant variable.
c. 35.9615.

4.7. Use linear equation.

Section 4.4

4.8. TEACHERS Data: $\hat{Y} = 4.14145 + 7.31022X$.
PRECIPITATION Data:
$\hat{Y} = 56.49358 + 0.00579X$.
SALES Data: $\hat{Y} = -0.32434 + 0.381416X$.
HEATING Data: $\hat{Y} = 0.73467 - 0.00799X$.

4.9. All results are the same as in Problem 2.14.

Section 4.5

4.10. a. $\hat{Y} = 1023.58358 - 6.03256X$.

4.11. a. $\hat{Y} = 2010.00276 - 35.30887X + 0.20955X^2$.

Section 4.6

4.12. Regression equations and scatter diagrams are the same as those found in Problems 2.8 and 2.9.

4.13. Results are the same as in Problem 2.14.

Section 4.7

4.14. Refer to Problems 2.11 and 4.5 c.

4.15. Refer to Problems 2.11, 4.6 c and 4.5.

4.16. Refer to Problem 4.7.

Section 4.8

4.17. Compare to Problem 3.22.

4.18. Compare to Problem 3.22.

4.19. $\begin{bmatrix} 0.6716384 & & (\text{sym}) \\ -0.1630060 & 0.0744829 & \\ -0.0137501 & 0.0006531 & 0.005788 \end{bmatrix}$.

Chapter 5

5.1. a. 25,100 teachers.

   b. For states in which the average salary is $4000, the average number of teachers is 25,100.

   c. The estimated 25,100 teachers is between the actual numbers for Alabama and West Virginia. The estimated value is only an average and is not intended to coincide exactly with the actual number of teachers in any state.

   d. Pennsylvania is not a Southern state, and no state in the sample paid an average salary as high as $7000.

   e. Data were collected for a given year and for a range of salaries not larger than $5100.

   f. It is a constant affecting number of teachers determined by something other than salary.

   g. It might seem reasonable that a state paying $0 in salaries would have no teachers.

   h. $\hat{Y} = 6.2908X$.

   i. If one state pays annual salaries averaging $1000 more than in another state, then on the average the state paying more will have 7310 more teachers than the lower-paying state.

   j. Probably other factors cause both the average salary paid and the number of teachers.

5.2. a. 56.8292 degrees.

   b. The typical city receiving 58 inches of rainfall will have average annual temperature of 56.8 degrees.

   c. The estimate is only an average and is not intended to coincide exactly with any actual. temperature.

   d. We have no evidence that the same relationship obtained for cities having between 51.2 and 69.2 inches of rainfall will hold for those having less than 51.2 inches.

   e. Yes.

   f. The regression constant is needed to orient the relationship between temperature and precipitation, but is determined by some factor or factors other than precipitation.

   g. It seems unlikely that locations having zero precipitation would have average annual temperature of zero.

   h. $\hat{Y} = 0.924X$.

   i. If one city has one more inch of rainfall than another, then on the average its temperature is 0.0058 degrees higher than in the other.

   j. Evidence of a relationship in a sample does not give evidence that one variable causes the other.

5.3. a. 584.98 robberies.

   b. The average number of robberies in cities employing 67 police officers is 584.98.

   c. Since the estimate is only an average, we would not expect it to coincide with the actual number of robberies in either city.

   d. We have no evidence that the same relationship holds for cities employing more than 90 policemen.

   e. No, out of range.

   f. Yes.

   g. The existence of a relationship does not imply a cause-effect relationship.

   h. The intercept is a constant affecting the number of robberies determined by something other than the number of police officers.

5.4. a. $6205.80.

   b. The typical store employing 17 people has weekly sales of approximately $6206.

   c. The estimate is only an average and is not intended to coincide with any actual values.

   d.–f. We only have information concerning the relationship for stores employing between 10 and 48 people.

g. $a = -0.3256$ is a constant affecting sales determined by something other than the number of employees.

h. It would seem reasonable that a store with no employees might have no sales.

i. $\hat{Y} = 0.3749X$.

j. If one store has one more employee than another, then its average weekly sales are $384.20 higher than in the other.

k. Probably larger stores tend to have both large number of employees and high sales.

5.5. a. 47¢.

b. The average daily bill when the temperature is 33 degrees is 47¢.

c. The estimate is only an average and is not intended to coincide with any actual bill.

d.–f. Estimates should be made only for days with temperatures between -2 and 50 degrees since this is the only range for which we have information.

g. The average bill on zero-degree days is 73¢.

h. We saw that the average bill on zero-degree days is 73¢, not 0¢.

i. $\hat{Y} = 0.0122X$.

j. If one day the temperature was one degree warmer than on another, the bill on the warmer day was 0.8¢ lower, on the average, than on the colder day.

5.6. a. 75.2222 bu.

b. The average yield from plots to which 25 lb of fertilizer was applied is 75.2222 bu.

c. The estimate is only an average and is not intended to coincide with any actual yield.

d.–e. The data contain information about yield at only 15, 20, and 25 lb fertilizer applied so estimates should be restricted to these levels.

f. If $z = 0$, or 20 lb fertilizer was applied, the average yield was 60.22 bu.

g. 0.2222 is a constant affecting yield determined outside of the relationship of yield to fertilizer.

h. It seems unlikely that yield would be zero if no fertilizer were applied.

i. Plots receiving five lb fertilizer more than others have, on the average, 15 bu greater yield.

j. Plots receiving five lb fertilizer more than others have, on the average, $3(5) = 15$ bu greater yield.

5.7. a. X could cause Y, Y could cause X, or perhaps larger firms (Z) both spend more and have greater sales.

b. Probably X causes Y.

c. X and Y are probably related by chance.

d. Probably size of the store (Z) determines both X and Y.

e. Probably spurious.

Chapter 6

Section 6.2

6.1. Scatter not homogeneous; the 15 different variances cannot be estimated from the data in this sample.

6.2. $n - 3$.

Sections 6.3 and 6.4

6.3. a.–b. When using the regression equation to estimate number of teachers from average annual salary, you will be off, on the average, by 6256 teachers.

c. 12.587 to 37.6118.

d. 25.11%.

6.4. a.–b. When using the regression equation to estimate temperature from precipitation, the estimates will be off by 1.2831 degrees, on the average, from the actual temperatures.

c. 54.263 to 59.3954.

d. 2.26%.

6.5. a.–b. When predicting sales from numbers of employees, our estimates will be off, on the average, by $1439.70 from the actual sales.

c. 3.3264 to 9.0852.

d. 12.48%.

6.6. a.–b. Estimating oil bill from temperature, we will be off, on the average, by 4.24¢.

c. 0.386 to 0.5556.

d. 7.27%.

6.7. a.–b. Estimating yield from amount of fertilizer, we will be off by 3.0132 bu from actual yield, on the average.

c. 69.1958 to 81.2486.

d. 5%.

Section 6.6

6.8. a. 1.725.   d. 2.2787.

b. -2.473.   e. ± 2.447.

c. -1.356.

6.9. a. Between 20,315 and 28,421.8 teachers.
      b. Between 55.3 and 58.3 degrees.
      e. The precision of the information decreases the farther X_i is from $\bar{X}$.

6.10. a. Between 8852.5 and 37,884.3 teachers.
      b. Between 52.7 and 61 degrees.
      c. Between \$15,405.90 and \$22,362.90.
      d. Between -0.47¢ and 32.23¢.

Section 6.9

6.11. From BMD, look under "standard error of estimate." For SPSS, look under "standard error." For SAS, look under "standard deviation."

6.12. *CV* must be calculated from $\bar{Y}$ and $s_{y \cdot x}$ for the BMD and SPSS. Look under C.V. in SAS output.

6.13. Correcting for the mean does not affect $s_{y \cdot x}$.

Chapter 7

Section 7.3

7.1. a. $t = 2.3896 > 1.182$, so states paying higher salaries tend to have more teachers than do states paying lower salaries.
      b. $t = 2.3896 > 2.228$, so number of teachers is related to salary.
      c. $F = 5.7104 > 4.96$, so number of teachers is related to salary.

7.2. a. $t = 0.0976 \not> 1.771$, so there is insufficient evidence to claim that the greater the precipitation, the higher the temperature.
      b. $t = 0.0976 \not> 2.160$ and $\not< -2.160$, so we have insufficient evidence to conclude that temperature is related to precipitation.
      c. $F = 0.0095 \not> 4.66$, so we have insufficient evidence to conclude that temperature is related to precipitation.

7.3. a. $t = 11.8853 > 2.650$ so you are more than 99% sure that stores having a large number of employees also tend to have high average weekly sales.
      b. $t = 11.8853 > 3.012$ so you are more than 99% sure that average weekly sales is related to number of employees.
      c. $F = 141.2446 > 9.07$, so you are more than 99% sure that sales are related to number of employees.

7.4. a. $t = -13.2329 < -1.771$, so you are more than 95% sure that the oil bill tends to be high on cold days.
      b. $t = -13.2329 < -2.160$, so you are more than 95% sure that oil bill is related to temperature.
      c. $F = 174.9444 > 4.66$, so you are more than 95% sure that oil bill is related to temperature.

7.5. $t = 12.1938 > 1.895$; you are more than 95% sure that the greater the amount of fertilizer applied, the greater the yield.

Section 7.4

7.6. a. $t = -0.3369 \not< -1.812$; you could be justified in fitting the line through the origin.
      b. $t = 15.5247 > 1.771$; you would not fit the line through the origin.
      c. $t = -0.3058 \not< -2.160$ and $\not> 2.160$; you cannot be 95% sure that the line should not be fitted through the origin.

7.7. a. $t = 1.9791 > 1.812$, so you are more than 95% sure that the greater the salary paid, generally the more teachers in the state.
      b. $t = 10.3198 > 1.771$; you are more than 95% sure that the greater the number of employees, the higher the sales.

Section 7.6

7.8. a. You are 95% sure that if one state pays \$1000 more than another, then the higher-paying state will have between 494 and 14,126 more teachers than the other.
      b. $-0.1225 < \beta < 0.1341$. Interpretation nonsensical; indicative of no relationship.
      c. You are 99% sure that if one store has one more employee than another, then its average retail sales are between \$286.80 and \$481.60 higher than in the other.
      d. You are 95% sure that if one day was one degree warmer than another, the bill was between 0.68¢ and 0.92¢ less on the warmer day than on the colder day.

7.9. a. $-31.5308 < \alpha < 23.2480$; no specific interpretation because no data on states paying \$0 salary.
      b. $45.5323 < \alpha < 67.4533$; no physical interpretation because no information on locations having zero population.

c. $-2.2113 < \alpha < 1.5601$; data included no stores with zero employees.

d. $0.6928 < \alpha < 0.7768$; we are 98% sure that the average bill on zero-degree days is between 69.28¢ and 77.68¢.

Section 7.7

7.10.

Part		F Calculated	F Critical	Signifi- cant?
a.	BMD	5.710	4.96	yes
	SPSS	5.710		
	SAS	5.71043		
b.	BMD	0.009	4.67	no
	SPSS	none calculated		
	SAS	0.00949		
c.	BMD	141.064	4.67	yes
	SPSS	141.063		
	SAS	141.06284		
d.	BMD	172.874	4.67	yes
	SPSS	172.88029		
	SAS	172.88030		

7.11.

Data Set	t Calculated	t Critical	Signifi- cant?
TEACHERS			
BMD	2.39	1.812	yes
SPSS	7.31022/3.05912 = 2.39		
SAS	2.38965		
PRECIPITATION			
BMD	0.097	1.771	no
SPSS	program terminated		
SAS	0.09743		
SALES			
BMD	11.877	1.711	yes
SPSS	0.38416/0.03234 = 11.879		
SAS	11.87699		
HEATING			
BMD	-13.148	-1.771	yes
SPSS	$-0.00799/0.00061$ = -13.098		
SAS	-13.14840		

7.12. a. $t = 0.33689 \not< -1.812$ so there is insufficient evidence that the intercept should be negative.

b. $t = 15.52502 > 1.771$, so you are more than 95% sure that the intercept is positive.

c. $t = -0.30442$, $\text{PR}\emptyset\text{B} > |T| = 0.7656 \not< 0.05$, so we are not 95% sure that the intercept is different from zero.

7.13.

	F-Values	
Data Set	X Not Corrected	X Corrected
TEACHERS		
BMD	5.710	5.710
SPSS	5.710	5.710
SAS	5.71043	5.71043
PRECIPITATION		
BMD	0.009	0.009
SPSS	none calculated	none calculated
SAS	0.00949	0.00949
SALES		
BMD	141.064	141.064
SPSS	141.06284	141.06286
SAS	141.06284	141.06284
HEATING		
BMD	172.874	172.879
SPSS	172.88029	172.88022
SAS	172.88030	172.88030

Chapter 8

Section 8.2

8.1. a. 36.35% of variation in number of teachers can be explained by salary differences.

b. 0.07% of the variation in temperature can be accounted for by differences in precipitation.

c. Differences in numbers of employees explains 91.56% of the variation in sales.

d. 93.02% of the variation in daily bills can be attributed to temperature differences.

Sections 8.3 and 8.4

8.2. a. $r = 0.6029$ indicates a relatively strong direct relationship.

b. $t = 2.3897 > 1.812$; you are more than 95% sure that states paying higher average salaries tend to have more teachers.

8.3. a. $r = 0.0270$ indicates a very weak direct relationship.

b. $t = 0.0974 \not> 2.160$ and $t \not< -2.160$; insufficient evidence to be 95% sure of a relationship.

8.4. a. $r = 0.9569$ indicates a very strong direct relationship.

b. $t = 11.8759 > 2.650$; you are more than 95% sure that the larger the number of employees, the greater the sales.

8.5. a. $r = -0.9644$ indicates a very strong inverse relationship.

b. $t = -13.1614 < -1.771$ so you are more than 95% sure that the colder the day, the higher the bill.

Section 8.5

8.6.

Data Set	r		r^2	
TEACHERS				
BMD	0.609	(given)	0.3635	(given)
SPSS	0.60289	(given)	0.36348	(given)
SAS	0.6029		0.36348017	(given)
PRECIPITATION				
BMD	0.0270	(given)	0.0007	(given)
SPSS	0.02701	(given)	0.00073	(given)
SAS	0.0270		0.00072969	(given)
SALES				
BMD	0.9569	(given)	0.9156	(given)
SPSS	0.95688	(given)	0.91562	(given)
SAS	0.9569		0.91561885	(given)
HEATING				
BMD	−0.9644	(given)*	0.9301	(given)
SPSS	−0.96440	(given)	0.93006	(given)
SAS	0.9644		0.93006252	(given)

* Sign must be determined from sign of b.

8.7.

Data Set	X Not Corrected	X Corrected
TEACHERS		
BMD	0.3635	0.3635
SPSS	0.36348	0.36348
SAS	0.36348017	0.36348017
PRECIPITATION		
BMD	0.0007	0.0007
SPSS	0.00073	0.00073
SAS	0.00072969	0.00072969
SALES		
BMD	0.9156	0.9156
SPSS	0.91562	0.91562
SAS	0.91561885	0.9156185
HEATING		
BMD	0.930	0.930
SPSS	0.93006	0.93006
SAS	0.93006252	0.93006252

Chapter 9

Section 9.2.

9.1. a. $\hat{Y} = 625.8359 + 36.7290X$.

c. Ms. Moore's salary appears to be an outlier.

d. $\hat{Y} = 624.4177 + 35.7308X$.
$703.1074 < \hat{Y}_{2.3} < 802.9898$.
Ms. Moore's salary is an outlier.

9.2. b. Small salaries: $s_{y\cdot x}^2 = 60.0112$.
Large salaries: $s_{y\cdot x}^2 = 269.7741$.
$F = 0.2224 \not< 0.1565$ and $\not> 6.39$.
Assume homogeneous variances.

9.3. b. Small: $s_{y\cdot x}^2 = 43.2833$.
Large: $s_{y\cdot x}^2 = 247.9799$.
$F = 0.1745 < 0.2020$; scatter not homogeneous.

c. $\hat{Y}' = 7.9831 + 4.6808X'$.

d. Small: $s_{y\cdot x}^2 = 10.6355$.
Large: $s_{y\cdot x}^2 = 9.4376$.
$F = 1.1269 \not< 0.2020$ and $\not> 4.39$; assume homogeneous scatter in transformed data.

Section 9.3

9.4. a. $F = 0.5199$; no linear relationship.

b. Simple linear model gives poor fit.

c. $F = \dfrac{2097.15/3}{64/5} = 54.6133$; simple linear model not appropriate.

d. $Y = 34.156353 - 2.652 + 8.320854Z^2$.

e. $F = 32.5779$; significant polynomial regression.

f. $F_1 = 4.4032$; no significant simple linear regression.
$F_2 = 60.7527$; quadratic term improves fit.

g. Results are the same as in part f.

h. $F = \dfrac{159.2836/2}{64/5} = 6.220$; second-degree polynomial does not provide an adequate description.

9.5. a. $\hat{Y} = 68.356 + 11.0491X$; plot indicates curvilinear model.

b. $F = \dfrac{386.9972/12}{211/11} = 1.6813$. Cannot conclude lack of fit.

c. $\hat{Y} = 116.3789 - 14.0954X + 3.0200X^2$.

d. $F = 7554.0495$; significant polynomial regression.

e. $F_1 = 13{,}257.259$; significant linear term.
$F_2 = 1850.84$; quadratic term improves fit.

9.6. a. $F = 38.6787$; significant regression.
     b. $F_1 = 66.8153$; significant linear term.
         $F_2 = 10.5420$; addition of quadratic term improves fit.
     c. Not enough degrees of freedom for pure error.

9.7. a. $\hat{Y} = -27.5486 + 0.1673X$.
     b. Effect of fatigue is present.

9.8. a.–b. $\hat{Y} = 20.2404 + 0.6037X$.
     c. Teacher A's students are almost all above the line; Teacher B's below.

Section 9.4.

9.9   a. $r_{t,t-1} = 0.7739$; $d = 0.4788$; conclude positive serial correlation.
     b. $\hat{Y} = 0.1120 + 0.1762X$, $r_{t,t-1} = -0.2366$, $d = 2.2015$, $d^- = 1.7985$; no negative serial correlation.

9.10. $r_{t,t-1} = 0.0715$, $d = 1.8613$; no serial correlation.

Section 9.5

9.11.

	Range	%			Range	%
a.	$-1, 1$	66.67	b.		$-1, 1$	9	60.00
	$-2, 2$	100.00			$-2, 2$	15	100.00
	$-3, 3$	100.00			$-3, 3$	15	100.00

Chapter 10

Section 10.3

10.1. a. $\hat{Y} = -3.6183 + 1.7896X_2$.
      b. $\hat{Y} = -2.447907 + 0.269825X_1 + 0.667393X_2$.

10.2. $\hat{Y} = 0.311472 + 1.023199X_1 + 0.072631X_2$.

10.3. $\hat{Y} = 35.221155 + 1.143994X_1 - 0.533289X_2$.

10.4. $\hat{Y} = -0.844886 + 0.525260X_1 + 0.039944X_2$.

10.5. $\hat{Y} = -15.293983 + 2.620384X_1 + 7.556194X_2 + 1.900656X_3$.

10.6. $\hat{Y} = -136.52606 + 9.527225X_1 + 3.348974X_2 - 0.596901X_3$.

10.7. Code temperature to $-1, 1$ and humidity to $-1, 0$, 1; then $\hat{Y} = 195.833333 + X_1 + 6X_2$, in terms of *coded* values.

10.8. $\hat{Y} = 2.916667 - 0.916667X_1 - 0.562500X_2$, in terms of *coded* values.

10.9. a. $\hat{Y} = 322.259259 + 51.166667 - 47.388889X_2 + 8.055556X_3$ in terms of coded values.

b. Regression of Y on X_1:
    $\hat{Y} = 322.259259 + 51.16667X_1$.
    Regression of Y on X_2:
    $\hat{Y} = 322.259259 - 47.388889X_2$.
    Regression of Y on X_3:
    $\hat{Y} = 322.259259 + 8.055556X_3$.

c. Regression of Y on X_1 and X_2:
    $\hat{Y} = 322.259259 + 51.166667X_1 - 47.388889X_2$.
    Regression of Y on X_1 and X_3:
    $\hat{Y} = 322.259259 + 51.166667X_1 + 8.055556X_3$.
    Regression of Y on X_2 and X_3:
    $\hat{Y} = 322.259259 - 47.388889X_2 + 8.055556X_3$.

d. Since X_1, X_2, X_3 are mutually uncorrelated, b_1, b_2, and b_3 are unaffected by the absence or presence of the other predictors.

Section 10.4

10.10. a. $a = -2.447909$ is a constant determined outside of the relationship. For two stores the same size, if one has one more employee than another, then on the average its sales will be $269.82 more than in the other. Among stores with the same number of employees, a store 1000 square feet larger than another will have $667.39 greater sales, on the average.

b. $a = 0.311472$ is determined outside of the relationship. For restaurants with the same liquor sales, if one has $1000 greater food sales than another, then it will average $1023.20 greater total profits than the other. If two restaurants have the same food sales but one has liquor sales $1000 greater than another, then it will average $72.63 higher profits than the other.

c. The intercept is determined by something other than the effect of dosage and age on time to relief. Among patients receiving the same dosage, if one is one year older than another, he will experience relief about half a minute sooner than will the younger. For patients the same age, one receiving one gram more than another will experience relief 1.14 minutes sooner than the other.

d. The constant represents an effect on cost determined by something other than age and mileage. If one car is one year older than another but both were driven the same number of miles, then the older car averaged $52.53 more in maintenance and repair costs. For two cars the same age, if one were driven 1000 miles farther than another, then its cost of upkeep averaged $3.99 more.

e. $a = 195.833333$ is determined outside the relationship. If humidity is held constant, then holding strength is one pound greater, on the average, at the higher temperature level than at the lower. If temperature is held constant, the holding strength is 6 pounds greater for each higher level of humidity.

f. a was determined outside of the relationship. If temperature is held constant, then an average of 0.92 fewer impurities are found at longer sterilization times than at the next shorter times. For equal sterilization times, an average of 0.56 fewer impurities are found at higher temperatures than at the next lower temperature level.

10.11. a. The constant is determined outside of the relationship. If two stores spend the same amount on radio and newspaper advertising but one spent $1000 more on TV ads, then on the average the store spending more on TV ads will have $2620.38 greater sales. For stores spending the same amount on TV and newspaper, one spending $1000 more on radio will have $7556.19 greater sales than another. If one store spends $1000 more on newspaper ads than another but both spend the same amount on radio and TV, then the higher-spending store will have $1900.66 greater sales.

b. The regression constant is determined outside of the relationship. For children the same height and weight, the average score of one child a year older than another is 9.5272 points higher. For two children the same age and weight, if one is one inch taller than another his average score will be 3.349 points higher. For children the same age and height, if one is one pound heavier than another, his score will be 0.5969 points lower than the lighter child's, on the average.

c. The average yield for plots receiving medium amounts of nitrogen, phosphorus, and potassium was 322.26 lb. For plots receiving the same amounts of potassium and phosphorus,

those receiving higher levels of nitrogen averaged 51.17 lb greater yield than those at the next lower levels. For plots receiving the same application of nitrogen and phosphorus, those receiving one level higher application of potassium averaged 47.39 lb less yield than those at the lower level. If two plots received the same amounts of nitrogen and potassium but one received one level higher application of phosphorus than another, the yield from the plot receiving more phosphorus was 8.06 lb greater than from the other.

10.12. a. $6810.90. It is only an average and is not intended to coincide with any actual value.

b. $2268.80. Other factors affecting total profits have not been included.

c. The experiment included no trials on which patients were given three grams.

d. $27.95.

e. Sample included no information on expenditures as large as $6000 on newspaper advertisements.

f. 74.1715. The estimate is only an average and is not expected to coincide exactly with any particular score.

g. 194.833333. Factors other than temperature and humidity, including random variation, also affect holding strength.

h. 3.270834; other factors omitted.

i. 334.092593 lb.

Section 10.5

10.13. a. b_1 and b_2 changed from the simple regressions to the multiple regression, so we know that multicollinearity is present.

b. No multicollinearity is present.

10.14. a. $r_{12} = 0.8965$; regression coefficients will be very unreliable.

b. $r_{12} = 0.4494$; regression coefficients will be rather unreliable.

c. $r_{12} = -0.0179$; regression coefficients slightly unreliable.

d. $r_{12} = 0$; regression coefficients are reliable measures of the effects of the independent variables.

10.15. a. $r_{12} = -0.0778$, $r_{13} = -0.3464$, $r_{23} = 0.0898$; regression coefficients unreliable.

b. $r_{12} = 0.8233$, $r_{13} = 0.7918$, $r_{23} = 0.6696$; regression coefficients unreliable.

Section 10.6

10.16. a. Estimating sales from numbers of employees and sizes of stores, we will be off, on the average, by $1172.30.

   b. Using food and liquor sales to estimate total profits, we will be off by $448.44, on the average.

   c. Our estimates will be off by an average of 5.844 minutes when estimating time to relief from dosage and age.

   d. Upkeep costs can be estimated from age and mileage within $53.99, on the average.

   e. Revenues can be estimated from TV, radio, and newspaper advertising within $19,606.40, on the average.

   f. Physical fitness score can be estimated from age, height, and weight within 15.6896 points, on the average.

   g. When estimating holding strength from temperature and humidity, the estimates will be off by 1.4011 pounds, on the average.

   h. The number of impurities can be estimated from temperature and time of sterilization within 0.8825, on the average.

   i. Yield can be estimated from amounts of the three elements applied within 109.2790 lb, on the average.

10.17. a. $2111.80.

   b. Predictions are more accurate using the number of employees than when using the sizes.

   c. More accurate predictions can be made by using both predictors than by using either predictor alone.

Chapter 11

Section 11.2

11.1. a. $r_{1Y}^2 = 0.9156$; differences in the numbers of employees alone explains 91.56% of the variation in the sales of the stores. $r_{2Y}^2 = 0.8187$; differences in the sizes of the stores explain 81.87% of the variation in sales.

   b. Variation in food sales explains 83.44% $= r_{1Y}^2$ of the variation in total profits, and variation in liquor sales explains $r_{2Y}^2 = 69.22\%$ of the variation in total profits.

   c. Differences in drug dosages explain 0.87% of the variation in time to pain relief, and

differences in ages explain 58.8% of the variation in time to pain relief.

   d. 67.07% of the variation in upkeep costs can be explained by differences in the ages of the cars, and 7.88% of the variation in cost can be accounted for by mileage differences.

   e. Differences in amounts spent on television, radio, and newspaper advertising explain 8.2%, 21.2% and 0.01%, respectively, of the variation in revenues.

   f. Variation in ages of the children explains 36.13% of the variation in scores, variation in heights explains 42.44% and variation in weights explains 13.55%.

   g. Differences in temperature explain 3.78% of the variation in holding strength, and differences in humidity account for 90.66% of the variation in strength.

   h. Differences in sterilization times account for 70.19% of the variation in the number of impurities remaining, and 7.05% of the variation can be explained by differences in sterilization temperatures.

   i. Differences in the amounts of nitrogen, potassium, and the phosphorus applied explain 12.56%, 10.77%, and 0.31% of the variation in yield, respectively.

Section 11.3

11.2. a. Together, differences in sizes and numbers of employees explain 94.84% of the variation in sales.

   b. Differences in food and liquor sales together explain 83.50% of the variation in profits.

   c. Differences in dosages and patient ages together explain 71.63% of the variation in time to relief.

   d. Age and mileage differences together explain 75.80% of the variation in cost of upkeep.

   e. Using amounts spent on radio, television, and newspaper advertising, we can explain 52.02% of the variation in revenues.

   f. Differences in ages, heights, and weights together account for 47.45% of the variation in scores.

   g. Using both temperature and humidity we can explain 94.44% of the variation in strength.

   h. 77.24% of the variation in the number of impurities can be accounted for by differences in sterilization times and temperatures.

i.   Differences in the amounts of nitrogen, potassium, and phosphorus applied account for 23.64% of the variation in yield.

11.3. a. 71.27% of the variation in the number of employees can be explained by the differences in sizes of the stores; the two predictors are giving very redundant information.

b. 80.37% of the variation in food sales can be explained by variation in liquor sales; the two predictors are giving very much the same information.

c. Since there was a tendency for older patients to receive larger doses, we cannot separate the percentages of variation explained by dosage alone and by ages alone.

d. There was a slight tendency for newer cars to be driven farther than older cars, so the percentages of variation explained by upkeep and mileage are not disjoint.

11.4. a. The amounts spent on advertising in the three media are not independent and cannot explain variation in revenue independently.

b. Since ages, heights, and weights of children are not independent, they cannot explain independent percentages of variation in scores.

11.5. In all cases the predictor variables are uncorrelated; thus they explain independent percentages of variation in the response and the total percentage of explained variation is equal to the sum of the percentages explained by each predictor separately.

Section 11.4

11.6. a. After sizes of the stores have been taken into account, variation in the number of employees explains 71.52% of the variation in sales. For stores with the same number of employees, variation in sizes explains 38.84% of the variation in sales.

b. For stores with the same liquor sales, variation in food sales explains 46.5% of the variation in profits. After differences in food sales has explained all it can, variation in liquor sales accounts for 0.52% of the variation in profits.

c. Once variation in ages of patients has explained all it can, variation in dosages explains 19.23% of the remaining variation in time to relief. Once variation in dosage has explained all it can of variation in time to relief,

differences in ages of the patients accounts for 66.42% of the remaining variation.

d. For cars with the same mileage, differences in ages explain 73.74% of the variation in upkeep costs; for cars the same age, 26.50% of the variation in upkeep costs can be explained by mileage differences.

11.7. a. After differences in amount spent on radio and newspaper advertising have explained all they can, variation in amounts spent on TV advertising explains 13.47% of the remaining variation in revenues. After TV and newspaper advertising have explained all they can, radio advertising explains 25.24% of the remaining variation. After radio and TV advertising have told all they can, newspaper advertising explains 0.58% of the remaining variation.

b. For children the same height and weight, age differences account for 8.76% of the variation in scores. For children the same age and weight, height differences account for 13.4% of the variation in scores. For children the same age and height, weight differences explain 6.47% of the variation in scores.

11.8. a. For children the same age and height, there is a slight tendency for heavier children to score lower than do lighter children.

b. The older, taller, and heavier child tends to score relatively high on the test.

11.9.

$$r_{1Y \cdot 234} = \frac{r_{1Y \cdot 24} - r_{13 \cdot 24} r_{3Y \cdot 24}}{\sqrt{1 - r_{13 \cdot 24}^2} \sqrt{1 - r_{3Y \cdot 24}^2}}$$

$$= \frac{r_{1Y \cdot 34} - r_{12 \cdot 34} r_{2Y \cdot 34}}{\sqrt{1 - r_{12 \cdot 34}^2} \sqrt{1 - r_{2Y \cdot 34}^2}}.$$

11.10.   0.0760.

Chapter 12

Section 12.2

12.1. a. Since $F = 110.1992 > 3.89$, conclude that either the number of employees, the size of the store, or both are significant predictors of sales.

b. Since $F = 17.7213 > 4.74$, either food or liquor sales, or both, are significant predictors of profits.

c. $F = 21.4640 > 3.59$. At least one predictor is significant.

d. $F = 18.7901 > 3.89$. At least one predictor is significant.

e. $F = 1.2558 \not> 4.07$. We cannot be 95% sure that amounts spent on television, radio, or newspaper advertising effectively predict revenues.

f. $F = 4.8150 > 3.24$; at least one predictor is significant.

g. $F = 76.4137 > 4.26$; at least one predictor is significant.

h. $F = 35.6241 > 3.47$; at least one predictor is significant.

i. $F = 2.3731 \not> 3.03$; we cannot be 95% sure that the amounts of nitrogen, potassium, or phosphorus is an effective indicator of yield.

Section 12.4

12.2. a. $F_1 = 6.1131$, $F_2 = 146.7142$; use
$\hat{Y} = 195.8333 + X_1 + 6X_2$.

b. $F_1 = 64.747$, $F_2 = 6.5012$; use
$\hat{Y} = 2.1967 - 0.9167X_1 - 0.5625X_2$.

c. $F_1 = 3.78172$, $F_2 = 3.24391$, $F_3 = 0.09374$; no significant predictors.

Section 12.5

12.3. a. $F_1 = 30.1577$, $F_2 = 7.6022$; use
$\hat{Y} = -2.447907 + 0.269825X_1 + 0.667383X_2$

b. $F_1 = 6.0686$, $F_2 = 0.0373$; use
$\hat{Y} = 0.3477 + 1.0950X_1$.

c. $F_1 = 5.1992$, $F_2 = 42.403$; use
$\hat{Y} = 35.221155 + 1.143994X_1 - 0.533289X_2$.

d. $F_1 = 33.6725$, $F_2 = 4.3405$; use
$\hat{Y} = -0.0878 + 0.5219X_1$.

e. $F_1 = 1.2393$, $F_2 = 2.6998$, $F_3 = 0.0439$; no significant predictors.

f. $F_1 = 1.2481$, $F_2 = 2.4867$, $F_3 = 1.1075$; no F's are significant. [Contradicts results of Exercise 12.1 (f.)]

Section 12.6

12.4. a. $0.162766 < \beta_1 < 0.376884$
$0.139968 < \beta_2 < 1.194818$.

b. $0.085392 < \beta_1 < 2.20261$
$-0.706090 < \beta_2 < -0.360488$.

c. $0.084917 < \beta_1 < 1.915083$
$4.879154 < \beta_2 < 7.120746$.

d. $-0.916667 < \beta_1 < -0.679703$
$-1.021369 < \beta_2 < -0.103731$.

Chapter 13

Section 13.2

13.1. Refer to answers of Exercises 12.1, 12.2, 12.3.

Section 13.3

13.2. Refer to answers to Exercises 12.1, 12.2, 12.3.

Section 13.4

13.3. Refer to answers of Exercises 12.1, 12.2, 12.3.

Chapter 14

Section 14.2

14.1. Only height is conditionally significant.

14.2. *Forward selection:* final model with all predictors significant at $\alpha = 0.25$ contains predictors population change, agricultural employment, percent with telephones, percent rural, and median age. If $\alpha = 0.05$, then use only population change and percent with telephones.

Stepwise: uses population change and percent with telephones, both significant at $\alpha = 0.05$.

Backward: predictors population change, agricultural employment, percent rural, median age. All significant at $\alpha = 0.05$.

$MAXR^2$: "best" models with all predictors significant at $\alpha = 0.05$:

k	Predictors
1	percent with telephones
2	percent with telephones
	population change
3	population change
	agricultural employment
	percent rural
4	population change
	agricultural employment
	percent rural
	median age

$MINR^2$: "best" models with all predictors significant at $\alpha = 0.05$ are the same as found by $MAXR^2$.

Section 14.3

14.3. Height is the only conditionally significant predictor.

14.4. All procedures find PØPCHG, INCØME, RURAL, AGE, and NEGRØ in final step; not all are significant at $\alpha = 0.05$.

Section 14.4

14.5. Height is the only conditionally significant predictor.

14.6. Only population change and percent with telephones are significant at $\alpha = 0.05$.

Section 14.5

14.7. *Forward*: If Negro population must be used, supplement it with population change and percent with telephones; both of these are significant.
  Stepwise: At Step 4, population change and percent rural are significant at $\alpha = 0.05$ and Negro population is significant at $\alpha = 0.0646$. At Step 5, population change, percent rural, and median age are all significant at $\alpha = 0.05$, and Negro population is significant at $\alpha = 0.1263$.

14.8. Five predictor model is found, with all predictors except the one forced in, significant at $\alpha = 0.15$.

Chapter 15

Section 15.2

15.1. $d = 1.4863 < d_L = 1.57$; conclude positive serial correlation.

15.2. Reasonably well-behaved plot.

15.3.

Interval	Percent
$-1, 1$	67
$-2, 2$	97
$-3, 3$	100

Section 15.3

15.4. $X_1 = \begin{cases} 1 \text{ if sophomore,} \\ 0 \text{ if otherwise;} \end{cases}$

$X_2 = \begin{cases} 1 \text{ if junior,} \\ 0 \text{ if otherwise;} \end{cases}$

$X_3 = \begin{cases} 1 \text{ if senior,} \\ 0 \text{ otherwise.} \end{cases}$

15.5. $X_1 = \begin{cases} 1 \text{ if no h.s. diploma,} \\ 0 \text{ if otherwise;} \end{cases}$

$X_2 = \begin{cases} 1 \text{ if graduated h.s.,} \\ 0 \text{ if otherwise;} \end{cases}$

$\cdots; \quad X_5 = \begin{cases} 1 \text{ if graduate work,} \\ 0 \text{ if otherwise.} \end{cases}$

15.6. If two patients have the same income and are the same age, but one is male and the other is female, then on the average the female's medical expenses will be \$20 less than the male's.

15.7. $\mathbf{y} = \begin{bmatrix} 40 \\ 72 \\ \vdots \\ 69 \end{bmatrix}$; $\quad \mathbf{X} = \begin{bmatrix} 1 & 14 & 0 & 1 & 0 \\ 1 & 38 & 1 & 0 & 1 \\ 1 & 25 & 0 & 0 & 0 \\ 1 & 24 & 1 & 0 & 1 \\ 1 & 19 & 1 & 0 & 1 \\ 1 & 10 & 0 & 1 & 0 \\ 1 & 23 & 1 & 0 & 1 \\ 1 & 12 & 1 & 0 & 0 \\ 1 & 26 & 1 & 1 & 0 \\ 1 & 24 & 1 & 0 & 1 \end{bmatrix}$

Section 15.4

15.8.

Model No.	PRESS
1	669.66
2	493.54
3	514.03
4	482.57

15.9.

Model	Cp	$Cp-p$	$q[F_{.05,p,v}-1]$
1	22.64	20.64	8.70
2	7.67	4.67	7.85
3	6.02	2.02	6.96
4	3.95	-1.05	

INDEX